Biogas Production

Nagamani Balagurusamy • Anuj Kumar Chandel
Editors

Biogas Production

From Anaerobic Digestion to a Sustainable Bioenergy Industry

Springer

Editors
Nagamani Balagurusamy
Laboratorio de Biorremediación, Facultad de Ciencias Biologicas
Universidad Autonoma de Coahuila
Torreón, Coahuila, Mexico

Anuj Kumar Chandel
Department of Biotechnology, Engineering School of Lorena (EEL)
University of Sao Paulo (USP)
Lorena, São Paulo, Brazil

ISBN 978-3-030-58826-7 ISBN 978-3-030-58827-4 (eBook)
https://doi.org/10.1007/978-3-030-58827-4

This Springer imprint is published by the registered company Springer Nature Switzerland AG.
The registered company address is: Gewerbestrasse 11, 6330 Cham, Switzerland

Preface

Although biogas production from animal manures is one of the oldest technologies in use, it has not gained much attention in the modern world as it is viewed as a rural technology. This is also supported by frequent reports of performance deficiencies. Dwindling fossil fuels, price fluctuations, accumulation of organic wastes, and the consequent climate change problems are the challenges threatening the balance of the world's natural resources. Biogas technology offers the scope of an integrated waste management technology in combination with recovery of resource from the organic wastes as renewable energy in the form of methane, organic fertilizers, and developing a biocircular economy.

With these ideas in focus, this book has been planned in three parts. Seven chapters on the current scenario status and perspectives on the biogas production are discussed with respect to India, Latin America, the USA, and Europe in the first part on "Trends in Biogas Production Technologies." Further a chapter on comparative analysis of biogas with other renewable fuels in relation to their carbon footprint is presented.

Though biogas technology is being practiced over a period of time, the process is still considered as "black box," since not much is deciphered clearly on the microbial groups and their interactions. Of late, there are lots of work that explains the microbiological, biochemical, and genomic aspects of the process and the ways and means to improve the methane yield and efficiency of the process. In the second part on "Improving Biogas Production: Progress, Challenges and Perspectives," two chapters on alternate feedstocks and their efficiency and limitations are discussed. Additionally, a chapter each on the use of microbiological and metabolic engineering tools to monitor and improve the efficiency is presented. The other three chapters focus on factors controlling large-scale production, purification of biogas as biomethane, storage, and value chain analysis. A chapter on the potential of digestate (effluent from biogas digester) as organic fertilizer is presented, which closes the loop of the biogas production technology.

Four chapters on the socioeconomic and technological impact of the process on society and how the governmental policies in developing world can aid in developing a biocircular economy are devoted in the last part on "Economics of Biogas Technology."

We the editors thank the contributing authors from various parts of the world, namely Mexico, India, the USA, Brazil, Denmark, the Netherlands, Greece, Turkey, and South Africa for their time and for sharing their knowledge for the benefit of students as well as professionals.

NB profoundly thanks Prof. Dr. S. Anthoni Raj and Prof. Dr. K. Ramasamy, eminent Professors of Tamil Nadu Agricultural University, Coimbatore, India, for sharing their expertise on the anaerobic microbial world and for teaching the importance of understanding them to explain the working of nature and, in particular, biogas technology. Editors also thank Ms. Martina Himberger, Mr. Srinivasan Manavalan, Ms. Narmadha Nedounsejiane, Dr. Miriam Sturm, Ms. Ulrike Daechert, and Springer publishing team for their constant support and guidance from the initial conception to the completion of the book.

Torreón, Mexico — Nagamani Balagurusamy
Lorena, Brazil — Anuj Kumar Chandel
December, 2020

Contents

Part III Economics of Biogas Technology

About the Editors

Nagamani Balagurusamy is Professor at the Faculty of Biological Sciences, Universidad Autonoma de Coahuila, Torreon, Mexico. Previously, he worked as Associate Professor at the Tamil Nadu Agricultural University, Coimbatore, India. Over the last 30 years, his research interest has been in the biology of anaerobic digesters, with particular emphasis on the interactions between anaerobic bacteria and archaea.

He teaches biochemistry, microbial biochemistry, biotechnology, environmental engineering, microbial physiology, bioremediation, and microbial molecular ecology to undergraduate and postgraduate students of Biochemical Engineering Programs. He is recognized as Level 1 Member of the National Research Scientists System of Mexico by the National Council of Science and Technology of the Government of Mexico.

Anuj Kumar Chandel is an Assistant Professor at the Department of Biotechnology, Engineering School of Lorena, University of Sao Paulo (USP), Brazil. Dr. Chandel's research interests are biofuels and biochemical production based on the biorefinery concept. As a frequently invited speaker at international conferences and forums, he regularly lectures on technical and strategic aspects of biomass energy.

He has published his research work in leading journals, and coedited books on xylitol, sustainable degradation of lignocellulosic biomass, Brazilian and Indian biofuels development, sustainable sources of energy, and others. Additionally, he is a reviewer for numerous prominent journals and industrial projects of various funding agencies and a consultant for several biotechnology startups.

Part I
Trends in Biogas Production Technologies

Technological Routes for Biogas Production: Current Status and Future Perspectives

Salvador Sánchez-Muñoz, Fernanda Gonçalves Barbosa, Jesús Jiménez-Ascencio, Edith Mier-Alba, Akhilesh Kumar Singh, Júlio César dos Santos, Nagamani Balagurusamy, Silvio Silvério da Silva, and Anuj Kumar Chandel

Abstract Biogas is produced from organic substrates by anaerobic digestion (AD) process. Methanogenic microorganisms anaerobically degrade the organic materials primarily into (50–75%), CO_2 (25–50%), N_2 (0–10%), H_2 (0–1%), H_2S (0.1–0.5%), and O_2 (0–0.5%). AD process could be performed by mesophilic or thermophilic microorganisms. Like natural gas, biogas can also be compressed, so-called compressed natural gas, for using as a transportation fuel in motor vehicles. A variety of biomass such as animal manure, agroresidues, lignocellulosic biomass, food waste, and municipal refuse may be exploited toward the biogas production in a closed reactor or tank or bioreactor, so-called anaerobic digester or biodigestor. Typically, the AD process may be categorized into four phases, i.e., hydrolysis, acidogenesis, and acetogenesis followed by methanogenesis. Pretreatment is an unavoidable process to make the lignocellulosic substrate amenable for fermenting microorganisms for biogas production. During the fermentation step, all these four steps occur to make biogas from substrates. Technological innovations are necessary in reducing the biomass pretreatment cost, enzyme cost and enrichment of methane gas. This chapter summarizes the various technological options for biogas production from a variety of organic substrates.

Keywords Biogas · Pretreatment · Lignocellulosic biomass · Anaerobic digestion

---

S. Sánchez-Muñoz · F. G. Barbosa · J. Jiménez-Ascencio · E. Mier-Alba · J. C. dos Santos · S. S. da Silva · A. K. Chandel (✉)
Department of Biotechnology, Engineering School of Lorena (EEL), University of Sao Paulo (USP), Lorena, São Paulo, Brazil
e-mail: anuj10@usp.br

A. K. Singh
Department of Biotech and Genome, Mahatma Gandhi Central University Bihar, Camp Office, Raghunathpur, Bihar, India

N. Balagurusamy
Laboratorio de Biorremediación, Facultad de Ciencias Biologicas, Universidad Autonoma de Coahuila, Torreón, Coahuila, Mexico

N. Balagurusamy, A. K. Chandel (eds.), *Biogas Production*,
https://doi.org/10.1007/978-3-030-58827-4_1

1 Introduction

Biogas production from anaerobic digestion (AD) is a well-known process toward the production of renewable energy and organic refuses. Biogas as a renewable energy has received great attention as it reduces greenhouse gas emissions considerably. Biogas competently substitutes natural gas and thus opened new channels for the development of advance techniques (Angelidaki et al. 2018; Atelge et al. 2018). In current scenario, the techniques available so far depict not only high production costs but also energy intensiveness as well. These factors limit the commercial production and applications of biogas (Sahota et al. 2018).

In the other side, potential marketing of biogas in a near future could be attractive, because of the wide-scale availability of residual biomass. Residual biomass is interesting due to the development of goals for future sustainability, significant economic potential, and positive environmental impact, recycling carbon in major of biological process (Cadenas and Cabezudo 1998; Sahota et al. 2018).

The implementation of biofuel technologies largely depends on the availability of local feedstock, requirement of energy and the policy in the region (Larson 2008; Elbehri et al. 2013).

2 Potential Feedstock for Biogas Production

Biogas is generated in the course of AD of organic substrates. Generally, most of these organic substrates are by-products or residual product of other processes, for example, animal manure, sewage sludge, etc. Biogas can also be generated in the course of anaerobic degradation of organic materials in landfills (Petersson and Wellinger 2009).

Recently, biogas production has increased significantly, particularly in the European Union (UE) countries. UE energy and climate policies have promoted the development of biogas production in the form of energy by giving incentives for the use of renewable resources (Scarlat et al. 2018). It is essential to purify the biogas for the removal of small impurities present in its composition, such as H_2S, water, CO_2, and some other trace elements (Scarlat et al. 2018). The source of organic substrate and its composition are important in the yield and chemical composition of the biogas (Rasi et al. 2007; Horváth et al. 2016; Khan et al. 2017; Scarlat et al. 2018) (Fig. 1).

Generally, four biochemical processes are involved in biogas production via AD: hydrolysis, acidogenesis, acetogenesis, and methanogenesis (Lam and Lee 2011). According to Kitani et al. (1989), AD uses low-cost substrates, favoring the production of biogas. They are classified according to their origin, as products of domestic origin, agricultural residues, and industrial waste activities.

Urban solid waste corresponds to food waste, paper, wood, plastic, metal, and glass and is usually converted to landfills (Donkin et al. 2013). Industrial wastes are waste from the food as well as beverage industries, the agro-industrials, such as paper and cellulose industry. Organic waste from food and beverage industry

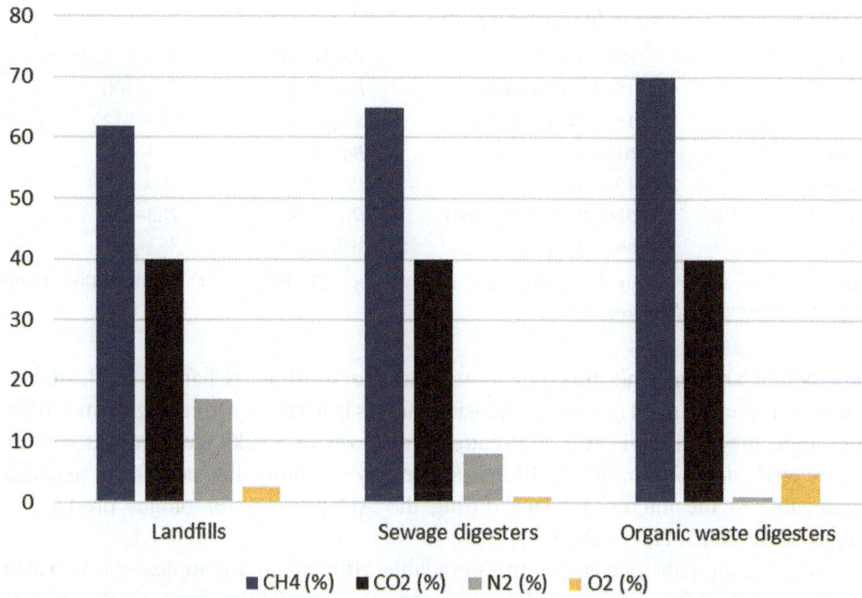

Fig. 1 Composition of biogas in relation to the feedstock. Source: Modified from Rasi et al. (2007)

produced in breweries, distilleries, fat processing, coffee, fruit, slaughterhouses, and refineries are also promising feedstock for biogas production (Demuynck et al. 1984). Agricultural residues are animal residues, which include cow manure, swine manure, chicken manure, horse manure, elephant manure, fishery residue, and slaughterhouse waste (Hobson et al. 1981).

Some residues of domestic and agricultural origin are not considered substrates with greater potential in anaerobic digestion, because these materials have already lost much of their energy content for the animal that produced it (Bhatia 2014). Therefore, as described by Ofoefule and Uzodinma (2006), co-digestion with two or more substrates can occur, that is, the organic residues can be mixed leading to enhance the yield of the biogas generation (Ofoefule and Uzodinma 2006). Ezeonu et al. (2002) reported an enhancement in biogas production of >400% from the brewery's grain mixture in the ratio of 4:1.

Many of these feedstocks are limited in relation to their availability, and to the yield of biogas produced, so there is currently use of plant residues, which are rich in lignocellulose (Chandra et al. 2012). The lignocellulosic feedstock is primarily composed of cellulose, hemicellulose as well as lignin. These polymers are bound together in the form of a network (Fengel and Wegener 1984). The high carbohydrate content present in this raw material, as well as its availability around the world, makes these materials important sources for fuel production (Zhang 2008; Chandra et al. 2012).

The challenge for the use of this type of feedstock is its structure. Hydrolysis of insoluble complex organic material in soluble monomers and oligomers is the first step in biogas generation from lignocellulosic material. For this, it is necessary that the responsible enzymes be produced by the microorganisms and that there is direct

Table 1 Different variety of feedstocks with biogas yield

Feedstock	*Biogas yield (m3/t)*	*Feedstock*	*Biogas yield (m3/t)*
Cattle slurry	15–25 (10% DM)	*Potatoes*	276–400
Grass silage	160–200 (28% DM)	*Sorghum*	295–372
Wheat grain	610 (85% DM)	*Barley*	353–658
Sunflower	154–400	*Peas*	390
Crude glycerine	580–1000 (80% DM)	*Rye grain*	283–492
Fats	up to 1200	*Wheat grain*	384–426

Source: "Renewable energy from crops and agro-wastes (CROPGEN)," C. Banks, 2007, Final report, *University of Southampton*

interaction amongst the enzymes as well as the substrate (Chandel et al. 2019). However, pretreatment of lignocellulosic biomass is necessary for using them further in biogas production via AD. Pretreatment removes or breaks the lignin as well as hemicellulosic portion of the biomass, thereby enabling the cellulosic material accessible to the microorganisms during the AD process, for biogas production (Karp et al. 2013; Fan et al. 2016).

The organic substrate varies in degradable effluents and complex solids waste (Steffen et al. 1998). During AD process, the raw material decomposition occurs at different kinetics. AD may occur more rapidly, if the substrates are short-chain hydrocarbons or simpler sugars. On the other hand, process could be slow if substrates are quite complex such as cellulose and hemicellulose (Bhatia 2014). Alkaline pretreatment of the feedstocks, nutrient addition, and co-digestion have increased biogas yield and productivities, eventually affecting the overall process performance (Ivo Achu 2012). Table 1 depicts the types of feedstock and the yield of biogas production.

3 Key Technological Process for Biogas Production (Pretreatments)

Pretreatment is an important step in biogas generation from lignocellulosic biomass. However, the effectiveness of pretreatment process depends on the physical nature and cell wall composition of lignocellulosic biomass, eventually influencing the success of biogas based biorefineries (Montgomery and Bochmann 2014; Achinas et al. 2017). Several promising pretreatment processes, for example, steam explosion, etc., have been quite successful in large-scale operations (Chandel et al. 2019).

Saha and collaborators (2018) studied three alkali reagents using NaOH, KOH, and $Ca(OH)_2$ at various dosages and found cumulative biogas production of 560 mL/gVS using 2% NaOH pretreated wheat straw (two times higher biogas produced than untreated substrates). In another study of Taherdanak et al. (2016), maximal methane production of 302.4 mL/g volatile solids (VS) was recorded from the substrate pretreated for 2 h that was 15.5% greater over untreated substrate.

Bauer et al. (2014) observed the effect of various steam explosion conditions on digestibility of hay. Enzyme-mediated hydrolysis of steam exploded at 220 °C for 15 min showed maximal glucose yields, while higher xylose yields were obtained at steam explosion (175 °C for 10 min). About 16% increase in methane yield was found from the steam-exploded hay than the untreated hay. Pengyu et al. (2017) studied two kinds of milling pretreatment (dry and wet) techniques. The anaerobic fermentation test showed the utmost methane production of 358.07 and 315.87 mL g^{-1} VS at 3 h dry milling and 6 h wet milling pretreated grass, respectively, that was 41.04 and 24.42% greater over untreated *Pennisetum* hybrid (253.88 mL g^{-1} VS).

Continuous pretreatment is an absolute method for the successful operation of biogas production plants using lignocellulose biomass. However, current pretreatment methods are not economically competitive and are energy intensive. Improvement in pretreatment process minimizing use of electric energy, heat, and catalyst load is important for economic biogas production from lignocellulosic biomass. Table 2 sums up the major benefits as well as drawbacks of various pretreatment methods.

4 Technological Process for Biogas Recovery

Biogas is the main source of renewable energy obtained from AD from biodegradable organic materials like biological waste, animal manure, landfills, wastewater treatment, and industrial waste (del Rosario Rodero et al. 2019). Depending upon the process environmental conditions, the biogas could be composed of CH_4, CO_2, halogens, etc. (Gilassi et al. 2019).

Biogas production represents a biotechnological advantage for agriculture, energy, transport, and waste disposal by greatly reducing greenhouse gas emissions (GHG) (Lyng et al. 2018). Nevertheless, the exploitation of biogas as fuel or production of heat and electricity is determined by its chemical composition. Because CO_2 represents 30–50% (Muñoz et al. 2015) of the biogas content, it is the main objective to be eliminated in order to recover and purify the biogas. The presence of CO_2 decreases the heat energy of the biogas, reducing the output power of the engine and volume of the storage cylinders, which would result in a higher energy consumption for biogas compression (Kapoor et al. 2019). In addition, it is necessary to remove other components such as H_2S and H_2O to avoid corrosion of the boilers or combustion heat and energy combustion systems, and because these compounds are in a low percentage, it can be eliminated more easily (Tilahun et al. 2017). On the other hand, some applications have permissible limits of some compounds within the biogas. It requires the elimination of H_2S below 250 ppm and the removal of moisture to use the biogas in gas heating boilers (Khan et al. 2017). Tilahun et al. (2017) reported with the same level of H_2S (<250 ppm) that the use of biogas in combined heat as well as power systems, avoiding corrosion that may imply higher maintenance costs, wherein for fuel and gas injection applications

Table 2 Overview on the benefits as well as shortcomings of different pretreatment technologies for biogas production

Pretreatment method	Pros	Cons
Milling	No inhibitors generation, increased methane yield (~25%)	Energy intensive, requirement of high maintenance cost
Extrusion	Augmentation of surface area of pretreated substrate	Energy intensive, requirement of high maintenance cost
Steam explosion	Effective method in removal of hemi-cellulose, fast method	Inhibitors generation, requirement of high capital investment, lignin relocation
Liquid hot water	High amount of cellulose retains in pretreated biomass, hemicellulose solubilization with low amount of inhibitors	Effectiveness of the process, inhibitors generation
Microwave	Effective process, reduces time and yield high biogas production	Scale up challenges, safety concerns
Acid pretreatment	Effectively solubilization of hemicellulose, leaving cellulose and lignin together	Inhibitors generation, equipment corrosion
Alkaline delignification	Effective lignin solubilization, high amount of cellulosic and hemicellulosic materials in the pretreated biomass for methanogenic bacteria	Loss of hemicellulose, solubilized lignin may act as inhibitors, corrosiveness to the reactors

in the consumer network for the home, must remove CO_2 and H_2S contents, humidity, and most of the trace impurities, which can vary according to the quality standards of each country (Sun et al. 2015).

Several biogas upgrading technologies like physical and chemical absorption techniques, etc. are used at the industrial level to eliminate the presence of CO_2, H_2S, H_2O, O_2, N_2, siloxanes, and halogens (Songolzadeh et al. 2014, Muñoz et al. 2015), depending on the chemical composition and biogas application (Fig. 2). Most used upgrading technologies for biogas recovery in relation to the quality required for its various applications are summarized in Fig. 2.

Pressure swing adsorption (PSA) is one of the utmost important adsorption techniques in which the most used absorbent materials such as activated carbon, silica gel, etc. are used (Ferella et al. 2019). The selective adsorption of CO_2 on CH_4 depends on the available specific surface area of porous adsorbents (Muñoz et al. 2015). The adsorption systems consist of solute transfer by gaseous flow to an adsorbent surface by using the concept of molecular size exclusion and adsorption affinity (Khan et al. 2017). The PSA systems allow the gas elimination such as CO_2, O_2, and N_2 (Verotti et al. 2016). This technology also adsorbs H_2S; therefore, the biogas achieved is free from H_2S (Kadam and Panwar 2017).

High-pressure water scrubbing (HPWS) is a process where CO_2 and H_2S are adsorbed at the same period, enriching the biogas with CH_4, being of great importance to know the solubility in H_2O of each biogas constituents (Wylock and

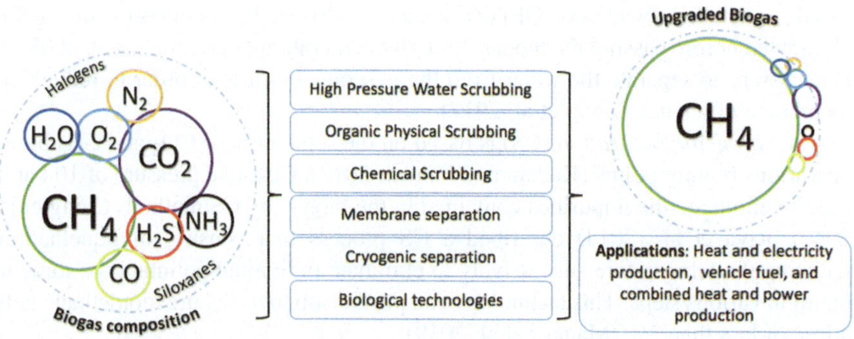

Fig. 2 Schematic illustration of upgrading technologies used at industrial level in the biogas recovery

Budzianowski 2017). This technique uses packed bed columns, where under optimal conditions result in greater concentration of CO_2 removal (Cozma et al. 2015). Eze and Agbo (2010) conducted a minimum supervision in the operation and a single step of the biogas through the column, obtaining biomethane with purity ranging from 95 to 97%. In addition, Läntelä et al. (2012) increased the pressure in the column of 20–25 bars, achieving a reduction of 99.1% of siloxanes and 99.9% of halogenated compounds.

Organic physical scrubbing (OPS) consists of the use of an organic solvent, (polyethylene glycol) based on an absorbent (selexol or genosorb) with a greater affinity for CO_2 as well as H_2S (Zhou et al. 2017). In this process, the biogas and the organic solvents are cooled to 20 °C before absorption, presenting as key benefit an anticorrosive nature of the solvents (Maurya et al. 2019). Furthermore, if H_2S, H_2O, and O_2 have not been formerly eliminated in the biogas cleaning process, they can also be removed together with CO_2 (Baena-Moreno et al. 2019).

In chemical scrubbing, absorbed substance and absorbent component are used. Within organic substances, amines are usually the utmost exploited toward the elimination of gasses like CO_2 or H_2S (Muñoz et al. 2015). Inorganic components are made up of an aqueous solution of an alkaline salt like Na, K, NH_4^+, or $Ca(OH)_2$ (Khan et al. 2017). In the chemical washing, intermediate chemical products such as CO_3^{2-} and HCO_3^- are produced exothermically when the CO_2 adsorbed reacts with the chemicals present in the washing solution, which results in a higher absorption capacity (Maurya et al. 2019).

Membrane-based separation (MS) is a biogas purification technology that works in selective separation of the biogas component (CO_2, H_2S, and H_2O) through semipermeable membranes eventually concentrating the CH_4 (Petersson and Wellinger 2009). In general, the membranes of mixed matrix membranes (MMM) are consisted of polymeric materials, such as cellulose acetate, cellulose triacetate, polyimides (PI), polyetherimide (PEI), and so on (Basu et al. 2010). Three diverse types of special fillers are exploited toward the preparation of MMMs, i.e., ordered mesoporous silicas (OMS), high aspect ratios (HAR), and silica-based particles and metal-organic frameworks (MOF) (Zornoza et al. 2011). To avoid corrosion, H_2S is eliminated before passing the biogas through the membranes (Kentish et al. 2008). It is necessary to separate the water from the gaseous mixture, in order to reduce the membrane efficiency (Zhou et al. 2017).

The *cryogenic separation* (CS) is based on the separation of CO_2 and CH_4 based on various boiling points (Kadam and Panwar 2017). Constant pressure of 10 bar is used to eliminate the impurities contained in the biogas by CS methods (Song et al. 2019). Riva et al. (2014) described a CS process that consists of liquefaction, reducing the temperature successively to eliminate individual pollutant (or some of them) in various steps. This technology can provide up to 97% pure biomethane with a loss of less than 2% (Maurya et al. 2019).

Biological upgrading technologies for biogas allow the reduction of CO_2 and H_2S by biological means with an ecological approach in the recovery of biogas. These technologies include H_2-assisted CO_2 bioconversion, microalgae-based CO_2 fixation, as well as biological H_2S removal.

In the *H_2-assisted CO_2 bioconversion*, the CO_2 contained in the biogas can serve as an electron acceptor and the H_2 as an electron donor for the bioconversion of CO_2 to CH_4 through the hydrogenotrophic methanogens (Zabranska and Pokorna 2018). Even the gasification processes of the biomass or synthesis gas, containing CO, H_2, and CO_2, can be used for the production of CH_4 depending on the biological activity of the methanogens such as *Methanobacterium* sp., *Methanococcus* sp., *Methanosaeta* sp., and so on (Muñoz et al. 2015).

The microalgae-based CO_2 fixation technology is based on the use of CO_2 present in the biogas to be reduced through the water photolysis (López et al. 2013). Most of the microalgae present growth inhibition at a concentration of 5% CO_2 in the biogas, but some strains of microalgae capable of tolerating concentrations of up to 60% CO_2 have been isolated (Wang et al. 2008). Another factor to consider in this technology is the higher H_2S concentrations more than 100 ppm, being an inhibitor in the growth of microalgae (Kao et al. 2012). However, with the use of H_2S-oxidizing bacteria and the chemical oxidation of H_2S in microalgae photobioreactors, it has been possible to oxidize the elemental sulfur to sulfate, preventing the inhibition of microalgae growth (Maurya et al. 2019). It has also been reported that up to 80% CO_2 have been removed and a 90% biomethane recovery (Muñoz et al. 2015).

For the *biological H_2S removal*, this method is represented by sulfur-oxidizing bacteria (SOB) that use H_2S as an electron donor, CO_2 as a carbon source, and O_2, NO_2, NO_2^- as electron acceptors to convert H_2S in sulfate (SO_4^{2}) together with elemental sulfur as an intermediate (Maurya et al. 2019). These SOBs belong to species such as *Acidithiobacillus ferrooxidans* and *Acidithiobacillus thiooxidans* (Montebello et al. 2014). SOBs can remove H_2S from the head space of a bioreactor through microaerobic and lithoautotrophic growth in the reactor wall, producing elemental sulfur (Muñoz et al. 2015). Microalgae can be employed to eliminate CO_2 and H_2S at the same time, where microalgae produce O_2 by photosynthesis that will later be assimilated by lithoautotrophs to oxidize H_2S to elemental sulfur (Maurya et al. 2019).

The development of technologies to improve the quality of biogas currently includes biological and hybrid technologies. The constant updating of traditional technologies, as well as emerging ones, has a better performance. Although several of them are still in development to obtain a higher performance, efficiency, and low cost in the process, it is opening the possibility of being widely used at commercial scale in the future.

5 Comparative Technological Profile of Biogas with Other Renewable Fuels

5.1 Bio-digester Improvements

For the utmost microbial management, design and operation of bio-digester have to be performed (Carballa et al. 2015). Looking at the historical point of view, the first Indian digester was built in 1859 (Kigozi et al. 2014). Thereafter, several modifications in reactor technologies have been done. More recently, a new digester type is developed by the Shenzhen Puxin Science and Technology Company (Puxin) of China developing Puxin digester (BiogasSA 2012). Table 3 presents a comparison between different bioreactors for biogas and ethanol production.

5.2 Technologies Improving Biogas Quality

Removal of CO_2 leads to enrichment of biogas having greater content of CH_4 and thus increasing energy content per unit volume (Brendeløkken 2016). Raw biogas is mainly consisted of CH_4 (55 to 65%) as well as CO_2 (35 to 45%). After removal of CO_2, calorific value of CH_4 increases up to 39,000 $kJ \cdot m^{-3}$ (Yin et al. 2009). There are some technological advancements that have been done for enrichment of biogas.

Water washing system was carried out to investigate the impact of various factors like including liquid/gas ratio, pressure, temperature, and CO_2 content. The lowest CO_2 content that was recorded after absorption was 2.6% at 1.2 MPa with 400 $L \cdot h^{-1}$ gas flow and 200 $L \cdot h^{-1}$ water flow, matching the need of CO_2 content in natural gas for vehicle fuel (Xiao et al. 2014).

Water scrubbing led the reduction of H_2S and CO_2 in biogas by 32.8 and 21.2%, correspondingly (Islamiyah et al. 2015).

Capturing CO_2 at low temperature showed the 99.7 mol.% purity of the captured CO_2 in liquid form as a by-product for transport at 110 bar (Yousef et al. 2016).

Activated carbon (PINPEL20) obtained from biomass waste (wood pellets) was activated with CO_2 at high temperature and showed the excellent properties as a selective adsorbent of CO_2/CH_4 (Vivo-Vilches et al. 2017).

Table 3 Bioreactor comparison between two main biofuels (methane and ethanol)

Biomethane				
Reactor type	Feed	Conditions	Methane	Reference
Chinese dome digester with a self-agitating mechanism	Cow manure	500 L capacity of digester, 15% influent TS concentration, 27–33 °C	HRT 40 days: 0.25 ± 0.05 L CH_4/g VS HRT 30 days: 0.23 ± 0.04 L CH_4/g VS	Jegede et al. (2019)
Jacketed fermenter (Biostat B)	Cow dung	Fermenter volume of 10 L, organic loading of up to 1.7 kg volatile solids (VS)/L d and an hydraulic retention time (HRT) of 10 days, 53°C	Biogas yields 0.15 L/kg VS_{added} and methane content of 47%	Abubakar and Ismail (2012)
Plug flow reactor (PFR)	Cattle manure	HRT of 25 days, total solids (7–10%), temperature of 37–40 °C, working volume of 3.85×10^4 m^3 and dispose 1504 m^3/d of waste for 3 years	Biogas yields 0.39 m^3/kg VS and methane content of 59%	Dong et al. (2019)
Upflow anaerobic sludge blanket-anaerobic filter (UASB-AF)	Cattle wastewater	Semi-continuous mode of operation, HRT of 6, 5, 3, and 2 days and organic loading rates of 3.8, 4.6, 7.0, and 10.8 kg CODt $m^{-3}d^{-1}$, 37 °C	Biogas volumes of 0.6–0.8 m^3 m^{-3} d^{-1} (3.8–4.6 kg CODt m^{-3} d^{-1}) and 1.2–1.4 m^3 m^{-3} d^{-1} (7.0–10.8 kg CODt m^{-3} d^{-1}), with CH_4 concentrations between 69 and 75%	de Mendonça et al. (2017)
Ethanol				
Reactor type	Feed/inoculum	Conditions	Ethanol	Reference
Immobilized cell reactor (ICR)	Sugarcane molasses, *Saccharomyces cerevisiae* (PTCC 5010)	Concentration of the sugarcane molasses (50, 100, and 150 g/l), dilution rates (0.064, 0.096, 0.144, and 0.192 h^{-1}) and HRT (5.21, 6.94, 10.42, and 15.63 h), pH of 4.5	19.5 g/L	Ghorbani et al. (2011)

(continued)

Table 3 (continued)

Biomethane					
Reactor type	Feed	Conditions		Methane	Reference
Continuously stirred tank reactor (CSTR)	Sugarcane molasses/ *S. cerevisiae*	Dilution rate of 0.12, 0.25, and 0.5 h^{-1}	30 °C and pH 5	Ethanol productivity (Qp) of 6.8 $g.L^{-1}\ h^{-1}$	Bouallagui et al. (2013)
ICR				Qp 48% improved in CSTR	
Membrane reactor (MBR)		Dilution rate of 0.5 h^{-1}		Qp of 19.2 $gL^{-1}\ h^{-1}$	

6 Conclusion

It is evident that biogas is one of the very essential renewable as well as clean energy resources. Presently, biogas is exploited toward the generation of electricity as well as heat. The main advantage of for biogas production is that a relatively wide range of renewable raw materials could be exploited for its production. Currently worldwide, the transportation gas (compressed natural gas) production is from crude oil refinery and, therefore, has a vast effect on the environment. Renewable raw materials (especially lignocellulosic biomass) can be a sustainable source for biogas production at competitive prices. However, biogas production has many challenges that need to be overcome for the biogas production at commercial scale. Biogas production employing the biorefinery platform can offer unique advantages to obtain value-added products along with biogas eventually augmenting environmental and economic sustainability of bioprocesses.

Acknowledgments AKC is grateful to the CAPES-Brazil for the financial assistance through visiting professor and researcher program (Process USP number: 15.1.1118.1.0). Authors are grateful to FAPESP and CNPq, Brazil, for the financial support to the laboratory.

References

Abubakar BSUI, Ismail N (2012) Anaerobic digestion of cow dung for biogas production. ARPN journal of engineering and applied sciences 7(2):169–172

Achinas S, Achinas V, Euverink GJW (2017) A technological overview of biogas production from biowaste. Engineering 3(3):299–307

Angelidaki I, Treu L, Tsapekos P, Luo G, Campanaro S, Wenzel H, Kougias PG (2018) Biogas upgrading and utilization: current status and perspectives. Biotechnol Adv 36(2):452–466

Atelge MR, Krisa D, Kumar G, Eskicioglu C, Nguyen DD, Chang SW et al (2018) Biogas production from organic waste: recent progress and perspectives. Waste and Biomass Valorization:1–22

Baena-Moreno FM, Rodríguez-Galán M, Vega F, Vilches LF, Navarrete B (2019) Recent advances in biogas purifying technologies. International Journal of Green Energy 16(5):401–412

Banks C (2007) Renewable energy from crops and agro-wastes (CROPGEN), Final report, University of Southampton. S017 1BJ, UK

Basu S, Khan AL, Cano-Odena A, Liu C, Vankelecom IF (2010) Membrane-based technologies for biogas separations. Chem Soc Rev 39(2):750–768
Bauer A, Lizasoain J, Theuretzbacher F, Agger JW, Rincón M, Menardo S et al (2014) Steam explosion pretreatment for enhancing biogas production of late harvested hay. Bioresour Technol 166:403–410
Bhatia SC (2014) Advanced renewable energy systems, (Part 1 and 2). WPI Publishing
BiogasSA (2012) biogassa: Digesters. In BiogasSA, ed.
Bouallagui H, Touhami Y, Hanafi N, Ghariani A, Hamdi M (2013) Performances comparison between three technologies for continuous ethanol production from molasses. Biomass Bioenergy 48:25–32
Brendeløkken HW (2016) Upgrading technologies for biogas production plants. Overview and life cycle cost analysis of available technologies
Cadenas A, Cabezudo S (1998) Biofuels as sustainable technologies: perspectives for less developed countries. Technol Forecast Soc Chang 58(1–2):83–103
Carballa M, Regueiro L, Lema JM (2015) Microbial management of anaerobic digestion: exploiting the microbiome-functionality nexus. Curr Opin Biotechnol 33:103–111
Chandra R, Takeuchi H, Hasegawa T (2012) Methane production from lignocellulosic agricultural crop wastes: A review in context to second generation of biofuel production. Renew Sust Energ Rev 16:1462–1476
Chandel AK, Albarelli JQ, dos Santos DT, Chundawat SPS, Puri M, Meireles MAA (2019) Comparative Analysis of Key Technologies for Cellulosic Ethanol Production from Brazilian Sugarcane Bagasse at the Commercial-Scale. Biofuels, Bioproducts and Biorefining 13, 994–1014
Cozma P, Wukovits W, Mămăligă I, Friedl A, Gavrilescu M (2015) Modeling and simulation of high pressure water scrubbing technology applied for biogas upgrading. Clean Techn Environ Policy 17(2):373–391
de Mendonça HV, Ometto JPHB, Otenio MH, dos Reis AJD, Marques IPR (2017) Bioenergy recovery from cattle wastewater in an UASB-AF hybrid reactor. Water Sci Technol 76 (9):2268–2279
del Rosario Rodero M, Lebrero R, Serrano E, Lara E, Arbib Z, García-Encina PA, Muñoz R (2019) Technology validation of photosynthetic biogas upgrading in a semi-industrial scale algal-bacterial photobioreactor. Bioresour Technol 279:43–49
Demuynck M, Nyns EJ, Palz W (1984) Biogas plants in Europe: A practical handbook. D. Reidel for the Commission of the European Communities
Dong L, Cao G, Guo X, Liu T, Wu J, Ren N (2019) Efficient biogas production from cattle manure in a plug flow reactor: A large scale long term study. Bioresour Technol 278:450–455
Donkin SS, Doane PH, Cecava MJ (2013) Expanding the role of crop residues and biofuel co-products as ruminant feedstuffs. Anim Front 3(2):54–60
Elbehri A, Segerstedt A, Liu P (2013) Biofuels and the sustainability challenge: a global assessment of sustainability issues, trends and policies for biofuels and related feedstocks. Food and Agriculture Organization of the United Nations (FAO)
Eze JI, Agbo KE (2010) Maximizing the potentials of biogas through upgrading. Am J Sci Ind Res 1(3):604–609
Ezeonu FC, Udedi SC, Okaka ANC, Okonkwo CJ (2002) Studies on Brewers Spent Grains (BSG) Biomethanation: I-Optimal Conditions for Digestion. Nigerian Journal of Renewable Energy 10 (1, 2):53–57
Fan S, Zhang P, Li F, Jin S, Wang S, Zhou S (2016) A review of lignocellulose change during hydrothermal pretreatment for bioenergy production. Curr Org Chem 20(26):2799–2809
Fengel D, Wegener G (1984) Wood: chemistry, ultrastructure. *Reactions*, vol 613, pp 1960–1982
Ferella F, Cucchiella F, D'Adamo I, Gallucci K (2019) A techno-economic assessment of biogas upgrading in a developed market. J Clean Prod 210:945–957

Ghorbani F, Younesi H, Sari AE, Najafpour G (2011) Cane molasses fermentation for continuous ethanol production in an immobilized cells reactor by Saccharomyces cerevisiae. Renew Energy 36(2):503–509

Gilassi S, Taghavi SM, Rodrigue D, Kaliaguine S (2019) Optimizing membrane module for biogas separation. International Journal of Greenhouse Gas Control 83:195–207

Hobson PN, Bousfield S, Summer R (1981) Methane Productionfrom Agricultural and Domestic Wastes. *Applied Science London*

Horváth IS, Tabatabaei M, Karimi K, Kumar R (2016) Recent updates on biogas production-a review. Biofuel Res J 10:394–402

Islamiyah M, Soehartanto T, Hantoro R, Abdurrahman A (2015) Water scrubbing for removal of CO2 (carbon dioxide) and H2S (hydrogen sulfide) in biogas from manure. KnE Energy 2 (2):126–131

Ivo Achu N (2012) Anaerobic digestion of crop and waste biomass: Impact of feedstock characteristics on process performance. Lund University

Jegede AO, Zeeman G, Bruning H (2019) Development of an Optimised Chinese Dome Digester Enables Smaller Reactor Volumes; Pilot Scale Performance. Energies 12(11):2213

Kadam R, Panwar NL (2017) Recent advancement in biogas enrichment and its applications. Renew Sust Energ Rev 73:892–903

Kao CY, Chiu SY, Huang TT, Dai L, Hsu LK, Lin CS (2012) Ability of a mutant strain of the microalga Chlorella sp. to capture carbon dioxide for biogas upgrading. Appl Energy 93:176–183

Kapoor R, Ghosh P, Kumar M, Vijay VK (2019) Evaluation of biogas upgrading technologies and future perspectives: a review. Environ Sci Pollut Res 26:1–31

Karp SG, Woiciechowski AL, Soccol VT, Soccol CR (2013) Pretreatment strategies for delignification of sugarcane bagasse: a review. Braz Arch Biol Technol 56(4):679–689

Kentish SE, Scholes CA, Stevens GW (2008) Carbon dioxide separation through polymeric membrane systems for fluė gas applications. Recent Patents on Chemical Engineering 1 (1):52–66

Khan IU, Othman MHD, Hashim H, Matsuura T, Ismail AF, Rezaei-DashtArzhandi M, Azelee IW (2017) Biogas as a renewable energy fuel–A review of biogas upgrading, utilisation and storage. Energy Convers Manag 150:277–294

Kigozi R, Muzenda E, Aboyade AO (2014) Biogas technology: current trends, opportunities and challenges

Kitani O, Hall CW, Wagener K (1989) Biomass handbook. Gordon and Breach Science Publishers

Lam MK, Lee KT (2011) Renewable and sustainable bioenergies production from palm oil mill effluent (POME): win–win strategies toward better environmental protection. Biotechnol Adv 29(1):124–141

Läntelä J, Rasi S, Lehtinen J, Rintala J (2012) Landfill gas upgrading with pilot-scale water scrubber: performance assessment with absorption water recycling. Appl Energy 92:307–314

Larson ED (2008) Biofuel production technologies: status, prospects and implications for trade and development. United Nations Conference on Trade and Development, New York and Geneva

López JC, Quijano G, Souza TS, Estrada JM, Lebrero R, Muñoz R (2013) Biotechnologies for greenhouse gases (CH4, N2O, and CO2) abatement: state of the art and challenges. Appl Microbiol Biotechnol 97(6):2277–2303

Lyng KA, Stensgård AE, Hanssen OJ, Modahl IS (2018) Relation between greenhouse gas emissions and economic profit for different configurations of biogas value chains: A case study on different levels of sector integration. J Clean Prod 182:737–745

Maurya R, Tirkey SR, Rajapitamahuni S, Ghosh A, Mishra S (2019) Recent Advances and Future Prospective of Biogas Production. In: Advances in Feedstock Conversion Technologies for Alternative Fuels and Bioproducts. Woodhead Publishing, pp 159–178

Montebello AM, Mora M, López LR, Bezerra T, Gamisans X, Lafuente J, Baeza M, Gabriel D (2014) Aerobic desulfurization of biogas by acidic biotrickling filtration in a randomly packed reactor. J Hazard Mater 280:200–208

Montgomery LF, Bochmann G (2014) Pretreatment of feedstock for enhanced biogas production. IEA Bioenergy, Ireland, pp 1–20
Muñoz R, Meier L, Diaz I, Jeison D (2015) A review on the state-of-the-art of physical/chemical and biological technologies for biogas upgrading. Rev Environ Sci Biotechnol 14(4):727–759
Ofoefule AU, Uzodinma EO (2006) Optimization of the qualitative and quantitative biogas yield from Poultry waste: In Proceedings. World renewable energy congress IX. 19–25. August 2006, University of Florence, Italy
Pengyu D, Lianhua L, Feng Z, Xiaoying K, Yongming S, Yi Z (2017) Comparison of dry and wet milling pre-treatment methods for improving the anaerobic digestion performance of the Pennisetum hybrid. RSC Adv 7(21):12610–12619
Petersson A, Wellinger A (2009) Biogas upgrading technologies–developments and innovations. IEA bioenergy 20:1–19
Rasi S, Veijanen A, Rintala J (2007) Trace compounds of biogas from different biogas production plants. Energy 32(8):1375–1380
Riva M, Campestrini M, Toubassy J, Clodic D, Stringari P (2014) Solid–liquid–vapor equilibrium models for cryogenic biogas upgrading. Ind Eng Chem Res 53(44):17506–17514
Sahota S, Shah G, Ghosh P, Kapoor R, Sengupta S, Singh P et al (2018) Review of trends in biogas upgradation technologies and future perspectives. Bioresource Technology Reports 1:79–88
Scarlat N, Dallemand JF, Fahl F (2018) Biogas: Developments and perspectives in Europe. Renew Energy 129:457–472
Song C, Liu Q, Deng S, Li H, Kitamura Y (2019) Cryogenic-based CO2 capture technologies: State-of-the-art developments and current challenges. Renew Sust Energ Rev 101:265–278
Songolzadeh M, Soleimani M, Takht Ravanchi M, Songolzadeh R (2014) Carbon dioxide separation from flue gases: a technological review emphasizing reduction in greenhouse gas emissions. Sci World J 2014
Steffen R, Szolar O, Braun R (1998) Feedstocks for anaerobic digestion. Institute of Agrobiotechnology Tulin, University of Agricultural Sciences, Vienna
Sun Q, Li H, Yan J, Liu L, Yu Z, Yu X (2015) Selection of appropriate biogas upgrading technology-a review of biogas cleaning, upgrading and utilisation. Renew Sust Energ Rev 51:521–532
Taherdanak M, Zilouei H, Karimi K (2016) The influence of dilute sulfuric acid pretreatment on biogas production from wheat plant. International journal of green energy 13(11):1129–1134
Tilahun E, Bayrakdar A, Sahinkaya E, Çalli B (2017) Performance of polydimethylsiloxane membrane contactor process for selective hydrogen sulfide removal from biogas. Waste Manag 61:250–257
Verotti M, Servadio P, Bergonzoli S (2016) Biogas upgrading and utilization from ICEs towards stationary molten carbonate fuel cell systems. International journal of green energy 13 (7):655–664
Vivo-Vilches JF, Pérez-Cadenas AF, Maldonado-Hódar FJ, Carrasco-Marín F, Faria RP, Ribeiro AM et al (2017) Biogas upgrading by selective adsorption onto CO2 activated carbon from wood pellets. J Environ Chem Eng 5(2):1386–1393
Wang B, Li Y, Wu N, Lan CQ (2008) CO2 bio-mitigation using microalgae. Appl Microbiol Biotechnol 79(5):707–718
Wylock CE, Budzianowski WM (2017) Performance evaluation of biogas upgrading by pressurized water scrubbing via modelling and simulation. Chem Eng Sci 170:639–652
Xiao Y, Yuan H, Pang Y, Chen S, Zhu B, Zou D et al (2014) CO2 removal from biogas by water washing system. Chin J Chem Eng 22(8):950–953
Yin B, Chen LM, Kong QP (2009) Research on purification technology for vehicle biogas. Modern Chemical Industry 29(11):28–33
Yousef AM, Eldrainy YA, El-Maghlany WM, Attia A (2016) Upgrading biogas by a low-temperature CO2 removal technique. Alex Eng J 55(2):1143–1150
Zabranska J, Pokorna D (2018) Bioconversion of carbon dioxide to methane using hydrogen and hydrogenotrophic methanogens. Biotechnol Adv 36(3):707–720

Zhang YHP (2008) Reviving the carbohydrate economy via multi-product lignocellulose biorefineries. J Ind Microbiol Biotechnol 35(5):367–375

Zhou K, Chaemchuen S, Verpoort F (2017) Alternative materials in technologies for Biogas upgrading via CO2 capture. Renew Sust Energ Rev 79:1414–1441

Zornoza B, Gorgojo P, Casado C, Téllez C, Coronas J (2011) Mixed matrix membranes for gas separation with special nanoporous fillers. Desalin Water Treat 27(1–3):42–47

Prospects and Challenges in Biogas Technology: Indian Scenario

Geetha Thanuja Kalyanasundaram, Anupriya Ramasamy, Blesy Godwin, Ramesh Desikan, and Karthikeyan Subburamu

Abstract Renewable energy technologies have started to play a major role in the global energy mix. India also has been encouraging renewable energy very aggressively and has large renewable energy expansion programmes. The Ministry of New and Renewable Energy has set a target of 10 GW of bioenergy by 2022 and 100,000 domestic size biogas plants. India has been promoting biogas technology for the past three decades with several governmental programmes. Though India has favourable environmental and sociological conditions such as optimal temperature, high cattle population and a strong agricultural economy, the attempts were unsuccessful. Hence there exists a need to explore the reasons for failure of biogas in India and as well interrogate the specific interventions required to revive back the biogas market in India.

Keywords Anaerobic digestion · Agricultural residues · Methanogens · Biogas plants · Biogas · Biodigested slurry

1 Introduction

In concurrence to the development of industries, energy, and electricity consumption has now become inevitable. The realism steered continuous demand for energy and made to rely on the sustainable energy source. Degeneration of fossils fuels coupled with their ill effects on the environment has paved the way for renewable energy sources coupled to the disposal of wastes. However, India's contribution to global biofuel production accounts for less than 1% (Purohit and Dhar 2018), and attempts

G. T. Kalyanasundaram · A. Ramasamy · B. Godwin · K. Subburamu (✉)
Department of Renewable Energy Engineering, Tamil Nadu Agricultural University, Coimbatore, Tamil Nadu, India
e-mail: skarthy@tnau.ac.in

R. Desikan
Horticultural College and Research Institute for Women, Tamil Nadu Agricultural University, Tiruchirapalli, Tamil Nadu, India

N. Balagurusamy, A. K. Chandel (eds.), *Biogas Production*,
https://doi.org/10.1007/978-3-030-58827-4_2

are being made to pick up the pace. Anaerobic digestion (AD) referred as biomethanation technology is an excellent technology used to convert the biodegradable organic wastes into gaseous biofuel. AD results in enriched organic biodigested slurry as by-product. The gaseous biofuel produced in anaerobic digestion is called biogas, which can be used for cooking, lighting, and dual-fuel engine running. Biogas is one of the promising candidates to replace fuelwood and meet out the cooking fuel requirement in rural and semi-urban areas. The biodigested slurry can be applied as manure in agricultural fields due to its manurial value. The salient benefits of biogas technology are simultaneously producing biofuel and value-added manure from organic wastes and thus lead to improve the sanitation conditions and eco-friendly solution for these wastes.

2 Biochemistry and Microbiology of Biogas Digestion

In the era of effective waste management for the conversion of waste into energy, anaerobic digestion/anaerobic co-digestion is regarded as effective technology since it provides renewable energy (Weiland 2003; Luo et al. 2013). The mechanism underlying with intricate pathways is still underexplored. To understand the thorough structure and metabolic functions involved in the digestion, several attempts have been made (Riviere et al. 2009; Werner et al. 2011). The intensive biodegradation of organic fractions by anaerobic route yields noteworthy advantages over other forms of anaerobic digestion. It is a complicated reduction process harnessing a variety of biochemical reactions and microbial guilds. In general, they involve four consecutive steps, viz. hydrolysis, acidogenesis, acetogenesis, and methanogenesis, which broadly include acid-forming and methane-forming microorganisms (Wirth et al. 2012).

As a consequence of this process, volume and weight of sludge are reduced and permit easy handling and disposal (Gerardi 2003). They are characterized by a difference in their physiology, nutrition requirement, kinetics, and adaptability to the environment. They are influenced by several parameters like pH, temperature, microbial community composition, the concentration of ammonia, heavy metals, and so on (Eduok et al. 2018). The study on metagenomics involving taxonomy and function of biogas-producing digester was analysed in comparison with industrial wastewater digester. The deep insight into major microbial groups revealed similar potential microbial functions and key metabolic pathways (Cai et al. 2016). Processing of feedstocks recruits hydrolytic activity which is favoured by members of *Proteobacteria.* They are also found to involve in all the stages except methanogenesis. However the members of *Clostridium* have reported role in hydrolysis (Dassa et al. 2014) and syntrophic acetate oxidation (SAO) (Müller et al. 2013); they are prime players in acidogenesis yielding major fermented products like formate, acetate, butyrate, and lactate. SAO bacteria account for providing H_2 and CO_2 to syntrophic methanogenic partner. Methanogenesis is dominated by the archaeal order *Methanomicrobiales,* while archaeal rRNA gene sequencing by

Wilkins et al. (2015) have reported *Methanosarcinales* as the dominant group. Methanogenesis gets varied by chemotaxic nature of the organisms and is explored by several studies. For instance, the Wolfe cycle gives a clear description of the unique metabolic nature of methanogens and their obligate synthesis of methane for their ATP synthesis irrespective of their substrate (Thauer 2012).

The anaerobic digestion of organic wastes by various groups of organisms results in methane production, which is exclusively due to methanogens. Biomethane produced by anaerobic digestion is colorless, odorless, and non-poisonous gas (Okonkwo et al. 2018).

3 Characterization of Feedstocks for Biogas Production

Every substrate has its own characteristics, bringing challenges combined with the chosen parameters for the digestion process. The main criteria to be considered are energy, environmental, and economic value combined with sustainability. Plant-based materials typically comprise of polysaccharides in the form of lignocellulose rendering potential for biogas formation. Material containing carbohydrates, proteins, and fats can be used as a feedstock as such or can be combined with appropriate co-substrates in order to enhance efficiency. They should be devoid of the pathogen and possess a balanced C/N ratio to mitigate ammonia accumulation (Braun 2007). Animal manures are endowed with rich microflora and have a moisture content of 75–92% and 72–93% of volatile solids evidencing them as ideal substrate (Fujino et al. 2005; Müller et al. 2004). They also characterized with low digestion and the presence of toxic compounds. Biomass with carbohydrate and proteins exhibits a higher conversion rate than fat; the latter is described to yield higher biogas (Zubr 1986). Lignocellulosic biomass serves as a readily available and cheap substrate for the anaerobic digestion (Paul and Dutta 2018). About 500 MT of crop residues are being generated every year in India (Punia et al. 2017), which is a tremendous potential for the surplus biomass to be used as a substrate.

Biogas yield and its compositions mainly depend upon feedstock's type and reaction conditions used in the biogas plant. So, the feedstocks must be analysed to estimate the important parameters of feedstocks or to find out their suitability for biogas production. The several parameters that have to be considered for selecting feedstocks for biogas production are pH, total solids, dry matter, volatile solids, organic dry matter, TKN and biomethane potential, etc. Analytical procedures used for estimating essential parameters for their suitability for biogas production are given in Table 1.

Table 1 Analysis followed for different feedstocks for biogas production (Drosg et al. 2013)

S. No.	Parameter to be determined	Name of standards used
1	Sample collection and preparation	VDI 4630 and ISO 566713
2	pH	Standards EN 12176 and APHA 4500-H + B
3	Total solids	Standards EN 12880 and APHA 2540 B
4	Volatile solids (VS) and organic dry matter (ODM)	EN 12879 and APHA 2540 E
5	Chemical oxygen demand (COD)	Standards DIN 38414 and APHA 5220 B
6	Nitrogen content	ISO 5663; ISO 11261; APHA 4500–Norg B
7	Biochemical methane potential	EN 11734, DIN 38414 (S8) and VDI 4630
8	Sulphur content	ISO 118851
9	Phosphorous content	ISO 6878, DIN 38414 (S12) & APHA 4500-P
10	Total organic carbon (TOC)	EN 1484 or APHA 5310

4 Biogas Potential in India

According to an estimate for fuelwood requirement in India, 853.9 million people utilize a quantity of 58.8 million tonnes of fuelwood for their fuel requirement (Indiastat 2011). Any organic waste digested under anaerobic conditions using a group of microorganisms is called as biomethanation technology. Based on cattle census of 1981–1982, the Ministry of New and Renewable Energy, Government of India, estimated the potential of family size biogas plants in the country was 1.2 cores. Nearly half crore biogas plants (48.35 lakh) with different capacities (1–10 m^3) were installed up to 2018 in India under the National Biogas and Manure Management Programme with MNRE fund. From the installed biogas plants in India, total biogas production is about 2.07 billion m^3 per year (Mittal et al. 2018), which is equivalent to 0.89 billion kg of LPG. Two significant benefits are obtained by this biogas technology. Firstly, reduce the import bill in terms of replacement of LPG (liquid petroleum gas) cylinders or alternate for fuelwood. Secondly, promote organic farming in terms of generation of organic fertilizer under biogas technology. Applications of biogas produced from domestic size biogas plants are used for cooking, lighting, and biogas-fuelled engine coupled with pump for irrigation purposes. For the adoption of this technology in rural villages, the Government of India is providing supports in terms of subsidy for plant construction.

5 Biogas Developments

Anaerobic digestion is not a new technology for India and introduced at the end of the last century. Numerous researchers from academic and private organizations and NGOs (non-governmental organizations) have contributed to the development of a variety of models of biogas plants suitable for organic wastes. The brief history of biogas developments in India is presented in Table 2. After the formulation of MNRE, several biogas programmes are implemented along with different schemes of renewable energy technologies under this ministry. Biogas schemes are popularized through different stakeholders such as academic institutions, different state-level renewable energy agencies, Biogas Development and Training Centres (BTDC), and NGOs. In order to streamline and popularize successful model biogas

Table 2 Chronological order of biogas developments in India

1897	Biogas used for lighting purpose at Matunga Leper Asylum, Mumbai
1900	A biogas plant built at Bombay, India, and plant did not functioned well
1937	Sewage treatment plant using anaerobic digestion process installed at Mumbai
1946	The first biogas plant designed by N.V Joshi at the Indian Agricultural Research Institute
1952	Development of the floating dome model, Grama Laxmi III by Joshbai Patel
1961	Establishment of Gobar Gas Research Station at Ajitmal, Planning Research and Action Division, Government of Uttar Pradesh
1962	Khadi Village and Industrial Commission (KVIC) developed KVIC model biogas plants and were standardized
1977	The first pilot plant (fixed dome type – Janata model) developed by Gobar Gas Research Station at Ajitmal, Etawah, Planning Research and Action Division
1978	Janata Model Biogas Plant by Gobar Gas Research Station, Ajitmal
1981	Launching of the National Project on Biogas Development (NPBD)
1982	The National Project on Biogas Development is starting to operate under the Department of Non-Conventional Energy Sources (DNES)
1984	Deenbandhu model biogas plant developed by Action for Food Production (AFPRO)
1985	Massive-scale implementation of the National Programme through subsidies, multi-organization, and multi-design approach
1986	Deenbandhu model biogas plant (1–6 m^3) was approved by DNES and popularized through NPBD
1985–1992	Improving designs, improving the organization, and results from the dissemination
1992–2002	Reduction in subsidies, promotion, dissemination, and extension
2005–2018	The National Biogas and Manure Management Programme (NBMMP)
2017–2020	Biogas-Based Power Generation (Off-Grid) Programme (BPGP) launched and scheme modified as Biogas-Based Power Generation and Thermal Application Programme (BPGTP)
2018–2020	The National Biogas and Organic Manure Programme (NBOMP) launched

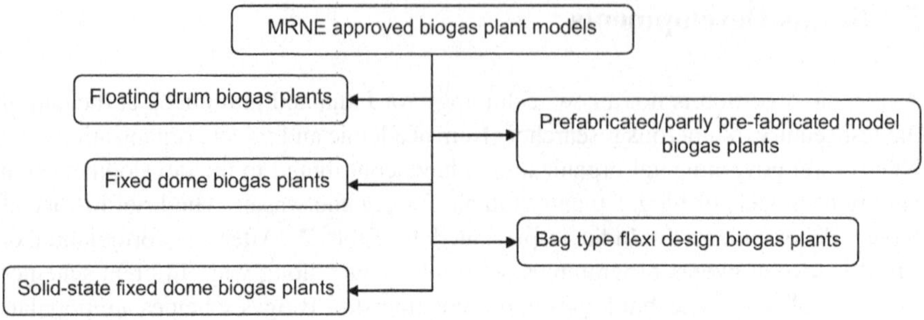

Fig. 1 The MRNE-approved models of biogas plants (1–25 m^3) in India

plants, the MNRE approved five different types of biogas plant models for subsidy and promotion (Fig. 1). The details of biogas plant potential (1–10 m^3) and the cumulative physical number of biogas plants installed in India through different biogas programmes is given in Table 3. An estimate shows that total installation potential biogas plant is about 1.23 crores of plants, and out of them only 48.35 lakhs numbers of biogas plants are installed (Table 3). Recently, biogas-based power generation is promoted, and totally 96 power plants installed between 2015 and 2018 and state-wise distribution of these power plants are presented in Table 4.

6 Popular Feedstocks

Several organic feedstocks are used for biogas production such as livestock wastes, agro-processing industrial wastes, energy crops, agricultural wastes, agro-industrial wastes, organic fraction of municipal solid wastes, sewage sludge, and forest residues (Kumar et al. 2015). It is a critical parameter governing the process of providing nutrients. Some can be used as a sole substrate, while few should be co-digested to ensure its contribution of favourable conditions for microbial growth. Co-substrates are added especially to increase the gas yield, and any bio-waste is commonly used. In India, cow dung is more popular feedstock used in the biogas plants installed at rural and semi-urban areas. Because of this reason, the biogas is also called as "Gobar gas." The use of different organic feedstocks influences the process besides their distinct physicochemical properties.

7 Popular Biogas Plant Models in India

The main components of any biogas plant (fixed dome or floating drum) are inlet and mixing tank, digester, gas holder, outlet tank, and pipe accessories for connecting inlet and outlet tanks to the digester. Based on the plant capacity and its installation

Table 3 Family size biogas plants potential and its installed numbers in India (Indiastat 2018a)

Name of the states/UTs	Potential, numbers	Installed biogas plants, numbers
Andaman and Nicobar Islands	22,000	137
Andhra Pradesh	1,065,000	532,311
Arunachal Pradesh	7500	3475
Assam	307,000	114,119
Bihar	733,000	129,826
Chandigarh	1400	97
Chhattisgarh	400,000	51,241
Dadra and Nagar Haveli	2000	169
Delhi	12,900	681
Goa	8000	4109
Gujarat	554,000	430,025
Haryana	300,000	60,753
Himachal Pradesh	125,000	47,424
Jammu and Kashmir	128,000	3072
Jharkhand	100,000	7326
Karnataka	680,000	478,958
Kerala	150,000	144,396
Madhya Pradesh	1,491,000	353,502
Maharashtra	897,000	871,494
Manipur	38,000	2128
Meghalaya	24,000	9996
Mizoram	5000	5020
Nagaland	6700	7903
Odisha	605,000	265,975
Puducherry	4300	578
Punjab	411,000	171,765
Rajasthan	915,000	70,139
Sikkim	7300	8874
Tamil Nadu	615,000	222,283
Telangana	–	9900
Tripura	28,000	3368
Uttar Pradesh	1,938,000	438,817
Uttarakhand	83,000	18,478
West Bengal	695,000	366,595
India	*12,339,300*	*4,834,934*

site, biogas plants are classified into three subclasses: (i) family-size biogas plants (< 6 m^3), (ii) institutional biogas plants (> 15 m^3), and (iii) community biogas plants (> 15 m^3). Various designs of the biogas plants being propagated under the programme may be broadly classified as floating drum popularly known as KVIC type and fixed dome popularly known as Janata and Deenbandhu type. The biogas plant can be

Table 4 Details of biogas power plants installed in India (2015–2018) (Indiastat 2018b)

States/UT	Organic waste-based biogas plants (nos.)
Andhra Pradesh	7
Karnataka	7
Madhya Pradesh	4
Maharashtra	7
Punjab	14
Tamil Nadu	13
Telangana	15
Uttar Pradesh	26
Uttarakhand	3
India	*96*

categorized based on the characteristics of organic waste used as wet (<10% total solids, TS), semi-dry (10–20% TS), and dry (>20% TS).

The family-size biogas plants are being promoted by the MNRE and use cattle dung slurry with 10% of TS. The biodigested slurry has more amount of water, which can be applied to agricultural fields in the form of wet or dried cake. The extension of biomethanation technology for domestic size biogas plants viz., floating drum, and fixed dome types has already progressed. These domestic biogas plants have to use the mixing ratio of cattle dung and water as 1:1. Several organizations developed a variety of biogas plants suitable for cow dung. Only few biogas plants got approved and promoted via biogas scheme. The biogas plants approved by MNRE for popularizing biogas technology through NBOMP scheme is shown in Fig. 1.

8 KVIC Model Biogas Plant

The main components of the KVIC biogas plant are inlet cum mixing tank, outlet tank, digester with a partition wall, control valve for biogas, gas holder, and inlet and outlet pipes. The KVIC model biogas plant is also known as floating drum-type biogas plant due to a metal drum used as a gas holder for this plant (Fig. 2). The well-shaped digester was constructed below ground level and cylindrical metal drum with a conical top (gas holder) placed in the digester. The digester and gas holder is separate in this type of biogas plant. Both inlet and outlet pipes are used to connect the inlet cum mixing tank and outlet tank, respectively. A partition wall was constructed at the bottom and the centre of the biogas plant to increase the hydraulic retention time. In other words, undigested slurry should be avoided by this partition wall arrangement. The floating drum supported by a guide pipe arrangement could help the gas holder move vertically up and down at the time of biogas production and consumption. The weight of the floating drum acting like a dead load on biogas accumulated in the drum, which would lead to release the biogas at constant pressure. This kind of biogas plant is costlier due to the metal gas holder, and drum cost alone accounts for 30–40% of the total cost.

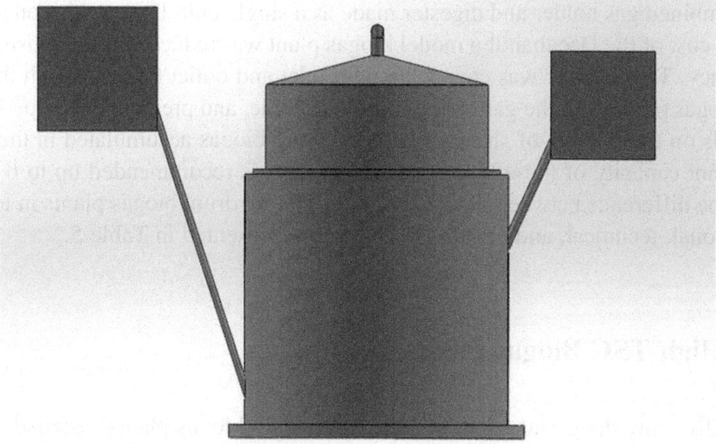

Fig. 2 Floating drum model biogas plant

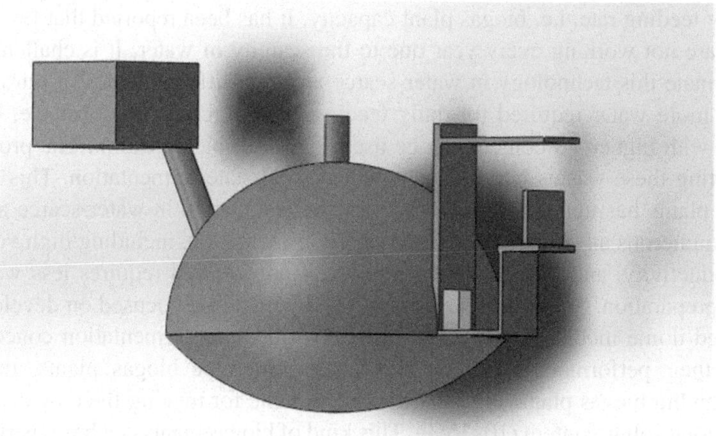

Fig. 3 Fixed dome model biogas plant

9 Fixed Dome Model Biogas Plant

The main drawbacks of KVIC biogas plant are a higher initial investment and shorter lifespan of gas holder. To overcome these issues, Deenbandhu model biogas plant (Fig. 3) was developed by AFPRO, New Delhi, in 1984, and it falls under a fixed dome biogas plant. This plant works as similar to Janata model, except the configuration of inlet entrance and digester. The digester was constructed in the segment of spherical shape, and dome was constructed with the same base diameter. The dome can act as a gas holder. As compared with the KVIC model biogas plant; this model has combined gas holder and digester made as a single unit. Due to this concept, the overall cost of the Deenbandhu model biogas plant was reduced without losing plant efficiency. The digester was connected with inlet and outlet tanks through the pipe. The biogas pressure at the gas outlet point is variable, and pressure developed utterly depends on the volume of slurry displaced by the biogas accumulated in the dome. The plant capacity of these types of biogas plants is recommended up to 6 m^3 per day. The difference between fixed dome and floating drum biogas plants in terms of operational, technical, and economic aspects are presented in Table 5.

10 High TSC Biogas Plants

Generally, cow dung was used as feedstock in the biogas plants installed in rural areas, and cow dung contains about 20% of total solids. For better biogas yield, cow dung with 10% of TS was fed into biogas plants. For example, 2 m^3 capacity biogas plants require 50 litres of water for cow dung slurry preparation to maintain optimal TS content. The daily requirement for slurry preparation increases with an increase in daily feeding rate, i.e. biogas plant capacity. It has been reported that few biogas plants are not working every year due to the scarcity of water. It is challenging to disseminate this technology in water-scarce areas in different states of our country due to more water required for daily feeding. To overcome this obstacle, organic wastes with higher TS content can be used for anaerobic digestion. The process of converting these wastes into biogas is called solid-state fermentation. This type of biogas plants has tremendous potential for the installation in water-scarce areas. It offers numerous advantages over submerged fermentation, including high volumetric productivity, and digestate with less water content and requires less water for slurry preparation. Several researchers worked with a goal focused on development for fixed dome model biogas plant based on solid-state fermentation concept and tested their performance. Among the fixed dome-type biogas plants, modified Deenbandhu biogas plant (Fig. 4) was well suitable for treating the cow dung with higher total solids content (10–15%). This kind of biogas plants can be constructed at low cost and requires less maintenance. Figure 5 illustrates the sequence of activities planned for the modified Deenbandhu model biogas plant.

11 Operational Conditions of Biogas Plants

The digester's optimum operational conditions are significant for efficient biofuel production. Besides the nutrient composition, there are a number of operating parameters that should be paid attention like pretreatment procedures, quality and quantity of input material, temperature, pH, retention time with certain qualitative evaluations like biological and chemical oxygen demand (van Lier et al. 2008),

Table 5 Comparative analysis of fixed dome and floating drum biogas plants

S. No.	Parameters	Fixed dome biogas plant	Floating drum biogas plant
A.	*Technical aspects*		
1	Plant construction	Requires trained and skilled persons required for dome construction	Requires skilled welders required for drum fabrication
2	Biogas storage	Dome	Floating drum
3	Example for each case	Janata model and Deenbandhu model	KVIC model
4	Digester and gas holder	Both are combined and made into a single unit	Digester and gas holder are separate
5	Space utilization	Entire digester constructed below ground level. The above space can be used for other purposes	May not be utilized for other purposes due to floating drum arrangement
6	The method used to avoid short circuit, i.e. fresh slurry noticed at the outlet tank due to lesser residence time/ without digestion	The digester shapes itself to support to complete digestion of the organic waste that takes place	Partition wall constructed at the centre of the digester
B.	*Operational aspects*		
7	Repair and maintenance	Identify the location of defects that is difficult	Identification is easy and easily repairable
8	Biogas pressure	Variable pressure. It depends on biogas production and pressure developed and acting on the slurry surface by produced biogas	Constant pressure. The drum weight increased the pressure the biogas
9	Effect of winter temperature on biogas production	Biogas production at the time of winter season is less	Biogas production at the time of winter season is less than summer
10	Breaking of scum	Difficult. Long-sized bamboo pole was used	Easy. Two or three times half rotation of the drum in clock–/anticlockwise direction
11	The simplest way to identify the biogas production in the plant	Slurry level in the outlet tank	Vertical movement of the floating drum in the plant
C.	*Economical aspects*		
12	Life span	25 years (for biogas plant)	10 years (for floating drum)
13	Investment	Capital investment is low due to masonry work	Higher due to higher material cost and labour cost for fabrication the metal gas holder
14	Cost of maintenance	Less as other	More due to painting cost

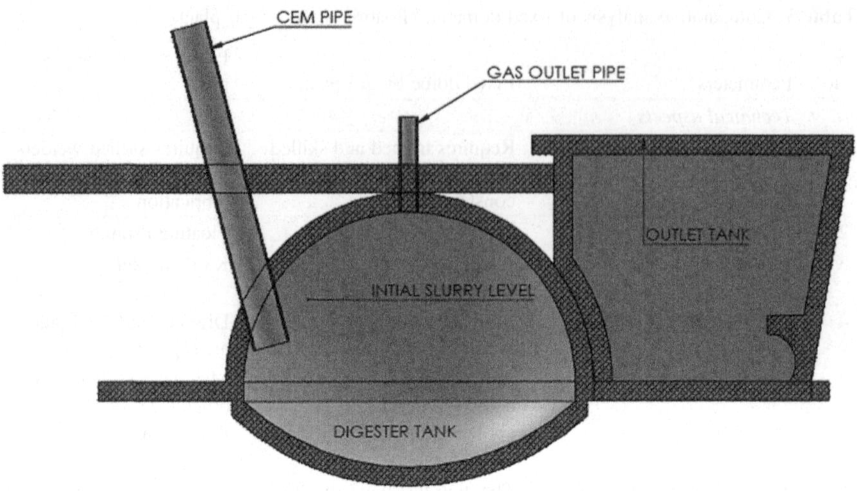

Fig. 4 Floating drum model biogas plant

carbon/nitrogen ratio, methane yield, and volatile solids. The prerequisite on the digester design is focused to maximum degradation of volatile solids and higher methane yield with short hydraulic retention time (Ward et al. 2008). Altogether the design and operation should warrant higher microbial activity and gas yield, minimizing the risk of inhibition. Co-substrate selection, their pretreatment, process optimization, and desulphurization are regarded as important steps in downstream process, which ultimately enhance biogas quality, digestate dewaterability, and biogas odour emission. Only 40–60% of carbon in the organic material is converted into methane, while the rest remains in digestate.

Nasir et al. (2012) have suggested the benefits of co-digestion of poultry litter wastes for improving biogas yield, separation of hydrolysis, and methanogen in good stability, use of poultry manure for higher biogas production potential. Usually, the introduction of H_2 and CO_2 into an anaerobic digester which is dominated by methanogenic environment creates a hydrogenotrophic dominant environment which is well studied by batch and continuous experiments (Szuhaj et al. 2016; Bassani et al. 2015). Enhanced performance was analysed by optimizing two critical operational conditions, viz. the reaction temperature and duration of stirrers in operation (Önen et al. 2018). Increase in optimum temperature from 40 to 43 °C and stirrer operating time yielded developed specific electricity of 11.7% with reduced electricity intake. The digester design should be useful in providing the yield with the resource available in the area, minimum monitoring, and suit practicality.

Fig. 5 The sequence of activities planned for construction of high solid-state biogas plant

12 Production Level

Eduok et al. (2018) have achieved higher biogas production and lower volatile fatty acids with the improved proliferation of methanogens. The anaerobic fermentation, along with combined urine as the wetting agent serves as a likely option for enhanced biogas production. Fantozzi and Buratti (2009) have designed single-stage, batch, mixed digester and assessed the biogas potential of different feedstocks. The maximum biogas production and energy observed for piggery wastes were about 0.35 Nm^3/kg and 1.35 kWh/kg of VS, respectively. Pilot studies on anaerobic digesters were conducted by Migliori and coworkers, where batch and continuous cultures have been compared under varied humidity and solid content pertaining to biogas quality and quantity. Using an organic fraction of municipal solid wastes as a substrate, the outcome of the study revealed that the biogas production increased with continuous process compared to the batch process. This is due to the increased dry matter production in the continuous process. A decrease in biogas quality (5%) has been observed on shifting from batch to continuous cultures due to production of volatile fatty acids (VFA). Reduction in the quality of biogas quality from 58.4 to 56.8 and the increase in production rates from 7.8 m^2 to 9.2m^2 raised the need for a detailed study on the factors influencing biogas quality as well as quantity (Migliori et al. 2019).

In the study conducted by Fernández et al. 2014 for methane production and the CO_2 utilization rates in two batch type anaerobic digesters fed with food wastes and sewage sludge respectively, different concentration of CO_2 (0.3, 0.6 and 0.9 molar fractions of CO_2) was injected to the digester. CO_2 has been injected for 20 minutes to reach equilibrium in the anaerobic digesters provided that change in the period of injection resulted in pH change. Reduction in CO_2 concentration has been observed at a rate of 3 to 11% and 8 to 34% for food waste and sewage sludge, respectively.

13 Energy Savings with Biogas

Energy is one of the most important global commodities to be conserved. The Energy Statistics (2018) reported that the compound annual growth rate (CAGR) of energy consumption is comparatively higher than the production rate over decades irrespective of their source. Nearly 48 lakhs of biogas plants were constructed in India to date, and the net production of natural gas has increased around 2.9% (Energy statistics 2019). Energy balance in biogas digester plays a crucial role in energy conservation. During biogas production, the process of heating, especially in the case of thermophilic anaerobic digesters, consumes comparatively more energy than other processes. Experimental results suggest that increased temperature reduces the digestion efficiency. For instance, experiments conducted by Navickas et al. (2013) resulted in a change in the energy conversion ratio of about 3.2 to 4.9 at 57 °C and 52 °C, respectively. One of the best ways for energy conservation is efficient production and utilization of energy. In the Indian scenario, LPG (liquid petroleum gas) is used widely, which has several disadvantages over biogas. It has been suggested that biogas has comparatively lowest environmental impacts over other fuels in India (Singh et al. 2014). India has a total estimated energy potential of about 25.7 GW of which the waste to energy conversion potential is about 1.2 GW (Rao et al. 2010) which brings to light that biogas can be the best alternative over other fuels.

14 Utilities for Biogas

Biogas and the digestate acquired from the process of anaerobic digestion entails further processing for the intended use and safe disposal. As the by-product of biogas digestion, digestate serves as high-value fertilizer and substitute for mineral fertilizer. This is accomplished only by enhancing the quality of biogas by removing unnecessary gases like ammonia, H_2S, siloxanes, carbon dioxide, and water vapour. In order to get pure methane, various CO_2 removal methods are followed like adsorption, absorption, cryogenic separation, and membrane separation (Vrbová and Ciahotný 2017). Each technique has its own merits and limitations. The steps for improving the digestate can be classified into two types, namely, pre-digestion and post-digestion.

15 Digestate Upgradation

Pre-digestion methods are done to increase gas production and digestate quality either by enzymes or chemicals or any physical or biological methods. The post-digestion includes two steps, viz. digestate conditioning and treatment. Digestate is solid-liquid suspension endowed with effluent macro- and micronutrients. Digestate management is the crucial step not only in preventing pollution but also in efficient storage and usage. Their characters are dependent on factors such as feedstocks, microbial community, reaction conditions, types of anaerobic digester used, and processing methods of digestates. Various digestate enrichment techniques are being followed to lessen environmental hazard and better applicability (Bauer et al. 2009). De-watering is the cost-effective and foremost step for efficient digestate management in solid/liquid separation. They are followed by processing of solid fractions which involve composting and drying. Liquid fraction processing is also done for the recovery of nutrients with a specific choice of interest. Edwards et al. (2017) and Li et al. (2017) have elucidated the accomplishment of biosolids with 15–30% solid content by typical dewatering process. They are lined by several factors like composition of the digestate and comprises of preconditioning and physical separation technique. The quality of the digestate is ensured chemically and biologically pertaining to the presence of heavy metals, organic contaminants, or pathogens.

16 Methane Upgradation

16.1 Carbon Dioxide Scrubbing

In order to upgrade the methane efficiency, the removal of CO_2 is essential. Since CO_2 impurity in the biogas reduces the calorific value (CV) of the methane produced, removal of CO_2 results in the increased CV and decreased the relative density of biomethane, and hence increase in the Wobbe index is observed (Ryckebosch et al. 2011). Various methods are being followed to remove CO_2 in biomethane production, which involves physical and chemical absorption, pressure swing adsorption (PSA), membrane and cryogenic separation, and biological CH_4 enrichment. Of these, biological methane enrichment is found to be sustainable since it is inexpensive and does not have any unwanted end products (Ryckebosch et al. 2011). The synthetic biogas (CH_4, 50–60%; CO_2, 30–40%; H_2S, 1–2%) was used to increase the methane content of biogas by using biological scrubbing method. The hollow fibre packed with organisms was used for this purpose. Test results show that the biogas was sent through hollow fibres yields biogas with higher methane content (CH_4, 96%; CO_2, 4%) (Strevett et al. 1995).

Several other techniques like high-pressure water scrubbing, a combination of adsorption and absorption methods (Al Mamun and Torii 2017), carbon molecular sieves, and membrane purification are being used according to the economic feasibility.

17 Socio-economic Impacts

Since 1970, the promotion of biogas technologies is on the progress to ease the energy crisis. Development of biogas plants encourages mitigation of greenhouse gases, reduction, and processing of wastes and helps in attaining self-sufficiency and sustainability of energy. Kurudampalayam Panchayat of Coimbatore District, Tamil Nadu state, India, has displayed biomethanation plant with the technical assistance of the Bhabha Atomic Research Centre (BARC), funded by the District Rural Development Agencies (DRDA) with the cost of 30 lakhs Indian rupees. It has the potential to produce four cylinders of gas per day, which benefits more than 250 families. The community plant utilizes cow dung and other biodegradable wastes like vegetables, fruits, agro-wastes, and slaughter wastes as substrate. Tamil Nadu Agricultural University (TNAU) has a plug flow biogas plant with the average biogas yield of 0.8–0.9 m^3/d. It is found to be appropriate for various feedstocks like cattle dung, vegetable residue, and several weeds and plant wastes. NBMMP is one of the Indian government schemes to promote and installation of biogas plants at rural and semi-urban areas. The family-size biogas plant (up to 6 m^3) utilizes the cow dung as feedstocks to produce biogas, which might fulfil their daily fuel requirements. Because of feedstocks availability, awareness, and government supports (through subsidies, promotion schemes, and training), nearly 48 lakhs of biogas plants have been constructed at different states of the country up to 2018 (Table 3).

18 Conclusion

The biogas technology is a well-proven technology to produce both biofuel and valuable manure from organic wastes simultaneously. This technology supports the safe disposal of organic waste and improves excellent sanitation practices or environment. The Indian Government is involved in promoting the biogas programs through various implementing agencies to achieve the target. Since, there are a sufficient number of bovines available in India, the effective implementation of the biogas programme to at least fulfil the fuel requirement of the homes in the rural and semi-urban areas. Among the different states, the northeast states show the very progressive implementation of the biogas programme, and the number of biogas plants installed in these states is more than that of estimated potential biogas plants. There will be a massive demand for the adoption of this technology in rural villages and the Government of India also providing supports in terms of subsidy for plant construction.

References

Al Mamun MR, Torii S (2017) Enhancement of methane concentration by removing contaminants from biogas mixtures using combined method of absorption and adsorption. Int J Chem Eng 2017:9

Bassani I, Kougias PG, Treu L, Angelidaki I (2015) Biogas upgrading via hydrogenotrophic methanogenesis in two-stage continuous stirred tank reactors at mesophilic and thermophilic conditions. Environ Sci Technol 49:12585–12593

Bauer A, Mayr H, Hopfner-Sixt K, Amon T (2009) Detailed monitoring of two biogas plants and mechanical solid–liquid separation of fermentation residues. J Biotechnol 142:56–63

Braun R (2007) Anaerobic digestion: A multi-faceted process for energy, environmental management and rural development, *Improvement of crop plants for industrial end uses*. Springer, pp. 335–416

Cai M, Wilkins D, Chen J, Ng S-K, Lu H, Jia Y, Lee PKH (2016) Metagenomic reconstruction of key anaerobic digestion pathways in municipal sludge and industrial wastewater biogas-producing systems. Front Microbiol 7:778

Dassa B, Borovok I, Ruimy-Israeli V, Lamed R, Flint HJ, Duncan SH, Henrissat B, Coutinho P, Morrison M, Mosoni P (2014) Rumen cellulosomics: Divergent fiber-degrading strategies revealed by comparative genome-wide analysis of six ruminococcal strains. PLoS One 9: e99221

Drosg B, Braun R, Bochmann G, Al Saedi T (2013) Analysis and characterisation of biogas feedstocks. In: The Biogas Handbook. Woodhead Publishing, pp 52–84

Eduok S, John O, Ita B, Inyang E, Coulon F (2018) Enhanced biogas production from anaerobic co-digestion of lignocellulosic biomass and poultry feces using source separated human urine as buffering agent. Front Environ Sci 6:67

Edwards J, Othman M, Crossin E, Burn S (2017) Anaerobic co-digestion of municipal food waste and sewage sludge: A comparative life cycle assessment in the context of a waste service provision. Bioresour Technol 223:237–249

Energy statistics (2018) 25th issue. Ministry of statistics and programme implementation. Government of India

Energy statistics (2019) 26th issue. Ministry of statistics and programme implementation. Government of India

Fantozzi F, Buratti C (2009) Biogas production from different substrates in an experimental continuously stirred tank reactor anaerobic digester. Bioresour Technol 100:5783–5789

Fernández YB, Soares A, Villa R, Vale P, Cartmell E (2014) Carbon capture and biogas enhancement by carbon dioxide enrichment of anaerobic digesters treating sewage sludge or food waste. Bioresour Technol 159:1–7

Fujino J, Morita A, Matsuoka Y, Sawayama S (2005) Vision for utilization of livestock residue as bioenergy resource in japan. Biomass Bioenergy 29:367–374

Gerardi M (2003) The microbiology of anaerobic digesters. Wastewater microbiology series, New Jersey, Wiley

Indiastat (2011). https://www.indiastat.com/table/forest-and-wildlife-data/13/forest-produce/17875/619501/data.aspx

Indiastat (2018a). https://www.indiastat.com/table/power/26/biomas-biogas-1985-2018/452705/1142342/data.aspx

Indiastat (2018b). https://www.indiastat.com/table/power-data/26/non-conventional-energy/184/1207557/data.aspx

Kumar A, Kumar N, Baredar P, Shukla A (2015) A review on biomass energy resources, potential, conversion and policy in india. Renew Sust Energ Rev 45:530–539

Li H, Jin C, Zhang Z, O'Hara I, Mundree S (2017) Environmental and economic life cycle assessment of energy recovery from sewage sludge through different anaerobic digestion pathways. Energy 126:649–657

Luo G, Wang W, Angelidaki I (2013) Anaerobic digestion for simultaneous sewage sludge treatment and co biomethanation: Process performance and microbial ecology. Environ Sci Technol 47:10685–10693

Migliori M, Catizzone E, Giordano G, Le Pera A, Sellaro M, Lista A, Zanardi G, Zoia L (2019) Pilot plant data assessment in anaerobic digestion of organic fraction of municipal waste solids. PRO 7:54

Mittal S, Ahlgren EO, Shukla PR (2018) Barriers to biogas dissemination in India: A review. Energy Policy 112:361–370

Müller B, Sun L, Schnürer A (2013) First insights into the syntrophic acetate-oxidizing bacteria–a genetic study. Microbiology 2:35–53

Müller JA, Winter A, Struenkmann G (2004) Investigation and assessment of sludge pre-treatment processes. Water Sci Technol 49:97–104

Nasir IM, Ghazi TIM, Omar R (2012) Anaerobic digestion technology in livestock manure treatment for biogas production: A review. Eng Life Sci 12:258–269

Navickas K, Venslauskas K, Petrauskas A, Zuperka V, Nekrosius A (2013) Energy balances of biogas production from industrial wastes and energy plants. Energy 4:6

Okonkwo UC, Onokpite E, Onokwai AO (2018) Comparative study of the optimal ratio of biogas production from various organic wastes and weeds for digester/restarted digester. J King Saud University-Engineering Sci 30:123–129

Önen S, Nsair A, Kuchta K (2018) Innovative operational strategies for biogas plant including temperature and stirring management. Waste Manag Res 37:237–246

Paul S, Dutta A (2018) Challenges and opportunities of lignocellulosic biomass for anaerobic digestion. Resour Conserv Recycl 130:164–174

Punia R, Marlar R, Kumar S, Tyagi S (2017) Potential of bioelectricity production in india through thermochemical conversion of lignocellulosic biomass, *Sustainable biofuels development in india*. Springer:189–206

Purohit P, Dhar S (2018) Lignocellulosic biofuels in India: Current perspectives, potential issues and future prospects. Aims Energy 6:453–486

Rao PV, Baral SS, Dey R, Mutnuri S (2010) Biogas generation potential by anaerobic digestion for sustainable energy development in india. Renew Sust Energ Rev 14:2086–2094

Riviere D, Desvignes V, Pelletier E, Chaussonnerie S, Guermazi S, Weissenbach J, Li T, Camacho P, Sghir A (2009) Towards the definition of a core of microorganisms involved in anaerobic digestion of sludge. ISME J 3:700

Ryckebosch E, Drouillon M, Vervaeren H (2011) Techniques for transformation of biogas to biomethane. Biomass Bioenergy 35:1633–1645

Singh N, Kumar A, Rai S (2014) Potential production of bioenergy from biomass in an Indian perspective. Renew Sust Energ Rev 39:65–78

Strevett KA, Vieth RF, Grasso D (1995) Chemo-autotrophic biogas purification for methane enrichment: Mechanism and kinetics. The Chemical Engineering Journal and The Biochemical Engineering Journal 58:71–79

Szuhaj M, Ács N, Tengölics R, Bodor A, Rákhely G, Kovács KL, Bagi Z (2016) Conversion of H_2 and CO_2 to CH_4 and acetate in fed-batch biogas reactors by mixed biogas community: A novel route for the power-to-gas concept. Biotechnol Biofuels 9:102

Thauer RK (2012) The wolfe cycle comes full circle. Proc Natl Acad Sci 109:15084–15085

van Lier JB, Mahmoud N, Zeeman G. (2008) Anaerobic wastewater treatment. Biological Wastewater Treatment: Principles, Modelling and Design: 415–456

Vrbová V, Ciahotný K (2017) Upgrading biogas to biomethane using membrane separation. Energy Fuel 31:9393–9401

Ward AJ, Hobbs PJ, Holliman PJ, Jones DL (2008) Optimisation of the anaerobic digestion of agricultural resources. Bioresour Technol 99:7928–7940

Weiland P (2003) Production and energetic use of biogas from energy crops and wastes in germany. Appl Biochem Biotechnol 109:263–274

Werner JJ, Knights D, Garcia ML, Scalfone NB, Smith S, Yarasheski K, Cummings TA, Beers AR, Knight R, Angenent LT (2011) Bacterial community structures are unique and resilient in full-scale bioenergy systems. Proc Natl Acad Sci 108:4158–4163

Wilkins D, Lu X-Y, Shen Z, Chen J, Lee PK (2015) Pyrosequencing of *mcr a* and archaeal 16s rRNA genes reveals diversity and substrate preferences of methanogen communities in anaerobic digesters. Appl Environ Microbiol 81:604–613

Wirth R, Kovács E, Maróti G, Bagi Z, Rákhely G, Kovács KL (2012) Characterization of a biogas-producing microbial community by short-read next generation DNA sequencing. Biotechnol Biofuels 5:41

Zubr J (1986) Methanogenic fermentation of fresh and ensiled plant materials. Biomass 11:159–171

Anaerobic Digestion Technology for Management of Organic Wastes: Latin American Context

Jesús Rubén Rodríguez-Nuñez and Omar Surisadai Castillo Baltazar

Abstract Organic waste generated from agricultural, livestock and industrial activities, as well as the organic fraction of municipal and domestic waste can be revalued through anaerobic digestion. In the application of this treatment, biogas and anaerobic effluents rich in carbon, nitrogen, and phosphorus are obtained. The biogas, rich in methane, can contribute to reducing the dependence on fossil fuels, as a means of heating or obtaining electricity. The effluents can be used as agricultural fertilizers and animal feed. In Latin America the most used large-scale anaerobic digestions are used to treat wastewater using "up-flow anaerobic sludge blanket" reactors, as well as the treatment of animal excreta. For small or medium scale, the domestic biodigesters are widely used by rural communities and agricultural producers throughout Latin America. The currently challenge for Latin American region is to be most efficient recycling its organic waste that allows to reduce their energy dependence on fossil fuels and therefore decrease their contribution of greenhouse gases. In this sense, the objective of this work is to present an overview about the use of anaerobic digestion technology in Latin America, focused on its progress in rural communities, small and medium agricultural producers and marginalization areas, as an alternative to improve the quality of life and its influence on reducing the environmental impact.

Keywords Revaluation of organic wastes · Domestic biodigesters · Heating from biogás · Rural communities

1 Introduction

The limited access to electric services, drinking water, and drainage and the cost of energy in some regions of Latin American obstruct the development of rural and indigenous communities. Furthermore, in the large cities with overpopulations exists

J. R. Rodríguez-Nuñez · O. S. Castillo Baltazar (✉)
Programa de Ingeniería en Biotecnología, Universidad de Guanajuato, Celaya, Mexico
e-mail: omar.castillo@ugto.mx

N. Balagurusamy, A. K. Chandel (eds.), *Biogas Production*,
https://doi.org/10.1007/978-3-030-58827-4_3

areas of high marginalization extend in the peripheries, which lack basic services such as water, drainage, electricity, and an appropriate waste collection system (Grau et al. 2015). All this social problem causes a decrease in the quality of life of the population and in public health (UN Environment 2018).

As alternative, the recycling of organic waste of food produced by chain of food industry, restaurants, agricultural, domestic, and markets, which correspond to one third part of the food produced for human consumption globally, as well as the municipal organic wastes estimated only in Latin America more than 50% of all solid waste collected (FAO 2016; HLPE 2014; LaGiglia et al. 2014; UN Environment 2018). This waste represents an excellent source of biomass to be treated by anaerobic digestion and generate biogas and biofertilizers as value-added products (Deublein and Steinhauser 2011).

In Latin America since the 1970s, the recycling of organic waste by anaerobic digestion at small and medium scale has been tried in response to the energy crisis, which recently it has extended (Chernicharo et al. 2015a; Martí-Herrero 2019). In those years the small dome-digesters were commonly used; however, the cost of installation, operation, and maintaining was expensive (Garfí et al. 2016). Nowadays, the tubular digester elaborated of polyvinyl chloride bags is the most used due to low cost (Lansing et al. 2008, 2017). Nevertheless, in Latin America this technology currently lacks of social acceptance, institutional supports, and limited financing; this represents a challenge to increase the use of this technology in the region (Martí-Herrero 2019).

The objective of this work is to present an overview about the use of anaerobic digestion technology in Latin America, focused on its progress in rural communities, small and medium agricultural producers and marginalization areas, as an alternative to improve the quality of life and its influence on reducing the environmental impact.

2 Organic Wastes Using for Anaerobic Digestion

Organic wastes that are used as substrates in anaerobic digestion are broad spectrum. Historically, anaerobic digestion has been used to treat waste effluents rich in organic matter such as excreta, municipal and industrial wastewater, and biological treatment sludge (Deublein and Steinhauser 2011; Vögeli et al. 2014). Since 1960, the use of municipal waste and agro-industrial waste as a substrate for anaerobic digestion was intensified due to its high organic load and its potential for biogas production (Vögeli et al. 2014). Table 1 shows the different substrates according on their origin.

Table 1 Main source of substrates used for anaerobic digestion (Vögeli et al. 2014)

Municipal waste	Agriculture	Industry
• Organic fraction from municipal waste • Excreta waste	• Cultive residues • Manure • Algae biomass • Agro-industrial waste	• Slaughterhouses waste • Food industry waste • Paper industry waste

Table 2 Percentage of TS and VS of diverse substrates used for anaerobic digestion (Deublein and Steinhauser 2011)

Substrate	TS (% of raw waste)	VS (% of TS)
Spent grain, fresh or ensilaged	20–26	75–95
Spent fruits	25–45	90–95
Vegetable wastes	5–20	76–90
Grass cuttings from lawns	37	93
Market wastes	8–20	75–90
Diverse kinds of cereals	85–90	85–89
Straw from cereals	86	89–94
Rice straw	25–50	70–95
Potato mash, potato pulp, Potato peelings	6–18	85–96
Wheat bran, wheat powder bran	87–88	93–95
Dry bread	65–90	96–98
Leftovers, overstored food	14–18	81–97

[a]*VS* volatile solids *TS* total solids

The potential use of the different solid substrates in anaerobic digestion depends on the biodegradable organic fraction that contributes to the formation of biogas (called volatile solids, VS), with respect to the total dry matter content, referred to as total solids (TS). In general, appropriate substrates for digestion are those that have between 70 and 95% of the TS as a biodegradable percentage. Those solids with a percentage of 60% or less are not considered appropriate substrates for anaerobic digestion (Deublein and Steinhauser 2011; Vögeli et al. 2014). Deublein and Steinhauser (2011) report extensive data on the percentage of TS and VS for different liquid and solid substrates and its viability to obtaining biogas through anaerobic digestion. Table 2 shows some of the solid wastes that are considered appropriate for biogas production.

Most of the substrates presented in Table 2 are harmless and with low complexity to be used for anaerobic digestion, with the exception of market wastes since these have a complex composition of biomass with different degrees of biodegradability (Deublein and Steinhauser 2011; Vögeli et al. 2014).

The substrates potential and particularly the solids used for biogas production can be increased by silage or pre-drying (Mortier et al. 2016). Silage is an anaerobic process (fermentation) that promotes the organic acid production mainly of the lactic acid and the production of volatile fatty acids (VFA's), which promotes the decrease of pH and inhibits the proliferation of harmful bacteria such as *Clostridium* (Vervaeren et al. 2010). The silage method is used for forage conservation of grasses and legumes (De la Roza-Delgado 2005). Recently, silage has been used as a pretreatment of the anaerobic digestion, since the VFAs and ethanol produced during this process are precursors for the production of methane (biogas), showing high yields (Baldini et al. 2017).

Historically, the biogas production using liquid excreta (effluents) as substrate for anaerobic digestion is the most common, due to their organic load and high availability (Deublein and Steinhauser 2011; Mortier et al. 2016; Kinyua et al. 2016).

Nowadays, the substrates are combined with other organic waste called "co-substrates" in order to increase the biogas yield (Deublein and Steinhauser 2011; Surendra et al. 2014). The addition of co-substrates increases the organic load of the feed to the anaerobic digester; also the use of co-substrates shows advantages due that offer the VFA's required to improve the silage (Surendra et al. 2014). However, the addition of co-substrates should be use with care due that carry some risks (depending on the usefulness of the digestates). For example, the use of co-substrates is not recommended when the digestates obtained from anaerobic digestion will be used in agricultural applications, because the co-substrates increase the levels of nitrogen and may exceed the permissible limits (Deublein and Steinhauser 2011; López and Borzacconi 2017).

The recycling of food wastes as substrates or co-substrates for anaerobic digestion can be revaluated, due to its high organic load and good biodegradability (Mortier et al. 2016; Vögeli et al. 2014). Also, since the last decade, the anaerobic digestion of municipal organic waste has emerged and continues to increase as an alternative to landfill and composting (Mortier et al. 2016). These residues are classified in two categories:

1. Garden, vegetables, and fruit wastes are usually used in composting and by silage as co-substrates in anaerobic digestion.
2. Domestic organic waste, used in anaerobic digestion to obtaining biogas (as the main product), in addition reduce the residues destined for landfill.

3 Solid Waste as Anaerobic Digestion Substrate: A Proposal for Latin America

The region of Latin America and the Caribbean produce 1 kg/day of municipal waste per habitant, producing a total of 541,000 t/day and is expected to increase 25% by 2050 (Hettiarachchi et al. 2018; UN Environment 2018). Also, important to remark is that 50% of the wastes generated are organic, which can be used as substrates for anaerobic digestion (UN Environment 2018).

In addition, Latin America is the world region that shows the highest rates of urbanization with 80% and was estimated that by 2050 the 90% of the population will live in urban areas (Hettiarachchi et al. 2018). This high rate of urbanization together with the economic crisis of this region generates a crisis in the waste management. Municipal authorities are unable to provide services and infrastructure to urban populations for the appropriate waste management (Grau et al. 2015; UN Environment 2018). In Latin America the improper handling of the waste management produce 5% of CO_2 emissions, mainly due to the disposal of open landfills (LaGiglia et al. 2014).

In addition to this problem, a regular and reliable waste collection service is required in Latin America, since more than 35,000 t/day remain uncollected, which impacts more than 7% of the region's population. Also, in this region implement an integral system for the management of solids using anaerobic digestion is

Table 3 Latin American countries with legislation focused to proper waste management (Mortier et al. 2016)

Country	Name of the law and approval year
Brazil	Brazil's National Law on Solid Waste (2010)
Argentina	Integral Management of Solid Urban Waste (2005)
Perú	General Law on Solid Wastes (2008)
Paraguay	Solid Waste Management Law (2009)
Mexico	General Law for Prevention and Integral Management of Wastes (2004)
Venezuela	Waste Management Law (2010)
Costa Rica	Integrated Waste Management Law (2010)

Table 4 Main food groups that generate high amounts of waste in Latin America (HLPE 2014; FAO 2016)

Food groups	Percentage losses (%)
Cereals	25
Roots and tubers	40
Oilseeds and legumes	20
Fruits and vegetables	55
Meat	20
Dairy products	20
Fish and seafood	33

complicated due to lack of resources to install this infrastructure (Grau et al. 2015; UN Environment 2018). Until 2010 only seven countries in this region promoted national legislation focused at the recovery, classification, recycling, minimization, and energy generation (biogas) from its generated wastes (Mortier et al. 2016) (Table 3).

The organic waste produced by food remains represent one of the main source of waste globally. The FAO (2016) reported that 1.3 billion tons of food are lost or wasted every year (equivalent to one third of the food produced for human consumption) (HLPE 2014; FAO 2016). Specifically, in Latin America more than 127 million tons of food waste are produced every year, being equivalent to 348,000 tons of food waste/day, enough to feed 300 million people and 10 times the total capacity of the Mexico City supply center, considered one of the largest in the world (FAO 2016). Table 4 shows the main food groups in Latin America that generate solid waste (HLPE 2014; FAO 2016).

Food losses generate problems of high impact about the economic and environmental area. The carbon footprint of wasted food is estimated at 3300 million tons of carbon dioxide in 2007 (8% of global greenhouse gas emissions), which is twice of the emissions produced by land transport in the USA. Also, in 2007, the total amount of water used to produce the discarded food was approximately 25,0000 m^3 for agricultural production, which represent 3.6 times the total water consumption in the USA in the same year (FAO 2013; FAO 2015; FAO 2016).

4 The Role of Anaerobic Wastewater Treatment in Latin America

During the 1980s, emerged the technology of reactors called "up-flow anaerobic sludge blanket (UASB)" and some Latin American countries adopted this technology to be used in sewage treatment plants (STPs) (Chernicharo et al. 2015a; López and Borzacconi 2017; Monroy et al. 2000). Currently in Latin America, it has been adopted to apply an anaerobic treatment prior to aerobic treatment (combined anaerobic/aerobic process) for the wastewater treatment, highlighting Mexico, Brazil, and Colombia in the use of this combined method (Chernicharo et al. 2015b).

Noyola et al. (2012) reported three main technologies used for wastewater treatment, (1) stabilization ponds, (2) activated sludge, and (3) UASB reactors, with percentages of 38, 26, and 17%, respectively. This technology is used by six Latin American countries (Brazil, Colombia, Chile, Dominican Republic, Guatemala, and Mexico). However, the UASB reactors and stabilization ponds only are used in STPs of low capacity, except to Brazil that work with UASB reactors for the municipal wastewater treatment from cities with more than one million of people (Chernicharo et al. 2015a). In wastewater treatment plants, anaerobic treatment is regularly a pre-aerobic treatment stage. The main aerobic treatments are shown below (von Sperling and de Andrada 2006).

- *Polishing ponds*: They function as stabilization ponds, but these treat UASB reactors effluents compared with the stabilization ponds that treat sewage.
- *Subsurface flow wetlands*: The nutrients from the effluents are used for plant cultivation. The plants are sown in a porous substrate where the effluent is supplied sub-superficially, avoiding the flooding of the substrate, the proliferation of harmful fauna, and the evaporation of the effluent.
- *Drip filters*: They consist of a tank filled with highly permeable material where the effluent is in the form of drops. The effluent is then filtered to the bottom allowing the growth of bacteria on the surface forming a biofilter.
- *Submerged aerated biofilters*. An aerated biofilter consists of a tank full of porous material through which the effluent and air flow upward. Throughout the process, the porous medium remains completely submerged.

5 Utilization of Biogas in Latin America as Energy Source from Anaerobic Digestion

Large-scale anaerobic digestion systems have not been widely implemented in Latin America due mainly to their technical complexity, investment, and maintenance costs. However, some countries such as Argentina, Chile, Brazil, and Mexico have implemented UASB and continuous stirred tank reactors to treat wastewater and excreta from animals (Silva-Martínez and Sanches-Pereira 2018).

Table 5 Production of biogas using agro-industrial residues by anaerobic digestion (Meneses-Jácome et al. 2016)

Activity (country)	Effluent	Potential energy
Ethanol from sugarcane (Brazil)	Bulk effluent (vinasses)	Biogas availability range between 2000 and 12,000 m^3/d
Tequila spirits production from agave (Mexico)	Bulk effluent (vinasses)	
Palm oil mills (Colombia)	Bulk effluent after grease traps	Capacity to produce 20,000 to 26,000 kWh/d treating anaerobically a flow of 6 L/s
Poultry slaughterhouse (Mexico)	Bulk effluent after grease traps	Calorific power of physicochemical sludge (DM) range between 4000 and 5000 cal/g
Poultry-viscera processing (Colombia)	Bulk effluent after grease traps	Capacity to supply 2000 MJ/d of heat with biogas
Dairy/cheese production (Colombia-Mexico)	Bulk effluent free of whey after grease traps	Capacity to drive an engine or micro-turbine (~ 100 kW)

There are reports that estimate the amount of energy that could be obtained from the use of biogas from anaerobic digestion of organic waste in Latin America. Cutz et al. (2016) mentioned that the Central American region has the potential to produce 3270 GWh/year of electricity using biogas. Voegele (2018) reported the use of Napier grass to produce 2 MW of electricity using the anaerobic digestion technology in Puerto Rico. In 2016, 74 large-scale anaerobic digesters were installed in Chile, to treat pig, dairy, and wastewater residues. Of these digesters, exceptional cases are highlighted, such as "La Farfana" treatment plant that produces 24 million m^3 of biogas/year and "Santa Irene" and "Las Pampas" with a potential to produce 800 kW and generate electricity for 2500 families (Ávila-Grothusen 2016).

In 2017, the first anaerobic digestion plant for the treatment of solid organic waste from food waste with a biogas production capacity of 170 m^3, equivalent to 173 kW/h, was inaugurated in Mexico City (Agencia EFE 2017). Meneses-Jácome et al. (2016) estimated the production of biogas using agro-industrial residues by anaerobic digestion (Table 5).

The government regulations and the high waste from sugarcane industry in Brazil have promoted the best advances about anaerobic digestion technology application in Latino America (Albanez et al. 2016; Bernal et al. 2017; Fuess et al. 2017; Giraldo et al. 2007; Moraes et al. 2014, 2015). In 2014, the potential for methane production through anaerobic digestion of agricultural waste was 14.3 million m^3/day, equivalent to produce 3478 MW. According to the National Biogas Register (Cadastro Nacional do Biogás), in 2015 the 127 anaerobic plants generated 1.6 million m^3/day of biogas. For 2018, the number of plants increases to 366, of which 276 operated with a production of 3.1 million m^3/day (Mercosur Biogas and Biomethane Report 2017). Currently, Brazil has the potential to produce 82 billion m^3/year, according to the installed capacity and the amount of waste that can be used (CIBiogás 2019). If this production was converted into electricity, it would be enough to supply the domestic electricity demand of Brazil for 1 month (CIBiogás 2019; Silva dos Santos et al. 2018).

The Mercosur member countries as Uruguay, Paraguay, and Argentina also produce organic waste in high volume that could potentially be recycling for biogas production. In Uruguay, the BIOVALOR project directed by BIOPROA research group analyzed the potential of generating biogas using organic waste and reported that 113.3 and 162.3 million m^3 of methane could be obtained per year. However, the real biogas production in 2016 was only 6900 m^3/day (Mercosur Biogas and Biomethane Report 2017).

In Argentina, the use of anaerobic digestion to obtain biogas has been implemented for more than 20 years. However, this technology has not reached the enough potential. Only some successful projects in Argentina can be highlighted: (1) bioelectric project with capacity to produce 1 MW of electricity by anaerobic digestion of manure and corn silage, (2) ACA-Yanquetruz project that produces 8000 MW with the digestion of pig excreta and corn, and (3) La Micaela project which generates 70 kWh by digestion of manure and urine from animals (Mercosur Biogas and Biomethane Report 2017; Vögeli et al. 2014). Finally, Paraguay has been made efforts to introduce biodigesters for the production of biogas; however, the progress reported has not been promising until 2016 (Mercosur Biogas and Biomethane Report 2017).

6 Anaerobic Digestion as an Alternative for Small and Medium Agricultural Producers in Latin America

Globally, small and medium agricultural producers face a situation of vulnerability due to effects as climate change, fluctuations in fossil fuel prices, prices of agrochemicals, and unequal competition with transnational industries (Martí-Herrero 2019). Furthermore, it is necessary to implement novel alternatives to strengthen the productive model from agricultural producers, improving their conditions and helping them to obtain value-added products. In this context, the use of anaerobic digesters is a useful tool to strengthen these models of production; it also allows the recycling of waste generated in the communities that apply this technology (Vögeli et al. 2014; Martí-Herrero 2019). Anaerobic digestion systems on a small scale allow producers and rural communities to have an external energy source due to the production of biogas and the production of biofertilizer from digestates (Silva-Martínez and Sanches-Pereira 2018; Garfí et al. 2016). These benefits are obtained by recycling organic waste from agricultural, municipal residues, excreta, fruits, and vegetables using anaerobic digestion technology (Silva-Martínez and Sanches-Pereira 2018; Garfí et al. 2016).

The first designs of anaerobic digestion systems to domestic or small scale were the fixed dome digesters, which consist of buried cylindrical chambers built with cement and bricks (Fig. 1a) (Rajendran et al. 2012; Garfí et al. 2016). These designs

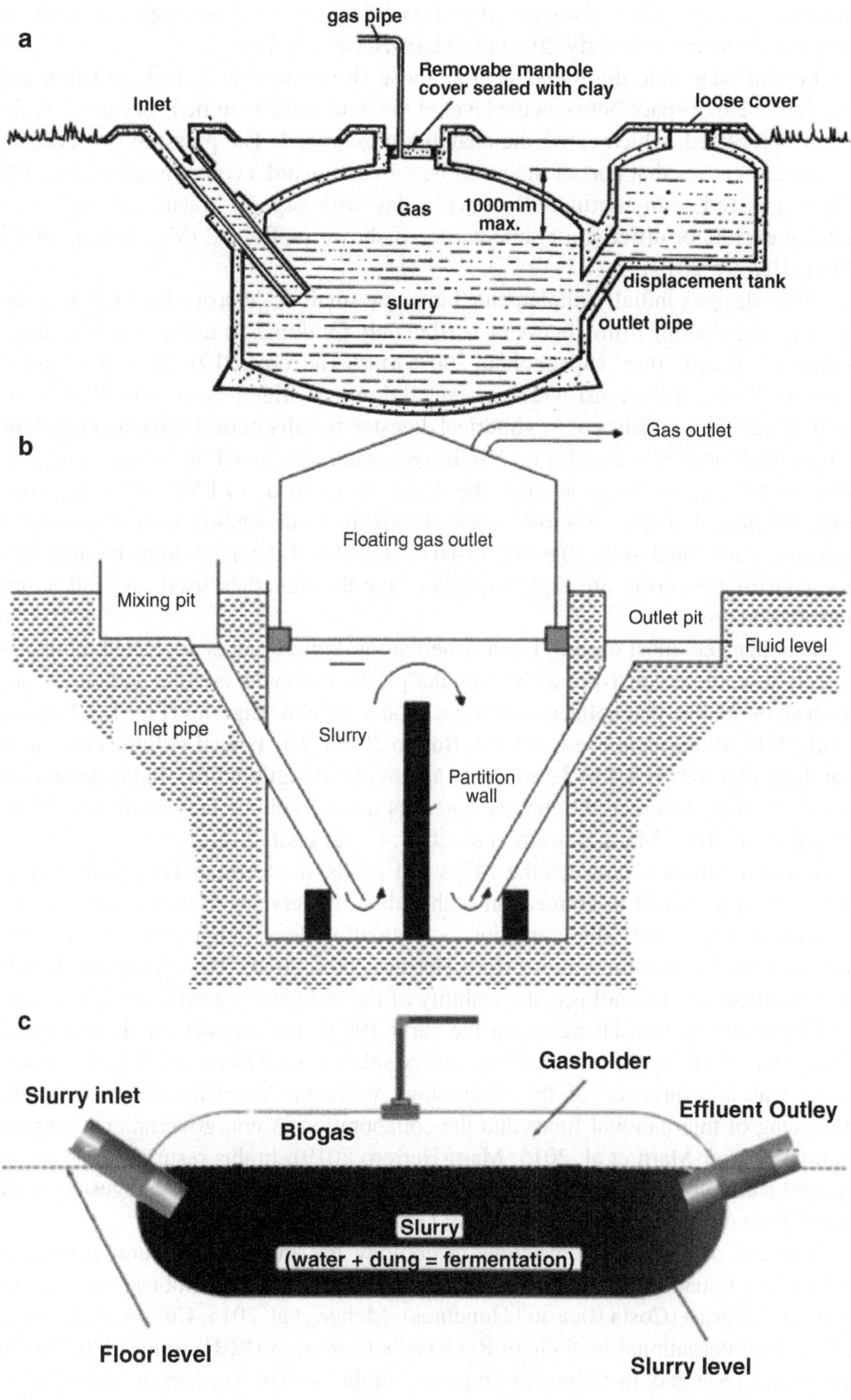

Fig. 1 Anaerobic digester designs. (**a**) Fixed dome digester (Fraenkel 1986); (**b**) floating drum digester (Surendra et al. 2014); (**c**) tubular digester (IMPACTLAB-Massachusetts Institute of Technology n.d.)

are costly for the materials required and qualified personnel for their construction, and its life is approximately 20 years (Martí-Herrero 2019).

Typical anaerobic digesters, such as those shown in Fig. 1, lack agitation and heating; the free space between the level of the mud and the dome is occupied by the biogas produced, which cause the chamber pressurized. The pressure produced by the biogas causes that part of the muds be moved toward a compensation tank. The digester is fed semi-continuously once a day with organic waste, regularly with diluted excreta or different organic waste, as shown in Table 2 (Vögeli et al. 2014; Martí-Herrero 2019).

Other designs initially implemented in some small-scale projects in Latin America were the floating drum digesters, used mainly for domestic use and with volumes between 1.6 and 10 m^3 (Spagnoletta 2007; Martí-Herrero 2019). Also, this kind of digester is one of the most widely accepted domestic digesters in India (Rajendran et al. 2012). This consists of a cylindrical digester, usually buried and a floating drum where the biogas is retained (Fig. 1b). Its construction is based on concrete and steel (the floating drum); in some cases the drum can be made of PVC. Also, the drum must be placed so that it allows vertical movement depending on the amount of biogas accumulated in the digester (Garfí et al. 2016). Like fixed dome biodigesters, the construction costs are high, especially for the materials used, as well as the installation cost.

The digesters most used in Latin America for domestic uses are called tubular or piston flow; these consist of a PVC tubular plastic bag (with an inlet and outlet) and with an outlet line that collects the biogas produced in the digester (Fig. 1c) (Lansing et al. 2008; Botero and Preston 1987; Botero 2011). The typical volume range used for these digesters is 2.4 to 7.5 m^3, with a ratio of 5 (length/radius), but in some rural areas of Latin America, the tubular digesters have a volume between 6 and 10 m^3 (Garfí et al. 2016; Martí-Herrero et al. 2014; Ferrer et al. 2011).

In Latin America, between the 1970s and 1980s, the German Technical Cooperation Agency (GTZ) in connection with public universities tried a project for the implementation and dissemination of small-scale biodigesters (fixed-dome digesters) in rural communities (Martí-Herrero 2008, 2019). These projects allowed to investigate the technology, the viability of the effluents as fertilizers, and the use of biogas in engines. However, in the early 1990s, the support for these projects disappeared leaving the installed digesters obsolete (Martí-Herrero 2019). Currently, there was a resurgence of this technology in Latin America, promoted by the financing of international funds and the collaboration of non-governmental organizations (Ferrer-Martí et al. 2015; Martí-Herrero 2019). In this resurgence, low-cost design technologies such as plastic tubular and membrane tubular digesters were used (Pound et al. 1981; Ni et al. 1993; Garwood 2010).

The use of anaerobic digestions technology has increased in rural regions of Colombia, Ecuador, Mexico, Nicaragua (Garfí et al. 2014), Caribbean islands, and Central America (Costa Rica and Honduras) (Mohee et al. 2015; Cutz et al. 2016). In 2012, the International Institute of Renewable Resources (IRRI) initiated the biogas program in Mexico, installing 256 digesters in the state of Yucatán in beneficial for more than 2000 people (Martí-Herrero et al. 2016).

The anaerobic digestion has been used under extreme weather conditions in regions as Bolivia and Andean countries (Martí-Herrero et al. 2014), Peru (Ferrer et al. 2009), Chile (Ávila-Grothusen 2016), Ecuador, and Argentina (Martí-Herrero et al. 2016). In 2012 and 2013, the Netherlands Development Organization (SNV) together to the non-governmental organization Hivos implemented a program for the dissemination and implementation of the use of domestic digesters called "National Biodigesters Plan," which operated in Bolivia and Peru (Martí-Herrero et al. 2013; Acosta-Bedolla et al. 2013). In 2017, the same financing agencies Hivos and SNV implemented the National Biogas Program in Nicaragua with the objective of developing a sustainable biogas market that allows small and medium agricultural producers and rural households to have access to a renewable energy (SNV Annual Report and Annual Accounts 2018; Martí-Herrero 2019).

In some Caribbean countries, small-scale anaerobic digesters have also been implemented. In Jamaica, about 200 biogas plants were installed between 1988 and 1993. In Cuba, the anaerobic digestion technology has been used since the middle of the last century, and until 2012, more than 700 small-scale plants were installed (Kranert et al. 2012). In this sense, the network called "Red de Biodigestores para Latino América y el Caribe (RedBioLac)" promotes the development, implementation, and dissemination of biodigesters in this regions, focused on stimulating the correct management of natural resources and promoting the socioeconomic well-being of Latin America and the Caribbean (RedBioLAC 2019; Martí-Herrero 2019). The RedBioLac has coordinated diverse institutions that promote investigation and diffusion about anaerobic digestion technology, some of them are presented in Table 6.

7 Conclusions

Latin America has a high generation of organic waste that can be recycled through anaerobic digestion, making this technology attractive for small-, medium-, and large-scale application at the domestic or industrial level. The benefits are reflected in obtaining energy such as biogas and nutrient-rich effluents with agricultural uses, which impact in rural and hard-to-reach communities. Also, it contributes to the use of environmentally friendly technologies, reducing the production of greenhouse gases due to use biogas obtained as an energy source.

Despite the benefits, this technology is currently not used efficiently in Latin America, with the exception of Brazil, where its small- and large-scale application has a favorable impact on energy consumption in the country. According with this panorama in the last 10 years, networks have emerged for coordinate private companies, educational institutions, and non-governmental organizations, with the purpose of encouraging the development and implementation of anaerobic digesters for the sustainable management of waste and promoting the socioeconomic well-being of the region.

The current challenge for Latin America is to reduce the costs of anaerobic digestion reactors and spread their benefits, so that cities, rural communities, and

Table 6 Participants of the RedBioLAC for Latin America (RedBioLAC 2019)

Country	Institution
Argentina	Universidad Nacional de Cuyo (http://www.uncuyo.edu.ar/) Las Camelias S.A. (http://www.lascameliassa.com.ar/inicio.php?idioma=es) MEYCO S.R.L. (http://www.meycosrl.com.ar/) Fundación Energizar (http://www.energizar.org.ar/) Grupo IFES (www.grupoifes.com) Instituto Nacional de Tecnología industrial-INTI (http://www.inti.gob.ar/e-renova/erBI/er26.php) Instituto Nacional de Tecnología Agropecuaria-INTA (http://inta.gob.ar/)
Bolivia	Proyecto EnDev-Bolivia (http://www.endev-bolivia.org/es/) Fundación HIVOS (https://www.hivos.org/biogas-esp/) Centro Internacional de Métodos Numéricos en Ingeniería-CIMNE (http://www.cimne.com/)
Brazil	Embrapa Suínos e Aves (https://www.embrapa.br/suinos-e-aves) CIBiogás (https://www.cibiogas.org/) University of São Paulo (www.usp.br) CETESB-Companhia Ambiental do Estado de São Paulo (cetesb.sp.gov.br) Universidade Federal de Itajubá, campus prof. José Rodrigues Seabra (www.unifei.edu.br) Viva Rio (vivario.org.br)
Chile	Universidad Católica del Norte (www.ucn.cl) Universidad de La Serena (www.userena.cl) Fen Biodinámico, Fundo Alhué – Curacaví, Chile – América del Sur (http://www.fen-bio-dinamico.cl/) Universidad de Santiago (www.usach.cl) Empleo Sustentable Compost Chile (http://www.compostchile.com/blog/inicio/nuestra-empresa/) Universidad de Chile (www.uchile.cl) CIFES-Ministerio de Energía, Gobierno de Chile Aiguasol (aiguasol.cl) Fundación Basura (www.fundacionbasura.org) Instituto del Medio Ambiente-IDMA (http://idma.cl/) Fundación Para la Innovación Agraria (http://www.fia.cl/) Treasure (http://traesure.com/) Universidad de Concepción (http://www.agronomiaudec.cl/home/) Red de Acción por los Derechos Ambientales-RADA (http://www.radaraucania.cl/) ONG Canales (http://www.ongcanales.cl/)
Colombia	Asociación de Cabildos Indígenas del Norte del Cauca-ACIN-ÇXHAB WALA KIWE (http://www.nasaacin.org/) Biotec (http://www.biotecsoluciones.co/) Fundación Cipav(www.cipav.org.co) Pro-Orgánica (http://www.pro-organica.org/index.php?id_cont=1) Universidad de los llanos (www.unillanos.edu.co) Asociación de Productores Indígenas y Campesinos de Riosucio Caldas Asproinca (http://agrobiodiversityplatform.org/refarm/case-study/asociacion-de-productores-indigenas-y-campesinos-de-riosucio-caldas-asproinca/) Filmtex (www.filmtex.com) CENSAT Agua viva (http://censat.org/) Universidad Distrital Francisco José de Caldas (www.udistrital.edu.co) Patrimonio natural (http://www.patrimonionatural.org.co/site/index.php)

(continued)

Table 6 (continued)

Country	Institution
	Fundación Guayacanal (http://guayacanal.org/) Universidad del Bosque (http://www.uelbosque.edu.co) Fundación Panamericana Para el Desarrollo (http://www.fupad.org/) UNIAGRARIA (http://www.uniagraria.edu.co/) Fundación Para la producción agropecuaria tropical sostenible-UTA Colombia (http://www.utafoundation.org/) Red Colombiana de Energía de la Biomasa-RedBioCol (http://www.redbiocol.org/) Universidad Pontificia Bolivariana (http://www.upb.edu.co/portal/page?_pageid=954,1&_dad = PORTAL&_schema = PORTAL) Fundación Salva Terra (http://fundacionsalvaterra.org/es/)
Costa Rica	Compañía Vega (http://www.companiavega.com/) VIOGAZ (http://www.viogaz.com/) Universidad EARTH (https://www.earth.ac.cr/es/) Centro de Investigación en Protección Ambiental y co (https://www.facebook.com/CentrodeInvestigacionenProteccionAmbiental/) Finca Escuela La Pilarica Guayabo (https://www.facebook.com/fincaescuelalapilarica/)
Ecuador	Escuela superior Politécnica del Litoral-ESPOL (www.espol.edu.ec) Universidad Laica Vicente Rocafuerte (www.ulvr.edu.ec) Coordinadora Ecuatoriana de Agroecología (www.agroecologia.ec) Instituto Nacional de Eficiencia Energética y Energías Renovables CARE (http://www.care.org.ec/) Ministerio de Agricultura, Ganadería, Acuacultura y Pesca-MAGAP (http://www.agricultura.gob.ec/)
Guatemala	Alterna (http://alternaimpact.org/en/)
Haití	SCAGITECH (https://web.facebook.com/profile.php?id=100007385146939)
Honduras	Escuela Agrícola Panamericana El Zamorano (www.zamorano.edu) Instituto Hondureño del Café-IHCAFE (www.ihcafe.hn) SNV Honduras (http://www.snv.org/sector/agriculture)
Mexico	Sistema Biobolsa (http://sistemabiobolsa.com/) Cooperativa Agrosol (http://ranchoagrosol.blogspot.cl/)
Nicaragua	Asociación Fénix (http://asociacionfenix.org/) Sunisolar (http://www.sunisolar.com/) Universidad Nacional de Ingeniería (http://www.uni.edu.ni/) Ministerio de Energía y Minas (www.mem.gob.ni) Universidad de Ciencias Comerciales (http://www.ucc.edu.ni/) Tecnosol (http://www.tecnosolsa.com.ni/) Universidad Politécnica de Nicaragua (https://www.upoli.edu.ni/) Universidad Nacional Agraria (http://www.una.edu.ni/)
Paraguay	Universidad Nacional de Asunción-Facultad de Ciencias Agrarias (www.agr.una.py)
Perú	Universidad Nacional de Tajamarca (http://www.unc.edu.pe/) Solaris Perú (http://www.solaris.org.pe/) Soluciones Prácticas (http://www.solucionespracticas.org.pe/) CIDELSA (http://www.cidelsa.com/es/about-us/) Universidad Peruana de Ciencias Aplicadas Monterrico-UPC Monterrico (http://www.upc.edu.pe/) Instituto Nacional de Innovación Agraria-INIA (http://www.inia.gob.pe/)

(continued)

Table 6 (continued)

Country	Institution
Uruguay	NETUM SRL (www.netum.com.uy)

industry apply this technology in the treatment of its organic waste, agro-industrial, municipal, domestic, and wastewater generated.

References

Acosta-Bedolla F, Martí-Herrero J, Gonzales L, Oppenoorth H, Veen M, Valori M (2013) Plan de Programa Nacional de Biodigestores en Perú, Hivos-SNV-Soluciones Prácticas http://www.snv.org/public/cms/sites/default/files/explore/download/snv-sp-hivos_2013_-_pnb-peru.pdf

Agencia EFE (2017) Inauguran en Ciudad de México biodigestor para tratamiento de residuos (journalistic news). https://www.efe.com/efe/america/mexico/inauguran-en-ciudad-de-mexico-biodigestor-para-tratamiento-residuos/50000545-3276567

Albanez R, Chiaranda BC, Ferreira RG, França AL, Honório CD, Rodrigues JA, Ratusznei SM, Zaiat M (2016) Anaerobic biological treatment of vinasse for environmental compliance and methane production. Appl Biochem Biotechnol 178:21–43

Ávila-Grothusen M (2016) Descripción y análisis de alternativas tecnológicas para la producción de biogás, In: Biogás de Residuos Agropecuarios en la Región de los Ríos. Instituto de Desarrollo Agropecuario (INDAP) & Gobierno Regional de los Rios (GORE Los Ríos), Chile. https://www.goredelosrios.cl/wp-content/uploads/2018/01/Libro-Biogas-PRDSAP-Regi%C3%B3n-de-Los-R%C3%ADos.pdf

Baldini M, da Borso F, Ferfuia C, Zuliani F, Danuso F (2017) Ensilage suitability and bio-methane yield of *Arundo donax* and *Miscanthus* × *giganteus*. Ind Crop Prod 95:264–275

Bernal AP, dos Santos IFS, Moni Silva AP, Barros RM, Ribeiro EM (2017) Vinasse biogas for energy generation in Brazil: an assessment of economic feasibility, energy potential and avoided CO2 emissions. J Clean Prod 151:260–271

Botero R (2011) El biodigestor de bajo costo, su aporte a la mitigación del cambio climático y su potencial para reducir la pobreza rural en América latina y el Caribe. Colombia: Futura Tecnología Renovable y Sostenible 1(11) http://www.kelcolombia.com/futura-11.pdf

Botero R, Preston TR (1987) Biodigestor de bajo costo para la producción de combustible y fertilizante apartir de excretas. Manual para su instalación, operación y utilización. CIPAV, Cali

Chernicharo CAL, Van Lier JB, Noyola A, Ribeiro TB (2015a) Anaerobic Sewage Treatment in Latin America, In Anaerobic Biotechnology-Environmental Protection and Resource Recovery. Fang HHP, Zhang T, Editors. Imperial College Press, pp 269–296

Chernicharo CAL, Van Lier JB, Noyola A, Ribeiro TB (2015b) Anaerobic sewage treatment: state of the art, constraints and challenges. Rev Environ Sci Biotechnol 14:649–679

CIBiogás (2019) April 10th Brazil's estimated potential for biogas production is 82 billions cubic metres/year (journalistic news). https://www.biogaschannel.com/en/news-and-events/brazils-estimated-potential-biogas-production-82-b/241/

Cutz L, Haro P, Santana D, Johnsson F (2016) Assessment of biomass energy sources and technologies: the case of Central America. Renew Sust Energ Rev 58:1411–1431

De la Roza-Delgado B (2005) El ensilado en zonas húmedas y sus indicadores de calidad. IV Jornadas de Alimentación Animal, Laboratorio de Mouriscade

Deublein D, Steinhauser A (2011) Biomass from waste and renewable resources, 2nd edn. Wiley-VCH Verlag GmbH & Co. KGaA, Germany

FAO (2013) Food wastage footprint: Impacts on natural resources. http://www.fao.org/3/i3347e/i3347e.pdf

FAO (2015) Food wastage footprint & Climate Change. http://www.fao.org/3/a-bb144e.pdf
FAO (2016) Newsletter #3, February. Food losses and waste in Latin America and the Caribbean. http://www.fao.org/3/a-i5504e.pdf
Ferrer I, Gamiz M, Almeida M, Ruiz A (2009) Pilot project of biogas production from pig manure and urine mixture at ambient temperature in Ventanilla (Lima, Peru). Waste Manag 29:168–173
Ferrer I, Garfí M, Uggetti E, Ferrer-Martí L, Calderón A, Velo E (2011) Biogas production in low-cost household digesters at the Peruvian Andes. Biomass Bioenergy 35:1668–1674
Ferrer-Martí L, Garfí M, Ferrer I (2015) Cooperation and human development projects as bachelor, master and PhD thesis: evaluating an internship program. Proc-Soc Behav Sci 196:63–68
Fraenkel PL (1986) Water Lifting Devices, FAO irrigation and drainage paper no.43, Food and Agriculture Organization, Rome. http://www.fao.org/3/ah810e/AH810E00.htm
Fuess LT, de Araújo Júnior MM, Garcia ML, Zaiat M (2017) Designing full-scale biodigestion plants for the treatment of vinasse in sugarcane biorefineries: how phase separation and alkalinization impact biogas and electricity production costs? Chem Eng Res Des 119:209–220
Garfí M, Cadena E, Pérez I, Ferrer I (2014) Technical, economic and environmental assessment of household biogas digesters for rural communities. Renew Energy 62:313–318
Garfí M, Martí-Herrero J, Garwood A, Ferrer I (2016) Household anaerobic digesters for biogás production in Latin America: a review. Renew Sust Energ Rev 60:599–614
Garwood A (2010) Network for biodigesters in Latin America and the Caribbean: case studies and future recommendations. Technical notes No. IDB-TN-207 2010 https://publications.iadb.org/en/publication/11443/network-biodigesters-latin-america-and-caribbean-case-studies-and-future
Giraldo E, Pena M, Chernicharo C, Sandino J, Noyola A (2007) Proceedings of the Water Environment Federation, DOI: https://doi.org/10.2175/193864707787969469
Grau J, Terraza H, Rodríguez-Velosa DM, Rihm A, Sturzenegger G (2015) Solid Waste Management in Latin America and the Caribbean. Núñez AM, Osuji O, Ed. Inter-American Development Bank
Hettiarachchi H, Ryu S, Caucci S, Silva R (2018) Municipal solid waste Management in Latin America and the Caribbean: issues and potential solutions from the governance perspective. Recycling 3(2):1–15
HLPE (2014) Food losses and waste in the context of sustainable food systems. A report by the High Level Panel of Experts on Food Security and Nutrition of the Committee on World Food Security, Rome. http://www.fao.org/3/a-i3901e.pdf
IMPACTLAB-Massachusetts Institute of Technology (n.d.). https://impactlabs.mit.edu/bag-biogas-digester
Kinyua MN, Rowse LE, Ergas SJ (2016) Review of small-scale tubular anaerobic digesters treating live stock waste in the developing world. Renew Sust Energ Rev 58:896–910
Kranert M, Kusch S, Huang J, Fischer K (2012) Anaerobic digestion of waste. In: Karagiannidis, A. (Ed.), Waste to Energy: Opportunities and Challenges for Developing and Transient Economies. Springer, pp. 128–129
LaGiglia M, López Legarreta P, Vanamoli A, Wang L (2014) Policy Report: Promoting Net-Zero emmissions from the waste sector in Latin America through NAMAS, Center for Clean Air Policy
Lansing S, Botero R, Martin J (2008) Waste treatment and biogás quality in small-scale agricultural digesters. Bioresour Technol 99:5881–5890
Lansing S, Maile-Moskowitz A, Eaton A (2017) Waste treatment and energy production from small-scale wastewater digesters. Bioresour Technol 245:801–809
López I, Borzacconi L (2017) Anaerobic digestion for agro-industrial wastes: a Latin American perspective. Int J Eng Appl Sci 4(8):71–76
Martí-Herrero J (2008) Biodigestores Familiares. ISBN978–99954–0-339-3. Guía de diseño y manual de instalación. Cooperación Técnica Alemana-GTZ, La Paz
Martí-Herrero J (2019) Latin America Experiences in the Democratisation of Biodigesters. Contributions to Ecuador. Climate Technology Centre and Network (CTCN)-UNFCCC. Ecuador

Martí-Herrero J, Acosta-Bedoya F, Gonzales L, Oppenoorth H, Veen M, Valori M (2013) Plan de Programa Nacional de Biodigestores de Bolivia, Hivos-SNV-Soluciones Prácticas-CIMNE http://www.snv.org/public/cms/sites/default/files/explore/download/snv-cimne-hivos_2013_-_plan_nacional_de_biodigestores_de_bolivia.pdf

Martí-Herrero J, Alvarez R, Rojas MR, Aliaga L, Céspedes R, Carbonell J (2014) Improvement through low cost biofilm carrier in anaerobic tubular digestion in cold climate regions. Bioresour Technol 167:87–93

Martí-Herrero J, Chipana M, Cuevas C, Paco G, Serrano V, Zymla B, Heising K, Sologuren J, Gamarra A (2014) Low cost tubular digesters as appropriate technology for wide spread application: results and lessons learned from Bolivia. Renew Energy 71:156–165

Martí-Herrero J, Pino M, Gallo-Mendoza L, Pedraza GX, Rodríguez L, Víquez J (2016) Oportunidades para el desarrollo de un sector sostenible de biodigestores de pequeña y mediana escala en LAC. Redbiolac. (ISBN: 978–956–362-921-7)

Meneses-Jácome A, Diaz-Chavez R, Velásquez-Arredondo HI, Cárdenas-Chávez DL, Parra R, Ruiz-Colorado AA (2016) Sustainable energy from agro-industrial wastewaters in Latin-America. Renew Sust Energ Rev 56:1249–1262

Mercosur Biogas and Biomethane Report (2017) GAHB - Mercosur Ad Hoc Group on Biofuels - Vol. 1, n. 1 CIBiogás: Foz do Iguaçu, 2017

Mohee R, Mauthoor S, Bundhoo ZMA, Somaroo G, Soobhany N, Gunasee S (2015) Current status of solid waste management in small island developing states: a review. Waste Manag 43:539–549

Monroy O, Famá G, Meraz M, Montoya L, Macarie H (2000) Anaerobic digestion for wastewater treatment in Mexico: state of the technology. Water Res 34(6):1803–1816

Moraes BS, Junqueira TL, Pavanello LG, Cavalett O, Mantelatto PE, Bonomi A, Zaiat M (2014) Anaerobic digestion of vinasse from sugarcane biorefineries in Brazil from energy, environmental, and economic perspectives: profit or expense? Appl Energy 113:825–835

Moraes BS, Zaiat M, Bonomi A (2015) Anaerobic digestion of vinasse from sugarcane ethanol production in Brazil: challenges and perspectives. Renew Sust Energ Rev 44:888–903

Mortier N, Velghe F, Verstichel S (2016) Organic recycling of agricultural waste today: composting and anaerobic digestion. In: Poltronieri P, D'Urso OF (eds) Biotransformation of agricultural waste and by-products. The food, feed, fibre, fuel (4F) economy. Elsevier, Netherlands

Ni JQ, Naveau H, Nyns EG (1993) Biogas: exploitation of a renewable energy in Latin America. Renew Energy 3:763–779

Noyola A, Padilla-Rivera A, Morgan-Sagastume JM, Güereca LP, Hernández-Padilla F (2012) Typology of municipal wastewater treatment technologies in Latin America. Clean Soil Air Water 40(9):926–932

Pound B, Bordas F, Preston TR (1981) Characteristics of production and function of a 15 m3 red-mud PVC biogas digester. Tropical Anim Prod 6:146–153

Rajendran K, Aslanzadeh S, Taherzadeh MJ (2012) Household biogás digesters - A review. Energies 5:2911–2942

RedBioLAC (2019) Network for Biodigesters in Latin America and the Caribbean. http://redbiolac.org/

Silva dos Santos IF, Braz Vieira ND, de Nóbrega LGB, Barros RM, Tiago Filho GL (2018) Assessment of potential biogas production from multiple organic wastes in Brazil: impact on energy generation, use, and emissions abatement. Resour Conserv Recycl 131:54–63

Silva-Martínez RD, Sanches-Pereira A (2018) Organic waste to energy in Latin America and the Caribbean (LAC); state of the art literature review

SNV Annual Report and Annual Accounts (2018) Netherlands Development Organization. http://www.snv.org/public/cms/sites/default/files/explore/download/snv_annual_report_2018-incl_annual_accounts-web.pdf

Spagnoletta SA (2007) Viability study for the application of small size biodigesters in the Andean rural zone of Cajamarca (Peru). Available from: http://grecdh.upc.edu/docencia/treballs-dirigits/altres/spagnoletta-sebastiano-alessandro

Surendra KC, Takara D, Hashimoto AG, Khanal SK (2014) Biogas as a sustainable energy source for developing countries: opportunities and challenges. Renew Sust Energ Rev 31:846–859
UN Environment (2018) Waste management outlook for Latin America and the Caribbean. In: United nations environment Programme, Latin America and the Caribbean office. Panama, Panama City
Vervaeren H, Hostyn K, Ghekiere G, Willems B (2010) Biological ensilage additives as pretreatment for maize to increase the biogas production. Renew Energy 35(9):2089–2093
Voegele E (2018) Organic power to install 2 MW AD plant in Puerto Rico., Biomass Magazine, http://biomassmagazine.com/articles/15045/organic-power-to-install-2-mw-ad-plant-in-puerto-rico
Vögeli Y, Lohri CR, Gallardo A, Diener S, Zurbrügg C (2014) Anaerobic digestion of biowaste in developing countries: practical information and case studies. Swiss Federal Institute of Aquatic Science and Technology (Eawag), Dübendorf, Switzerland
von Sperling M, de Andrada JGB (2006) Simple wastewater treatment (UASB reactor, shallow polishing ponds, coarse rock filter) allowing compliance with different reuse criteria. Water Sci Technol 54(11):199–205

Biogas Technology for Animal Manure Management in the USA: State of the Art, Opportunities, Challenges, and Perspectives

David Henry Huber

Abstract Anaerobic digestion is an established technology for treating organic wastes and producing renewable energy as biogas. This technology is well suited for US livestock operations which have been moving toward increasingly larger facilities with large quantities of manure. However, out of an estimated 8000 plus dairy and swine farms that could use AD, there are currently only 248 operational farm digesters in the USA. The environmental problems associated with livestock manures are summarized, and the benefits of AD to alleviate the problems are discussed. Increasing the versatility of products derived from farm-based AD may make biogas technology more attractive. Some of these products could be enhanced methane production through power-to-gas (PtG) technology, combined cooling, heat and power (CCHP), bio-crude oil, and the enhanced production of medium-chain fatty acids with modified AD microbiomes.

Keywords Anaerobic digestion · Bioenergy · Methane · Livestock manure

1 Introduction

Agricultural sustainability and environmental sustainability have become increasingly coupled objectives in the USA as well as worldwide (Tilman et al. 2002; Robertson et al. 2014). This is particularly evident with regard to the management of livestock production and livestock residues. In the 50 states, 17% of the total land area (391 million acres) is classified as cropland, and 29% (655 million acres) is dedicated to livestock as either grassland, pasture, or range (USDA ERS 2017). This land supports the annual production of more than 1 billion tons of livestock manure which provides significant environmental and health challenges (Sommer et al. 2013). One of the best methods to address these problems is biogas technology

D. H. Huber (✉)
Department of Biology, Gus R. Douglass Institute, West Virginia State University, WV, Institute, USA
e-mail: huberdh@wvstateu.edu

N. Balagurusamy, A. K. Chandel (eds.), *Biogas Production*,
https://doi.org/10.1007/978-3-030-58827-4_4

which can simultaneously minimize detrimental effects while converting manure into useful resources (Sakar et al. 2009). Biogas is produced by anaerobic digestion which is a remarkably versatile and robust microbial technology that is used worldwide for stabilizing organic wastes and producing bioenergy. This review summarizes the current state of AD applications on American livestock farms and highlights opportunities for further incorporating AD into sustainability objectives.

2 Trends in US Livestock Farms

US livestock farms have been shifting away from integrated farming systems where livestock operations and crop production occur together toward specialized larger size concentrated animal feeding operation (CAFO) facilities. This move has occurred to improve economics and efficiency, and some CAFO facilities house several thousand animals (MacDonald and McBride 2009). The five largest livestock in the USA in terms of numbers are chickens (broilers) with 1.62×10^9 animals, chickens (layers) with 3.52×10^8, hogs with 7.22×10^7, beef cows with 3.17×10^7, and dairy cows with 9.54×10^6 (USDA 2017). The distribution of these livestock by state is shown in Fig. 1. Since 2002, the number of dairy farms with 500 to 999 cows has increased by 6.7% (5696 to 6073), and farms with 1000–2499 cows have increased by 18% (2081 to 2448). But the most dramatic increase has been of extremely large farms. In 2017, there were 677 farms with herds of 2499 to 4999 animals and 241 farms with 5000 or more animals (USDA, NASS). From 1997 to 2017, extremely large swine operations (> 5000 animals) also doubled from 1851 to 3600. While larger herds provide an improved scale of economics and efficiency, other problems are magnified, particularly manure management.

3 Current Status of Anaerobic Digestion Deployment in the USA

The current status of anaerobic digestion deployment on US livestock farms is summarized in the EPA AgSTAR database (2019). The AgSTAR Program is jointly sponsored by the US Environmental Protection Agency, US Department of Agriculture, and US Department of Energy. This program provides excellent online resources (https://www.epa.gov/agstar) that shows the current status of biogas technology in the USA and provides useful information for commercial farms that want to evaluate whether AD technology is appropriate for their situation.

The deployment of AD systems on US farms has been slowly progressing since the 1970s. The first farm digester was built in 1972, and several additional digesters were added in other states in the early 1980s. A boom in digester construction occurred between 2004 and 2012, and the number of operational farm digesters

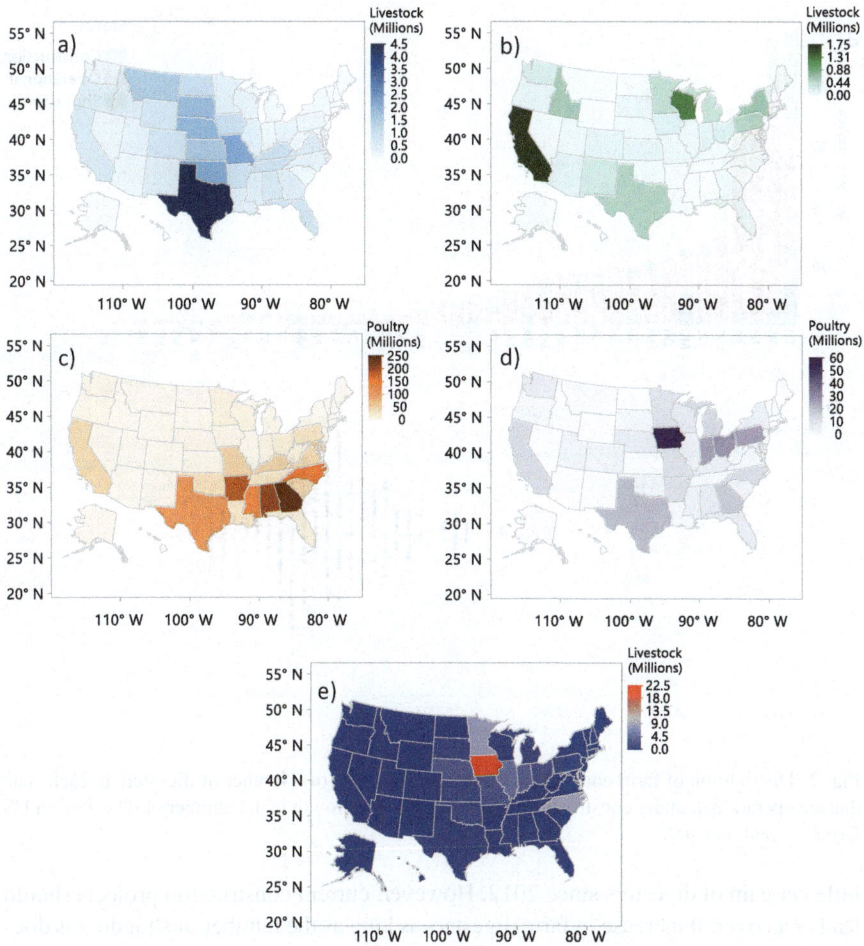

Fig. 1 Distribution of the primary livestock that are used for farm-based anaerobic digesters: (a) beef cattle, (b) dairy cattle, (c) chickens (broilers), (d) chickens (layers), (e) hogs

has grown about tenfold in the USA since 2000 (Fig. 2). However, due to digester shut downs, the total number has not increased over the past 4 years. Shut downs have occurred in most states, but the greatest number have been in California and Wisconsin (Fig. 2). There were 248 operational farm digesters by January 2019. In terms of feedstock, these digesters used animal waste from 198 dairy farms, 43 hog farms, 8 poultry, and 8 beef farms; some digesters used more than one type of manure. In 2018, the major AD technologies employed in the USA were 40% plug flow, 35% complete mix, and 19% covered lagoon. Minor use of anaerobic sequencing batch reactor, induced blanket reactor, fixed film, and dry technologies was also present. The largest number of operational farm digesters occurs in Wisconsin (37), New York (33), Pennsylvania (29), California (20), and Vermont (16) (Fig. 2). Farm digester shut downs have occurred almost every year since 2000, and there has been

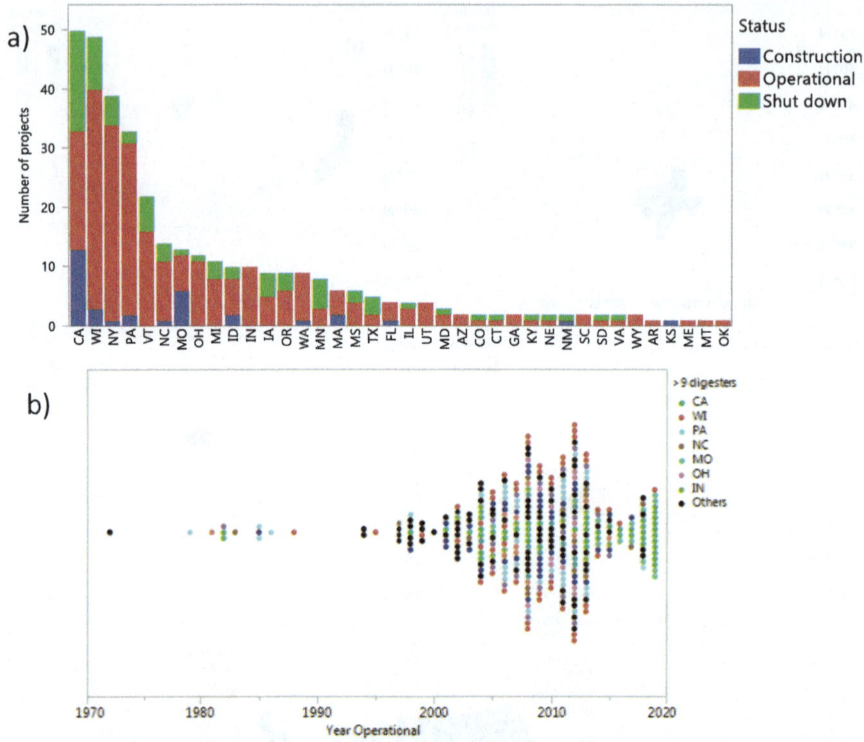

Fig. 2 Distribution of farm anaerobic digesters in the USA. (**a**) Number of digesters in each state that are operational, under construction, or shut down. (**b**) Growth in the number of digesters on US farms by year and state

little net gain of digesters since 2012. However, current construction projects should lead to an overall increase in farm digesters as long as the number of shut downs does not keep pace.

4 Environmental Problems Associated with Livestock Manures

The environmental problems that can arise from livestock manure production have been extensively addressed (Sommer et al. 2013). The increase in size of CAFO facilities has exacerbated some problems associated with maintaining large numbers of animals in close proximity as well as the production and storage of large quantities of manure. The primary issues concern nutrient release into watersheds and groundwater, greenhouse gas (GHG) and odor emissions, and the spread of pathogens of animals and humans. The total quantity of manure produced by state is shown in Fig. 3. The obvious stand-out is Iowa which boasts by far the largest number of hogs

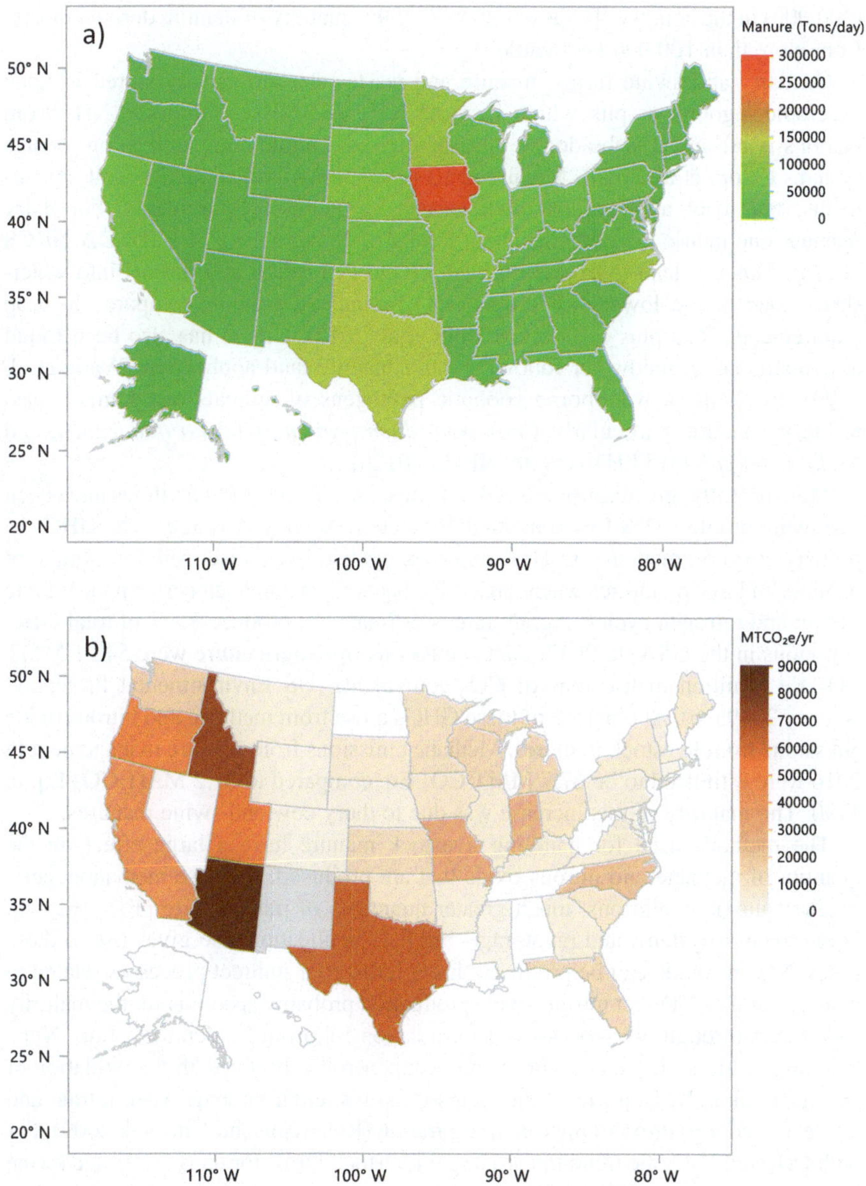

Fig. 3 (**a**) Total manure production per state including cattle cows, dairy cows, broiler chickens, layer chickens, goats, and sheep. (**b**) Total GHG emission reductions (MTTCO$_2$Eq./yr) due to farm-based anaerobic digesters

(200,000) in the country. Texas has about half the quantity of manure derived mostly from more than 100,000 beef cattle.

On dairy and swine farms, manure and wastewater are usually stored in open anaerobic lagoons or pits which increase NH_3 and GHG emissions. NH_3 from lagoons is estimated to be double the quantity per animal compared to pit storage systems (Szogi et al. 2015). Manure is also commonly land-applied as soil amendments, but local applications often exceed agronomic requirements. For dairy manure, one metric ton contains about 6.6 kg of N and 1.1 kg of P (USDA-NRCS 2012a). This can lead to excess nutrient (N and P) loading and run-off into watersheds. Due to the lower N:P ratio (<4:1) for animal manure compared to crop requirements, P surplus can occur (Szogi et al. 2015). Nitrate has also been found to contaminate groundwater following dairy manure land application (Wang et al. 1999). In addition, waterborne zoonotic pathogens of animals and humans also reside in manure, particularly *Cryptosporidium parvum*, *Giardia duodenalis*, and *Escherichia coli* O157:H7 (USDA-NRCS 2012b).

More recently, greenhouse gas (GHG) emissions from CAFO facilities have been a growing problem (US Environmental Protection Agency 2019a, b). The GHGs of primary concern, methane (CH_4) and nitrous oxide (N_2O), arise from the storage of manures in large quantities where anaerobic bacteria and archaea perform the anoxic carbon and nitrogen cycles. Agriculture is estimated to produce 8.4% of total GHG emissions in the USA. In 2017, GHG emissions from agriculture were 542.1 MMT CO_2 Eq. (million metric tons of CO_2 equivalent) (US Environmental Protection Agency 2019a, b). About 15% of these GHGs arose from methane and nitrous oxide emissions from livestock manures. Methane emissions from manure management in 2016 were estimated to be 67.7 MMT CO_2 Eq. compared to 37.2 MMT CO_2 Eq. in 1990. The majority of this increase was due to dairy cow and swine manures.

The methods used for handling livestock manure have a large effect on the quantity of methane and nitrous oxide that are produced. Because methanogenesis requires anoxic conditions, much greater quantities of methane are produced with liquid manure systems and pit storage, but field application also gives rise to these gases. Nitrous oxide can be produced through direct or indirect processes related to manure handling. Direct nitrous oxide production probably accounts for the majority and occurs through the processes of nitrification followed by denitrification. Nitrification produces nitrate and nitrite and occurs aerobically through the oxidation of NH_3 or organic N. Denitrification occurs in anoxic environments when nitrate and nitrite are reduced through microbial respiration (Robertson and Vitousek 2009). For both CH_4 and N_2O, the trend toward larger livestock farms for dairy cattle and swine has increased these GHGs.

Perhaps less appreciated at individual farms is the additional environmental impact that arises from antibiotics administered to livestock. Antibiotics are administered for three general reasons: treatment of disease, prophylactically to prevent disease, and subtherapeutically for growth promotion (Chee-Sanford et al. 2009). The concern is that large quantities of antibiotics will lead to a proliferation of antibiotic-resistant bacteria that could affect human health (Silbergeld et al. 2008). Antibiotics are known to be excreted in the feces and urine of livestock. When manure is land-applied, particularly from CAFO farms, antibiotics will be directly spread in the environment. While antibiotics can degrade in soil, the half-lives vary greatly; tetracyclines and quinolones are among the most persistent with half-lives of nearly 100 days. Recently, further unintended consequences of manure applications to soils were identified. Udikovic-Kolic et al. (2014) found that dairy manure stimulated blooms of antibiotic-resistant bacteria in soil. These bacteria were not present in manure but represented several species of native soil bacteria that already carried β-lactamase genes.

The enormous use of antibiotics in livestock operations has also led to proliferation of antibiotic resistance genes (ARGs) not only in manure storage areas but also in agricultural fields where manure is spread (Chee-Sanford et al. 2009). This has led to significant antibiotic footprints around farms where the abundance of antibiotic resistance genes has increased. For example, Marti et al. (2013) found that soil and vegetables fertilized with swine and dairy manure contained a greater abundance of antibiotic-resistant bacteria and ARGs. Indeed, studies have shown that workers on poultry farms harbor significantly more antibiotic-resistant bacteria in their gut microbiome than non-farm workers (van den Bogaard et al. 2002). Even more ominous, a recent study found that farm workers at industrial livestock operations carried methicillin-resistant *Staphylococcus aureus* (MRSA), while workers at antibiotic-free livestock operations did not (Rinsky et al. 2013).

5 Environmental Benefits of Manure Treatment with Anaerobic Digestion

The primary benefits of treating manure with AD have been well discussed and relate to reducing the environmental problems as well as using the byproducts of biogas generation (Khanal 2008). First, GHG emissions from manures have been greatly reduced in the USA through AD. For example, in 2017 direct and indirect GHG emissions on livestock farms were reduced by 3.95 MMT CO_2 Eq. (US Environmental Protection Agency 2019a, b). Since 2000, total GHG emission reduction from livestock farms has been estimated to be 34.6 MMT CO_2 Eq. If all dairy and swine farms where AD is feasible actually employed the technology, about 85% of total methane emissions could be eliminated (AgSTAR 2019). However, the greatest reductions in manure-generated methane do not always occur in the states with the most animals and reflect the limited adoption of the technology on livestock farms (Fig. 3).

Second, the problems arising from the release of antibiotics, antibiotic-resistant bacteria, and ARGs into the environment from livestock manure can be mitigated by AD. Recent research has shown that the effectiveness of AD for removing antibiotics and ARGs depends on the type of antibiotic, type of AD, and digester operating conditions. Feng et al. (2017) evaluated removal of several antibiotics in a study comparing thermophilic and psychrophilic digestion of pig manure. They found that

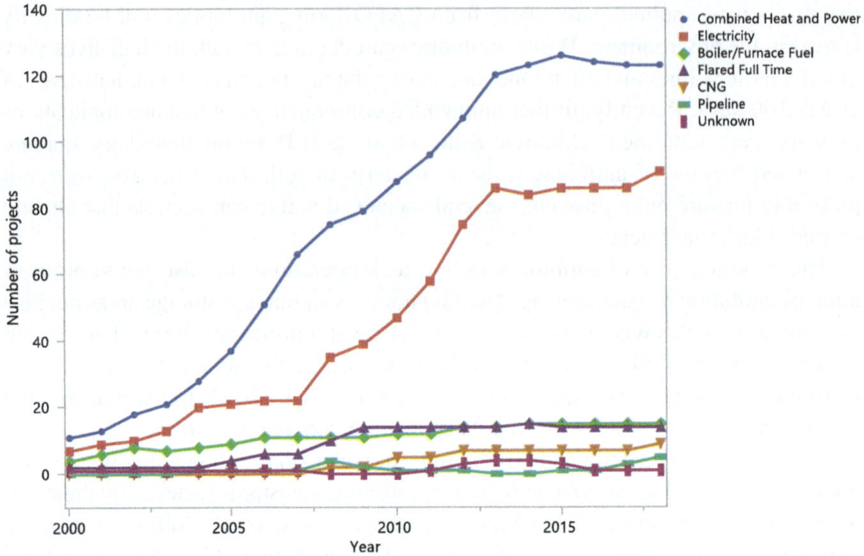

Fig. 4 End use of biogas for US farm digesters from 2000 through 2018. (Source: EPA AgSTAR 2019)

some antibiotics (clarithromycin, sulfadiazine, sulfamethizole) were not removed by AD, while erythromycin reduction varied by AD type: 99% removal with thermophilic digestion but only 20% removal with psychrophilic. In a mesophilic system treating cow manure, Turker et al. (2018) found that the operating conditions, particularly mixing rate and volatile solids content, affected the elimination of oxytetracycline and ARGs. Another issue is that antibiotics can inhibit AD and reduce methane production. However, digester microbiomes have been found to have the potential to adapt to at least some antibiotics. For example, monensin is given to dairy cows to increase milk production, and it is retained in their manure. Spirit et al. (2018) found that higher concentrations of monensin inhibited AD but a gradual increase in monensin dosing permitted adaptation by the microbiome.

Third, biogas is a renewable energy source that can be used locally. In 2018, 35% of the US farm anaerobic digesters were used for electricity generation, while 47% were used for combined heat and power (CHP). Both uses have been steadily increasing since 2000 (Fig. 4). CHP refers to the generation of electricity in conjunction with the capture and use of the excess heat generated by the process allowing greater efficiency. Lantz (2012) reported that electric power can recover 30–40% of total energy from biogas, while CHP which includes low heat recovery can recover 35–55%. Total electricity generation from farm AD biogas has produced approximately 1.14 million megawatt-hours of electricity during 2018 (EPA AgStar 2019). The other most common uses for biogas were boiler/furnace, flared, and compressed natural gas (Fig. 4).

An outstanding example of energy recovery from biogas is Fair Oaks Farms, Indiana (AgSTAR 2019). This farm cooperative is exceptionally large and maintains 12 dairies and 2 digesters. A DVO two-stage mixed plug flow digester receives manure from more than 9000 cows, while a vertical plug flow digester receives manure from 3000 head. The DVO digester delivers 200,000 ft^3 of biogas per day. Methane from this digester is purified to 99% and converted into compressed natural gas that is used to fuel a fleet of 42 milk trucks. This has reduced diesel fuel use by more than 2 million gallons per year. The second digester is used to generate electricity. Fair Oaks Farms has also created an agricultural science center that promotes environmental sustainability and receives 500,000 visitors per year.

6 Codigestion for Improved AD Performance and Economics

Livestock manures as sole feedstock for AD do not provide optimal nutrition for the microorganisms and consequently do not maximize the efficiency of biogas production. It is generally accepted that C:N ratios of 20–30:1 are ideal for AD (Ward et al. 2008). However, dairy manures have a C:N ratio of about 9:1, while swine and poultry manures have ratios of about 7:1. To improve the nutrient balance and increase methane production, codigestion is often used. Typically, digesters are started and stabilized using a particular organic waste or waste mixture. Codigestion refers to the addition of a second type of waste in the already-established digester. The co-substrate could be applied continuously or intermittently. Major benefits of codigestion can be improvements in the C:N ratio of the feedstock, thereby improving overall biodegradation and methane production and improving economics through waste diversification (Khanal 2008, Ashekuzzaman and Poulsen 2011). The use of manure as co-substrate can also be beneficial because it provides buffering that compensates for the acidogenesis step of AD and it provides a wide range of additional nutrients needed by the microbiome. Of the 248 operating farm digesters in the USA, all of them at least occasionally use codigestion (AgSTAR 2019). Specific information about co-substrates are available for only 104. Of these, 35% used food processing wastes, 15% used dairy processing wastes, 15% used agricultural residues, nearly 7% used FOG (fats, oils, greases) wastes, and 27% used a variety of other feedstocks (e.g., undefined process water).

Considerable research on optimizing codigestion methods has been ongoing for years (Ward et al. 2008). Many different types of organic wastes have been evaluated as co-substrates in conjunction with various manures. Some recent examples include codigestion with food wastes (Zhang et al. 2013; Comino et al. 2012), crop residues (Wang et al. 2018), and slaughterhouse waste (Pitk et al. 2014). But studies have also shown that new substrates can have a broad range of effects on biogas production ranging from little impact to inhibition (Ward et al. 2008). The optimal feeding regimen to incorporate co-substrates should be evaluated empirically because of the unique characteristics of particular digesters and digester microbiomes (Wang et al. 2012).

Pretreatment methods to enhance the digestibility of agricultural residues, particularly fruits and vegetables, cereals, and grassland biomass, are also being developed. Plant residues contain lignocellulose which is not quickly decomposed with anaerobic processes. Pretreatments would be used to increase enzyme access to the fermentable components of lignocellulose prior to AD (Sawatdeenarunat et al. 2015). A variety of chemical, mechanical, thermal, or biological processes have been evaluated (reviewed in Paudel et al. 2017; Carrere et al. 2016). However, each of these methods has both benefits and limitations. Chemical pretreatments that use alkalis or acids leave chemical residues that can inhibit AD processes. Mechanical and thermal pretreatments involve costly mechanization, while biological methods lead to loss of cellulose and require longer treatment times. Currently, pretreatment methods for lignocellulose require further development and simplification if they are to be considered for ordinary farm use.

Linking manure digestion with other organic wastes produced as byproducts from bioenergy industries could also provide economic benefits for biogas operations. Ethanol production from corn grain fermentation results in considerable quantities of stillage waste. Using stillage as a co-substrate in manure digesters has been shown to increase biogas output in mesophilic (Westerholm and Hansson 2012) and thermophilic (Sharma et al. 2013) systems. Furthermore, thin stillage is generated at a high temperature and could thereby be coupled to the maintenance of reactor temperatures for mesophilic or thermophilic digesters if the two facilities are in close proximity. Biodiesel manufacturing has been growing quickly in the USA as well as worldwide. In 2013, biodiesel production in the USA surpassed 63 million metric tons (NBB 2014). But for every 100 kg of biodiesel, 10 kg of crude glycerol waste is produced. Since glycerol can be fermented and crude glycerol has an extremely high C:N ratio, this waste has been evaluated in codigestion with manures. For example, crude glycerol has been found to be a useful co-substrate to increase biogas in mesophilic digesters treating pig manure (Astals et al. 2012) and thermophilic digestion with cattle manure (Castrillon et al. 2013). Mixtures of multiple wastes in conjunction with manures have been studied extensively as well. Microalgae are being pursued as potential sources of bio-oil, and large-scale waste biomass may eventually be available. A recent bench-scale experiment showed that a triple codigestion of oil-extracted microalgae biomass, crude glycerol, and chicken litter produced higher methane than either co-substrate alone (Meneses-Reyes et al. 2018).

7 Potential for Expansion of Biogas Technology on US Farms

EPA AgSTAR (2019) analysis has estimated that an additional 8113 US livestock farms could utilize anaerobic digestion. This represents 5409 swine farms and 2704 dairy farms. The criteria used to determine whether a farm is suitable for AD considered herd size and the type of manure management system. For swine farms, the criteria were either herd sizes of 2000 animals and manure stored in anaerobic lagoons and liquid slurry systems or herds of 5000 animals with deep pit manure management systems. For dairy farms, the requirement was a herd of at least 500 cows and use of anaerobic lagoon or slurry manure systems. However, while new farm digesters did become operational each year for the last 5 years (2013–2018), there has been practically no net gain due to closures.

Considering the large numbers of dairy and hog farms that have the potential to support AD, the slow adoption of the technology is an issue. Studies have found that the large capital costs for building digesters have been prohibitive for most animal farms (DeVuyst et al. 2011; Bishop and Shumway 2009). Cowley and Brorsen (2018) addressed this issue from the point of view of those who have already adopted AD and asked whether users make adoption decisions for reasons other than economics. They found that farm characteristics (size, type, location) were also important factors. Larger farms were more likely to adopt AD, and the frequency of in-state neighbors with digesters also encouraged adoption. They further compared the attitudes of farmers who considered using AD versus those who actually implemented the technology. Although neighborhood effects and environmental beliefs stimulated consideration, these provided insufficient motivation for adoption. It was suggested that increased government grants for setting-up AD systems would encourage further use. Stokes et al. (2008) also cited the need for grant funding to encourage adoption because of user uncertainty vis-à-vis the actual value of biogas technology.

Anaerobic digestion is also greatly underutilized for poultry farm wastes where more than ten million tons of poultry litter are produced annually in the USA (Perera et al. 2010). This may be partly due to the perception that poultry manure, which is high in ammonia, is not a good substrate. However, pilot-scale and full-scale digesters have demonstrated that this waste can be digested even as sole substrate. In 2017, there were 320 farms with more than 100,000 chickens (layers), but few had digesters (USDA 2017). An excellent example, though, of using AD for poultry manure management is the Wenning Poultry Farm in eastern Ohio which has 600,000 animals and has been using a plug flow digester since 2008. Another example of a successful poultry litter digester is a single-stage, thermophilic (56 °C), pilot-scale (40m^3) sCSTR (semi-continuous stirred tank reactor) digester that was operated at West Virginia State University for 10 years using only poultry litter (Espinosa-Solares et al. 2009). This digester was reliable and produced biogas with 55–62% methane using litter with a C:N ratio of 7:1. Mixing was accomplished in this digester through recirculation of headspace gas. Of particular interest was the long-term stability of this system in spite of large changes in the feeding schedule each year: feeding ceased each winter and was restarted each spring (Bombardiere et al. 2007).

8 Further Developments of Biogas System Technology

While anaerobic digestion is itself considered to be a mature technology, supportive technologies for feedstock diversification, the expansion of economically useful bioproducts, and the integration of AD into anaerobic biorefineries present exciting opportunities for the (near) future.

Thermophilic digestion. Thermophilic anaerobic digestion (TAD) has several benefits over lower-temperature AD, including much better pathogen removal, higher process rates requiring a smaller system footprint, and higher methane production (Labatut et al. 2014). However, TAD is much less common than mesophilic in the USA. The primary reason is probably due to the perception that TAD is less stable than mesophilic AD and has higher maintenance costs. In fact, direct comparisons of the processes have shown that this can be true. Substrates that were digested without causing stress in a mesophilic AD were found to introduce moderate stress in TAD (Labatut et al. 2014). The reasons for reduced stability toward new substrates are not understood but may be related to microbial interdependencies (syntrophy). In our research, we have evidence that prolonged recovery time for TAD following substrate-induced stress was associated with large reductions in the population sizes of syntrophic acetate-oxidizing bacteria (SAOB) (unpublished results). The food web within TAD typically relies upon hydrogenotrophic methanogenesis that requires SAOB to oxidize acetate into CO_2 and H_2. This linkage in the food web may be more susceptible to environmental perturbations than the mesophilic food web. However, full-scale thermophilic digesters do have proven track records.

Expanding anaerobic digestion applications and products. Anaerobic digestion on US farms has been primarily used for treating high-strength organic wastes and producing electricity. But in most cases, simply relying on biomethane production for electricity generation does not provide sufficient economic return. However, the diversification of AD products has advanced considerably in the last decade and could be brought to farm operations, improving the utility and economics of the process. Angenent et al. (2018) have summarized several additional process units that could be added to large-size digesters on farms that would provide greater versatility: power-to-gas (PtG) technology, combined cooling, heat and power (CCHP), bio-oil, and carboxylic acid production. The PtG concept seeks to convert the CO_2 in biogas into additional methane. The conversion would be done using hydrogenotrophic methanogens (ex situ biomethanation) and hydrogen provided by electrolysis. This process has already been scaled-up in Germany (Götz et al. 2016). The goal of CCHP is to add cooling capabilities to the standard CHP process. The low-grade heat provided by AD can be used to drive the cooling cycle of absorption refrigeration which could have many uses on farms. These cooling systems are currently widely used in industry. In addition, bio-crude oil can be synthesized from the wet organic matter in AD effluent using hydrothermal liquefaction.

Another potentially valuable bioproduct from AD is medium-chain carboxylic acids. The anaerobic microbial food web in AD converts complex organic molecules (polysaccharides, lipids, proteins) into fermentable substrates. The metabolic intermediates in this food web include short- and medium-chain fatty acids (acetate, propionate, butyrate, valerate, caproate). The medium-chain fatty acids have value as platform chemicals for a variety of industrial syntheses. A growing research thrust has been to modify the metabolic pathways of AD in order to increase the production of fatty acids rather than methane. This concept has been called the carboxylate platform (Agler et al. 2011). Laboratory-scale increases in carboxylate production have been achieved, but full-scale operations have not yet been attained. Challenges remain for steering the microbial metabolic pathways toward the medium-chain fatty acids and also for recovering these carboxylates from water in an economical manner.

9 Conclusions

The application of anaerobic digestion for manure management in the USA has been slowly growing since the 1970s. However, it has been estimated that an additional 8000 dairy and hog farms have suitable herd sizes and manure management operations that could benefit from biogas technology. Clearly, the expansion of AD adoption on farms has not kept pace with the development of ever-larger CAFO facilities. Promising new process unit technologies have been proposed that could increase the economic return for farm-based AD, but these are still in development. Converting biogas into compressed natural gas has already been shown to be successfully implemented on some farms. AD may be further expanded to form the central platform technology for future anaerobic biorefineries that produce a variety of bioproducts from waste biomass. Ideally, as anaerobic digestion technology grows to produce multiple products, it will become integrated into the lifecycle of many more US livestock farms without reliance on government subsidies for support.

Acknowledgments I thank USDA NIFA for several years of grant funding that has supported our anaerobic digestion research. Thanks also to Jesus E. Chavarria-Palma for assistance with data collection and graphics.

References

Agler MT, Wrenn BA, Zinder SH, Angenent LT (2011) Waste to bioproduct conversion with undefined mixed cultures: the carboxylate platform. Trends Biotechnology 29:70–78

AgSTAR, Environmental Protection Agency (2019) Market Opportunities for Biogas Recovery Systems at U.S. Livestock Facilities. EPA-430-R-18-006

Angenent LT, Usack JG, Xu J, Hafenbradl D, Posmanik R, Tester JW (2018) Integrating electrochemical, biological, physical, and thermochemical process units to expand the applicability of anaerobic digestion. Bioresour Technol 247:1085–1094

Ashekuzzaman SM, Poulsen TG (2011) Optimizing feed composition for improved methane yield during anaerobic digestion of cow manure based waste mixtures. Bioresour Technol 102:2213–2218

Astals S, Nolla-Ardevol V, Mata-Alvarez J (2012) Anaerobic co-digestion of pig manure and crude glycerol at mesophilic conditions: biogas and digestate. Bioresour Technol 110:63–70

Bishop CP, Shumway CR (2009) The economics of dairy anaerobic digestion with coproduct marketing. Rev Agric Econ 31(3):394–410

Bombardiere J, Espinosa-Solares T, Domaschko M, Chatfield M (2007) Thermophilic anaerobic digester performance under different feed-loading frequency. App Biochem Biotechnol 137-140:765–775

Carrere H, Antonopoulou G, Affes R, Passos F, Battimelli A, Lyberatos G, Ferrer I (2016) Review of feedstock pretreatment strategies for improved anaerobic digestion: from lab-scale research to full-scale application. Bioresour Technol 199:386–397

Castrillon L, Maranon E, Fernandez-Nava Y, Ormaechea P, Quiroga G (2013) Thermophilic co-digestion of cattle manure and food waste supplemented with crude glycerin in induced bed reactor (IBR). Bioresour Technol 136:73–77

Chee-Sanford JC, Mackie RI, Koike S, Krapac IG, Lin YF, Yannarell AC, Maxwell S, Aminov RI (2009) Fate and transport of antibiotic resistance residues and antibiotic resistance genes following land application of manure waste. J Environ Qual 38:1086–1108

Comino E, Riggio VA, Rosso M (2012) Biogas production by anaerobic co-digestion of cattle slurry and cheese whey. Bioresour Technol 114:46–53

Cowley C, Brorsen BW (2018) The hurdles to greater adoption of anaerobic digesters. Agricultural and Resource Economics Review 47(1):132–157

DeVuyst EA, Pryor SW, Lardy G, Eide W, Wiederholt R (2011) Cattle, ethanol, and biogas: does closing the loop make economic sense? Agric Syst 104:609–614

Espinosa-Solares T, Valle-Guadarrama S, Bombardiere J, Domaschko M, Easter M (2009) Effect of heating strategy on power consumption and performance of a pilot plant anaerobic digester. App. Biochem. Biotechnol. 156:35–44

Feng L, Casas ME, Ottosen LDM, Moller HB, Bester K (2017) Removal of antibiotics during the anaerobic digestion of pig manure. Sci Total Environ 603-604:219–225

Götz M, Lefebvre J, Mörs F, McDaniel Koch A, Graf F, Bajohr S, Reimert R, Kolb T (2016) Renewable power-to-gas: a technological and economic review. Renew Energy 85:1371–1390

Khanal SK (2008) Anaerobic biotechnology for bioenergy production: principles and applications. Wiley, Blackwell

Labatut RA, Angenent LT, Scott NR (2014) Conventional mesophilic vs. thermophilic anaerobic digestion: a trade-off between performance and stability? Water Res 53:249–258

Lantz M (2012) The economic performance of combined heat and power from biogas produced from manure in Sweden – a comparison of different CHP technologies. Appl Energy 98:502–511

MacDonald JM, McBride WD (2009) The transformation of U.S. livestock agriculture: scale, efficiency, and risks. United States Department of Agriculture, Electronic Information Bulletin Number 43

Marti R, Scott A, Tien YC, Murray R, Sabourin L, Zhang Y, Topp E (2013) Impact of manure fertilization on the abundance of antibiotic-resistant bacteria and frequency of detection of antibiotic resistance genes in soil and on vegetables at harvest. Appl Environ Microbiol 79:5701–5709

Meneses-Reyes JC, Hernández-Eugenio G, Huber DH, Balagurusamy N, Espinosa-Solares T (2018) Oil-extracted *Chlorella vulgaris* biomass and glycerol bioconversion to methane via continuous anaerobic co-digestion with chicken litter. Renew Energy 128:223–229

NBB (2014) National Biodiesel Board. http://nbb.org

Paudel SR, Banjara SP, Choi OK, Park KY, Kim YM, Lee JW (2017) Pretreatment of agricultural biomass for anaerobic digestion: current state and challenges. Bioresour Technol 245:1194–1205

Perera R, Perera P, Vlosky RP, Darby P (2010) Potential of using poultry litter as a feedstock for energy production. Louisiana State University Forest Products Development Center

Pitk P, Palatsi J, Kaparaju P, Fernandez B, Vilu R (2014) Mesophilic co-digestion of dairy manure and lipid rich solid slaughterhouse wastes: process efficiency, limitations and floating granules formation. Bioresour Technol 166:168–177

Robertson GP, Vitousel PM (2009) Nitrogen in agriculture: balancing the cost of an essential resource. Annu Rev Environ Resour 34:97–125

Robertson GP, Gross KL, Hamilton SK, Landis DA, Schmidt TM, Snapp SS, Swinton SM (2014) Farming for ecosystem services: an ecological approach to production agriculture. Bioscience 64:404–415

Sakar S, Kaan Y, Kocak E (2009) Anaerobic digestion technology in poultry and livestock waste treatment: a literature review. Waste Management Res 27:3–18

Sawatdeenarunat C, Surendra KC, Takara D, Oechsner H, Khanal SK (2015) Anaerobic digestion of lignocellulosic biomass: challenges and opportunities. Bioresour Technol 178:178–186

Sharma D, Espinosa-Solares T, Huber DH (2013) Thermophilic anaerobic co-digestion of poultry litter and thin stillage. Bioresour Technol 136:251–256

Silbergeld EK, Graham J, Price LB (2008) Industrial food animal production, antimicrobial resistance, and human health. Annu. Rev. Public Health 29:151–169

Sommer SG, Christensen ML, Schmidt T, Jensen LS (2013) Animal manure recycling: treatment and management. Wiley

Spirit CM, Daly SE, Werner JJ, Angenent LT (2018) Redundancy in anaerobic digestion microbiomes during disturbances by the antibiotic monensin. Appl Environ Microbiol 84(9): e02692–e02617

Stokes JR, Rajagopalan RM, Stefanou SE (2008) Investment in a methane digester: an application of capital budgeting and real options. Rev Agric Econ 30:664–676

Szogi AA, Vanotti MB, Ro KS (2015) Methods for treatment of animal manures to reduce nutrient pollution prior to soil application. Curr Pollution Rep 1:47–56

Tilman D, Cassman KG, Matson PA, Naylor R, Polasky S (2002) Agricultural sustainability and intensive production practices. Nature 418:671–677

Turker G, Akyol C, Ince O, Aydin S, Ince B (2018) Operating conditions influence microbial community structures, elimination of the antibiotic resistance genes and metabolites during anaerobic digestion of cow manure in the presence of oxytetracycline. Ecotoxicol Environ Saf 147:349–356

Udikovic-Kolic N, Wichmann F, Broderick NA, Handelsman J (2014) Bloom of resident antibiotic-resistant bacteria in soil following manure fertilization. Proc Natl Acad Sci 111(42):15202–15207

US Environmental Protection Agency (2019a) EPA 430-R-19-001. Inventory of US Greenhouse Gas Emissions and Sinks: 1990-2017

US Environmental Protection Agency (2019b) EPA 430-R-18-006. June 2018. Market opportunities for biogas recovery Systems at U.S. Livestock Facilities

USDA-NRCS (2012a) Agricultural Waste Management Field Handbook, vol 2012. USDA, Washington DC

USDA-NRCS (2012b) Introduction to waterborne pathogens in agricultural watersheds. Nutrient management technical note no. 9. USDA, Washington DC

USDA ERS (2017) Major uses of land in the United States [data file]

USDA (2017) Census of agriculture

van den Bogaard AE, Willems R, London N, Top J, Stobberingh EE (2002) Antibiotic resistance of faecal enterococci in poultry, poultry farmers and poultry slaughterers. J Antimicrob Chemotherapy 49:497–505

Wang SJ, Fox DG, Cherney DJR, Klausner SD, Bouldin DR (1999) Impact of dairy farming on well water nitrate levels and soil content of phosphorus and potassium. J Dairy Sci 82:2164–2169

Wang X, Li Z, Bai X, Zhou X, Cheng S, Gao R, Sun J (2018) Study on improving anaerobic co-digestion of cow manure and corn straw by fruit and vegetable waste: methane production and microbial community in CSTR process. Bioresour Technol 249:290–297

Wang X, Yang G, Feng Y, Ren G, Han X (2012) Optimizing feeding composition and carbon-nitrogen ratios for improved methane yield during anaerobic co-digestion of dairy, chicken manure and wheat straw. Bioresour Technol 120:78–83

Ward AJ, Hobbs PJ, Holliman PJ, Jones DL (2008) Optimisation of the anaerobic digestion of agricultural resources. Bioresour Technol 99:7928–7940

Westerholm M, Hansson M, Schnurer (2012) Improved biogas production from whole stillage by co-digestion with cattle manure. Bioresour Technol 114:314-319

Zhang C, Xiao G, Peng L, Su H, Tan T (2013) The anaerobic co-digestion of food waste and cattle manure. Bioresour Technol 129:170–176

Anaerobic Digestion in Europe: Key to Waste Management, Challenges, and Perspectives

Efraín Reyes Cruz, Alicia Guadalupe Talavera-Caro, and Aldo Almeida

Abstract Anaerobic digestion is a natural occurring process developed by a wide microbial consortium. From it, CH_4, a valuable biofuel, is produced from agro-industrial wastes. The European Union has an enormous potential to continue the development of this technology. Not only the countries within it are leading the discoveries and implementing novel strategies for the process, but also they have a large amount of usable resources as raw material. Furthermore, current policies for the adaptation of the circular economy and green technologies enhance the opening of biogas plants. It is possible to search for the improvement of the process by different means. For example, reactor design has a main task, which is to provide the adequate conditions to enhance microbial metabolism. On the other hand, there are technologies to improve biogas to biomethane, which can be injected to the grid in the form of gas or electricity. However, if the process is not sustainable from an energetic or economic view, the process has no sense. To know this, technoeconomical analyses are performed. This chapter describes the current political conditions that enhance the circular economy within the EU. In addition, examples of potential modifications in reactors' design, response from co-digestion experiments, and potential substrates for the process are provided. At last, technoeconomical examples are explained too.

Keywords Biomass Europe · Anaerobic digestion · Biomethane · European bioenergy · Techno economical assessment

E. R. Cruz · A. G. Talavera-Caro
Laboratorio de Biorremediación, Facultad de Ciencias Biologicas, Universidad Autonoma de Coahuila, Torreón, Coahuila, Mexico

A. Almeida (✉)
Department of Plant and Environmental Sciences, University of Copenhagen, Copenhagen, Denmark

N. Balagurusamy, A. K. Chandel (eds.), *Biogas Production*,
https://doi.org/10.1007/978-3-030-58827-4_5

1 Introduction

Climate change has accelerated the need to shift our main source of energy, the fossil fuels, to a sustainable energy production by eco-friendly technologies. The constant research of these technologies had achieved the improvement of wind turbines, photovoltaic systems, solar power technologies, tidal and wave systems, hydropower plants, and more (Tchanche et al. 2013).

There exists another technology that has earned high expectations in Europe, which is the process of anaerobic digestion (AD) which allows the exploitation of diverse feedstocks such as agricultural residues (animal manure and crop wastes), industrial residues (sewage sludge, food waste, etc.), and municipal solid waste, to produce biogas rich in methane (~65% CH_4) (Scarlat et al. 2018).

CH_4 is a renewable source of energy considered a valuable biofuel, and its combustion generates 36.5 MJ m^{-3} (Guo et al. 2015). However, raw biogas contains, in less proportion, other components as CO_2 (~35%), water vapor, hydrogen sulfide (H_2S), siloxanes, halogenated hydrocarbons, ammonia (NH_3), and nitrogen (N_2) (Batlle-Vilanova et al. 2019; Ryckebosch et al. 2011). Therefore, a purification process to remove the mentioned gasses and upgrade the biogas into biomethane (<95% CH_4) is required. In this way, it can be injected to the gas grid, be liquefied for transportation fuel, or be burned for heat and electricity production (Achinas and Willem Euverink 2020; Scarlat et al. 2018; Ryckebosch et al. 2011).

In addition to CH_4, an organic fertilizer (digestate) rich in nutrients is obtained (Uçkun Kiran et al. 2016). Because animal manure can function as a substrate for AD, it can substitute its direct implementation on farmlands as a fertilizer, which will reduce odor problems, greenhouse gas emissions, and soil and water contamination. Therefore, biowaste can function as a suitable source for energy production, which is favored by the current practices for waste management. However, AD and waste collection improvement is indispensable to impulse the application of this technology (Achinas and Willem Euverink 2020).

The process of AD is based on the degradation of complex organic materials by a set of biological processes (hydrolysis and/or fermentation, acetogenesis, and methanogenesis), performed by the constant interaction of a wide microbial consortium in anaerobic conditions (Alvarado et al. 2014). Therefore, AD is susceptible to inhibition from a variety of compounds and conditions (Chen et al. 2014), which makes the constant monitoring of the process an unquestionable requirement.

Countries belonging to the European Union (EU) (Fig. 1) have had an important role in the development and application of AD, which is reflected in the implementation and operation of biogas plants. For example, Germany has reported a total of 9444 biogas plants in operation in 2018 (German Biogas Association 2019). Furthermore, a recent study presented a variety of modeled scenarios of the biomass and biogas energy potential from agricultural residues in the EU for year 2030, which highlighted the potential of this technology even in the worst conditions (Meyer et al. 2018, b).

In another aspect, the supporting schemes and governmental initiatives, provided by the circular economy from the EU, promote the application of renewable energies

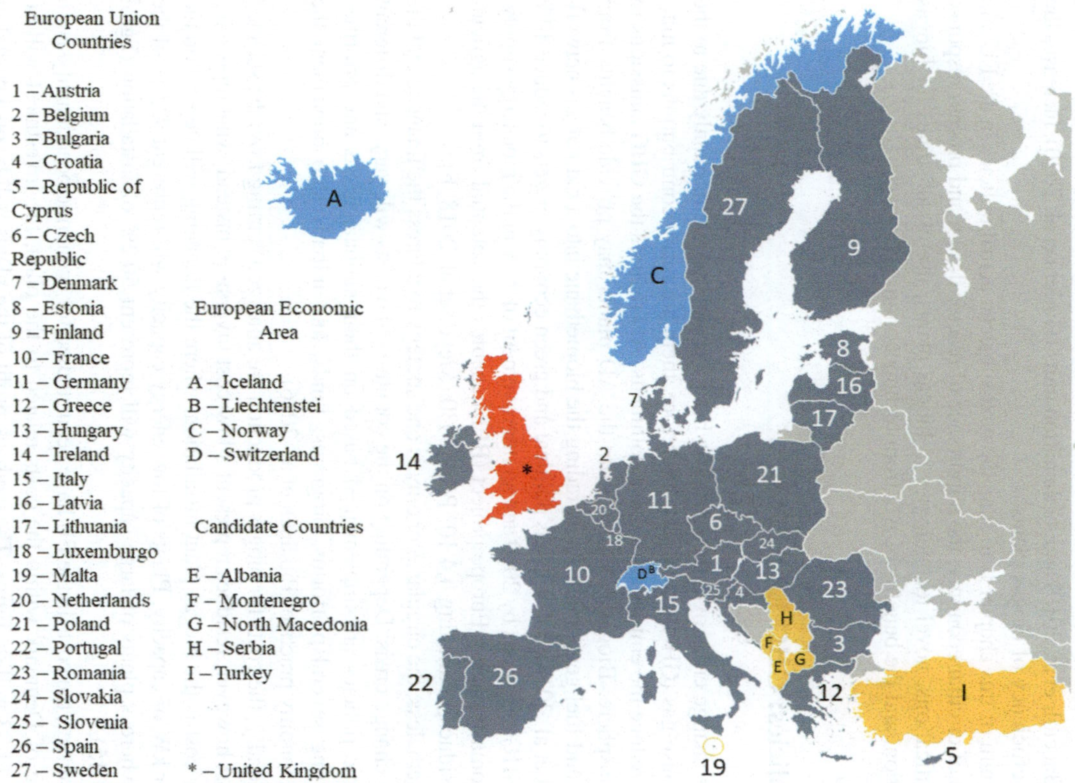

Fig. 1 Countries belonging to the European Union, the European Economic Area and the Candidates to form part of EU (European Union 2020). * Due to the recent exit of United Kingdom from the EU, it was considered within it for the elaboration of this chapter

and impulse the development of biogas plants (Achinas et al. 2017; Scarlat et al. 2018).

Improvement of anaerobic digesters can be achieved by the modification of process variables, such as temperature (Feng et al. 2018), digester configuration (Baldi et al. 2019), mixing time (Kress et al. 2018), and mixing components (Lebranchu et al. 2017). Furthermore, addition of extra systems, i.e., organic rankine cycle (Baccioli et al. 2019) and methanation units (Guilera et al. 2020), increases the overall efficiency of biogas plants.

The aim of this chapter is to state the next subjects: actual policies in the EU and the role of circular economy, the importance of reactor design and process response to modifications, novel technologies in biogas plants, useful substrates for biogas production, and the benefits of AD from technoeconomical aspects.

2 Policies: GHG Emissions

Climate change crisis over the years has been a critical problem, mainly due to the greenhouse gas (GHG) emissions. Therefore, there are many contributions to mitigate or reduce the emissions of these pollutants. AD reduces the GHG emissions to the atmosphere. Biogas produced from the AD plants may provide benefits from energy and fuel generation by upgrading the biomethane into a natural gas network (Scarlat et al. 2018). The EU has political and green economy targets to reduce 40% of the GHG emissions by 2030 and an improvement of 27% in the bioenergy supply. The Biomass Energy Europe project (BEE) enables the potential from the agricultural residues, estimating $4.4 \cdot 10^3$ PJ in 2030 (Meyer et al. 2018, b).

Nonetheless, the circular economy is one strategy to address the financial and the climate change crisis. Depending on the countries, the costs will vary, and different economic policies are implemented based on their national needs and priorities developing several platforms, programs, funds, and initiatives to implement the green economy practice (Pitkänen et al. 2016).

Generally, the costs for biogas production are analyzed starting from the selected biomass; however, residues represent a low-cost inversion. Instead, what represents a major cost is the biogas plant installation, where the financing will depend on the cost per kW of capacity. Plants of low energy capacity will represent 92.0% of the costs, whereas plants of large capacity will represent 50.2% of installation (Sgroi et al. 2015).

Moreover, the costs in the EU have been studied for economic benefit analysis. In the case of Denmark within the costs related to the plant, the investment will be around 11.33 million Euros. The payback profit of the plants will be on a period of 10 years after installation of 2.67 million Euros. Even so, the plants for the EU countries are mostly supported by the funding investment for the integration of circular economy placed by the European Structural and Investment Fund (Lybæk and Kjær 2017).

The value of the energy production from biogas produced from renewable sources indicates 25% of electrical efficiency. Then biomethane produced cost in

Europe is between 38 €/MWh for 100 m^3/h and 10 €/MWh for 2000 m^3/h on average. Moreover, the costs of the biomethane to be distributed as natural gas may include the processing of injection, compression, and pipeline costs. Thus, the maximum cost is about 60 €/MWh for 100 m^3/h sharing 13% to 39% of the total costs (Flotats 2019).

For circular economy from biogas plants integrated to the market in local communities, a different policy implementation based on a project for biogas research and plant installation facilitates the support and the impact of the AD. Mainly, in the EU there is an Environmental Impact Assessment (EIA) policy for covering the biogas facilities and photovoltaic power plants for evaluation and assessing the impact for support (Larsen et al. 2018).

The EU has a highly promoting system for initiatives through renewable energy production, primarily from organic waste management. Europe runs 9000 plants, within a total of biogas production which involves landfills, sewage wastewater treatment, and digestion plants, being Germany the head leader of the biogas market (Fig. 2). Nevertheless, in the EU the upcoming markets in other countries are well established and subsidized such as Italy. On the other hand, France is another potential market growing strong, as well as Sweden which is already transporting natural gas as part of their economy by city buses (Yousuf et al. 2016).

Furthermore, biogas for natural gas application, contributing with more than 30% in the European energy source, is expected to be the major energy demand in 2040 in the New Policies scenarios. In order to estimate the costs to develop biogas plants, the economic analysis is necessary to evaluate the variability of operation costs for the employment of the process, also considering some of the incentives and financial programs. In Italy, for example, the cost estimation is assessed by the NETL methodology, where five levels should be considered to define capital costs, being the bare erected cost (BEC), engineering procurement and construction cost (EPCC), total plant cost (TPC), total overnight cost (TOC), and total as spent capital (TASC). Therefore, the costs will be estimated for the biogas production on a system before the installation and getting advantage from the Italian incentive programs introduced for the legal framework; thus, the costs will be reduced. These frameworks and incentives for biogas promotion are also adopted in Germany, Sweden, and the UK (Rotunno et al. 2017).

Since the renewable energy directive from the European Commission promotes the biogas plant installation to contribute to the domestic application. To increase the energy supply, opportunities are promoted to provide the development of technology and innovation, thus in the future strengthening economic growth. Identification of biogas plant types is essential to stablish innovations.

The European classification of biogas plants according to the International Energy Agency (IEA) of different European countries (Denmark, Sweden, Switzerland, UK) is tested in a Complemented Framework for Categorization depending on each country's necessities. The categorization is supposed to state possibilities in several approaches when plants are classified. On result, the framework will then promote technology and policy application if needed (Lindkvist and Karlsson 2018).

Some other studies and funded projects in EU such as BIOSURF, FaBbiogas, ESBF, GreenGasGrid, BiogasIN, and many others aimed to the diffusion on the

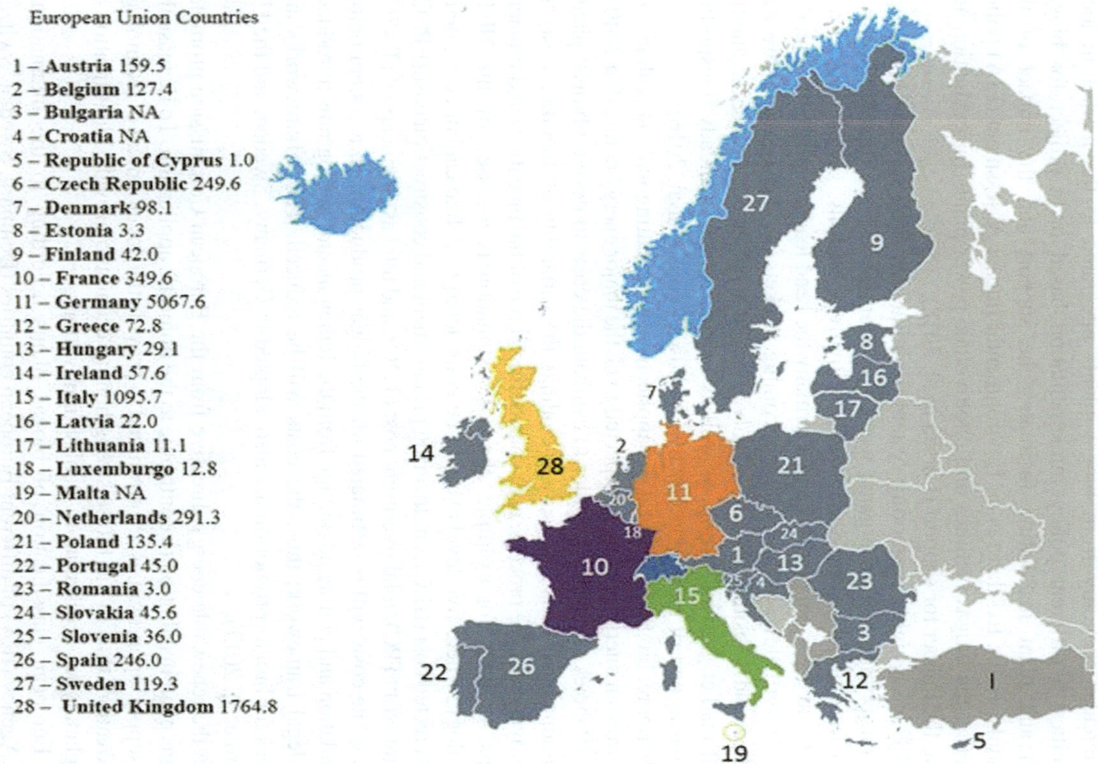

Fig. 2 Overview of the biogas production for energy from landfills, wastewater treatment plants and digester of different organic matrices in the European Union (ktone·y^{-1}). Modified from Urbini and Raboni 2014

large scale of biomass-derived energy including institutional capacity building, improvement of framework conditions for biogas production including the transportation systems and optimal models, and financing the projects of installation implementing policies, tariffs, and incentives (Capodaglio et al. 2016).

The aim of establishing a circular economy for biomethane production is to deal with the biomass waste with a low cost and high supply of energy for many European countries. The key factor is to reduce the costs and state an environmental alternative for the biogas sector by strengthening the investment in plant installation and promote the governmental initiatives, such as feed-in tariffs or tax policies, in order to succeed on the full positive introduction of a renewable electricity and biofuels taking advantage of funds (Achinas et al. 2017).

3 Design

It is known that AD consists of a series of biological reactions that occurred continuous and simultaneously. Therefore, to maintain the best conditions inside the digester, it must be carefully designed to provide the required environment for the participant microbial groups. Among the variables that are related to this aspect are the shape and volume of the digester, its configuration (continuous or batch, single, or multistage process), the mixing systems along with their components, and the installation of novel equipment in the biogas plant, just to mention some examples (Radetic 2018).

3.1 Configuration

Multistage process is based on separating AD stages to improve their development and have a better control of the possible occurring constraints compared with a one-stage process. This facilitates the digestion of harsh organic material.

For example, Baldi et al. (2019) applied a two-stage process for co-digestion of food waste (FW) with activated sludge (AS), substrate with low and inefficient biodegradability that neither allows a high organic loading rate (OLR) or high biogas production, and compared it against a one-stage process. The two-stage process had a fermentative reactor (3 L), where only hydrolysis and fermentation stages were developed with the consequent production of H_2, a valuable biofuel as CH_4. Digestate expulsed from this digester directly passed to methanogenic reactor (12 L) for CH_4 production. In the case of one-stage process, it only had the methanogenic reactor. Their results showed that biogas and its CH_4 content didn't differ in one-stage against two-stage process. Nevertheless, the fermentative reactor of the last configuration had the advantage of H_2 production. Another benefit was the reduction in the hydraulic retention time (HRT), because compared with the one-stage, the two-stage process time decreased by 5 days (from 17 to 15 days). Even if the two-stage process co-digestion efficiency was a bit lower compared with

FW as the only substrate, it is a great strategy to take advantage of both wastes, produce two valuable biofuels, reduce the HRT, and manage more volumes.

3.2 Mixing

To ensure the homogenization of substrates, and their contact with microorganisms inside digesters, mixing systems are used. Each anaerobic digester will require a certain mixing system to function properly, which can be obtained by empiricism. However, the application of science must be the main strategy to reach this goal (Dapelo and Bridgeman 2018). According to the digester hydrodynamics and the rheological properties of the substrate, impeller and agitation rate are selected. Common substrates for AD, e.g., manure, are considered as non-Newtonian fluids, a property which makes them harder to homogenize (Lebranchu et al. 2017; Kress et al. 2018). Non-efficient mixing can result in temperature and concentration gradients that will impact the microbial populations and hamper the process development (Lebranchu et al. 2017). However, mechanical mixing systems require high energy input, which leads to reduced mixing time to cut operational costs. This strategy can lead to limitations in biodigester efficiency (Kress et al. 2018). Therefore, choosing the most suitable mixing system, along with its components, is of utmost importance for AD plants.

Lebranchu et al. (2017) studied the impact of mixing in a lab-scale digester (2 L). In their work, three agitation rates of a six-bladed Rushton turbine (22, 66, and 110 rpm) and a nonstandard double helical ribbon (10, 50, and 90 rpm) were compared to determine the best impeller for biogas production from CM with constant cellulose additions. From a total experimentation time of approximately 1440 h, there was a biogas production of 106 and 159 L for the turbine and the helical ribbon agitation rates, respectively. Therefore, the latter produced 50% more biogas and had a suitable CH_4 content (~ 64%). Enhancement was related with the start-up of biogas production, the helical ribbon started gas production at hour 10, whereas the Rushton turbine started at hour 70, approximately. Furthermore, less time passed between each addition of cellulose, which allowed five extra additions with the helical ribbon. On the other hand, high mixing frequencies negatively affected the process. Therefore, AD with a nonstandard double helical ribbon can overpassed CH_4 production from a six-bladed Rushton turbine, a common use impeller, from co-digestion of CM with cellulose.

Another research analyzed the triggered effects for reducing the mixing time in a full-scale CSTR. This digester was fed with a mixture of crop residues and animal manures. Furthermore, it was equipped with a submersible mixer and a propeller incline agitator. It was found that inlet mixture accumulated near the feeding system, but it reached an equal distribution among all the height levels inside the digester. Furthermore, all the mixing times didn't occasion a segregation effect on the substrate. With regard to the energy consumption of impellers, the submersible mixer consumed 40% more electricity compared with a propeller incline agitator.

In a mixing time of 10, 5, and 2 min per 30 min, the former consumed, in order, 97, 47, and 19 kWh/d, compared with the latter that consumed 43, 19, and 8 kWh/d, for the respective times. It was noted that biogas production increased with higher mixing rates. However, this trend completely shifted after an intense stirring (4 hours) at the end of each treatment, which released the trapped gas and modified average gas yields, which allow the lower mixing time to be the most efficient. Reduction in mixing time increased the viscosity of the substrate and didn't let biogas to escape. In conclusion, there was no effect in nutrient distribution by reducing mixing times. Furthermore, the authors suggested to not discard the strategy of push the limits before critical operational conditions appear. In addition, it's important to mention the effectivity of applied mixers because they avoided the presence of dead zones in the CSTR (Kress et al. 2018). Therefore, this work demonstrated a great scenario for AD with a stable and flexible biogas production, coupled with less electricity requirement for homogenization thanks to the brief application of combined mixers.

The actual trend in anaerobic digesters consists in the application of advanced computational software to model base scenarios with the aim to know the viability of the project or to evaluate the performance of the applied systems in an operating one (Dapelo and Bridgeman 2018). For the study of mixing and hydrodynamics in biodigesters, computational fluid dynamics (CFD) simulations are used, these are based on 3D simulations that show the behavior of liquid velocity fields when are homogenized by moving parts or not. However, simulations require a significant number of variables such as Reynolds number, power number, power dissipation, agitation rate, shear rate, shear stress, viscosity, density, and some more constants (Lebranchu et al. 2017).

A study made CFD simulations, with a commercial CFD software called ANSYS Fluent 16.0, for a deeper comparison in the mixing efficiency of two impellers. It was found that the Rushton turbine left unmixed zones, called "caverns," at all agitation rates. Furthermore, the substrate presented near-zero velocity and zero-shear viscosity that extended from the blades ends to the wall of the digester. Accordingly, substrate distribution inside the reactor was almost static. On the other hand, the digester equipped with a helical ribbon mixer at is lower agitation rate contained less than 1% of dead zones. Therefore, this type of mixers can be recommended for their capacity for great homogenization compared with common turbines. Furthermore, energy requirements for helical ribbon mixers didn't importantly differ from the Rushton turbines (Lebranchu et al. 2017).

Another parameter that was obtained with the CFD simulations was the volume-averaged shear stress (σ). To determine its impact in the floc breakage, the maximum value of σ (σ_{max}) in each treatment was used. The best performance was observed when agitation rates were near the maximal shear stress (30 Pa), 26 Pa (10 rpm) for helical ribbon, and 25 Pa (66 rpm) for Rushton turbine. At higher agitation rates, σ_{max} exceeded the maximal shear stress and caused slower acid consumption, less biogas production, and CH_4 content. Therefore, estimation of shear stress for the variety of impeller types and the rotation frequencies are primary tasks in the design and improvement of anaerobic digesters (Lebranchu et al. 2017).

3.3 Feeding Systems

Biogas plants require a flexible biogas production according to the energy demands (Hahn et al. 2014). For this, variations in feeding intervals, mass addition, or in the type of added substrate are helpful strategies. However, a constant feeding pattern is required to maintain the optimum microbiological processes and, thus, the development of AD (Feng et al. 2018).

Some of the mentioned fundaments were applied in the work of Feng et al. (2018). Lab-scale experiments with a single pulse feeding of maize silage (MS) or briquetted meadow grass (BMG) were analyzed through 7 days in a CSTRs mainly fed with CM. There was a remarkable boost in biogas production in the first 2 days in digester with MS, which increased 130 and 100% compared with the control, while in the remaining days, enhancement reached 20%. Moreover, it just took around 10 hours to increase CH_4 content to 60% from 40%. In contrast, pulse feeding of BMG increased average production by 30% the first 2 days and below 5% the following.

The same authors experimented with pilot-scale digesters at mesophilic (41 °C, 10 m^3) and thermophilic (51 °C, 30 m^3) temperatures. As in lab-scale digesters, MS and BMG enhanced biogas production. Thermophilic digestion of MS importantly enhanced biogas production in the first 2 days (90 and 57%). At the fourth day, biogas production was less than the control. Unlike the mentioned digester, mesophilic digester with MS had a moderated biogas production, with the highest values in the first 3 days (60, 32, and 17%). Despite those differences, both treatments enhanced biogas yield to 1.2 $m^3/m^3{*}d$, approximately, 20% above compared with its control. Regarding the CH_4 content, both conditions maintained around 60% of this parameter. On the other hand, the increment in biogas yield from thermophilic digester with BGM ranged between 17 and 34%, which allowed a total upgrade of 28% compared with its control (1.25 $m^3/m^3{*}d$) and maintained its CH_4 content between 58 and 60%. This stability was possibly due to retention of BMG as a floating layer inside the digester. Therefore, pulse feeding of BMG at pilot-scale digesters was better than that of MS. Although there was an improvement in the process, the authors reported that pilot-scale CSTRs didn't reach the same effectiveness compared with laboratory-scale experiments, probably by differences in the OLR, retention time, and inoculum (Feng et al. 2018). However, it's noteworthy the potential of pulse feeding to create oscillations in biogas production, with higher yields some hours after pulse feeding and a constant production for the next period, thanks to the main substrate addition.

Although adoption of novel technologies in pilot or full-scale digesters is based on results from lab-scale experiments, this approach may not be such reliable as Kress et al. (2018) argued. Therefore, research with full-scale digesters must be driven.

3.4 Additional Equipment

With internal combustion engines or micro-gas turbines, generated biogas from AD plants is transformed into heat and energy, which can cover the plant necessities or be sold (Baccioli et al. 2019). A method to recover part of the generated heat is the addition of an organic Rankine cycle (ORC) to the infrastructure of the plant. The system is composed of an evaporator, an expander, a condenser, and a pump. It's based on the use of heat from combustion gasses to evaporate a working fluid, commonly an organic compound such as oil, that immediately expands in a turbo-generator; after that, it is condensed and pumped back to the evaporator (Tchanche et al. 2013). Thereby, thermal energy can be used to produce electric energy.

For the implementation of novel technologies, the foresight of the possible scenarios is fundamental. At present, there also are sophisticated modeling systems that can accomplish this goal (Table 1). For example, Baccioli et al. (2019) worked with AMESim software to evaluate the addition of an ORC system for two anaerobic digesters, with a total volume of 4600 m^3, to treat 10.8 t/h of a wastewater mixture of sewage and municipal organic waste. The plant had a biogas production of 276.6 kg/h^- with CH_4 content of 65%, which burned with a micro-gas turbine (mGT) Capstone C600s. For the modeling, 1 year was taken as reference time, and variables such as ambient, sludge seasonal, and air temperatures were taken into account. In the reference scenario, the exhaust gasses from the mGT heated a water loop to maintain the digester at 37 °C and preheat the sludge mixture before entering the digester. Furthermore, there was a regeneration heat exchanger of 10 m^2 that used the heat from digestate to increase the temperature of substrate sludge too. Results from the model indicated that ambient temperature had an important impact in the efficiency of the mGT; when temperature increased efficiency of the mGT decreased, this was reflected in less energy production. The improved scenario consisted in the addition of an ORC system; it was an Infinity Turbine IT50, a radial outflow turbine generator that produced 55 kW_{el} by the use of 550 kW_{th}. The exhaust gasses were used to heat the evaporator of the ORC and the substrate sludge. The ORC evaporator required a water loop at 95 °C, which needed an increment in the regenerator size to 35 m^2. This modification allowed the ORC operation nearly mentioned conditions but reduced the heat extracted for the substrate sludge heating. However, this effect was compensated by the heat obtention from the condenser of the ORC, which was used to heat the sludge. After 1 year, the ORC provided 8.6% of the electric energy produced by the plant (average production of 12,700 kWh/d). As in the reference scenario, efficiency of mGT decreased when ambient temperatures were high but increased the heat available from exhaust gasses which was exploited by the ORC. This led to a turbine recovery efficiency of 74.6–77% compared with the <35% of the reference scenario. Therefore, implementation of ORC systems to biogas plants can improve the recovery and use of heat derived from exhaust gasses (Table 1).

As can be seen, the application of novel technologies or change from common use equipment can have a significant impact in AD. The gaps that remained in the

Table 1 Biogas production and CH_4 content AD processes with the implementation of different technologies

AD scenario	OLR	Biogas production	Methane content (%)	Reference
[L]One-stage process	2.5 kgTVS/m^3*d of FW	694.4 ± 24.6 NL/ kgTVS*d	65.2 ± 1.9	Baldi et al. (2019)
[L]Two-stage process	14.2 kgTVS/m^3*d of FW 2.5 kgTVS/m^3*d of FW	43.1 ± 12.8 NL/ kgTVS*d 704.6 ± 28.5 NL/ kgTVS*d	22.9 ± 5.5% of H_2 68.4 ± 1.1	
[L]One-stage process	2.5 kgTVS/m^3*d of FW + AS	485.9 ± 25.8	61.2 ± 2.2	
[L]Two-stage process	14.6 kgTVS/m^3*d of FW + AS 2.5 kgTVS/m^3*d of FW + AS	44.8 ± 12.6 NL/ kgTVS*d 611 ± 45.4 NL/ kgTVS*d	18.4 ± 6.3% of H_2 70.1 ± 1.6	
[L]Reactor equipped with a non-standard double helical ribbon	N/A CM + cellulose	123–175 mL/h	57–64	Lebranchu et al. (2017)
[L]Reactor equipped with a six-bladed Rushton turbine	N/A CM + cellulose	82–85 mL/h	59–64	
[F]CSTR, mixing time of 10 min/30 min	3.5 kg oDM/m^3*d^1 of crop residues and animal manure mixture	74 → 65 m^3/h	N/A	Kress et al. (2018)
[F]CSTR, mixing time of 5 min/30 min		65 → 68 m^3/h		
[F]CSTR, mixing time of 2 min/30 min		63 → 72 m^3/h		
[L]CSTR control	4 kgVS/m^3*d^{-1} of CM	0.76 m^3/m^3*d	~53	Feng et al. (2018)
[L]Pulse feeding of BMG to CM	8 kgVS/m^3*d^1 the first day, the rest were with 4 kgVS/ m^3*d of CM	0.94 m^3/m^3*d	~55	
[L]Pulse feeding of MS to CM		1.26 m^3/m^3*d	~58	
[P]Pulse feeding of BMG to CM (mesophilic)	5.2 kgVS/m^3*d the first day, the rest were with 2.6 kgVS/m^3*d of CM	~1.2 m^3/m^3*d	~60	
[F]Pulse feeding of BMG to CM (thermophilic)		~1.2 m^3/m^3*d	~60	
[P]Pulse feeding of MS to CM		~1.55 m^3/m^3*d	58–60	

[L] Lab scale, [P] pilot scale, [F] full-scale digester, N/A not available

development of full-scale technologies must be closed. Therefore, constant research and development of biogas technologies can't stop.

4 Process

Biogas plants require the correct development of serial steps to achieve biomethane production, one of them is the AD performance. In this section, digestion of different substrates was evaluated, and some of the currently implemented technologies are described.

4.1 Animal Manure

The aperture of new biogas plants requires the constant availability of feedstock for the process. The selected substrate must be sustainable in all aspects, covering its cultivation, recollection, transport, storage, and an efficient conversion in AD. Frequently, agricultural wastes or animal manure meet these requirements (Meyer et al. 2018, b). An exhaustive modeling work evaluated the potential of biogas production from animal manure in the EU for year 2030. The forecasted manure production (cattle, poultry, and pig) ranged from 83 to 122 Mt./y with a total energy potential for AD of 670–890 PJ/y (Meyer et al. 2018, b). Thus, these substrates must undergo AD to harness that possible biogas.

The potential of CM for biogas production, along with recent strategies for its improvement, has been constantly reviewed (Achinas and Willem Euverink 2020; Meyer et al. 2018, b; Scarlat et al. 2018). Therefore, mentioning them here would be repetitive.

Nevertheless, poultry manure (PM) hasn't received the same attention as CM. PM is a troublesome substrate by its high content of ammonium ions (NH_4^+-N) and free ammonia nitrogen (FAN), which can diffuse passively into the cell and cause proton imbalance and/or potassium deficiency (Chen et al. 2014; Świątek et al. 2019). This could lead to an unstable process with its consequent failure. Therefore, finding the conditions that trigger this effect would help to prevent the mentioned scenario.

Świątek et al. (2019) experimented PM mono-fermentation in a CSTR (15 L) for 50 days. OLR increased weekly 0.5 g/L∗d, and from an initial value of 1, it reached 2.5 g/L∗d. Top process performance was on the 26th day with 20 L of biogas (with 66.8% of CH_4). After this, production started to decrease constantly. On the 45th day, biogas produced was just 35% of the maximal production (20 L), with 42.6% of CH_4. Released nitrogen increased NH_4^+-N and FAN to 3.86 and 0.17 g/L, respectively, values known to inhibit microorganisms. Furthermore, acetic acid concentration increased to 9.26 g/L on the 46th day, but acidification didn't occur by the high alkalinity of the medium provided by NH_4^+-N, and pH ranged between 7.41 and 7.98.

With regard to *bacterial* groups, *Firmicutes* (mainly *Lactobacillales* and *Clostridiales*) were the most abundant in the substrate, while *Bacteroidetes* and candidate division WWE1 (from family *Cloacamonaceae*) were in the inoculum. In the process, *Bacteroidetes* (16–25%) and *Firmicutes* (31–58%) were the dominant groups. On the other hand, initial *Archaeal* genera were *Methanosaeta*, *Methanosarcina*, and *Methanobrevibacter*, but after PM addition the first group overcome the second. Generally, *Methanosarcina* dominates at high ammonia concentrations. Therefore, *Methanosaeta* demonstrated an abnormal behavior. Predominance of the mentioned group prevented syntrophic acetate oxidation between hydrogenotrophic methanogens and syntrophic acetate-oxidizing bacteria (SAOB). This would explain the lack of biogas in the second half of the experiment. Therefore, mono-fermentation of PM can be suitable for a period. However, at high OLRs, NH_4^+-N and FAN concentrations will reach dangerous levels (Świątek et al. 2019).

Another study focused on the water extraction of water-soluble inorganic and organic nitrogen compounds from PM, coupled with its C/N ratio amendment by co-digestion. After soaking, PM was centrifuged to obtain treated PM (T-PM). By this separation, C/N ratio increased to 19.8 from 7.5 by the decrement of nitrogen content from 53.75 to 21.99 mg/kg. Then, T-PM was digested alone or in co-digestion with MS. CH_4 yield from batch experiments determined that co-digestion of T-PM with MS or T-PM alone produced 260.2 and 209.5 mL-CH_4/g-oTS, respectively. The next test consisted in monitoring 16 weeks CSTRs, which managed an OLR of 0.5 g-oTS/L$*$d the first half and increased to 1 g-oTS/L$*$d the second half. Biogas production didn't decrease but wasn't as high as in batch experiments. Probably, the microbial consortium required more time before the exit of substrate. Cumulative CH_4 production from MS alone produced ~143 L, higher value than co-digestion of T-PM with MS (~95 NL) or T-PM alone. Albeit biogas production of the last treatments was almost the same; at the end of the experiment, NH_4^+-N content reached 3.6 g/L inside digester only with T-CM, whereas it just reached 2.3 g/L in co-digestion of T-CM and MS (Böjti et al. 2017). Therefore, co-digestion of PM with substrates rich in carbon maintains process stability with great biogas productions.

4.2 Microalgae Biomass

Production of microalgae biomass has a wide set of advantages. For example, they can be easily cultivated, remove contaminants from aqueous effluents, and show higher photosynthetic efficiencies and growth rates (Wirth et al. 2019).

In the last paragraph of the previous section, it was mentioned that water treatment can remove organic, nitrogen, and phosphate compounds from PM. However, if this process is reproduced in a full-scale digester, large quantities of wastewater will result. This fact gives the opportunity for microalgae application, as an additional process, to remove organic compounds from wastewaters, recirculate cleaned water, and produced biomass (Böjti et al. 2017).

The potential of the microalgae *Chlorella vulgaris* as feedstock for AD was evaluated as fresh or heat pre-treated mass. Each substrate was fed in its respective CSTR (1 L working volume, OLR of 1.5 g-COD/L$*$d). Fresh mass had a CH_4 yield of 85 mL-CH_4/g-COD, and the heat pre-treated mass produced 126 mL-CH_4/g-COD. Furthermore, biogas production from the latter resulted in higher COD and VS removal. Microbial analysis indicated that the most abundant phyla were *Proteobacteria*, *Bacteroidetes*, *Chloroflexi*, and *Firmicutes*. According to the treatment, there were differences in microbial abundance. For example, *Proteobacteria* predominated in fresh mass because these hydrolytic bacteria must degrade the hard cell wall of the microalgae. Therefore, if biomass was treated with heat, it will contain more soluble solids; this was observed by the relevant presence of *Chloroflexi*, whose members participate in the production of acetate, lactate, and hydrogen. In the case of *Firmicutes*, their presence was similar; these groups participate in the fermentative metabolism and hydrolysis of macromolecules (Sanz et al. 2017).

4.3 Wastewater Treatment

At present, biogas plants are searching for innovative technologies to upgrade biogas resultant from AD. To reach this goal, four main technologies are used: water scrubbing, membrane separation, chemical scrubbing, and pressure swing adsorption. With Germany, the UK, Sweden, and France are leading in biogas upgrading plants (International Energy Agency (IEA) 2018).

On the other hand, Spain's new research is focusing on the upgrading of water scrubbing systems to improve biomethanation processes. Batlle-Vilanova et al. (2019) proposed a case scenario for biogas upgrading, based on lab-scale results, in a WWTP fed with primary and waste activated sludge. Here, a novel water scrubbing-based technology, developed by De Godos et al. (2015), would be implemented and combined with a bioelectrochemial system (BES). The first would harness the same wastewater used as a substrate to remove CO_2 from biogas, initially destined for flaring (345 Nm^3/d), which will produce 689 m^3 of wastewater ([CO_2] increased from 108 to 360 mg CO_2/L). Then, the resultant wastewater would be introduced to the BES, with an efficiency of 30%, to produce 41 Nm^3/d of extra CH_4. Thus, total biomethane production would reach 17.5%, reduce by 42.8% CO_2 emissions from biogas upgrading, and produce 1830 kg of Cl_2 equivalents/d resultant for electrolysis for wastewater cleaning.

Another study in a WWTP aimed for the synthetic production of natural gas from AD of primary and secondary slurry. This plant had a complex function scheme. Initially, AD inside reactors produced the biogas. Then, its composition was determined and found the main components (~65% CH_4 + ~35% CO_2). Immediately, biogas went through two carbon filters to remove impurities. Next, a compressor increased the biogas pressure to separate CO_2 from CH_4 within a three-step membrane process. Part of gas was upgraded and contained 97.5% of CH_4, which was injected to the grid without problem. However, a combination of biogas, the permeable resultant gas, and the partially cleaned biogas were subjected to another step, the methanation process. In this unit, CH_4 was produced from the stoichiometric combination of CO_2 and H_2 streams ($CO_2 + 4H_2 \rightarrow CH_4 + H_2O$), the latter stream was provided by an alkaline electrolyzer that used tap water. The methanation unit consisted in two micro-heat exchangers; the first functioned as the catalytic reactor, while the second was the equilibrium reactor. Both exchangers were composed of 180 micro-channels that prevent hot spots and temperature gradients and had short start-up times. Effluents of this unit reached a CH_4 content above 95%, optimum for gas grid injection. However, the pilot methanation unit was limited because its capacity could only treat half of the hourly production (50 Nm^3) (Guilera et al. 2020).

As it was observed, AD process can be very unstable; if the constant microbial activity can't regulate components within the process at a certain limit, failure occurs. However, it is also shown that there exist a great number of solutions to enhance the correct development of AD, even to the extent to perfectionate certain process.

5 Substrates and Applications

The concept of a centralized economy by energy recovery from anaerobic digestion, for large plant or farm, requires methane production optimization. Anaerobic digestion is able to digest urban biowastes, energy crops, and agri-residues, being most abundant wastes in the EU, that may be transported by renewable energy subsidies granted by the government (Vlyssides et al. 2015).

Almost 1.25 billion tons of animal manure and slurries, also, nearly 1 billion tons per year of agricultural residues, are generated and accumulated. The European food residues are about 2.5 billion tons per year, among other few more wastes, which potentially could be utilized for biogas production. Some EU countries (e.g., Germany and Belgium) take food waste (FW) for valorization and application of new technologies like AD. France, on the other hand, has implemented a source separation program for FW management and valorization, whereas Scotland is now prohibiting the direct landfills of FW.

The European Community 2002–2012 developed a key strategy for waste prevention and management on the Sixth Environmental Action Programme, because these biowastes can be processed for by-product recovery and valorization, such as in AD for several applications (Fig. 3) (Fava et al. 2013; Fisgativa et al. 2016). Another source of biomass for biomethane production is forest residues, which is a less abundant waste, but it has an efficiency of 62–64% for AD.

Moreover, the Göteborg biomass gasification project (GoBiGas) demonstrates the capacity of this residue to produce biogas and provide it for cars every year, with a 65% of efficiency for transport sector supply (Li et al. 2017). Mainly, the properties of the biomass residues are determined as mass and energy for environmental impact via the attributed life cycle analysis (aLCA) according to the European guidance for waste management (Pierie et al. 2015).

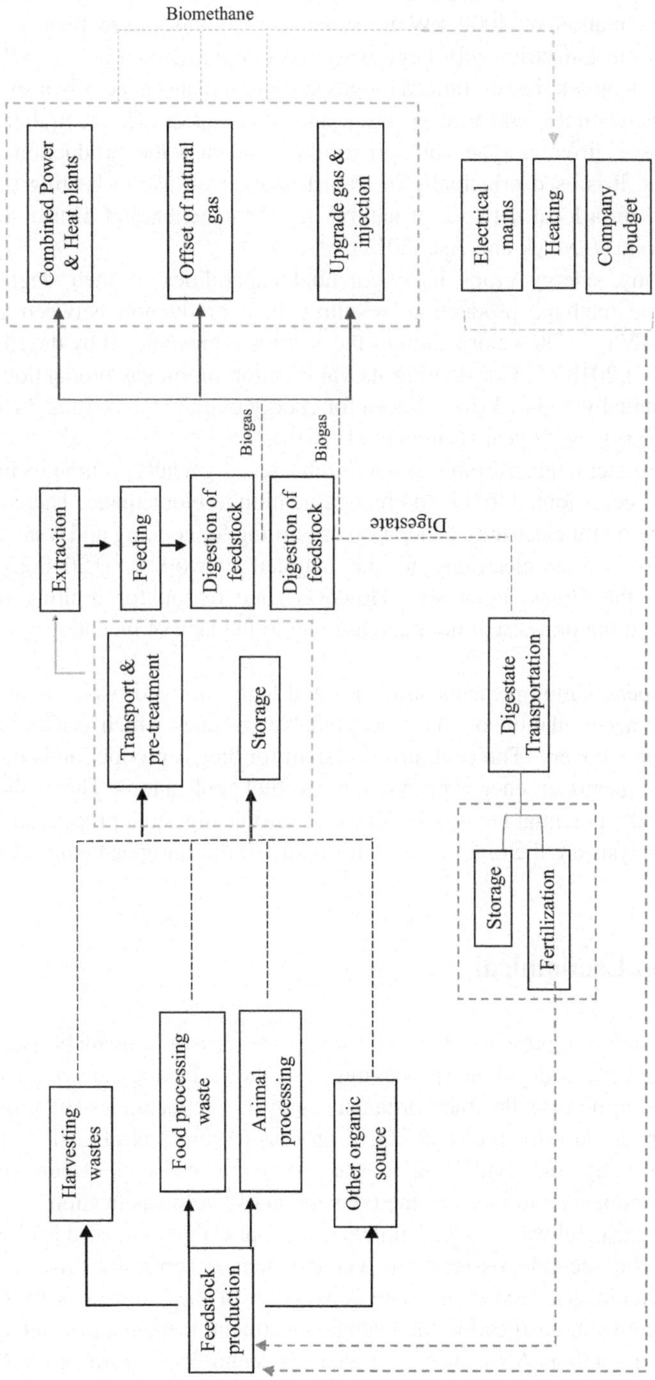

Fig. 3 Biogas production system of centralized plants using different feedstock sources for several applications in EU (Pierie et al. 2015; Hijazi et al. 2016)

Currently, the EU-27 the biogas energy generation could cover up to households, having an estimation of 4000 kW h, which can provide more than 18 million households from industries with large waste production (Lorenz et al. 2013). Furthermore, in Denmark the centralized biogas system of plants generally uses rape oil, soybean, corn residues, wheat, deep litter, and food wastes. The centralized plants digesting these different types of agri-residues increase the production of farm biogas plants. It is used principally in Danish market of district heating (DH) and is distributed on a local network of natural gas for injection and distribution (cost 0.133 Euro cent) (Lybæk and Asai 2017).

In Germany, energy crops have potential capabilities of their degradability improving the methane production, resulting in a production between 253 and 351 L $CH_4 \cdot TVS^{-1}$. The valorization of these crops is encouraged by the EU Policy Directive (EU) 2018/85. Considering its valorization on biogas production, it may contribute annually 2584 PJ (61.7 Mtoe) for energy output representing 34.1% total renewable energy every year (Garcia et al. 2019).

The by-product interest from residual biomasses, especially from agro-industrial sector in Greece, is joined to the AD biodegradation to biomethane. Therefore, it is mainly an engine for electricity generation or burning for cooking and water heating. The price for selling electricity to the national network is 0.20–0.22 €/kWh according to the Greek legislation. However, one reason for limiting the high production and the prices will not decrease due to the lack of incentives (Vlyssides et al. 2015)

The European Union presents strategies and legislations for waste management reducing the accumulation of the biodegradable residues which can be used for by-production recovery. The centralized system for biogas production is one of the technological means for energy production and fuel applications. The evaluation of the energetically potential feedstocks for the anaerobic digestion process and the full economic analysis are the main focuses for many of the European countries.

6 Techno Economical

Biogas generation represents many factors, like feedstock availability, process efficiency, and the end-product processing. For the technological optimization of the process, starting from the microorganism analysis, pretreatments and purification of the biogas are key for a cost-effective process (Achinas et al. 2017). The first optimization technology will have to start from the feedstock, where there are different developments to increase the conversion of waste gasification.

The biological pathway is a well-functioning application in more than 200 plants in Europe. The second possibility is via the thermochemical pathway, applied usually for lignin-rich feedstocks. The conversion of the biomass is by different stages, pretreatment, gasification, raw syngas treatment, methanation, and synthetic natural gas upgrading. Also, there is a starting technology, power-to-gas, which converts exhaust power available into hydrogen (Billig and Thrän 2016).

Another option of technological optimization to increase the biogas production is the co-digestion system. The municipal solid waste is generated globally, which can be digested with animal manure. Using this co-digestion system in plants can reduce de start-up time during the AD and the total solids decrease. These residues are a potential source for maximizing the biogas recovery. Additionally, these types of residues generated are effective for the solid waste management according to the specification of the European Parliament legislation (Directive 2008/98/CE) (Stan et al. 2018).

Even if there is availability of the feedstocks, there are many technologies for anaerobic digesters, such as garage-type digesters, silo-type digester, and bag-type digesters, among others, which are only in prototype stage, and they have not been tested on large scale. Indeed, is a technological and scientific hurdle related then to the economic viability evaluation (André et al. 2018). The technological models of digesters are well studied in lab scale; nevertheless, the large-scale systems remain an essential optimization for the primary parameters.

Some authors consider biogas purification as one of the most important steps, since after biogas production the unnecessary remaining components must be removed. The technology for CO_2 removal is commonly used in the chemical absorption by alkanolamines. Another component that must be removed is hydrogen sulfide by using the physical methods (charcoal), absorption (water), chemical absorption (oxygen reduction), and membranes as well.

The upgrading technologies are gas separation membranes, organic solvent scrubbing, amine scrubbing, water scrubbing, and pressure swing adsorption (Makareviciene and Sendzikiene 2015; Lora Grando et al. 2017). In Europe, the most significant technologies employed in the biomethane plants are the water scrubbing (WATS) in 40% of the plants and chemical scrubbing (CHEMS) around 25% of the 200 biogas plants functioning currently (International Energy Agency (IEA) 2018).

Water scrubbing represents the best efficient technology to remove the CO_2 based on the absorption employing a solvent, in this case water. The yielding range of CH_4 purified by using this technology reaches 94%, and the purity is around 98%. The chemical scrubbing, on the other hand, is based on CO_2, using solvents such as monoethanolamine (MEA), diethanolamine (DEA), or diglycolamine (DEA) (Niesner et al. 2013). This upgrade of biogas is the main important technologies for the operation in the energy sector; however, the investment cost will vary depending on the quality of the raw biogas and on the biogas plant capacity. For the WATS and CHEMS selecting technology for purification performance will be less expensive with a range of 0.20–0.43 kWh/Nm^3 and 0.4–0.5 kWh/Nm^3, respectively (Ullah Khan et al. 2017).

Since the energy supply in Europe is constantly requiring different renewable sources, the installation of biogas plants and the new upcoming technologies should be cost-effective and with a high efficiency to be optimally applied. The technoeconomic overviews must be also considered on the statement of favorable subsidies; thus, the energy price will be reduced.

7 Perspectives

The generation of biogas as an added-value by-product is generally applied in the energetic sector of Europe, which requires the extension of policy instruments for support on infrastructure for efficient performance. The main scope of biowaste methanization is to reduce the GHG emissions and to contribute to the socioeconomic sector. The influence of the financial retrieves impacts directly into the development of the biogas production and plant installation.

The biobased circular economy is one of the major ambitions for the EU. Therefore, some of the upcoming technologies are focused to improve degradation of biowastes, find the correct combination of feedstocks for co-digestion, reduce process-time, manage more volume, apply combined heat and energy systems, understand microbial behavior, increase methanation efficiency, and decrease GHG emissions from biogas plants, Furthermore, new regulations and funds encourage the reliability of this green energy.

In favor of increasing a profitable power engineering power system applied in the society. The innovative and competitive market of the EU countries and their constant improvement may play as a role model for many other countries which are not part of Europe, to state the plant engineering and biogas application, based on the well-developed policies and incentives for the renewable energy production.

Furthermore, it is noteworthy to mention the recent separation of the UK from the EU, because a lot of modeling research considers and places it as one of the main participant countries to reach the required percentages of clean energy in future schemes. Therefore, the posture of the UK must be known as soon as possible to determine the impact in the development of AD in Europe.

As it was reflected by this work, after an overview of biogas production, the AD, in EU mainly consists in the application of CSTRs at lab scale. Therefore, the aim of developed new technologies must reach the fields of full-scale or at least the pilot-scale experiments. We are aware that a lot of data from full-scale processes can't be found freely on the Internet, but most of the research articles managed lab-scale experiments.

References

Achinas S, Achinas V, Euverink GJW (2017) A technological overview of biogas production from biowaste. Engineering 3(3):299–307

Achinas S, Willem Euverink GJ (2020) Rambling facets of manure-based biogas production in Europe: a briefing. Renew Sust Energ Rev 119:109566

Alvarado A, Montañez-Hernández LE, Palacio-Molina SL, Oropeza-Navarro R, Luévanos-Escareño MP, Balagurusamy N (2014) Microbial trophic interactions and mcrA gene expression in monitoring of anaerobic digesters. Frontiers in Microbiology 5(NOV):1–14

André L, Pauss A, Ribeiro T (2018) Solid anaerobic digestion: state-of-art, scientific and technological hurdles. Bioresour Technol 247:1027–1037

Baccioli A, Ferrari L, Vizza F, Desideri U (2019) Feasibility analysis of coupling an ORC to a mGT in a biogas plant. Energy Procedia 158:2311–2316

Baldi F, Pecorini I, Iannelli R (2019) Comparison of single-stage and two-stage anaerobic co-digestion of food waste and activated sludge for hydrogen and methane production. Renew Energy 143:1755–1765

Batlle-Vilanova P, Rovira-Alsina L, Puig S, Balaguer MD, Icaran P, Monsalvo VM, Rogalla F, Colprim J (2019) Biogas upgrading, CO2 valorisation and economic revaluation of bioelectrochemical systems through anodic chlorine production in the framework of wastewater treatment plants. Sci Total Environ 690:352–360

Billig E, Thrän D (2016) Evaluation of biomethane technologies in Europe – technical concepts under the scope of a Delphi-survey embedded in a multi-criteria analysis. Energy 114:1176–1186

Böjti T, Kovács KL, Kakuk B, Wirth R, Rákhely G, Bagi Z (2017) Pretreatment of poultry manure for efficient biogas production as monosubstrate or co-fermentation with maize silage and corn Stover. Anaerobe 46(February 2018):138–145

Capodaglio AG, Callegari A, Lopez MV (2016) European framework for the diffusion of biogas uses: emerging technologies, acceptance, incentive strategies, and institutional-regulatory support. Sustainability 2016(8):298. https://doi.org/10.3390/su8040298

Chen JL, Ortiz R, Steele TWJ, Stuckey DC (2014) Toxicants inhibiting anaerobic digestion: a review. Biotechnol Adv 32(8):1523–1534

Dapelo D, Bridgeman J (2018) Assessment of mixing quality in full-scale, biogas-mixed anaerobic digestion using CFD. Bioresour Technol 265(June):480–489

De Godos I, Cano R, Santiago JR, Lara E, Llamas B (2015) Device and method for simultaneous removal of hydrogen sulphide and carbon dioxide from biogas. EP 3061515A1

European Union (EU) (2020) About the EU: countries. Link: https://europa.eu/european-union/about-eu/countries_en. Accessed: 6 February 2020

Fava F, Totaro G, Diel L, Reis M, Duarte J, Carioca OB, Poggi-Varaldo HM, Ferreira BS (2013) Biowaste biorefinery in Europe: opportunities and research & development needs. New Biotechnol

Feng L, Ward AJ, Guixé PG, Moset V, Møller HB (2018) Flexible biogas production by pulse feeding maize silage or briquetted meadow grass into continuous stirred tank reactors. Biosyst Eng 174(2013):239–248

Fisgativa H, Tremier A, Dabert P (2016) Characterizing the variability of food waste quality: a need for efficient valorisation through anaerobic digestion. Waste Manag 50:264–274

Flotats X (2019) Biogas: perspectives of an old technology. Biorefinery:313–349

Garcia NH, Mattioli A, Gil A, Frison N, Battista F, Bolzonella D (2019) Evaluation of the methane potential of different agricultural and food processing substrates for improved biogas production in rural areas. Renew Sust Energ Rev 112:1–10

German Biogas Association (2019) Biogas sector statistics 2018/2019. Link: https://www.biogas.org/edcom/webfvb.nsf/id/EN-German-biogas-market-data/$file/19-07-12_Biogasindustryfigures-2018-2019_english.002.pdf. Accessed: 26 January 2020

Guilera J, Andreu T, Basset N, Boeltken T, Timm F, Mallol I, Morante JR (2020) Synthetic natural gas production from biogas in a waste water treatment plant. Renew Energy 146:1301–1308

Guo J, Peng Y, Ni B-J, Han X, Fan L, Yuan Z (2015) Dissecting microbial community structure and methane-producing pathways of a full-scale anaerobic reactor digesting activated sludge from wastewater treatment by metagenomic sequencing. Microb Cell Factories 14(1):33

Hahn H, Krautkremer B, Hartmann K, Wachendorf M (2014) Review of concepts for a demand-driven biogas supply for flexible power generation. Renew Sust Energ Rev 29:383–393

Hijazi O, Munro S, Zerhusen B, Effenberger M (2016) Review of life cycle assessment for biogas production in Europe. Renew Sust Energ Rev 54:1291–1300

International Energy Agency (IEA) (2018) Task 37: Energy from Biogas. Plant Lists Upgrading Plant List 2017

Kress P, Nägele HJ, Oechsner H, Ruile S (2018) Effect of agitation time on nutrient distribution in full-scale CSTR biogas digesters. Bioresour Technol 247(September 2017):1–6

Larsen SV, Hansen AM, Nielsen HN (2018) The role of EIA and weak assessments of social impacts in conflicts over implementation of renewable energy policies. Energy Policy 115:43–53

Lebranchu A, Delaunay S, Marchal P, Blanchard F, Pacaud S, Fick M, Olmos E (2017) Impact of shear stress and impeller design on the production of biogas in anaerobic digesters. Bioresour Technol 245(June):1139–1147

Li H, Mehmood D, Thorin E, Yu Z (2017) Biomethane production via anaerobic digestion and biomass gasification. Energy Procedia 105:1172–1177

Lindkvist E, Karlsson M (2018) Biogas production plants; existing classifications and proposed categories. J Clean Prod 174:1588–1597. https://doi.org/10.1016/j.jclepro.2017.10.317

Lora Grando R, de Souza Antune AM, da Fonseca FV, Sánchez A, Barrena R, Font X (2017) Technology overview of biogas production in anaerobic digestion plants: a European evaluation of research and development. Renew Sust Energ Rev 80:44–53

Lorenz H, Fischer P, Schumacher B, Adler P (2013) Current EU-27 technical potential of organic waste streams for biogas and energy production. Waste Manag 33(11):2434–2448

Lybæk R, Asai M (2017) Future pathways for deploying agricultural based biogas Plantsin Denmark and Japan: a comparative analysis. GMSARN Int Jl 11:129–138

Lybæk R, Kjær T (2017) Enhancing identified circular economic benefits related to the deployment of the Solrød biogas plant. Eng Appl Sci Res 44(2):97–105

Makareviciene V, Sendzikiene E (2015) Technological assumptions for biogas purification. Environ Technol 36(14):1745–1750

Meyer AKP, Ehimen EA, Holm-Nielsen JB (2018) Future European biogas: animal manure, straw and grass potentials for a sustainable European biogas production. Biomass Bioenergy 111:154–164

Niesner J, Jecha D, Stehlik P (2013) Biogas upgrading techniques: state of art review in european region. Chem Eng Trans 35:517–522

Pierie F, Van Someren CEJ, Benders RMJ, Bekkering J, Van Gemert WJT, Moll HC (2015) Environmental and energy system analysis of bio-methane production pathways: a comparison between feedstocks and process optimizations. Appl Energy 160:456–466

Pitkänen K, Antikainen R, Droste N, Loiseau E, Saikku L, Aissani L et al (2016) What can be learned from practical cases of green economy? –studies from five European countries. J Clean Prod 139:666–676

Radetic B (2018) Anaerobic Digestion, Important Aspects Regarding Digester Design and Sludge Mixing Systems. Handbook of Water and Used Water Purification, 1–13

Rotunno P, Lanzini A, Leone P (2017) Energy and economic analysis of a water scrubbing based biogas upgrading process for biomethane injection into the gas grid or use as transportation fuel. Renew Energy 102:417–432

Ryckebosch E, Drouillon M, Vervaeren H (2011) Techniques for transformation of biogas to biomethane. Biomass Bioenergy 35(5):1633–1645

Sanz JL, Rojas P, Morato A, Mendez L, Ballesteros M, González-Fernández C (2017) Microbial communities of biomethanization digesters fed with raw and heat pre-treated microalgae biomasses. Chemosphere 168:1013–1021

Scarlat N, Dallemand J-F, Fahl F (2018) Biogas: developments and perspectives in Europe. Renew Energy 129:457–472

Scarlat N, Fahl F, Dallemand JF, Monforti F, Motola V (2018) A spatial analysis of biogas potential from manure in Europe. Renew Sust Energ Rev 94(June):915–930

Sgroi F, Foderà M, Trapani AMD, Tudisca S, Testa R (2015) Economic evaluation of biogas plant size utilizing giant reed. Renew Sust Energ Rev 49:403–409

Stan C, Collaguazo G, Streche C, Apostol T, Cocarta D (2018) Pilot-scale anaerobic co-digestion of the OFMSW: improving biogas production and startup. Sustainability 10(6):1939

Świątek M, Lewicki A, Szymanowska D, Kubiak P (2019) The effect of introduction of chicken manure on the biodiversity and performance of an anaerobic digester. Electron J Biotechnol 37:25–33

Tchanche BF, Loonis P, Petrissans M, Ramenah H (2013) Organic Rankine cycle systems: Principles, opportunities and challenges. 2013 25th International Conference on Microelectronics, ICM 2013, (December)

Uçkun Kiran E, Stamatelatou K, Antonopoulou G, Lyberatos G (2016) Production of biogas via anaerobic digestion. In: Handbook of biofuels production: processes and technologies: second edition. Elsevier, Ltd.

Ullah Khan I, Hafiz Dzarfan Othman M, Hashim H, Matsuura T, Ismail AF, Rezaei-Dasht Arzhandi M, Wan Azelee I (2017) Biogas as a renewable energy fuel – a review of biogas upgrading, utilisation and storage. Energy Convers Manag 150:277–294

Urbini G, Raboni M (2014) Production and use of biogas in Europe: a survey of current status and perspectives. Ambiente e Agua - An Interdisciplinary J of Appl Sci 9(2)

Vlyssides A, Mai S, Barampouti EM (2015) Energy generation potential in Greece from agricultural residues and livestock manure by anaerobic digestion technology. Waste and Biomass Valorization 6(5):747–757

Wirth R, Böjti T, Lakatos G, Maróti G, Bagi Z, Rákhely G, Kovács KL (2019) Characterization of core microbiomes and functional profiles of mesophilic anaerobic digesters fed with *Chlorella vulgaris* green microalgae and maize silage. Front Energy Res 7(October):1–18. https://doi.org/10.3389/fenrg.2019.00111

Yousuf A, Khan MR, Pirozzi D, Ab Wahid Z (2016) Financial sustainability of biogas technology: barriers, opportunities, and solutions. Energy Sources, Part B: Economics, Planning, and Policy 11(9):841–848

Advances and Challenges of Anaerobic Digestion of Wastes and Wastewaters from Different Industrial Sectors

Reyna Roxana Guillén-Enríquez, Marcela Estefanía Téllez Reséndiz, Inty Omar Hernández-De Lira, and Jesús Morlett Chávez

Abstract Anaerobic digestion is an attractive strategy in which a consortium of microorganisms produces biogas from different types of waste and oxygen-free environment. Biogas is an alternative fuel used in electricity and heat production, and contributes to greenhouse gas and pollullants reduction. Biogas is composed methane (50–80%), carbon dioxide (30–50%), and may have traces of hydrogen sulfide and ammonia. The conditions of the process (temperature, pH, and pressure), digester design, and substrate characteristics influence biogas production. Organic matter used is obtained from various sources and industrial sectors (food waste, wastewater, vegetable waste, sewage sludge, etc.). For best degradation of substrates, methods such as pretreatment and co-digestion are used. Pretreatment facilitates the digestion and improves the accessibility of the source carbon utilizable by the microbial community, and mixing sources (co-digestion), working together as substrates, provides several advantages that improves biogas yields, methane production, and various other benefits. In conclusion, the importance of biogas production from diverse sources, its impact on different industries, and its representation as alternative energy which is compatible with our surroundings, daily activities, and industrial residues, were observed.

Keywords Anaerobic digestion · Biogas · Methane · Industrial wastes · Wastewater

R. R. Guillén-Enríquez · M. E. T. Reséndiz
Laboratorio de Biorremediación, Facultad de Ciencias Biológicas, Universidad Autónoma de Coahuila, Torreón, Coah, Mexico

Laboratorio de Biología Molecular, Facultad de Ciencias Químicas, Universidad Autónoma de Coahuila, Saltillo, Coah, Mexico

I. O. H.-D. Lira
Laboratorio de Biorremediación, Facultad de Ciencias Biologicas, Universidad Autonoma de Coahuila, Torreón, Coahuila, Mexico

J. M. Chávez (✉)
Laboratorio de Biología Molecular, Facultad de Ciencias Químicas, Universidad Autónoma de Coahuila, Saltillo, Coah, Mexico
e-mail: antoniomorlett@uadec.edu.mx

N. Balagurusamy, A. K. Chandel (eds.), *Biogas Production*,
https://doi.org/10.1007/978-3-030-58827-4_6

1 Introduction

The use of fossil products as the principal strategy to satisfy human necessities, in terms of energy, has contributed to environmental problems such as pollution and global warming (Maurya et al. 2019). The energy demand and high concentration of wastes (organic residues) from different sectors worldwide contributes to the need for an alternative energy source compatible with the environment (Divya et al. 2015). Methane produced under anaerobic digestion (AD) can be used as such an energy alternative, consistent with the environment (Divya et al. 2015). Lora et al. (2017) reported methane contributes to climate change because its effect is 24 times greater than carbon dioxide (Lora et al. 2017; Pessuto et al. 2016).

AD is an attractive strategy where a consortium of microorganisms produces biogas from different types of diverse substrates as wastes (organic residue and wastewater) and oxygen-free environment (Divya et al. 2015; Hagos et al. 2016; Pessuto et al. 2016). AD has been improved and is currently a factor in many aspects such as pretreatment of biomass, reactor design, configuration and development, conditions and monitoring process, biological treatment (anaerobic and microbiology area) and looking for an eco-friendly, cost-effective and high-quality yield process (Maurya et al. 2019). The process is carried out in the following stages: hydrolysis, acidogenesis, acetogenesis, and methanogenesis (Divya et al. 2015; Hagos et al. 2016; Pessuto et al. 2016). Two groups of microorganisms participate in those processes: (a) acidogens and acetogens and (b) methanogens. In acidogenesis, hydrolysis occurs mediated by bacteria and produces hydrogen, carbon dioxide, and organic acids. Acetogenesis is the second step, in which the oxidation of acidogenesis products to obtain acetates, hydrogen, and carbon dioxide occurs. During methanogenesis, two kinds of methanogenic bacteria are involved: one produces biogas methane from acetic acid or methanol, and the other classes of methanogenic bacteria produce biogas methane from hydrogen and carbon dioxide (Fagbohungbe et al. 2017; Pessuto et al. 2016).

This chapter addresses (1) the key factors that are involved and associated with challenges in the AD process, and (2) the advancements and application of AD for methane production with strategies (the pretreatments and co-digestion process) employing wastes and wastewater from different industrial sectors. Some general aspects are shown in Table 1.

2 Biogas and the Factors Involved During AD Process

Biogas is composed of methane (50–80%), carbon dioxide (30–50%), and may have traces of hydrogen sulfide and ammonia (Lora et al. 2017; Pessuto et al. 2016). Biogas is an alternative fuel and high-quality fertilizer used in electricity and heat production. Also, biogas contributes to greenhouse gas reduction and environmental protection from pollutants (Divya et al. 2015; Lora et al. 2017) (Table 1). The

Table 1 General aspects on the wastes and wastewater from different industrial

Waste or wastewater	Advance	Challenge	Reference
Residues organic produce form diverse wastes			
Animal manure and slaughter-house waste	The use of two-stage system. Thermophilic conditions.	High ammonium concentration.	(Cu et al. 2015; Dalkilic and Ugurlu 2015)
Food waste	Vertical pressure steam sterilizer to remove lipids. Co-digestion with animal manure or sewage sludge. Two-phase anaerobic digestion system.	High lipids level. Excessive concentration of ammonia and VFA.	(Deressa et al. 2015; Ren et al. 2018a, b; Xu et al. 2017; Zhang et al. 2016a, b, c)
Plant and vegetables	Water vapor treatment.	High concentration of lignocellulose.	(Forgács 2012; Li et al. 2018a, b)
Fat-Oil-Grease	FOG as a co-substrate. Stronger mixing condition.	Concentration of LCFAS.	(Chowdhury et al. 2019; Salama et al. 2019a, b)
Wastewater from different industrial sectors			
Agro-wastes	C/N optimisation Pretreatment Co-digestion	Uncontrol of discharges. Innapropied feedstock for AD. Lignocellulosic composition.	(Gontard et al. 2018)
Paper and pulp	New technology for treatmentment specific biomass.	To establish protocol for treatment waste and wastewater.	(Apruzzese et al. 2017).
Tannery	It is possible employ different reactors for the treatment.	Efficiency on effluent tratment water. Chemiclals employ for the leather process need to be more eco-friendly.	(Apruzzese et al. 2017; Kumar et al. 2019; Singh and Singh 2019).
Winery	Effluent can be recycled. Microorganisms can survive on wine residues making a environment for them.	High concentration of different pollutants in the water.	(Singh and Singh 2019).
Textile	Biological treatment results to be a good option.	Huge volume of effluent is a problem for employing the biological treatment.	(Kumar et al. 2019).
Food	Feedstock has wide range of sources for the AD. Treatment is cost-effective for short time.	Composition of wastewater from dairy process is heavily contaminated. According to the long startup periods have a high capital cost.	(Goli et al. 2019; Singh and Singh 2019).

conditions of the process (temperature, pH, and pressure), digester design, and substrate characteristics influence in biogas production (Divya et al. 2015; Fagbohungbe et al. 2017; Pessuto et al. 2016).

Among the factors are feedstock, which is the organic waste employed during AD for methane production (Divya et al. 2015). Years ago, during biogas production, only vegetable biomass and animal waste treatment were used. But today there is a wide range of biomass types derived from both agricultural, industrial activities, and municipal wastes (Divya et al. 2015). However, the use of pre-treatment with biological, mechanical, or physicochemical methods is required for cellulose, hemicelluloses, and lignin for increasing the substrates availability to AD (Lora et al. 2017). Second, temperature is an essential parameter that can affect microbial growth; the ideal temperature depends on the type of microorganism employed (Divya et al. 2015). However, temperature for mesophilic digestion (MD) is 37 °C, and thermophilic digestion (TD) is 55–70 °C (Mao et al. 2015). A process with MD requires 10–40 days, while TD needs only 14 days (Divya et al. 2015). Prior, Kougias et al. (2017) demonstrated that thermophilic conditions produce more methane and CO_2 than mesophilic conditions do (Kougias et al. 2017). Resende et al. (2016) report that the average daily biogas produced in the summer (~20 °C–50 °C) and winter months (~10°c–40) was 18.7 and 16 L day−1 (Resende et al. 2016). Third, the pH interferes in the enzymatic activities of biogas producer microorganisms (Divya et al. 2015; Pessuto et al. 2016). Also, the variation of pH levels derives from metabolites production during biogas formation as well as a pH range of 5.5–8.5 given optimum substrate degradation (Divya et al. 2015). Next, the carbon/nitrogen (C/N) ration represents the nutritional requirements of microorganisms involved in AD (Divya et al. 2015; Mao et al. 2015). Besides, increasing the C/N ratio induces a low protein solubilization rate, causing a decrease in total ammonium nitrogen (TAN) and FA concentrate ions (Divya et al. 2015; Mao et al. 2015). Hence, optimizing the C/N ratio in the AD may avoid the ammonia inhibition (Divya et al. 2015; Mao et al. 2015). For example, manure has a low C/N ratio, causing an increment of ammonia concentration, which may inhibit methanogenic bacteria (Lora et al. 2017). The optimum conditions for anaerobic metabolic activity are shown in Table 2.

Table 2 Optimum conditions for anaerobic metabolic activity

Parameters	Optimum conditions	Reference
Temperature	Mesophilic (37 °C)Thermophilic (55–70 °C)	(Mao et al. 2015)
pH	5.5–8.5	(Divya et al. 2015)
Carbon to nitrogen ratio (C/N ratio)	10–35	(Divya et al. 2015)
Organic loading rate (OLR) and nutrient concentration	According to the substrate and inoculum	–

3 Organic Residues Produced from Diverse Wastes

Biomass as a renewable energy source is estimated to contribute to the world's energy supply (10–14%) (Taghizadeh et al. 2017). Organic matter used is obtained from various resources as food waste, industrial waste, fruit and vegetable waste, sewage sludge, agricultural residues, manure of livestock, and different manure animals, etc. (Abdeshahian et al. 2016; Taghizadeh et al. 2017).

3.1 Animal Manure and Slaughterhouse Waste

The amount of manure is a continuous production from different industries. China has produced 209.3 million tons of pig manure annually, of which more than 100 million tons is discharged into the environment (Zhou et al. 2016). The United Arab Emirates (UAE) produce around 1.5 million bulk metric tons (MT) manure per day from dairy cows (Abdallah et al. 2018). The United States produces 41.5 million MT of ready-to-eat meat (chicken, pork and beef), and discharges an equal amount of slaughterhouse waste (SHW) (Wang et al. 2018a, b). However, biogas production depends on manure composition, including such factors as the kind of animal, diet, and digestion (Abdallah et al. 2018; Cu et al. 2015). Previously, a study compared different categories of cows from the same livestock: milking cows, dry cows (breeding cows), fresh cows (to be milked for the first time after lamb), young cows (18 months old), and young-2 (9 months), and those cows fed different diets (Table 3). The authors recorded different levels of methane production, with fresh cow manure (215 ml CH4/g VS), after Young-2 (207 ml CH4/g VS), high (195 ml CH4/g VS), dry (160 ml CH4/g VS) and Young-1 (147 ml CH4/g VS); nevertheless, it has not established a protocol regarding the kind of manure (Abdallah et al. 2018). Also, the energy recovered in methane has represented a challenge for diverse authors who have explored reducing the multiple environmental impacts and employing the conversion of animal waste to produce a clean energy resource (Abdallah et al. 2018; Resende et al. 2016).

Another study has shown that the best Biochemical Methane Potential (BMP) is from piglet manure (443.6 NL CH4 kg VS-1) in comparison with cow, sow, chicken, rabbit, buffalo, and sheep manures (222, 177.7, 173, 172.8, 153, and 150.5 NL CH4

Table 3 Characteristics of cow manure

Sample Type	TS %	VS%	Ashes %	Carbohydrates %/ TS	Lipids %/TS	Proteins %/ TS
High	25.3	18.4	6.9	63.7	1.19	7.94
Dry	24.7	17.7	7.1	64.8	1.34	5.34
Fresh	22.6	15.5	7.1	62	1.02	5.35
Young	21.3	14.5	7.1	60.3	1.2	5.99
Young-2	24.5	17.5	7	65	1.14	5.22

(kg VS−1 correspondingly) (Cu et al. 2015). At 443.6 NL CH4 kg VS-1 as a result of the relation between biogas production and the chemical conformation of the substrate. Also, chicken manure showed to be better for producing methane under mesophilic conditions than the thermophilic acidogenic reactor. Piglet manure has a composition of matter dry, 19.40%; volatile solids, 82.88%; protein, 24.73%; lipid, 7.89%, hemicellulose, 17.88%; cellulose, 10.47%; lignin, 6.88% (Cu et al. 2015). However, Cu et al. (2015) reported as the highest matter dry of 37.9% in a study which inhibited the biogas production from chicken manure; this is due to the nitrogen-richness of chicken manure, which can inhibit conversion of organic material because in the degradation of protein and amino acid it forms ammonium (Cu et al. 2015; Dalkilic and Ugurlu 2015). Also, methane production from chicken manure has been proved to be better on mesophilic methanogenic reactor with the production between 75 and 85% than on single thermophilic acidogenic reactor (30–40%), while the use of a two-stage system has produced 74% (Dalkilic and Ugurlu 2015). In addition, the thermophilic condition affects the methane production, improving the reaction rate and lowering HRT (Dalkilic and Ugurlu 2015).

3.2 *Plant and Vegetables Residues*

The most abundant fruit cultivation in the world is citrus (principally in tropical and subtropical regions), with the global production of citrus fruit of 169.4 tons in 2013 (Taghizadeh et al. 2017). China produces 130 million tons per year of fruits and vegetable wastes, which are usually discarded (Li et al. 2016a, b). Taghizadeh et al. (2017) reported that methane produced from AD is between 45 to 116 m^3 per ton of wet citrus wastes (CWs); in view of the biomass of 682,987.97 t CWs, then, the production of methane will be 79 million m^3. Further, the most important of these is orange (61% of 100%). In 2013–2014, orange peel waste (OPW) was increased by 5% over 2012 to 51.8 million metric tons (Martín et al. 2018; Taghizadeh et al. 2017). It has been reported the methane yields from citrus wastes lies between 0.05 and 0.33 Nm^3/ kg VS (Taghizadeh et al. 2017; Wikandari et al. 2014). Previously, Forgács (2012) reported the methane production at 0.537 Nm3/kg VS using a water vapor treatment at 150 °C for 20 min; owing to this, the compressed structure of citrus wastes is destroyed (Forgács 2012).

The vegetables are considering, such as a principal resource of lignocellulose. Lignocellulose is a complex substrate, composed of cellulose (15–99%), hemicellulose (0–85%), and lignin (0–40%) (Li et al. 2018a, b). The pretreatment with acid affects the lignocellulose structure, producing biomass to produce methane (Li et al. 2016a, b). Li et al. (2018a, b) evaluated the methane production from different substrates including microcrystalline cellulose (MC), α-cellulose (α-CE) and alkali lignin (LI), xylan (XY), glucomannan (GM), and arabinogalactan (AG). Hence, the highest production was from the fortieth day, at 245.4 mL/g-VS by MC, followed by α-CE at 241.7 mL/g-VS. In contrast, the lowest show was GM at 178.6, probably for its stable distribution in acidic and alkalescent settings and the high thickness.

3.3 *Food Waste Industry*

Food waste (FW) from domestic kitchens, restaurants and cafeterias is one of the main wastes present in the municipal waste system (Kiran et al. 2015; Zhao et al. 2017; Ren et al. 2018a, b). Last year the annual production of FW was as follows: 38 million tons of FW in the U.S., 98 million tons in the European Union, and more than 90 million tons annually in China (Xu et al. 2017). FW presents in the landfills, was used in incineration plants, or to produce compost. However, the improper handling of these wastes has become the primary source of groundwater pollution and the emission of toxic and greenhouse gases (Kiran et al. 2015; Xu et al. 2017; Zhao et al. 2017; Ren et al. 2018a, b).

FW presents high levels of organic matter (volatile solids/total solids [VS/TS]: 0.8–0.9), high moisture, and excellent biodegradability; hence anaerobic digestion has been proposed as an efficient method for waste treatment to generate energy as hydrogen and methane (Fig. 1) (Kiran et al. 2015; Zhao et al. 2017; Li et al. 2018a, b; Peng et al. 2018; Wang et al. 2018a, b).

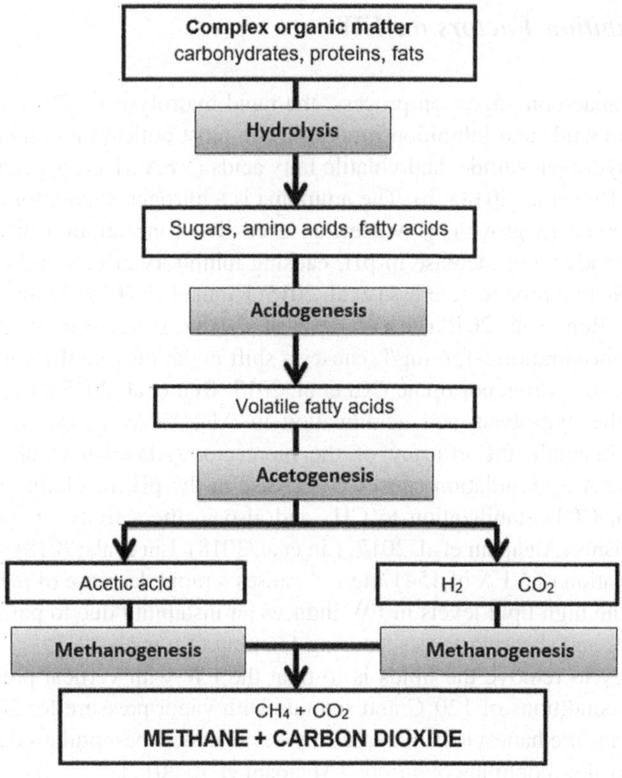

Fig. 1 Anaerobic digestion process

Table 4 Biogas production from different components of FW (Wang et al. 2018a, b)

Component	Biogas (1 g^{-1})	CH_4 (%)
Lipids	1.425	69.5
Carbohydrates	0.83	50
Proteins	0.921	68.8

3.4 *Methane Potential of Different Sources*

The different sources of FW have a specific degree of lipids, carbohydrates, and proteins, and a minor amount of cellulose and hemicellulose. Each carbon source therefore has a different degree of methane production: lipids have a higher methane production, followed by proteins, and finally, carbohydrates has the least methane yield. Table 4 shows the production of biogas by different components of FW (Xu et al. 2017; Li et al. 2018a, b; Wang et al. 2018a, b). AD treatment has to be designed according to the characteristics of each FW to obtain the highest yield methane in the biogas produced (Xu et al. 2017).

3.5 *Inhibition Factors on FW*

During the anaerobic digestion process, the rapid hydrolysis of FW produces both benefit compounds and inhibition products. The most critical inhibition factors are ammonia, hydrogen sulfide, and volatile fatty acids (VFA) (Deressa et al. 2015; Xu et al. 2017; Ren et al. 2018a, b). The ammonia is a nitrogen source for microorganisms to promote its growth. However, an excessive concentration of ammonia in the reactor leads to an increase in pH, causing inhibitory effects and affecting the biogas production process (Deressa et al. 2015; Kiran et al. 2015; Xu et al. 2017; Liu et al. 2018; Ren et al. 2018a, b; Wang et al. 2018a, b; Zhao et al. 2017). Also, ammonia concentration >156 mg/L causes a shift in the microbial community from acetotrophic to hydrogenotrophic (Xu et al. 2017; Ren et al. 2018a, b).

During the hydrolysis and acidification in AD, VFAs produced must be in balance to maintain the efficacy of the bioreactor (Algapani et al. 2017). The excessive VFA accumulation causes a decrease in the pH, resulting in a negative influence on COD stabilization to CH_4 and also in the activity of methanogenic microorganisms (Algapani et al. 2017; Liu et al. 2018). Liu et al. (2018) reported that the accumulation of VFA of 1541 mg L^{-1} causes a rapid decrease of pH, failing the AD. Also, the high lipid levels in FW induces an instability due to partial degradation of lipids and shearing forces, causing foaming (Xu et al. 2017).

A strategy to remove the lipids is to treat the FW with vertical pressure steam sterilizer at conditions of 120°C and saturate with vapor pressure for 20 min. With this treatment, methane yield increased by 4.5% under mesophilic conditions and 24.6% under thermophilic conditions (Algapani et al. 2017).

For stable food waste, digestion often has to be performed at low organic rates (OLR) of 2–3 g/L/d of COD, and the co-digestion with animal manure or sewage sludge is a common method because it can provide the alkalinity to avoid process failure (Xu et al. 2017). On the other hand, the balance between carbon and nitrogen, expressed in C/N ratio, affects the AD. The 15:30 is the optimal C/N ratio to promote the efficacy in the AD process. A high C/N ratio decreases the efficiency of the AD, whereas a low C/N ratio will have a lethal effect on methanogenic bacteria due to the pH (Srisowmeya et al. 2019).

3.6 Anaerobic Reactor Configurations for Food Waste Treatment

Many researchers have improved the structure and function of the fermentation reactor to obtain a higher methane yield (Ren et al. 2018a, b). To begin, the traditional reactors contain a single-phase anaerobic digestion system or a two-phase anaerobic digestion system. In the two-phase system, hydrolysis and acidogenesis are performed in the first reactor, and the methanogenesis takes place in the second reactor. Although the single-phase system is simpler than the two-phase system, and all reactions occur simultaneously, researchers have concluded that the two-phase anaerobic digestion system is more efficient (Ren et al. 2018a, b).

3.7 The Latest Anaerobic Digestion Systems and Technologies

A compact three-stage anaerobic digester (TSAD) was developed for AD of food waste. This system combines three independent chambers for hydrolysis, acidification, and methanogenic, respectively; TSAD had a 24–54% higher methane yield compared with single-phase or two-phase anaerobic digestion systems (Zhang et al. 2016a, b, c). A two-phase pressurized biofilm (TPPB) system was developed, including a conventional continuously stirred tank reactor and a pressurized biological anaerobic reactor. The pressure has significant effects on methane yield and quality, and the highest methane yield was detected under pressure of 0.3 MPa (Li et al. 2016a, b). In another method of two-phase anaerobic digestion system, the first reactor is operated at pH 4.0 with lactate as its principal product, and then the effluent from the first reactor is degraded in UASB to produce methane (Ren et al. 2018a, b). A three-stage process that consisted of saccharification, ethanol fermentation of the saccharified liquid, and anaerobic treatment of the saccharified residue to convert FW to ethanol and CH_4, achieved a 27.5% increase in the FW decomposition rate, a 51.8% reduction in the energy requirement for system operation, and

17.6% improvement in the total energy yield compared to the single-stage system (Wu et al. 2015). A two-phase anaerobic digestion reactor treating food waste with the reutilization of acidogenic off-gas in the methanogenic UASB reactor increased the methane recovery up to 38.6% (Yan et al. 2016). Digestate recirculation plays an important role because it facilitates mixing in the system and the purification of impurities in the gas; this could promote the transformation of carbon dioxide into methane (Ren et al. 2018a, b).

3.8 AD of Fat-Oil-Grease and Their Respective Impacts in Energy Production

Biowaste and high-strength wastewater generated from restaurants, food processing industries and domestic properties contain a significant amount of fat, oil, and grease (FOG). Anaerobic digestion presents an attractive option for biomethane recovery from FOG because the lipid content in FOG has a higher methane potential of 1014 L (kg VS)$^{-1}$ compared with carbohydrates (e.g., 370 L (kg VS)$^{-1}$ for glucose) and protein (740 L (kg VS)$^{-1}$) (Chowdhury et al. 2019). Lipids show less biodegradability as a single substrate, but FOG is considered to be a desirable substrate to enhance biomethane production through co-digestion; studies have reported increasing the methane yield by 250–350% with FOG co-digestion (Salama et al. 2019a, b). However, utilization of FOG as a feedstock for AD implies several challenges (Chowdhury et al. 2019). Hydrolysis of FOG produces glycerol and long-chain fatty acids (LCFAS). The LCFAS are degraded via β-oxidation to short-chain fatty acids (SCFAS), acetate, H_2, and biomethane. The presence of LCFAS in anaerobic systems is an operational problem due to the inhibition of methanogens, and the LCFAS causes substrate diffusion limitations (Salama et al. 2019a, b). A stronger mixing condition in the continuous stirred tank reactor (CSTR) increase the lipid degradation (63–68%) compared to that in the up-flow anaerobic sludge blanket reactor (UASB) (48–67%). High concentrations of FOG inhibit the growth of key microorganisms in AD. The addition of an optimal calcium concentration (0.5%) showed maximum COD removal and methane production, due to avoiding the growth-inhibitory and toxic effects in AD systems OR The addition of an optimal calcium concentration (0.5%), which prevents growth-inhibitory and toxic effects in AD systems, showed maximum COD removal and methane production (Salama et al. 2019a, b). The addition of an optimal calcium concentration (0.5%) showed maximum COD removal and methane production, due to avoiding the growth-inhibitory and toxic effects in AD systems (Salama et al. 2019a, b). It has been reported that addition of FOG at a fraction of 20–30%, corresponding to VS fraction of 70%, increased methane production up to 200% in co-digestion with sludge (Chowdhury et al. 2019). Food waste is another substrate for co-digestion with FOG due to its wide availability and because it contains a high concentration of readily degradable organics (carbohydrates and proteins) (Chowdhury et al. 2019).

4 Wastewater from Different Industrial Sectors

Contamination of water by residual sources from different industries is increasing. Moreover, recalcitrant compounds present in the water are not degrading by conventional technologies. Therefore, AD is a necessary process to remove the "easy" contaminants and recalcitrant compounds of industrial wastewater. As result, during anaerobic treatment of industrial wastewater, there is methane formation (Cu et al. 2018). For example, the olive oil industry generates large amount of residues, including plastic olive cake, olive pomace, and liquid stream, called "olive oil mill wastewater" (OMW); high concentrations of polyphenols in the AD of OMW affect the methane production rates. The average of polyphenols in the wastewater concentrations is in 0.5–24 g l^{-1} (Calabrò et al. 2018). A decrease in methane yield has been reported at concentrations of about 0.5–2 g l^{-1}. Therefore, co-digestion is recommended to raw OMW (Calabrò et al. 2018). Microorganisms are also an alternative for the increase and for improving the methane production due to their adaptation ability, enhancing inhibition toleration according to the substrate concentrations (1.0 and 2.0 g/L) (Calabrò et al. 2018).

Currently, there are around 9000 large-scale anaerobic plants for treatment of industrial wastewaters (Cuff et al. 2018). For example, the dairy industry has a high wastewater load of around 180,000 m^3 of waste effluent annually. Positively, the 99% of organic matter in the dairy industry is biodegradable whereby using AD could be useful (Kothari et al. 2016). However, the traditional biological treatment of such wastewater has an incomplete degradation of the organic portion, as the response causes odor and creates a hatchery for pathogens and insects (Kothari et al. 2016). Also, many anaerobic plants use wastewater treatment processes under different configurations, such as upflow anaerobic sludge blanket (UASB), upflow anaerobic bioelectrochemical (UABE), expanded granular sludge bed (EGSB), anaerobic filter, fixed film reactor, and anaerobic membrane bioreactor (AnMBR) (Cu et al. 2018; Ohimain and Izah 2017; Shi et al. 2017). Thus, a practical (cost-effective) treatment process with specific configurations and efficient degradation potential is vigorously sought for management of huge loads from industry wastewater (Kothari et al. 2016). Table 5 shows the most principal anaerobic processes used for each wastewater. These kinds of reactors help to minimize the effluent of the COD, thereby increasing both the efficiency of COD elimination and the biogas production (Cuff et al. 2018). Also, Table 5 shows the methane yield according to COD removal efficiency in different wastewaters.

4.1 *Recent Trends in AD of Wastewater*

Up-flow anaerobic sludge blanket (UASB) reactor is one of the most widely used high-rate anaerobic wastewater treatment systems due to its simple design, construction, and low operating costs. In addition, UASB has a short HRT and a long SRT,

Table 5 Principal anaerobic technologies used for wastewater kind

Wastewater kind	Principal anaerobic processes	Reference
Distillery	Anaerobic sludge blanket (UASB) Anaerobic filter Anaerobic fluidized bed reactor Anaerobic bioelectrochemical reactor (UABE)	(Anon n.d.)
Pharmaceutical	Anaerobic membrane bioreactor (AnMBR) Upflow anaerobic sludge blanket (UASB) Anaerobic sequencing batch reactor (AnSBR) Moving bed biofilm reactor (MBBR)	(Shi et al. 2017)
Soft drink industry	Expanded granular sludge bed (EGSB)	(Cu et al. 2018)
Palm oil mill	Ultrasonic-assisted membrane anaerobic system (UAMAS)	(Ohimain and Izah 2017)
Brewery	Upflow anaerobic sludge blanket (UASB)	(Karina et al. 2017)
Antibiotics (Tetracycline)	EGSB	(Zhang et al. 2018)

and therefore can process large amounts of wastewater in a short time (Dutta et al. 2018). Pharmaceutical wastewaters contain high levels of nitrogenous compounds, due to the frequent use of nitrogen-containing organics for the manufacturing process (Shi et al. 2017). An effective system to remove nitrogen is the addition of sequential biocatalyst to the UASB (SBA-UASB), which is able to remove nitrogen of up to 9.4 kg N/m^3/d (Shi et al. 2017).

Packing nano mediator into an anaerobic system is an attractive technology to strengthen refractory pollutant removal and methane production wastewater. In a study, granular activated carbon/nano zero valent iron mediator (GAC/NZVI) was added into an EGSB reactor for anaerobic digestion of tetracycline wastewater; the results indicated that the biogas production and methane content were increased by 21.2% and 26.9% respectively (Zhang et al. 2018).

The implementation of anaerobic membrane bioreactor (AnMBR) in municipal wastewater (MWW) treatment can achieve a perfect performance in COD removal, methane conversion, and sludge yield; thereby, implementations of AnMBRs in industrial wastewater have attained great success worldwide. However, there are still some challenges to be overcome, such as the high energy demand from membrane fouling, recovery of dissolved methane from effluent, low COD/sulfate ratios, and deficiency of alkalinity, before industrial deployment is possible (Lei et al. 2018). Therefore, many researchers have developed different AnMBR configurations, such as vibrating ANMBRs (V-AnMBRs), gas-lifting AnMBRs (Gl-AnMBRs), anaerobic bio-entrapped membrane bioreactors (AnBEMRs), anaerobic dynamic membrane bioreactors (AnDMBRs), and anaerobic membrane sponge bioreactors (AnMSBRs) for sustainable fouling mitigation strategies (Chen et al. 2017). Another promising technology for the fouling mitigation is the granular anaerobic membrane bioreactor (G-AnMBR), a technology that incorporates the granular technology with membrane-based separation; biomass retention is achieved by the spontaneous formation of granular sludge without the need for mechanical mixing, as it is performed in C-AnMBRs predominantly in the form of completely stirred tank reactor (Chen et al. 2017).

4.2 The Principal Industrial Wastewaters

4.2.1 Agro-industrial Waste

Agro-industrial sectors are the biggest water consumers reflected in an environmental emergency as a result of different productions of the wastewaters (Amor et al. 2019). The treatment of agro-industrial wastewater is currently of high importance due the impact of problematic pollutants related to fast industrialization (e.g., fruit processing...) and other diverse activities (Wong et al. 2018). For example, uncontrol discharge of wastewaters from urban sources is the principal factor for water contamination in developed countries. Likewise, 70% of industrial waste is released into the environment without treatment, and 90% of wastewater is discharged into zones like rivers, lakes, and coastal water, which represents a problem as much for the environment, humans, animals, and plants (Amor et al. 2019; Wong et al. 2018). The AD process is thus the natural technology for utilizing the agricultural residues to obtain alternative eco-friendly outcomes like biogas and fertilizers (Gontard et al. 2018). However, selection of specific energy crops as primary feedstocks instead of local agricultural residues can improve the adoption of biogas technology on a larger scale (Gontard et al. 2018). Furthermore, AD is low for lignocellulosic rich waste streams because this waste has a low economic value (low conversion yields) (Gontard et al. 2018). Besides, depending on how the selection of feedstock is determined, an indirect variation in acreage use, and moreover, from the agricultural residue after treatment is issued the absence for developing innovative building blocks, molecules, and materials. Frequently, the capital and operation cost for pretreatment are greater than the significant return in terms of biogas production as well (Gontard et al. 2018). Co-digestion have been reported to optimise the process by minising the costs involved in pretreatment and alos enhance the process efficiency (Gontard et al. 2018). A pretreatment with white-rot fungi has also obtained a significant increase in methane production applied to residues of wheat straw (Rouches et al. 2016). Therefore, the advance of innovative and effective strategies such as co-digestion and pretreatments support AD and raise the capacity for the use of diverse feedstock to increase the biogas production by about 20–30% (7–10-5 billion m^3 biogas per year) (Gontard et al. 2018). Among the advances in agro-industrial wastes, many authors have focused on obtaining an accelerated carbon: nitrogen (C/N) optimization and a high solid loading process (Zahan et al. 2018). However, overcoming these conditions require designing and evaluating treatment; developing kinetic models, characteristics, and composition of agro-industrial wastes enhanced the methane yield and improved the balance of the C/N ratio in the reactors (Zahan et al. 2018). Other agro-wastes have demonstrated the high biofuel and bioenergy potential of poultry droppings, press mud, as well as other large agro-based industries such as sugar mills (sugarcane bagasse). In development for most of the tested feedstock, the increased total solids content usually increased the biogas/methane production, but only with an optimum limit of total solids according to this study, with TS contents of 12.07, 13.18, and 12.27% for poultry droppings, press mud, and SB, respectively, impacting significantly on the methane production (Rahman et al. 2019).

The olive oil industry is one of the most common in the Mediterranean region. It produces plastic olive cake, very wet olive pomace, and liquid stream (olive oil mill wastewater) (Calabrò et al. 2018). The constitution of olive mill wastewater (OMW) depends on the kind of extraction process, the olive variety, the season, vegetation, and fruit (Amor et al. 2019). AD is recommended for removing the OMW residues. However, presence of high phenol concentration affects the microorganisms and their activity (Amor et al. 2019; Calabrò et al. 2018). Further, conditions such as low pH, nitrogen content, alkalinity, and the presence of other inhibitory compounds affect the process (Calabrò et al. 2018).

4.3 Paper and Pulp Industry

There are a lot of industries, such as the paper and pulp industries, dedicated to transforming raw materials that originate in agriculture and livestock (Apruzzese et al. 2017). Total paper production was around 403 million tons in 2013, and according to a global estimate, this could be approximately 500 million tons in 2025 (Singh and Singh 2019). Paper and pulp industries consume a high raw material like wood (natural resources), chemical, energy (fossil fuels, electricity), and water (Singh and Singh 2019). As a result, waste material includes bleached pulp (41.8%), solid waste (4.2%), dissolved organic matter (2.3%), suspended solids, and 20–250 m^3/ton of wastewater (Singh and Singh 2019). Looking for alternatives to use this biomass has brought about developing an advanced technology proposed with the next steps: first, suspension and separation of the industrial organic waste and separate the pulp from mineral residues; second, the residue mineral is removed via acidification and washing of the waste; third, anaerobic digestion is employed via enzymatic hydrolysis with the high solids of the demineralized residues to obtain fermentable sugars; and finally, fermentation into biofuel (Apruzzese et al. 2017). A challenge is to establish a protocol as the last present, but it should be necessary to characterize the waste's kind and total wastewater composition (Apruzzese et al. 2017). Since a characterization, it can implement measures according to non-biodegradable or difficult materials as lignocellulosic and inhibitors (sulfur, resin acids, ammonia, heavy metals, and organochloride compounds) (Kamali et al. 2019).

In the same way, on configuration/operational conditions and reactor election for an optimized process (Kamali et al. 2019). Nowadays, the main anaerobic reactor employed is the up-flow anaerobic sludge blanket (UASB) because it is cost-effective in terms of high organic loading rate capacity, low process expense, and abundant methane–biogas production (Kamali et al. 2019). However, the expanded granular sludge bed (EGSB), which has a quicker rate of upward-flow velocity for the wastewater with low suspended particles passing through the sludge bed without mixing and saving energy (Singh and Singh 2019). Another reactor used is the moving bed biofilm reactor, characterized by thorough removal of organic matter and the production of added-value products (Kamali et al. 2019).

4.4 *Tannery (Leather) Industry*

A high volume of water and chemicals is used for the conversion of raw animal skin into leather; in other words, a colossal amount of water and contaminants are released (Apruzzese et al. 2017; Verma and Jaiswal 2016). The process consumes approximately 25–80 m^3/tons of raw material, 83 million hides and 140 million skin pieces, and in India alone, chrome salts consumption is about ~3000 ton and more than 40% chromium (Singh and Singh 2019; Verma and Jaiswal 2016). Although there are a lot of advantages, this is lamentably one of the most wastewater-contaminating industries (Singh and Singh 2019). Thus, the efficiency of effluent treatment process represents a challenge because of the presence of toxic elements such as ammonia, sulfides, aldehydes, nitrogenous compounds, suspended solids, heavy metals, tannins, and volatile hydrocarbons in wastewater (Apruzzese et al. 2017; Kumar et al. 2019; Singh and Singh 2019). Attempts to meet this challenge of contaminants from the tannery effluent include placing better control on the use of several chemicals during leather process as well as advancement in in-plan and end-of-pipe treatment technologies (Singh and Singh 2019). For this kind of wastewater, the up-flow anaerobic sludge blanket reactor (UASB) is recommend due to its advantage for containing a high concentration of basic bacteria immobilized to remove organic matter, and because it is not necessary to add support accessories for high concentration of biomass, which reduces the cost of treatment (Singh and Singh 2019). Also, the anaerobic bio-filter reactor (ABFR) for the treatment of tannery effluent involved (1) wide space for the growth of the microorganisms, and filler increase hydraulic retention time; (2) a wide surface area for interaction between the wastewater and film (Singh and Singh 2019).

4.5 *Textile Industry*

The textile industry contributes to environmental problems by employing large volumes of water for dyeing and finishing processes. Also, these industries spill more than 8000 chemical products (Singh and Singh 2019). Currently, the biological treatments have an immense ability to degrade the textile dyes and resolve superfluous pollutants of effluents (Singh and Singh 2019). However, establishing conditions like temperature, pH, and specific nutrient constituents are challenging for researchers (Kumar et al. 2019). Further, it is necessary to include microorganisms capable of degrading or transforming the dyes in wastewater to improve the efficiency of the process (Kumar et al. 2019).

4.6 Food Processing

Almost 30% of the total global energy is consumed for food production and supply chains (Gontard et al. 2018). In addition, world population is projected to be around 9 billion people in 2050, which will lead to increased food production requiring a proportionate increase in primary agricultural residues (Gontard et al. 2018). Among food industries is the dairy chain, in which residues derive mostly from transportation, equipment in production units, cleaning of tanks, and washing products (Singh and Singh 2019). The wastewater from dairy processes comprise casein, lactose, inorganic salts, dissolved sugar, proteins, fats, additives, preservatives, detergents and sanitizers, as well other pollutants such as heavy metals (Goli et al. 2019). Advancement of biological treatment is considered to be one of the most positive methods for washing the effluents assimilates of the wastewater components, and has a low cost scale (Singh and Singh 2019). For the treatment of dairy wastewater, UASB reactors, hybrid digesters, and anaerobic sequencing batch reactors (ASBR) are also used (Singh and Singh 2019). Among the advantages of these are low energy requirements, less sludge, less space needed, high removal efficiency, cost-effectiveness, lack of pathogenic organisms. Nevertheless these have disadvantages: high capital cost, long startup periods, strict control operation, toxic compounds, and high energy needs (Goli et al. 2019).

4.7 Winery Industry

Worldwide wine production last year exceeded 250 MhL, within Europe, France and Spain as the main producer (Amor et al. 2019). The process requires almost 1100 gal of water per ton of grapes (Singh and Singh 2019). During this process, a huge amount of water is used at different steps (e.g., distillation) of production, barrel cleaning, tank, floor, and equipment washing (Amor et al. 2019; Da Ros et al. 2017). Diverse treatment alternatives have been proposed for an efficient technology with cost-effective and easy management (Amor et al. 2019). Frequently, the treatment of wastewater employs a conventional system including activated sludge reactors, sequencing batch reactors (SBR) and physico-chemical treatment by coagulation–flocculation (Amor et al. 2019). Therefore, the ability of microorganisms to survive in an environment created by wine residues is significant (Singh and Singh 2019). In this way, the treatment with AD is designed for high organic loading and reduction

Table 6 Methane yield according to COD removal efficiency in different wastewaters industry

Type of wastewater	pH	Type of AD process	Temperature	Methane yield	Reference
Distillery	5.6	UABE reactor	35 ± 2 °C	0.469 ± 0.005 (L/g COD)	(Anon n.d.)
Distillery	7.0	UABE reactor	35 ± 2 °C	0.463 ± 0.005 (L/g COD)	(Anon n.d.)
Soft drink	6.5–7.5	EGSB	30–35 °C	2200 $m^3 \cdot d^{-1}$	(Cu et al. 2018)
Palm oil	–	Modified anaerobic baffled bioreactor	–	0.32–0.421 (L/g COD)	(Ohimain and Izah 2017)
Brewery	6.9	UASB	26–32 °C	0.32 (L/g COD)	(Enitan et al. 2015)
Tetracycline	8.0 ± 0.3	EGSB	35 ± 1 °C	0.1994 ± 0.212 (L/g COD)	(Zhang et al. 2018)
Cassava	5.5	UASB	37 °C	115,230 (l/g COD)	(Intanoo et al. 2015)

of COD, employing reactors which required time to degrade organic matter (Singh and Singh 2019). Most often employed, USAB increasing its efficiency with the connection of anaerobic lagoons, other anaerobic reactor employed is sequence batch reactor with HRT for two days attempted more than 98% COD removal (Singh and Singh 2019) (Table 6).

5 Pretreatments as Alternative for a Better AD

Pretreatment facilitates the digestion and improves the accessibility of the source carbon utilizable by the microbial community (Montingelli et al. 2016; Patinvoh et al. 2017). Currently, pretreatments have a global classification known as biological (enzymatic, fungal), chemical (alkaline, acid, or inorganic salts) and physical (mechanical, microwaves or beating) (Table 7) (Patinvoh et al. 2017; Xihui et al. 2018). Prior authors reported that mechanical pretreatment has increased the biodegradation; for example, grape pomace, 13.1%; pulp, 4.8%; and seeds, 22.2% (El Achkar et al. 2016). Also, microwave pretreatment of winery waste, cotton gin waste, olive pomace, and juice industry waste increased methane production by 112% in 7 days (Pellera and Gidarakos 2017). If a different pretreatment is used, it has variation about the methane yields; for example, the Irish macroalgae has demonstrated a better yield using beating (37%) than microwave (7%) (Montingelli et al. 2016). Achkar, et al. (2016) reported that the effect of mechanical pretreatment depends of source type, physical and chemical conditions, and the biodegradability and solubility of the material.

Chemical pretreatment of lignocellulose under alkaline conditions has produced lignin, hemicelluloses, and cellulose of 46.03%, 50.43%, and 31.55%, respectively

Table 7 Classification of pretreatments used in the digestion process

Pretreatment	Source	Processes	Effect	CH4 yield	Reference
Mechanical (P) Microwave pretreatment Beating pretreatment	Winery waste Cotton gin waste Olive pomace Juice industry waste Macroalgae *Laminaria spp.*	Using a Mars 6 (CEM) microwave reaction system. A high-pressure beating action of the feedstock against an inclined plate	To have higher moisture contents Pulp of macroalgae in different consistencies.	256.8 + −11.8, 125 °C 234.5 + − 55.9, 200 °C 244.5 + −3.4, 200 °C 387.4 + − 5.1, 200 °C 335 Nml g-1VS, 25 days.	(Pellera and Gidarakos 2017) (Montingelli et al. 2016)
Oxidative (C)	Coffee husks	Produced of a liquid fraction	Solubilization of lignin and hemicelluloses	91.07 NmLCH4/g-COD	(Santos et al. 2018)
Acid (C)	*A.Tequila* bagasse	HCI or H2SO4 solutions at different concentrations	Hydrolysates of *A. tequilana* bagasse	0.19 NL-CH4/g-COD	(Breton-Deval et al. 2018)
Alkaline (C)	Pennisetum hybrid (Pennisetum americanum×P. purpureum)	NaOH as a catalyst with different concentration and temperatures	Variations in cellulose crystallinity, hemicellulose and lignin content and hydrolysate composition	53.6 mL g − 1 VS d − 1 at second day for 8% NaOH.	(Xihui et al. 2018)
Extraction (P)	Orange peel waste (OPW)	The OPW was homogenized and crushed until reaching a particle size of <2 mm.	The crushed OPW was then treated by distillation to eliminate the d-limonene content. The OPW pre-treatment removed 70% after 1 h.	1.5 g COD/L	(Martín et al. 2018)

(Santos et al. 2018). Another author has mentioned that alkaline pretreatment increases the removal of hemicellulose (80.2%) and lignin (about 84.8%) (Xihui et al. 2018). Also, prior authors showed that using both NaOH as pretreatment in anaerobic digestion and activated sludge as inoculum have increased methane production from 0.2512 L/L•d to 0.3622 L/L•d (Geng et al. 2016). Meanwhile, acid pretreatment uses acids such as hydrochloric (HCl) or sulfuric (H_2SO_4). This pretreatment supports the hydrolysis of fermentable sugars (Breton-Deval et al. 2018). During the pretreatment of *Agave tequilana* bagasse, HCl was more effective than H_2SO_4 (Breton-Deval et al. 2018). However, the pretreatment of hemicellulose with HCl showed poor yields. Otherwise, hemicellulose pretreated under alkaline conditions and degraded under AD showed an increase of the biogas production 31.6% and biogas yield per kg of waste 31.5% (Andrade et al. 2018).

Pretreatment (P) with rumen fluid in paper sludge (PS) improves the hydrolysis produced methane 3.4 times more than untreated PS (Takizawa et al. 2018). Hence, pretreatment with rumen fluid increases the conversion of PS into VFAs (Takizawa et al. 2018).

6 Anaerobic Co-Digestion for Biogas/Methane Production

Mono-digestion of recalcitrant feedstocks, protein-rich substrates, or harmful composites causes a slow process with low biogas yield. However, mixing substrates for digestion offers many advantages, including ecological, technological, and economic benefits, over the digestion of one-substrate processing (Patinvoh et al. 2017; Rodriguez et al. 2018a, b). Anaerobic co-digestion (AcoD) process used in anaerobic conditions is the simultaneous digestion of a mixture of two or more substrates (Maragkaki et al. 2017; Patinvoh et al. 2017; Prajapati and Singh 2018; Rodriguez et al. 2018a, b). Hence, AcoD involves mixing sources working together as substrates (Fig. 2) and provides several advantages that improve biogas yields, methane production, and various additional benefits (dilution of potential toxic compounds, C/N, microbial diversity, alkalinity, etc.) to the process of AD (Sebola et al. 2015; Xavier et al. 2015; Zheng et al. 2015; Rodriguez et al. 2018a, b). In summary, first, co-digestion can employ nutrients from various wastes and balance the bacterial community to optimize the digestion operation, increasing the yield of biogas by balancing the carbon/nitrogen (C/N) ratio necessary for the microbial growth (Zhang et al. 2016a, b, c; Wang et al. 2018a, b; Náthia-Nevesorcid et al. 2018). Also, AcoD has several critical factors for the digestion such as substrate, temperature, pH, organic loading rate, and retention time (Table 8). Table 9 shows different co-substrates used for enhancing the AD. For example, kitchen waste is an excellent option to AD, but it has a low pH (4.2), which represents a problem; an alternative could be the use of cow manure (pH 7.3) (Zhai et al. 2015). Likewise, tea powder waste with fruit, food, and vegetable co-digestion were used to increase methane production (Thanarasu et al. 2018). For some reason, however, FW digestion has negative results in the process: 1) high organic loading of FW increasing

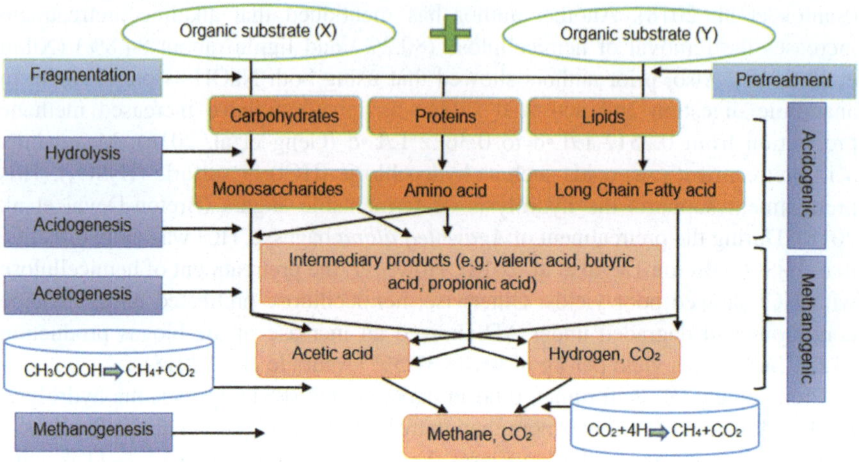

Fig. 2 Diagram of Anaerobic Co-digestion process

osmotic pressure, which causes bacterial dehydration; 2) poor balance of C/N; 3) decreased pH by the limitation of trace elements (Mehariya et al. 2018). Hence, to use FW in co-digestion with sewage sludge has a methane yield of 85.3% higher than FW mono-digestion; in turn, it provided better growth support, dilution of toxic elements. However, there is presently insufficient knowledge to establish an AcoD process and strategies (Mehariya et al. 2018).

Finally, nutrients balance is an essential factor that could affect conditions of the bioprocess. In this respect, microalgae like *P.canaliculata* is used as a source of nitrogen to balance C/N during the co-digestion with waste paper. This strategy has allowed an increase in methane production by 22% more than mono-digestion of microalgae (Rodriguez et al. 2018a, b). Also, Tunisian green macroalgae and sugar wastewater are an alternative to conventional AD (Karray et al. 2017). As has been reported, co-digestion of swine manure (high N) and crop straw (high C) present good conditions to indicate C/N balance (Mao et al. 2017). In addition, cornstalk and pig manure enhanced the degradability and methane production during AD process (Wang et al. 2017). Furthermore, agro-wastes contain a low percentage of nitrogen and a high percentage of carbon, enhancing methane yields, C/N ratio, and the risk of ammonia inhibition decreased by the co-digestion of sewage sludge and agro-wastes (Elsayed et al. 2016). For sewage waste, crude glycerol, food waste, cheese whey, grape residues and sheep manure have been used to improve methane production co-digestion (Maragkaki et al. 2018). On the other hand, cheese whey wastes and poultry manure denote an environmental problem, but they are an economical option together, and their co-digestion has proved to produce a more significant amount of methane while at the same time reducing the acidification of cheese whey, improving C/N ratio, and reducing the TS fed into the plant (Carlini et al. 2015). Notably, co-digestion is effectively an alternative to anaerobic digestion of mono-substrate

Table 8 Key factors of anaerobic co-digestion and operational conditions

Key factors	Characteristics	Reference
Substrate	Specific organic compounds may predominate, although contain with a wide range of simple and complex matters. However, depending on their sources will be the conformation of the substrates.	(Hagos et al. 2016)
Carbon based compounds	Commonly sugars. They are easy to biodegrade by methanogens with the formation of volatile fatty acids. Suppression of methanogenesis, lower pH, VFA accumulation may be caused by high levels of simple sugars.	(Hagos et al. 2016)
Nitrogen based compounds	Protein compounds are high ammonia. During the process is necessary at least 500 mg L-1 concentration. However, high concentration of ammonia causes an inhibition in methane production	(Hagos et al. 2016; Mehariya et al. 2018)
Fat rich based compounds	High concentration of fatty compounds cause blocking, adsorption to biomass and microbial inhibition, but they are used to AD by their easy degradability.	(Hagos et al. 2016)
Temperature	This parameter is important for the growth of microorganisms, hydrolysis kinetics and solubility of several compounds. There are three different ranges mentioned to AD: 25 °C (psychrophilic), ~35 °C (mesophilic) and ~ 55 °C (thermophilic), while mesophilic and thermophilic are the best temperature recommended to the grow of the moos. However, mesophilic process is more stable than thermophilic process because is harder to control and needs more energy.	(Hagos et al. 2016; Mehariya et al. 2018; Siddique and Wahid 2018)
pH	pH has greater influence in AcoD process. Maximum methane yield uses a pH in between 6.8 and 7.2. Also, pH optimum is a reason to separate the process in two-phase; hydrolysis and acidogenic between 5.5 and 6.5.	(Hagos et al. 2016; Mehariya et al. 2018; Siddique and Wahid 2018)
Organic loading rate (OLR)	Parameter which indicates the amount of organic material is loaded to an AcoD per a determined period (per unit time, per unit volume, as capacity per day) of a digestion. ORL is strictly linked with Hydraulic retention time (HRT). For example, higher the HRT, higher the ORL. The accumulation of material undigested results in fatty acids caused an acidic condition to destabilize the process.	(Matheri et al. 2017; Mehariya et al. 2018; Siddique and Wahid 2018)
Retention time (RT)	It is the period taken by biodegradable material and moos in the AD to reach depletion. This depends on the source composition, temperature, reactor system, as well as the processes. RT in mesophilic conditions is between 15 to 40 days, whereas in thermophilic conditions is around 12 to 14 days. HRT is the time required for any moos to degrade	(Matheri et al. 2017; Mehariya et al. 2018; Siddique and Wahid 2018)

(continued)

Table 8 (continued)

Key factors	Characteristics	Reference
	and synthesize the substrate digested, while Solids RT relates to the time taken by moos that are alive during the AD process.	
Volatile Fatty acids (VFAs)	VFAs are intermediates produced from substrate hydrolysis in AcoD. Organic matter is a factor to produce high concentration of VFAs, but it can be used high nitrogen substrates concentration to decreased the inhibition by the amount of acids prior it is important to obtain a control of C/N ratio.	(Li et al. 2016a; Mehariya et al. 2018)
Ration carbon/nitrogen	This factor has effects in all AcoD process. Sources with optimal C/N ratio provide sufficient nutrients for moos. However, lower C/N values leads to higher concentrations of ammonia, pH rises excessively and decrease microbial growth, while a high C/N value more than optimal value produce a high concentration of VFAs and result in lower gas production	(Cristina Rodriguez et al. 2018a, b; Siddique and Wahid 2018)

Table 9 Co-substrates used to improve the Anaerobic digestion

Substrates		pH	Methane yield	Reference
Kitchen waste	Cow manure	5.2	0.1214 L/g VS fed	(Zhai et al. 2015)
Food waste	Sewage sludge		0.215 LCH4 g − 1 VS	(Mehariya et al. 2018)
Wastepaper	Macroalgae		386,000 L g-1 VS	(Rodriguez et al. 2018a, b)
Corn straw	Swine manure	7.5	0.220 L g − 1 VS	(Mao et al. 2017)
Sugar wastewater	Green macroalgae		0.114 L g-1 VS	(Karray et al. 2017)
Cheese whey	Poultry manure		0.223 L /gVS	(Carlini et al. 2015)
Cow slurry	Olive pomace and apple pulp		216,100 CH4/g SV	(Riggio et al. 2015)

and obtains improved methane production and contributes to the development of conditions for a better biodegradability (Karray et al. 2017; Wang et al. 2017).

7 Conclusion

Over the years, various paradigms in the anaerobic digestion technology have been presented; for this reason, it is still important to develop further research in this field. Biogas production has been become a feasible energy source. The importance of biogas production has been observed in diverse sources, along with its impact on different industrial settings and its potential as alternative energy that is compatible with our surroundings, daily activities, and industrial residues.

Moreover, factors such as design of reactors, C/N ratio of the feedstock, pretreatment strategies, temperature, pH, etc control the efficiency of methane production.

Significantly, AD has often used mono-substrate, which has a lower yield, but after years of research, the AcoD method is seen as a viable technology.

References

Abdallah M, Shanableh A, Adghim M, Ghenai C, Saad S. (2018) "Biogas production from different types of cow manure." 2018 Advances in Science and Engineering Technology International Conferences, ASET 2018 1–4

Abdeshahian P, Lim JS, Ho WS, Hashim H, Lee CT (2016) Potential of Biogas Production from Farm Animal Waste in Malaysia. Renew Sust Energ Rev 60:714–723

El Achkar JH, Lendormi T, Hobaika Z, Salameh D, Louka N, Maroun RG, Lanoisellé JL (2016) Anaerobic Digestion of Grape Pomace: Biochemical Characterization of the Fractions and Methane Production in Batch and Continuous Digesters. Waste Manag 50:275–282

Algapani DE, Wang J, Qiao W, Su M, Goglio A, Wandera SM, Jiang M, Pan X, Adani F, Dong R (2017) Improving Methane Production and Anaerobic Digestion Stability of Food Waste by Extracting Lipids and Mixing It with Sewage Sludge. Bioresour Technol 244:996–1005

Amor C, Marchão L, Lucas MS, Peres JA (2019) Application of Advanced Oxidation Processes for the Treatment of Recalcitrant Agro-Industrial Wastewater: A Review. Water (Switzerland) 11 (2):205

Andrade WR, Santos TMB, Arruda LDO, Rosa EM, Cappi N (2018) Biogas production from dairy cattle manure subjected to chemical pre-treatment. Braz J Biosyst Eng 12(1):91–100

Anon (n.d.) Influence of Neutralization in Acidic Distillery Wastewater on Direct Interspecies Electron Transfer for Methane Production in an Upflow Anaerobic Bioelectrochemical Reactor. Pdf

Apruzzese I, Bhardwaj V, Degrassi G, Bhardwaj RK (2017) Biogas Production From Organic Wastes Of Paper And Leather Industries. Biotechnology and Biochemistry 3(4):8–14

Breton-Deval L, Méndez-Acosta HO, González-Álvarez V, Snell-Castro R, Gutiérrez-Sánchez D, Arreola-Vargas J (2018) Agave tequilana bagasse for methane production in batch and sequencing batch reactors: acid catalyst effect, batch optimization and stability of the semi-continuous process. J Environ Manag 224(July):156–163. https://doi.org/10.1016/j.jenvman.2018.07.053

Calabrò PS, Fòlino A, Tamburino V, Zappia G, Zema DA (2018) Increasing the Tolerance to Polyphenols of the Anaerobic Digestion of Olive Wastewater through Microbial Adaptation. Biosyst Eng 172:19–28

Carlini M, Castellucci S, Moneti M (2015) Biogas Production from Poultry Manure and Cheese Whey Wastewater under Mesophilic Conditions in Batch Reactor. Energy Procedia 82 (December):811–818

Chen C, Guo W, Hao H, Woong S, Duc D, Dan P, Thanh X, Yun W (2017) Impact of Reactor Configurations on the Performance of a Granular Anaerobic Membrane Bioreactor for Municipal Wastewater Treatment. Int Biodeterior Biodegradation 121:131–138

Chowdhury B, Lin L, Dhar BR, Nazrul M, Mccartney D, Kumar A (2019) Enhanced Biomethane Recovery from Fat, Oil, and Grease through Co- Digestion with Food Waste and Addition of Conductive Materials. Chemosphere 236:124362

Cu G, Turcios AE, Mohammad-pajooh E, Kujawski O, Weichgrebe D, Rosenwinkel K-h (2018) High-Rate Anaerobic Treatment of Wastewater from Soft Drink Industry : Methods, Performance and Experiences. 220(January):8–15

Cu TTT, Nguyen TX, Triolo JM, Pedersen L, Le VD, Le PD, Sommer SG (2015) Biogas Production from Vietnamese Animal Manure, Plant Residues and Organic Waste: Influence of Biomass Composition on Methane Yield. Asian Australas J Anim Sci 28(2):280–289

Cuff G, Turcios AE, Weichgrebe D, Kujawski O, Rosenwinkel K, Mohammad-pajooh E (2018) High-rate anaerobic treatment of wastewater from soft drink industry: methods, performance and experiences. J Environ Manag 220(May):8–15. https://doi.org/10.1016/j.jenvman.2018.05.015

Dalkilic K, Ugurlu A (2015) Biogas Production from Chicken Manure at Different Organic Loading Rates in a Mesophilic-Thermopilic Two Stage Anaerobic System. J Biosci Bioeng 120(3):315–322

Deressa L, Libsu S, Chavan RB, Manaye D, Dabassa A (2015) Production of Biogas from Fruit and Vegetable Wastes Mixed with Different Wastes. Environ Ecol Res 3(3):65–71

Divya D, Gopinath LR, Merlin Christy P (2015) A Review on Current Aspects and Diverse Prospects for Enhancing Biogas Production in Sustainable Means. Renew Sust Energ Rev 42:690–699

Dutta A, Davies C, Ikumi DS (2018) Performance of Upflow Anaerobic Sludge Blanket (UASB) Reactor and Other Anaerobic Reactor Configurations for Wastewater Treatment: A Comparative Review and Critical Updates. Journal of Water Supply: Research and Technology 1:27

Elsayed M, Andres Y, Blel W, Gad A, Ahmed A (2016) Effect of VS Organic Loads and Buckwheat Husk on Methane Production by Anaerobic Co-Digestion of Primary Sludge and Wheat Straw. Energy Convers Manag 117:538–547

Enitan AM, Adeyemo J, Swalaha FM. (2015) "Anaerobic Digestion Model to Enhance Treatment of Brewery Wastewater for Biogas Production Using UASB Reactor"

Fagbohungbe MO, Herbert BMJ, Hurst L, Ibeto CN, Li H, Usmani SQ, Semple KT (2017) The Challenges of Anaerobic Digestion and the Role of Biochar in Optimizing Anaerobic Digestion. Waste Manag 61:236–249

Forgács G (2012) Biogas Production from Citrus Wastes and Chicken Feather : Pretreatment and Co-Digestion

Geng Y, Zhang B, Du L, Tang Z, Li Q, Zhou Z, Yin X (2016) Improving Methane Production During the Anaerobic Digestion of Waste Activated Sludge: Cao-Ultrasonic Pretreatment and Using Different Seed Sludges. Procedia Environ Sci 31:743–752

Goli A, Shamiri A, Khosroyar S, Talaiekhozani A, Sanaye R, Azizi K (2019) A Review on Different Aerobic and Anaerobic Treatment Methods in Dairy Industry Wastewater. Journal of Environmental Treatment Techniques 6(1):113–141

Gontard N, Sonesson U, Birkved M, Majone M, Bolzonella D, Celli A, Angellier-Coussy H, Jang GW, Verniquet A, Broeze J, Schaer B, Batista AP, Sebok A (2018) A Research Challenge Vision Regarding Management of Agricultural Waste in a Circular Bio-Based Economy. Crit Rev Environ Sci Technol 48(6):614–654

Hagos K, Zong J, Li D, Liu C, Xiaohua L (2016) Anaerobic Co-Digestion Process for Biogas Production: Progress, Challenges and Perspectives. Renew Sust Energ Rev 76(March):0–1

Intanoo P, Chaimongkol P, Chavadej S (2015) Hydrogen and Methane Production from Cassava Wastewater Using Two-Stage Upflow Anaerobic Sludge Blanket Reactors (UASB) with an Emphasis on Maximum Hydrogen Production. Int J Hydrog Energy 41:1–8

Kamali M, Alavi-Borazjani SA, Khodaparast Z, Khalaj M, Jahanshahi A, Costa E, Capela I (2019) Additive and Additive-Free Treatment Technologies for Pulp and Paper Mill Effluents: Advances, Challenges and Opportunities. Water Resources and Industry 21(March):100109

Karina M, Jos H, Sequinel R, Pioneiro R. (2017) "ScienceDirect Treatment of Brewery Wastewater and Its Use for Biological Production of Methane and Hydrogen." 2

Karray R, Karray F, Loukil S, Mhiri N, Sayadi S (2017) Anaerobic Co-Digestion of Tunisian Green Macroalgae Ulva Rigida with Sugar Industry Wastewater for Biogas and Methane Production Enhancement. Waste Manag 61:171–178

Kiran EU, Trzcinski AP, Liu Y (2015) Enhancing the Hydrolysis and Methane Production Potential of Mixed Food Wastes by an Effective Enzymatic Pretreatment. Bioresour Technol 183:47–52

Kothari R, Kumar V, Pathak VV, Tyagi VV (2016) "ScienceDirect Sequential Hydrogen and Methane Production with Simultaneous Treatment of Dairy Industry Wastewater : Bioenergy Profit Approach." 1–10

Kougias PG, Treu L, Peñailillo D, Boe K, Campanaro S, Angelidaki I (2017) Bioresource Technology Ex-Situ Biogas Upgrading and Enhancement in Different Reactor Systems. Bioresour Technol 225:429–437

Kumar D, Pandit PD, Patel Z, Bhairappanavar SB, Das J (2019) Perspectives, Scope, Advancements, and Challenges of Microbial Technologies Treating Textile Industry Effluents. Elsevier Inc

Lei Z, Yang S, Li Y-y, Wen W, Wang XC, Chen R (2018) Application of Anaerobic Membrane Bioreactors to Municipal Wastewater Treatment at Ambient Temperature : A Review of Achievements , Challenges, and Perspectives. Bioresour Technol 267(June):756–768

Li K, Liu R, Sun C (2016a) A review of methane production from agricultural residues in China. Renew Sust Energ Rev 54:857–865. https://doi.org/10.1016/j.rser.2015.10.103

Li Y, Liu H, Yan F, Su D, Wang Y, Zhou H (2016b) High-Calorific Biogas Production from Anaerobic Digestion of Food Waste Using a Two-Phase Pressurized Biofilm (TPPB) System. Bioresour Technol 224:56–62

Li L, Peng X, Wang X, Di W (2018a) Anaerobic Digestion of Food Waste: A Review Focusing on Process Stability. Bioresour Technol 248(174):20–28

Li W, Khalid H, Zhu Z, Zhang R, Liu G, Chen C, Thorin E (2018b) Methane Production through Anaerobic Digestion: Participation and Digestion Characteristics of Cellulose, Hemicellulose and Lignin. Appl Energy 226(May):1219–1228

Liu XF, Song DZ, He XQ, Nie BS (2018) Effect of microstructures on methane adsorption characteristics of soft and hard coal. J China Univ Min Technol 47(1):155–161

Lora R, Maria A, De SA, Valéria F, Sánchez A, Barrena R, Font X (2017) Technology Overview of Biogas Production in Anaerobic Digestion Plants : A European Evaluation of Research and Development. Renew Sust Energ Rev 80(May):44–53

Mao C, Feng Y, Wang X, Ren G (2015) Review on Research Achievements of Biogas from Anaerobic Digestion. Renew Sust Energ Rev 45:540–555

Mao C, Zhang T, Wang X, Feng Y, Ren G, Yang G (2017) Process Performance and Methane Production Optimizing of Anaerobic Co-Digestion of Swine Manure and Corn Straw. Sci Rep 7 (1):1–9

Maragkaki AE, Fountoulakis M, Gypakis A, Kyriakou A, Lasaridi K, Manios T (2017) Pilot-Scale Anaerobic Co-Digestion of Sewage Sludge with Agro-Industrial by-Products for Increased Biogas Production of Existing Digesters at Wastewater Treatment Plants. Waste Manag 59:362–370

Maragkaki AE, Fountoulakis M, Kyriakou A, Lasaridi K, Manios T (2018) Boosting Biogas Production from Sewage Sludge by Adding Small Amount of Agro-Industrial by-Products and Food Waste Residues. Waste Manag 71:605–611

Martín MA, Fernández R, Gutiérrez MC, Siles JA (2018) Thermophilic Anaerobic Digestion of Pre-Treated Orange Peel: Modelling of Methane Production. Process Saf Environ Prot 117:245–253

Matheri AN, Ndiweni SN, Belaid M, Muzenda E, Hubert R (2017) Optimising biogas production from anaerobic co-digestion of chicken manure and organic fraction of municipal solid waste. Renew Sust Energ Rev 80(March 2016):756–764. https://doi.org/10.1016/j.rser.2017.05.068

Maurya R, Tirkey SR, Rajapitamahuni S, Ghosh A, Mishra S (2019) Recent Advances and Future Prospective of Biogas Production. In: Advances in Feedstock Conversion Technologies for Alternative Fuels and Bioproducts. Elsevier, pp 159–178

Mehariya S, Patel AK, Obulisamy PK, Punniyakotti E, Wong JWC (2018) Co-Digestion of Food Waste and Sewage Sludge for Methane Production: Current Status and Perspective. Bioresour Technol 265(March):519–531

Montingelli ME, Benyounis K, Stokes J, Olabi AG (2016) Pretreatment of Macroalgal Biomass for Biogas Production. Energy Convers Manag 108:202–209

Náthia-Nevesorcid G, de Alencar Neves T, Berni M, Dragone G, Mussatto SI, Forster-Carneiro T (2018) Start-up Phase of a Two-Stage Anaerobic Co-Digestion Process: Hydrogen and Methane Production from Food Waste and Vinasse from Ethanol Industry. Biofuel Research Journal 5 (2):813–820

Ohimain EI, Izah SC. (2017) "Crossmark." 70(November 2016):242–253

Patinvoh RJ, Osadolor OA, Chandolias K, Horváth IS, Taherzadeh MJ (2017) Innovative Pretreatment Strategies for Biogas Production. Bioresour Technol 224:13–24

Pellera F-m, Gidarakos E (2017) Journal of Environmental Chemical Engineering Microwave Pretreatment of Lignocellulosic Agroindustrial Waste for Methane Production. Biochem Pharmacol 5(1):352–365

Peng L, Appels L, Su H (2018) Combining microwave irradiation with sodium citrate addition improves the pre-treatment on anaerobic digestion of excess sewage sludge. J Environ Manag 213:271–278

Pessuto J, Scopel BS, Perondi D, Godinho M, Dettmer A (2016) Enhancement of Biogas and Methane Production by Anaerobic Digestion of Swine Manure with Addition of Microorganisms Isolated from Sewage Sludge. Process Saf Environ Prot 104:233–239

Prajapati KB, Singh R (2018) Kinetic Modelling of Methane Production during Bio-Electrolysis from Anaerobic Co-Digestion of Sewage Sludge and Food Waste. Bioresour Technol 263:491–498

Rahman MA, Saha CK, Feng L, Møller HB, Alam MM (2019) Anaerobic Digestion of Agro-Industrial Wastes of Bangladesh: Influence of Total Solids Content. Engineering in Agriculture, Environment and Food 12(4):484–493

Ren Y, Yu M, Wu C, Wang Q, Gao M, Huang Q (2018a) Bioresource Technology A Comprehensive Review on Food Waste Anaerobic Digestion : Research Updates and Tendencies. Bioresour Technol 247:1069–1076

Ren Y, Yu M, Wu C, Wang Q, Gao M, Huang Q, Yu L (2018b) A Comprehensive Review on Food Waste Anaerobic Digestion: Research Updates and Tendencies. Bioresour Technol 247 (September 2017):1069–1076

Resende JA, Godon J-J, Bonnafous A, Arcuri PB, Silva VL, Otenio MH, Diniz CG (2016) Seasonal Variation on Microbial Community and Methane Production during Anaerobic Digestion of Cattle Manure in Brazil. Microb Ecol 71(3):735–746

Riggio V, Comino E, Rosso M (2015) Energy production from anaerobic co-digestion processing of cow slurry, olive pomace and apple pulp. Renew Energy 83:1043–1049. https://doi.org/10.1016/j.renene.2015.05.056

Rodriguez C, Alaswad A, El-Hassan Z, Olabi AG (2018a) Improvement of Methane Production from P. Canaliculata through Mechanical Pretreatment. Renew Energy 119:73–78

Rodriguez C, Alaswad A, El-Hassan Z, Olabi AG (2018b) Waste Paper and Macroalgae Co-Digestion Effect on Methane Production. Energy 154:119–125

Da Ros C, Cavinato C, Pavan P, Bolzonella D (2017) Mesophilic and Thermophilic Anaerobic Co-Digestion of Winery Wastewater Sludge and Wine Lees: An Integrated Approach for Sustainable Wine Production. J Environ Manag 203:745–752

Rouches E, Zhou S, Steyer JP, Carrere H (2016) White-Rot Fungi Pretreatment of Lignocellulosic Biomass for Anaerobic Digestion: Impact of Glucose Supplementation. Process Biochem 51 (11):1784–1792

Salama E-s, Jeon B-h, Kurade MB, Patil SM, Usman M, Li X, Lim H (2019a) Enhanced Anaerobic Co-Digestion of Fat, Oil, and Grease by Calcium Addition: Boost of Biomethane Production and Microbial Community Shift. Bioresour Technol 296:122353

Salama E-s, Saha S, Kurade MB, Dev S, Woong S, Jeon B-h (2019b) Recent Trends in Anaerobic Co-Digestion : Fat, Oil, and Grease (FOG) for Enhanced Biomethanation. Prog Energy Combust Sci 70:22–42

dos Santos LC, Adarme OFH, Baêta BEL, Gurgel LVA, de Aquino SF (2018) Production of Biogas (Methane and Hydrogen) from Anaerobic Digestion of Hemicellulosic Hydrolysate Generated in the Oxidative Pretreatment of Coffee Husks. Bioresour Technol 263(May):601–612

Sebola MR, Tesfagiorgis HB, Muzenda E (2015) Methane Production from Anaerobic Co-Digestion of Cow Dung, Chicken Manure, Pig Manure and Sewage Waste. I(September)

Shi X, Leong KY, Ng HY (2017) Anaerobic Treatment of Pharmaceutical Wastewater : A Critical Review. Bioresour Technol 245(August):1238–1244

Siddique MNI, Wahid ZA (2018) Achievements and perspectives of anaerobic co-digestion: a review. J Clean Prod 194(1):359–371. https://doi.org/10.1016/j.jclepro.2018.05.155

Singh RL, Singh RP (2019) Advances in Biological Treatment of Industrial Waste Water and Their Recycling for a Sustainable Future. In: Singh RL, Rajat PS (eds) Applied Environmental Sciencie and Engineering for a Sustainable Future

Srisowmeya G, Chakravarthy M, Nandhini Devi G (2019) Critical Considerations in Two-Stage Anaerobic Digestion of Food Waste–A Review. Renew Sust Energ Rev 119:109587

Taghizadeh A, Hosseini SH, Ghobadian B, Motevali A (2017) Biofuel Production from Citrus Wastes: A Feasibility Study in Iran. Renew Sust Energ Rev 69(October):1100–1112

Takizawa S, Baba Y, Tada C, Fukuda Y, Nakai Y (2018) Pretreatment with Rumen Fluid Improves Methane Production in the Anaerobic Digestion of Paper Sludge. Waste Manag 78:379–384

Thanarasu A, Periyasamy K, Devaraj K, Periyaraman P, Palaniyandi S, Subramanian S (2018) Tea Powder Waste as a Potential Co-Substrate for Enhancing the Methane Production in Anaerobic Digestion of Carbon-Rich Organic Waste. J Clean Prod 199:651–658

Verma JP, Jaiswal DK (2016) Book Review: Advances in Biodegradation and Bioremediation of Industrial Waste. Front Microbiol 6(January):2015–2016

Wang M, Zhou J, Yuan YX, Mei Dai Y, Li D, Li ZD, Liu XF, Zhang XY, Yan ZY (2017) Methane Production Characteristics and Microbial Community Dynamics of Mono-Digestion and Co-Digestion Using Corn Stalk and Pig Manure. Int J Hydrog Energy 42(8):4893–4901

Wang S, Jena U, Das KC (2018a) Biomethane Production Potential of Slaughterhouse Waste in the United States. Energy Convers Manag 173(May):143–157

Wang X, Li Z, Xue B, Zhou X, Cheng S, Gao R, Sun J (2018b) Study on Improving Anaerobic Co-Digestion of Cow Manure and Corn Straw by Fruit and Vegetable Waste: Methane Production and Microbial Community in CSTR Process. Bioresour Technol 249(August 2017):290–297

Wikandari R, Millati R, Cahyanto MN, Taherzadeh MJ (2014) Biogas Production from Citrus Waste by Membrane Bioreactor. Membranes 4(3):596–607

Wong S, Ngadi N, Inuwa IM, Hassan O (2018) Recent Advances in Applications of Activated Carbon from Biowaste for Wastewater Treatment: A Short Review. J Clean Prod 175:361–375

Wu C, Wang Q, Xiang J, Yu M, Chang Q, Gao M (2015) Enhanced Productions and Recoveries of Ethanol and Methane from Food Waste by a Three-Stage Process. Energy Fuel 29 (10):6494–6500

Xavier CAN, Moset V, Wahid R, Møller HB (2015) The Efficiency of Shredded and Briquetted Wheat Straw in Anaerobic Co-Digestion with Dairy Cattle Manure. Biosyst Eng 139:16–24

Xihui K, Yongming S, Lianhua L, Xiaoying K, Zhenhong Y (2018) Bioresource Technology Improving Methane Production from Anaerobic Digestion of Pennisetum Hybrid by Alkaline Pretreatment. Bioresour Technol 255(September 2017):205–212

Xu F, Li Y, Ge X, Yang L, Li Y (2017) Anaerobic Digestion of Food Waste – Challenges and Opportunities. Bioresour Technol 247(September 2017):1047–1058

Yan BH, Selvam A, Wong JWC (2016) Innovative Method for Increased Methane Recovery from Two-Phase Anaerobic Digestion of Food Waste through Reutilization of Acidogenic off-Gas in Methanogenic Reactor. Bioresour Technol 217:3–9

Zahan Z, Othman MZ, Muster TH (2018) Anaerobic Digestion/Co-Digestion Kinetic Potentials of Different Agro-Industrial Wastes: A Comparative Batch Study for C/N Optimisation. Waste Manag 71:663–674
Zhai N, Zhang T, Yin D, Yang G, Wang X, Ren G, Feng Y (2015) Effect of Initial PH on Anaerobic Co-Digestion of Kitchen Waste and Cow Manure. Waste Manag 38(1):126131
Zhang J, Loh K-c, Li W, Wei J, Dai Y, Wah Y (2016a) Three-Stage Anaerobic Digester for Food Waste. Appl Energy 194:287–295
Zhang Q, Jianjun H, Lee D-j (2016b) Biogas from Anaerobic Digestion Processes : Research Updates. Renew Energy 98:108–119
Zhang Z, Gao P, Cheng J, Liu G, Zhang X (2018) Enhancing Anaerobic Digestion and Methane Production of Tetracycline Wastewater in EGSB Reactor with GAC / NZVI Mediator. Water Res 136:54–63
Zhang Z, Zhang G, Li W, Li C, Xu G (2016c) Enhanced Biogas Production from Sorghum Stem by Co-Digestion with Cow Manure. Int J Hydrog Energy 41(21):9153–9158
Zhao J, Liu Y, Wang D, Chen F, Li X, Zeng G (2017) Potential Impact of Salinity on Methane Production from Food Waste Anaerobic Digestion. Waste Manag 67:308–314
Zheng Z, Liu J, Yuan X, Wang X, Zhu W, Yang F, Cui Z (2015) Effect of Dairy Manure to Switchgrass Co-Digestion Ratio on Methane Production and the Bacterial Community in Batch Anaerobic Digestion. Appl Energy 151:249–257
Zhou J, Zhang R, Liu F, Yong X, Wu X, Zheng T, Jiang M, Jia H (2016) Biogas Production and Microbial Community Shift through Neutral PH Control during the Anaerobic Digestion of Pig Manure. Bioresour Technol 217:44–49

Comparative Analysis of Biogas with Renewable Fuels and Energy: Physicochemical Properties and Carbon Footprints

Edith Mier-Alba, Salvador Sánchez-Muñoz, Fernanda Gonçalves Barbosa, Vijay Kumar Garlapati, Nagamani Balagurusamy, Silvio Silvério da Silva, Júlio César dos Santos, and Anuj Kumar Chandel

Abstract Industrialization, infrastructure development, energy production, and their use are the engines of the economic growth of any country. These growth drivers permanently require the uninterrupted supply of energy and fuels. Currently, gasoline and its products are a primary source which is fulfilling the energy requirement. However, their excessive usage as an energy source is generating the greenhouse gasses (GHGs), which deteriorate the environment and eventually cause negative impact on the climate. Currently, more than 85% of the world's total energy needs are being fulfilled by crude petroleum resources. Therefore, it is necessary to look out for alternative fuels that are renewable and cause a low impact on the environment. Because of environmental concerns, there is a strong push to implement renewable energy sources. Renewable fuels and energies can play a crucial role in lowering GHG emissions and thus keeping the environment clean and green, which eventually meets the sustainability goals. Cellulosic ethanol in the USA and Brazil, biodiesel and biogas in Brazil and some European countries, and solar energy in India have shown promising growth. The carbon emissions of renewable fuels are very low compared to crude petroleum. This chapter compares the current scenario of biofuel implementation in various countries. Particular emphasis is placed on the current status of biogas development and commercialization. The carbon footprints of various biofuels/energies are also discussed.

E. Mier-Alba · S. Sánchez-Muñoz · F. G. Barbosa · S. S. da Silva · J. C. dos Santos · A. K. Chandel (✉)
Department of Biotechnology, Engineering School of Lorena (EEL), University of Sao Paulo (USP), Lorena, São Paulo, Brazil
e-mail: anuj10@usp.br

V. K. Garlapati
Department of Biotechnology & Bioinformatics, Jaypee University of Information Technology (JUIT), Waknaghat, HP, India

N. Balagurusamy
Laboratorio de Biorremediación, Facultad de Ciencias Biologicas, Universidad Autonoma de Coahuila, Torreón, Coahuila, Mexico

N. Balagurusamy, A. K. Chandel (eds.), *Biogas Production*,
https://doi.org/10.1007/978-3-030-58827-4_7

Keywords Biogas · Bioethanol · Biodiesel · Carbon footprints · Carbon emissions · Sustainability

1 Introduction

Due to the finite amount of fossil fuels and their negative impact on the environment, research of new renewable energy sources has been explored for a long time. Recent geopolitical factors, food versus fuel debate, and climate change concerns have asked for sustainable sources of energy and fuels with the competitive prices (Chandel et al. 2018; Silva and Chandel 2014). Renewable energy sources like solar and wind energy, geothermal energy, ocean energy, and hydropower seem to be established in some countries, but the large-scale production still have to be more economical to fulfill the societal needs (Banerjee et al. 2019; Sevda et al. 2019; Kumar et al. 2017; Montpart et al. 2014). Moreover, the production of biodegradable energy sources by biological agents reduce the toxic and irreversible petroleum derivative-based greenhouse gas emissions. In the last two or three decades, various bioenergy technologies that have been developed are at a stage of fundamental research, demonstration, and commercialization. Biological, thermal, and physical processes of conversion of biomass offer renewable energy/fuels and biochemicals under the biorefinery platform (Wu et al. 2010; Chandel and Silveira 2017).

In the last two decades, renewable energy sources like solar and wind energy have shown great potential, but their utilization is difficult due to their fluctuating and unstable nature. Further, the storage of these energies poses a big challenge. Hydrogen has shown promising results in terms of production at a laboratory scale, but their large-scale viability is still a question. However, long-term storage of hydrogen as a sustainable energy vector is possible with minimum efforts (Gahleitner 2013).

Bioethanol and biodiesel are the most common liquid transportation fuels which have shown significant progress; however, their economic production at commercial scale has to be resolved. The USA and Brazil are two crucial countries that are successful in implementing bioethanol as a viable option as an alternative of gasoline in fleets. Biodiesel also is quite successful in Brazil as a blending option diesel. Currently, bioethanol is being produced either from corn grains or sugarcane molasses. Production of cellulosic ethanol and biodiesel production is still not economically competitive at commercial scale in any country.

The minimum emission of GHG is an essential parameter in the selection of alternative fuels/energy. These alternative fuels/energy sources have been given subsidies by governments to promote. However, more efforts are required to push renewable fuels/energy to match the sustainability goals by governments. Biogas has shown an enthusiastic success curve in European countries. However, a lot has to be done in the rest of the world for biogas promotion and implementation as a sustainable energy source. Countries like the USA, Brazil, India, and Mexico have a potential in the implementation of biogas as a clean and economical energy source for the common people.

No matter what is the renewable energy/fuel source, the carbon footprint is always lower than petroleum sources. The production of biofuels is more eco-friendly and has lower carbon footprints than the refining of crude petroleum into diesel, petrol, or gas. Upon burning as fuels/energy, they produce significantly lower carbon emissions than conventional petroleum products. This chapter presents a comparative analysis of various renewable fuels/energy and their physicochemical properties and carbon footprints.

2 Potential Renewable Fuels: An Overview of Production Technology

2.1 Biogas

Anaerobic digestion of wastes, crop, and waste residues has to gain more popularity than the other bioenergy production schemes due to the sustainability and ease of the process with lesser GHG emissions. Moreover, the anaerobic digestion toward biogas production was reported as sustainable energy-efficient technology (Fehrenbach et al. 2008). However, anaerobic digestion can produce acidification and eutrophication 25 and 12 times higher than fossil fuel options (Whiting and Azapagic 2014). The utilization of anaerobic digestion, coupled with the utilization of local waste residues, results in lower GHG emissions (Weiland 2010). The main product of anaerobic digestion is biogas, which is a promising substitute for fossil fuels and heat and power generation. It can be used in a more efficient way than used today. On the other hand, the digestate from anaerobic digestion is a useful and valuable fertilizer due to its high nitrogen availability, minimum presence of pathogens, and its better short-term fertilization effect (Weiland 2010). Digestate is very important to apply to land because if digestate from anaerobic digestion is not used to replace artificial fertilizers and is accumulated or not properly stored, it would cause a higher negative impact on the environment. The environmental impact of anaerobic digestion is influenced by the type and source of feedstock, digestate storage, and its application on land (Whiting and Azapagic 2014).

As shown in Table 1, the substrate for biogas production usually comprises carbohydrates, proteins, lipids, and lignocellulosic constituents. The proportions of biogas and methane production through anaerobic digestion depend mostly on the substrate and adopted the digestion process and process parameters (Braun et al. 2008; Gemmeke et al. 2009).

Second-generation biogas avoids conflicts between food and feed. If biogas is used for cogenerating electricity and heat, then it can present approximately 50% lower global warming potential and can lead to a significant reduction in most environmental impacts than fossil fuels (Whiting and Azapagic 2014). Chinnici et al. (2018) reported that 1 ton of feedstock, comprising municipal, farm, and agricultural wastes, could yield approximately 65 Nm^3 of biogas, which is enough

Table 1 Biogas yields from anaerobic digestion from different substrates

Substrate	Biogas yield (Nm^3/ton of substrate)	Energy yield (KWh/ton of substrate)	Thermal energy yield (kWh/ton of substrate)	Reference
Municipal, farm, and agricultural wastes mixture	65	104	78	Chinnici et al. 2018
Pig manure	350	1350	–	Fantozzi and Buratti 2009
Mixture of olive husk and piggery manure	280	1070		
Pig manure	356*	~1100**		Gebrezgabher et al. 2010
Energy maize	390*	~1205**		
Poultry manure	410*	~1267**		
Food waste	500*	~1545**		
Flower bulbs	500*	~1545**		

*Methane yield (Nm^3/ton of substrate)**Estimated from methane yield using factor of conversion (3.09)

to produce 104 KWh and 78 KWh of electrical energy and thermal energy, respectively. Usually, farm animal manure along with the co-substrates results in higher methane yields in agricultural-based biogas setups. Table 1 shows the biogas production from different organic substrates via anaerobic digestion.

The various process can be employed in wet fermentation with total solids (TS) concentrations in the fermenter less than 10% and dry fermentation systems with a TS between 15% and 35% (Gemmeke et al. 2009). Generally, wet fermentations will operate at mesophilic temperatures (38 and 42 °C), but some will operate under thermophilic conditions (50 to 55 °C). At higher temperature, the degradation rate is faster, in turn causing short retention time and smaller reactor volumes without changing final methane productivity. Adopting lesser temperatures near about 50 °C helps in the reduction of ammonia toxicity along with the thermophile growth. However, the risk of a washout of the microbial population can occur at this point using thermophilic microorganisms (Angelidaki et al. 2003).

Another modification in the fermentation process is continuous dry fermentation processes where more than 25% dry matter substrates were used in the horizontal fermenter with a mechanical mixing system or vertical plug flow fermenters. In vertical digesters, mixing inside the fermenter is not required as substrate flows from the top to the bottom due to the gravitational force (Weiland 2010; De Baere and Mattheeuws 2008).

2.2 *Other Biofuels*

To promote sustainability and independence from fossil fuel, some biofuels like bioethanol, biomethanol, and biodiesel also have crucial role (Paulauskiene et al. 2019; Thangavelu et al. 2016; Wei et al. 2014). Bioethanol is a clean, renewable transportation fuel that has gained more attention as a blend or fossil petrol substitute since it is considered to be clean, renewable, and green (Bhatia et al. 2019; Wei et al. 2014; Thangavelu et al. 2016). The source of substrate for bioethanol varies from an edible source in first-generation to nonedible sources in second-generation biofuels (Aditiya et al. 2016; Kuila et al. 2017; Roy et al. 2015). Algal biomass is another renewable feedstock that is used as third-generation bioethanol and for biochemical production (Azhar et al. 2017; Jambo et al. 2016; Jha et al. 2017).

Some research centers at Malaysian universities (Universiti Teknologi Malaysia, Universiti Malaysia Sarawak, and Swinburne University of Technology) showed a remarkable improvement in combustion characteristics and enhanced combustion for second-generation bioethanol as a fuel. Fuel emissions like carbon monoxide and unburned hydrocarbon emissions were analyzed and reduced. However, emissions such as oxides of nitrogen, carbon dioxide, acetaldehyde, carbonyls, and aromatics were not significantly reduced (Thangavelu et al. 2016).

Bioethanol is commonly produced by wild yeasts, where *Saccharomyces cerevisiae* is the primary model microorganism for this process (Thangavelu et al. 2016). However, for the production of cellulosic ethanol, a recombinant or wild microorganism, which can utilize hemicellulosic and cellulosic sugars, is used to obtain desired ethanol production. Further other bioprocessing methods such as using immobilized cells in the recycling of media or even cells are used to produce ethanol. Other advantages of using immobilized cells are high cell density, high substrate conversion, short reaction time, less inhibition, and cell recycling for other processes (Azhar et al. 2017).

Biodiesel is another alternative to replace petroleum fuel because it is renewable and biodegradable and presents similar properties of fossil diesel fuel and it can be used in compression ignition engine and in marine engines when it is blend with diesel and biomethanol (Ghazali et al. 2015; Paulauskiene et al. 2019). However, its characteristics may vary depending on the feedstock nature and its properties and in some conditions have better engine performance than pure diesel (Mahmudul et al. 2017; Verma and Sharma 2016).

Biodiesel (fatty acid methyl esters) is produced by transesterification of animal fats, vegetable oils, algal lipids, or wastes, a mixture derived from renewable feedstock (Garlapati et al. 2017, 2015, 2013; Kumari et al. 2009; Meher et al. 2006). The primary process parameters for biodiesel production include the molar ratio of oil and alcohol, molar ratio of oil and solvent, catalyst concentration, reaction time and temperature, and FFA content of the oils. The commonly used alcohols in biodiesel production include methanol and ethanol and alkaline catalysts are sodium or potassium hydroxides.

Some vegetable oils are used or have the potential to be used as biodiesel feedstock since their triglyceride concentration supports biodiesel production, including canola, coconut, sunflower, and corn with a potential annual yield between 18 and 300 gallons/acre produced. Besides, algal lipids represent an excellent feedstock since microalgae can fix large amounts of CO_2 per land area than other plants and would reach more than 5000 gallons of biodiesel per acre of land (Hoekman et al. 2012).

For the cost economic and competitive production of biodiesel, parameters used in the process of transesterification play a crucial role (Verma and Sharma 2016). For example, algal feedstock derived biodiesel is a clean fuel but still expensive when compared with fossil fuels. This is necessary to use appropriate media formulation for the production of biodiesel and biogas in a single process from algal feedstock to avoid the toxicity of ammonium traces (Alaswad et al. 2015).

An investigation of automotive engineering groups of Malaysia has shown that biodiesel-diesel blends with up to 30% of biodiesel have very similar properties as pure diesel. Biodiesel-diesel blends have shown reduced heat release rate; reduction of HC, CO, and PM emissions; and a shorter ignition delay when compared to diesel; on the other hand, the NO_x emission becomes slightly higher (Hasan and Rahman 2017).

Another option for bioenergy is the production and use of biohydrogen. The clean production of biohydrogen, using renewable sources, has the potential to minimize carbon dioxide emissions (Wang et al. 2014). Actually, biohydrogen is produced by water electrolysis, microbial electrolysis cell, water dissociation, thermochemical water splitting, and photolysis or bio-photolysis by cyanobacteria and green algae (Dincer and Acar 2015).

Dark fermentation is a potential method of hydrogen production from anaerobic digestion utilizing the waste materials. A maximum yield of 4 mol H_2/mol hydrogen yield was reported in the case of dark fermentation where glucose is metabolized either to acetate or to acetone (Khanna and Das 2013). Nevertheless, several gross challenges in production, storage, and transportation have a decisive role in large-scale production (Khan et al. 2018).

The use of hydrogen as a source of energy offers several advantages such as high energy conversion efficiencies, easy transportation, abundance, ease of conversion to other forms of energy, and higher lower and higher heating values as compared with conventional fossil fuels (Acar and Dincer 2014).

3 Physical and Chemical Properties of Alternative Fuels and Energy

With growing concern about environmental problems, the demand for products from renewable sources is increasing. To have wide use of fuels and alternative energy is necessary to know the physical properties of these products. The main physicochemical properties of fuels and energy from renewable sources will be discussed below.

3.1 Biogas

Biogas is a fuel of high energy potential and corresponds to a mixture of carbon dioxide and methane. Other elements present in a smaller quantity are H_2 (hydrogen), NH_3 (ammonia), H_2S (hydrogen sulfide), O_2 (oxygen), CO (carbon monoxide), N (nitrogen), and H_2O (water) (Petersson and Wellinger 2009; Santos and Lima 2017) (Table 2).

The proportion of each gas in the mixture depends on parameters such as the type of digester and the substrate used. In addition to CH_4, all other constituents of biogas are considered pollutants; many treatments are currently performed to remove these undesirable compounds (Angelidaki et al. 2003; Zachow 2000). The physicochemical properties of biogas are related to their main constituents, methane and carbon dioxide; according to their proportions, the physicochemical properties may vary (Tables 3 and 4).

Biogas is a colorless gas, insoluble in water, and, in the absence of impurities, odorless. It has a low density, being lighter than air, dissipating rapidly in the atmosphere and presenting lower risks of explosion when compared to natural gas (Kim et al. 2015).

Calorific value is an important fuel parameter. It represents the amount of energy released in the complete combustion of a unit mass of the same (KJ/kg). According to the percentage of methane in the biogas composition, the calorific value can vary from 5.815 to 8.141 kWh/m^3. By reducing the amount of carbon dioxide in the gas mixture, the calorific value can reach 13,956 kWh/m^3 (Mazzucchi 1980; Parchen 1979). The calorific value of purified biogas is considered to be 9.5 kWh/m^3 (Coelho et al. 2000).

Table 2 Composition of biogas

Component	Content
Methane	50–70%
Carbon dioxide	30–45%
Hydrogen sulfide	0.001–2%
Hydrogen	0.01–2
Nitrogen	0–3%
Oxygen	0–1%
Argon	0.001
Carbon monoxide	0.001–2
Ammonia trace	Trace
Organics	Trace

Source: Retrieved from Pauss et al. (1987)

Table 3 Physicochemical properties of methane

Properties	Methane (CH_4)
Boiling temperature (°C)	−161,49
Melting temperature (°C)	– 186,48
High calorific value (kcal/m^3)	9.520,00
Lower calorific value (kcal/m^3)	8.550,00
Ignition temperature (° C)	650,00

Source: Vargaftik (1975)

Table 4 Density of methane gas

Methane contents %	50	60	80	90	100
Gas density relative to air	1.040	0,942	0,745	0,652	0,555

Source: Pauss et al. (1987)

Table 5 Physicochemical properties of biodiesel

Properties	ASTM D6751	EN 14214	Units
Kinematic viscosity (40 ° C)	1.9–6.0	3.5–5.0	mm^2/s
Number Cetane	Minimum 47	Minimum 51	–
Flash point	Minimum 130	>101	C
Cloud point	Report	–	° C
Density	–	860–900	Kg/m^3

Source: Retrieved from ASTM (1995). Annual Book of ASTM Standards. Metals Test Methods and Naytical Procedures. ASTM

3.2 *Biodiesel*

According to ASTM (American Society for Testing and Materials, 1995), biodiesel is defined as a "fuel is made of mono-alkyl esters of long-chain fatty acids originated from oils either from vegetable oils or animal fats." ASTM designated biodiesel as B100. The properties of biodiesel as fuel depend on the structural characteristics and types of esters of fatty acids (Knothe and Razon 2017). ASTM and the European Committee for Standardization (CEN) have established biodiesel quality specifications (B100), ASTM D6751, and EN 14214 (Knothe 2008). Table 5 shows some properties of biodiesel governed by fatty acid composition and properties.

Biodiesel's viscosity is primarily influenced by structural factors, such as chain length, number, and nature of the double bonds of alkyl esters of individual fatty acids. In addition, the viscosity is also influenced by the double bond configuration. Compounds with cis and trans double bonds have different viscosities (Knothe and Steidley 2005). According to ASTM D6751, the kinematic viscosity of biodiesel must meet the requirement between 1.9 and 6.0 centistokes (cSt) at 40 °C (Knothe and Steidley 2007).

The density is specified only in EN 14214 with a range of 860 to 900 kg/m^3 for biodiesel and depends on the composition of fatty acids and their purity. Density depends on the degree of unsaturation and the content of heavier atoms (Knothe and

Razon 2017). The fuel quality broadly depends on the cetane number (CN), which influences the ignition profile of the fuel. The higher the CN, the shorter the ignition delay time. The cetane number shows that CN decreases with decreasing chain length and increasing branching (Knothe and Razon 2017).

3.3 Biohydrogen

Biohydrogen has high energy density and forms water vapor during its combustion; hydrogen gas is considered a clean and alternative energy source (Gupta 2008; Rahman et al. 2016). Some physicochemical properties of H_2 contribute to its use as an alternative fuel, among them are high air diffusion, high floatage (better than methane, propane, and gasoline), low toxicity, higher flammability range (4–75% in volume), and low energy requirements in its ignition, among others (Gupta 2008). In addition, the energy content of H_2 is the highest among conventional fuels, with a value of 141.90 $J.kg^{-1}$, which is higher than the energy content of ethanol (29.9 $J.kg^{-1}$), biodiesel (37.0 $J.kg^{-1}$), and natural gas (50.0 $J.kg^{-1}$) (Gupta 2008). Some properties are listed in Table 6.

3.4 Ethanol

Ethanol is an alternative fuel produced by fermentation processes and distillation processes, the most used for transportation worldwide. It is produced from different agricultural raw materials, such as plain sugar, starch, and lignocellulose (Balat and Balat 2009). Ethanol is characterized as being water-soluble (polar solvent) with a flashpoint of 55 °F, boiling point of 173 °F, autoignition temperature of 793 °F, and a specific gravity of 0.79 (lighter than water). Its vapor density is 1.59, higher than the

Table 6 Physicochemical properties of the biohydrogen

Properties	Biohydrogen
Boiling temperature	20 K (−420 ° F, −253 ° C)
Melting temperature	14 K (−434 ° F, −259 ° C)
Odor, color, and taste	Odorless, colorless, tasteless
Toxicity	Nontoxic
Density	4.432 lb./ft3 (70.8 kg/m3)– (1 atm/ RT)
Reactivity	High
Flash point	< −423 ° F (< −253 ° C 20 K)
Autoignition temperature	1085 ° F (585 ° C)

Source: Retrieved from "Handbook of physical properties of liquids and gasses-pure substances and mixtures," N. B. Vargaftik 1975

Table 7 Physical and chemical properties of ethanol fuel

Properties	Ethanol
Flash point (° F)	55
Ignition temperature (° F)	793
Specific gravity	0,79
Vapor density	1,49
Vapor pressure (mmHg)	44
Boiling point (° F)	173
Conductivity	Yes
Toxicity	Lower
Solubility	Highly
Density (kg/m^3)	785,5
Viscosity (kg/m.s)	0,001007
Surface tension (N/m)	0,02314

Source: Retrieved from "Perry's Chemical Engineers' Handbook" D. W. Green and R. H. Perry 1997

air, being more substantial, so the ethanol vapors do not go up (Green and Perry 1997; Torres-Jimenez et al. 2009) (Table 7).

4 Commercialization Status of Biofuels and Bioenergy

Electricity is one of the essential services in the world, and it is required for most of the activities we realize every day. Most of the energy used in the world still comes from fossil resources despite the growing environmental concerns and the excessive use of fossil fuels. The search for renewable energy from organic and residual resources has increased (Hosseini and Wahid 2013; González et al. 2017) which results in at least 20% of electricity produced from renewable sources throughout the world by 2014 (Our World in Data, 2019). The renewable sources include biomass, hydropower, solar, wind, geothermal, and marine energy.

China, the USA, Brazil, and India are countries that had stood out as leaders in investment in renewable energy technologies, with a total global inversion of more than 2600 billion dollars per year with solar energy as the renewable energy with the highest inversion until 2016. This inversion amount will be affected by the country's scientific and technological development and their ecological compromise. Also, the technologies and types of renewable energy may vary according to the principal economic and agricultural activities and the natural sources of each country. Following the increasing interest of the utilization of renewable energy sources, the European Union presents initiatives to increase its production and utilization in various sectors. These countries aim to reduce greenhouse gas emissions by 80–95% by 2050. In addition, they expect to increase renewable energy use from 6.7 EJ in 2012 to 10.3 EJ in 2020, to increase about 10% (1.5 EJ) the use of renewable energy in transport (Scarlat et al. 2015).

The production and use of biogas are growing in the world market. Analyses of the "Global Biogas Market 2014-2018" report showed that the global market of European, American, Middle East, African, and Asia Pacific regions would have a composite annual growth rate of 8.64% in the period 2013–2018. According to the report, this increase in biogas production and use is stimulated by increased energy demand, availability and easy access to waste materials used as feedstock for production, the efficiency of the production process, and its properties (Yousuf et al. 2017).

According to data from EBA (European Biogas Association (http://european-biogas.eu/2017/12/14/eba-statistical-report-2017-published-soon/, recovered on April 8, 2019)), biogas is expected to contribute at least 1.5% of the EU's primary energy mix, which corresponds to approximately 5% of the EU's natural gas consumption. And by 2030, EBA estimates at least 30 billion m^3/year in biogas production from anaerobic digestion. Data taken from the EBA 2017 report show an increase in the number of biogas-producing plants in Europe; between 2009 and 2016, there was an increase from 6227 to 17,662 installations. The enhanced growth results from the increase in plant numbers on agricultural substrates, which increased from 4797 units in 2009 to 12,496 units in 2016.

The substrate utilization of biogas as a source of natural gas needs to adopt some purification techniques to enhance its quality (Kárászová et al. 2015; Khan et al. 2017). The purified biogas served as fuel like a fossil gas in natural gas-based vehicles. In this case, purified biogas can be injected into the natural gas grid, mixed with the natural gas present, and used to produce electricity/heat/cold, as well as its use as a vehicular fuel (Irena 2017). The usage of gaseous fuels such as biomethane in the transportation vehicles is growing swiftly in several countries with the European Union serving as the biggest market value for biomethane, which is worth 160 million m^3 in 2015 (Eurostat 2005). The major biogas producers as a transportation fuel in 2016 were Germany, Sweden, Switzerland, the UK, and the USA. According to the EBA 2017 report (http://european-biogas.eu/2017/12/14/eba-statistical-report-2017-published-soon/, recovered on April 8, 2019), there was an increase in biomethane output, from 752 GWh (2011) to 17,264 GWh (2016). Among different continents, Europe records a 40% enhancement in bioethanol production.

Reports from the World Biogas Association (http://www.worldbiogasassociation.org / january-2019-biogas-investment-update-from-acucomm/, recovered on April 8, 2019) show that the leading countries in biogas and anaerobic digestion investment in the year 2018 were the USA representing $ 986 million or 30% of the total and the European Union, representing US $ 1351 million or 42% of the total. The markets are Denmark, France, Germany, France, the Netherlands, and the UK (http://www.worldbiogasassociation.org/january-2019-biogas-investment-update-from-acucomm/, recovered on April 8, 2019).

4.1 *Biohydrogen*

Biohydrogen production technology must overcome a few challenges in its production before it could successfully compete in the fuel market and be deployed on a large scale (Sharma and Kaushik 2017). As all biofuels, the total production cost of biohydrogen depends on the processes and the feedstock used as shown in Table 3; nevertheless, it was estimated that the levelized price of biohydrogen energy content is less susceptible to the biomass feedstock cost than other biofuels, and its price was calculated as US $ 2.0–3.0/Kg which gives a great advantage to this type of bioenergy. Also, the economic feasibility of biohydrogen was endorsed by several financial bodies with a positive trend in the near future for investment and commercial success (Lee 2016). Even though technologies for hydrogen storage, production, and distribution have evolved over the years, there is a need to adopt the same for use in an energy system.

5 Emissions and Carbon Footprints of Alternative Fuels and Energy

In 2017, there is around 1.4% increase in CO_2 emissions, accounting for 460 million tons (Mt). The stats were recorded even after the issue is raised in the Paris Agreement on climate change too. Three factors such as economic growth, lower fossil fuel prices, and weaker energy efficiencies contribute largely to carbon emissions and carbon footprints as these contributed to global energy demand by 2.11% in 2017 (Garlapati et al. 2019). Countries such as the USA, the UK, Mexico, and Japan have shown a declination in carbon emissions. Switching over to renewables and coal-to-gas can potentially contribute to reducing emissions.

Renewables and nuclear power contributed around 17% and 20% in electricity generation, respectively (IEA 2018). Biofuels gained significant attention as a potential alternative to fossil fuels in addition to generate employment and protection of ecosystems and biodiversity and de-carbonizing the economy (Khanna et al. 2011; Chakrabarty et al. 2013).

Major biofuel-producing countries such as the USA, Brazil, and European countries are using food/feed crops for fuels. To have marked effects on carbon footprints, holistic use of crops must be done for fuel production. The utilization of lignocellulosic residues into fuels and chemicals can significantly reduce the net carbon footprints (Chandel et al. 2018). Bioenergy is carbon neutral as the CO_2 released from the combustion of biofuels has a global warming potential (GWP) of zero, as proved in several life cycle assessment (LCA) analyses.

Biogas from anaerobic digestion of residues/by-products of waste, crops, bioethanol, and biodiesel has also had substantial advantages like other biofuels in terms of resource availability and sustainability issues. Further, technological and business innovations aiming to cost-competitive ethanol and other biofuels are necessary to achieve the sustainability goals and fix the climate change issues (Börjesson and Mattiasson 2008; Chandel et al. 2019).

5.1 *Carbon Footprint of Different Biofuels*

The carbon footprint categorically analyzes the entire process for carbon emissions (Weber and Matthews 2008). The total amount of CO_2 and CH_4 emissions from different biofuels, system or activity, sinks, and storage within the spatial and temporal boundary of the population are measured by LCA analysis (Wright et al. 2011). Biofuels have potential to offer positive environmental benefits compared to fossil fuels. Ecological foot printing of biofuels can significantly support regional decision-making processes (Stoeglehner and Narodoslawsky 2009). Productive agricultural land use has the highest amount of the footprint, which follows the footprints of water and transport components. The global biofuel footprint was found to be enhanced from 0.248 billion (bn) global hectares (2010) to 0.449 bnghain (2019) (Hammond and Seth 2013).

5.1.1 Biogas

The stats from Scandinavian countries have shown carbon footprints of 7–108 CO_2e P.E.$^{-1}$ year^{-1} from the data obtained from 16 municipal WWT plants and found that the direct emission of nitrous oxides was the significant contributors for the specified ranges of carbon footprints (Gustavsson and Tumlin 2013). In another study, Szabó et al. (2014) reported a carbon footprint of 208,173 kg CO_2 equivalents (CO_2e) from a biogas power plant of Hungary which produces a power output of 0.637 MW in 2013 with a production capacity of 4347.21 MWh and 4607.89 MWh of electric and thermal energies, respectively. In Thailand, biogas production plants from starch demonstrated reduced carbon footprints of the Thai cassava starch industry from 0.9 to 1.0 million tons CO_2eq/year with a total carbon footprint of 609–966 kg CO_2eq/FU (Hansupalak et al. 2016).

5.1.2 Ethanol, Biodiesel, and Others

The carbon footprints of biofuels (ethanol, biodiesel, and others) vary on several factors such as feedstock, feedstock cultivation, production methods, and combustion. The USA and Brazil are the largest producers of biofuel with a contribution of >70% of total world bioethanol production from corn and sugar cane. In contrast, Europe's worldwide contribution to bioethanol production seems to be ~6% (Licht 2006). The GHG emissions from ethanol range from 0.7 to 1.5 kg CO_2eq per kg ethanol. Muñoz et al. (2014) reported that fossil-based ethanol generates GHG emissions of 1.3 kg CO_2eq per kg from cradle-to-gate.

Global warming potential (GWP) is another important indicator of the carbon footprint from biofuels. Total amounts of global warming potential (GWP) were 2740, 1791, and 1910 kg CO_2e $ha^{-1}y^{-1}$ for biofuel sugarcane, energy cane, and sweet sorghum, respectively. However, after applying tillage practices, the GHG emissions were reduced by 13%, 23%, and 8% for biofuel from sugarcane, energy cane, and sweet sorghum, respectively (Izursa et al. 2013).

Mekonnen et al. (2018) compared the carbon footprints of bioethanol industries of the US (corn-based) and Brazilian (sugarcane-based) counterparts in terms of energy balance and impinging carbon footprint. The study reported the superior results with the Brazilian bioethanol industries (17.7 MJ/L energy balance and 38.5 g CO_2e/MJ carbon footprint) rather than the US sugarcane bioethanol industries (11.2 MJ/L energy balance and 44.9 g CO_2e/MJ). With regard to the Brazilian diesel industry, a normal total transportation fleet runs around 60,000 km/year with a usage of 5 km/l (5000 km/m^3) by the emission of 87.95 Mton CO_2/5 years. The negative impacts with the high carbon emissions of diesel combustion can be significantly reduced by the usage of biodiesel blending in higher proportions (Coronado et al. 2009).

6 Conclusions

Fossil fuel-derived GHGs (CO_2, CH_4, N_2O) have high "global warming potential" (GWP). Life cycle analysis (LCA) evaluates the environmental indicators during the production of biofuels and the impact of the burning of fuels on climate change and GHG emissions. LCA is the key to observe sustainability biofuels. However, LCA of particular biofuel from a particular region cannot be implied for other geographical regions. The carbon footprints of biofuels depend on several factors such as feedstock cultivation, feedstock preparation, production methods, and combustion. Among biofuels, bioethanol and biodiesel have shown a great impact on the environment in Brazil and the USA, respectively. Biogas is a sustainable renewable energy which has higher acidification and eutrophication than fossil fuels but has a positive impact of reduced GHG emissions over the fossil fuels with the usage of locally available resources of residues/by-products of waste, crop, and other biofuel industries. Biogas is a promising alternative for fossil fuels and cogeneration of heat and power.

In conclusion, renewable fuels/energy has a positive impact on the environment in terms of lower GHG emissions, eventually improving the environment. However, large-scale production of biofuels to meet the vast demand of society is still a grand challenge. There is a lot of scope in process improvements for the cost-effective production of renewable fuels/energy. The biofuels industry growth is possible with the involvement of biofuels process developers, biofuels industries, stakeholders and governments.

Acknowledgments The authors thank FAPESP, CNPq, and CAPES for the financial assistance. AKC is grateful to the CAPES-Brazil for the financial support through visiting professor and researcher program (Process USP number: 15.1.1118.1.0).

References

Acar C, Dincer I (2014) Comparative assessment of hydrogen production methods from renewable and non-renewable sources. Int J Hydrog Energy 39(1):1–12

Aditiya HB, Mahlia TMI, Chong WT, Nur H, Sebayang AH (2016) Second generation bioethanol production: a critical review. Renew Sust Energ Rev 66:631–653

Alaswad A, Dassisti M, Prescott T, Olabi AG (2015) Technologies and developments of third generation biofuel production. Renew Sust Energ Rev 51:1446–1460

Angelidaki I, Ellegaard L, Ahring BK (2003) Applications of the anaerobic digestion process. In: Ahring BK et al (eds) Biomethanation II, Advances in biochemical engineering/biotechnology. Springer, Berlin, Heidelberg, pp 1–33

ASTM (1995) Annual book of ASTM standards: 1995. Metals Test Methods and Naytical Procedures, ASTM

Azhar SHM, Abdulla R, Jambo SA, Marbawi H, Gansau JA, Faik AAM, Rodrigues KF (2017) Yeasts in sustainable bioethanol production: a review. Biochemistry and Biophysics Reports 10:52–61

Balat M, Balat H (2009) Recent trends in global production and utilization of bio-ethanol fuel. Appl Energy 86(11):2273–2282

Banerjee R, Kumar SPJ, Mehendale N, Sevda S, Garlapati VK (2019) Intervention of microfluidics in biofuel and bioenergy sectors: technological considerations and future prospects. Renew Sust Energ Rev 101:548–558

Bhatia L, Garlapati VK, Chandel AK (2019) Scalable Technologies for Lignocellulosic Biomass Processing into cellulosic ethanol. In: Pogaku R (ed) Horizons in bioprocess engineering. Springer Nature, Switzerland AG, pp 73–90. https://doi.org/10.1007/978-3-030-29069-6_1

Börjesson P, Mattiasson B (2008) Biogas as a resource-efficient vehicle fuel. Trends Biotechnol 26 (1):7–13

Braun R, Weiland P, Wellinger A (2008) Biogas from energy crop digestion. In IEA bioenergy task (Vol. 37, pp. 1–20)

Chakrabarty S, Boksh FM, Chakraborty A (2013) Economic viability of biogas and green self-employment opportunities. Renew Sust Energ Rev 28:757–766

Chandel AK, Albarelli JQ, dos Santos DT, Chundawat SPS, Puri M, Meireles MAA (2019) Comparative analysis of key Technologies for Cellulosic Ethanol Production from Brazilian sugarcane bagasse at the commercial-scale. Biofuels Bioprod Biorefin 13(4):994–1014

Chandel A, Silveira MHL (2017) Advances in sugarcane biorefinery: technologies, commercialization, policy issues and paradigm shift for bioethanol and by-products. Elsevier Press, Amsterdam, The Netherland

Chandel AK, Garlapati VK, Singh AK, Antunes FAF, Silva SS (2018) The path forward for lignocellulose biorefineries: bottlenecks, solutions, and perspective on commercialization. Bioresour Technol 264:370–381

Chinnici G, Selvaggi R, D'Amico M, Pecorino B (2018) Assessment of the potential energy supply and biomethane from the anaerobic digestion of agro-food feedstocks in Sicily. Renew Sust Energ Rev 82:6–13

Coelho ST, Paletta CEM, Freitas MAV (2000) Medidas mitigadoras para a redução de emissões de gases de efeito estufa na geração termelétrica. In M. A. Freitas (Ed). Brasília: Dupligrafica Ed./ ANEEEL/CENBIO/BUN/MCT, PNUD

Coronado CR, de Carvalho Jr JA, Silveira JL (2009) Biodiesel CO2 emissions: a comparison with the main fuels in the Brazilian market. Fuel Process Technol 90(2):204–211
De Baere L, Mattheeuws B (2008) State of the art anaerobic digestion of solid waste. Waste management world 9(4):1–8
Dincer I, Acar C (2015) Review and evaluation of hydrogen production methods for better sustainability. Int J Hydrog Energy 40(34):11094–11111
Eurostat (2005) Europe in figures: Eurostat yearbook (Vol. 1). Office for Official Publications
Fantozzi F, Buratti C (2009) Biogas production from different substrates in an experimental continuously stirred tank reactor anaerobic digester. Bioresour Technol 100(23):5783–5789
Fehrenbach H, Giegrich J, Reinhardt G, Sayer U, Gretz M, Lanje K, Schmitz J (2008) Kriterien einer nachhaltigen Bioenergienutzung im globalen Maßstab. UBA-Forschungsbericht 206:41–112
Gahleitner G (2013) Hydrogen from renewable electricity: an international review of power-to-gas pilot plants for stationary applications. Int J Hydrog Energy 38(5):2039–2061
Garlapati VK, Gour RS, Sharma V, Roy LS, Samudrala PJK, Thakur AK, Banerjee R (2017) Current status of biodiesel production from microalgae in India. In: Advances in Biofeedstocks and biofuels. Production Technologies for Biofuels. Wiley–Scrivener Publishing House, Austin, USA, pp 129–154
Garlapati VK, Kant R, Kumari A, Mahapatra P, Das P, Banerjee R (2013) Lipase mediated transesterification of *Simarouba glauca* oil: a new feedstock for biodiesel production. Sustainable chemical processes 1:11
Garlapati VK, Tewari S, Ganguly R (2019) LCA of first-, second- generation, and microalgae biofuels. In: Advances in feedstock conversion Technologies for Alternative Fuels and Bioproducts. Woodhead Publishing, USA, Elsevier, Imprint, pp 355–371
Garlapati VK, Roy LS, Banerjee R (2015) An overview of reactor designs for biodiesel production. In: Bioenergy: opportunities and challenges. CRC Press, Taylor and Francis Group, USA, pp 221–240
Gebrezgabher SA, Meuwissen MP, Prins BA, Lansink AGO (2010) Economic analysis of anaerobic digestion - a case of Green power biogas plant in the Netherlands. NJAS-Wageningen Journal of Life Sciences 57(2):109–115
Gemmeke B, Rieger C, Weiland P, Schröder J (2009) Biogas-Messprogramm II, 61 Biogasanlagen im Vergleich. Fachagentur Nachwachsende Rohstoffe, Guelzow-Pruezen
Ghazali WNMW, Mamat R, Masjuki HH, Najafi G (2015) Effects of biodiesel from different feedstocks on engine performance and emissions: a review. Renew Sust Energ Rev 51:585–602
González MOA, Gonçalves JS, Vasconcelos RM (2017) Sustainable development: case study in the implementation of renewable energy in Brazil. J Clean Prod 142:461–475
Green DW, Perry RH (1997) Perry's Chemical Engineers' Handbook/edición Don W. Green y Robert H. Perry (No. C 660.28 P47 2008.)
Gupta RB (2008) Hydrogen fuel: production, transport, and storage. Crc Press
Gustavsson DJI, Tumlin S (2013) Carbon footprints of Scandinavian wastewater treatment plants. Water Sci Technol 68(4):887–893
Hammond GP, Seth SM (2013) Carbon and environmental footprinting of global biofuel production. Appl Energy 112:547–559
Hansupalak N, Piromkraipak P, Tamthirat P, Manitsorasak A, Sriroth K, Tran T (2016) Biogas reduces the carbon footprint of cassava starch: a comparative assessment with fuel oil. J Clean Prod 134:539–546
Hasan MM, Rahman MM (2017) Performance and emission characteristics of biodiesel–diesel blend and environmental and economic impacts of biodiesel production: a review. Renew Sust Energ Rev 74:938–948
Hoekman SK, Broch A, Robbins C, Ceniceros E, Natarajan M (2012) Review of biodiesel composition, properties, and specifications. Renew Sust Energ Rev 16(1):143–169
Hosseini SE, Wahid MA (2013) Feasibility study of biogas production and utilization as a source of renewable energy in Malaysia. Renew Sust Energ Rev 19:454–462

International Energy Agency (2018) Global Energy & CO2 Status Report 2017. https://www.iea.org/publications/freepublications/publication/GECO2017.pdf

IRENA (2017) Biogas for road vehicles: technology brief

Izursa JL, Hanlon E, Amponsah N, Capece J (2013) Global warming potential and eutrophication potential of biofuel feedstock crops produced in Florida. Measured Under Different Scenarios. U.S. Department of energy office of scientific and technical information, Hendry County, Florida

Jambo SA, Abdulla R, Azhar SHM, Marbawi H, Gansau JA, Ravindra P (2016) A review on third generation bioethanol feedstock. Renew Sust Energ Rev 65:756–769

Jha D, Jain V, Sharma B, Garlapati VK (2017) Microalgae-based pharmaceuticals and Nutraceuticals: an emerging field with immense market potential. ChemBioEng Reviews 4 (4):257–272

Kárászová M, Sedláková Z, Izák P (2015) Gas permeation processes in biogas upgrading: a short review. Chem Pap 69(10):1277–1283

Khan IU, Othman MHD, Hashim H, Matsuura T, Ismail AF, Rezaei-DashtArzhandi M, Azelee IW (2017) Biogas as a renewable energy fuel - a review of biogas upgrading, utilisation and storage. Energy Convers Manag 150:277–294

Khan MA, Ngo HH, Guo W, Liu Y, Zhang X, Guo J, Chang SW, Nguyen DD, Wang J (2018) Biohydrogen production from anaerobic digestion and its potential as renewable energy. Renew Energy 129:754–768

Khanna M, Crago CL, Black M (2011) Can biofuels be a solution to climate change? The implications of land use change-related emissions for policy. Interface Focus 1(2):233–247

Khanna N, Das D (2013) Biohydrogen production by dark fermentation. WIREs Ener Environ 2:401–421

Kim D, Lee K, Park KY (2015) Enhancement of biogas production from anaerobic digestion of waste activated sludge by hydrothermal pre-treatment. Int Biodeterior Biodegradation 101:42–46

Knothe G (2008) "Designer" biodiesel: optimizing fatty ester composition to improve fuel properties. Energy Fuel 22(2):1358–1364

Knothe G, Razon LF (2017) Biodiesel fuels. Prog Energy Combust Sci 58:36–59

Knothe G, Steidley KR (2005) Kinematic viscosity of biodiesel fuel components and related compounds. Influence of compound structure and comparison to petrodiesel fuel components. Fuel 84(9):1059–1065

Knothe G, Steidley KR (2007) Kinematic viscosity of biodiesel components (fatty acid alkyl esters) and related compounds at low temperatures. Fuel 86(16):2560–2567

Kuila A, Sharma V, Garlapati VK, Singh A, Roy LS, Banerjee R (2017) Present status on enzymatic hydrolysis of lignocellulosic biomass for bioethanol production. In: Advances in Biofeedstocks and biofuels. Volume 1: Biofeedstocks and their processing. Wiley – Scrivener Publishing LLC, Austin, USA, pp 85–96

Kumar SPJ, Garlapati VK, Dash A, Banerjee R, Scholz P (2017) Sustainable Green solvents and techniques for lipid extraction from microalgae: a review. Algal Research: Biomass, Biofuels and Bioproducts 21:138–147

Kumari A, Mahapatra P, Garlapati VK, Banerjee R (2009) Enzymatic transesterification of Jatropha oil. Biotechnol Biofuels 2:1

Lee D-H (2016) Cost-benefit analysis, LCOE and evaluation of financial feasibility of full commercialization of biohydrogen. Int J Hydrog Energy 41(7):4347–4357

Licht FO (2006) World ethanol markets: the outlook to 2015

Mahmudul HM, Hagos FY, Mamat R, Adam AA, Ishak WFW, Alenezi R (2017) Production, characterization and performance of biodiesel as an alternative fuel in diesel engines - a review. Renew Sust Energ Rev 72:497–509

Mazzucchi OAJ (1980) Biodigestor rural. São Paulo, CESP 29

Meher LC, Sagar DV, Naik SN (2006) Technical aspects of biodiesel production by transesterification - a review. Renew Sust Energ Rev 10(3):248–268

Mekonnen MM, Romanelli TL, Ray C, Hoekstra AY, Liska AJ, Neale CM (2018) Water, energy, and carbon footprints of bioethanol from the US and Brazil. Environ Sci Technol 52 (24):14508–14518

Montpart N, Ribot-Llobet E, Garlapati VK, Rago L, Baeza JA, Guisasola A (2014) Methanol opportunities for electricity and hydrogen production in bioelectrochemical systems. Int J Hydrog Energy 39(2):770–777

Muñoz I, Flury K, Jungbluth N, Rigarlsford G, Canals LM, King H (2014) Life cycle assessment of bio-based ethanol produced from different agricultural feedstocks. Int J Life Cycle Assess 19 (1):109–119

Parchen CAP (1979) Manual de biogás. Acarpa-Emater

Paulauskiene T, Bucas M, Laukinaite A (2019) Alternative fuels for marine applications: biomethanol-biodiesel-diesel blends. Fuel 248:161–167

Pauss A, Naveau H, Nyns EJ (1987) In D. O. Hall, R. P. Overend (eds.): Biomass: Regenerable Energy (pp. 273–291). Chichester

Petersson A, Wellinger A (2009) Biogas upgrading technologies–developments and innovations. IEA bioenergy 20:1–19

Rahman SNA, Masdar MS, Rosli MI, Majlan EH, Husaini T, Kamarudin SK, Daud WRW (2016) Overview biohydrogen technologies and application in fuel cell technology. Renew Sust Energ Rev 66:137–162

Roy LS, Garlapati VK, Banerjee R (2015) Challenges in harnessing the potential of Lignocellulosic biofuels and the probable combating strategies. In: Bioenergy: opportunities and challenges. CRC Press, Taylor and Francis Group, USA, pp 171–204

Santos RB, Lima AKDC (2017) Análise comparativa do biogás: processo em biodigestores e de aterro sanitário. RevistaEletrônica de Energia 6(1)

Scarlat N, Dallemand JF, Monforti-Ferrario F, Banja M, Motola V (2015) Renewable energy policy framework and bioenergy contribution in the European Union–an overview from National Renewable Energy Action Plans and Progress reports. Renew Sust Energ Rev 51:969–985

Sevda S, Garlapati VK, Sharma S, Bhattacharya S, Sreekrishnan TR (2019) Microalgae at niches of bioelectrochemical systems: a new platform for sustainable energy production coupled industrial effluents. Bioresource Technology Reports 7C:100290

Sharma M, Kaushik A (2017) Biohydrogen economy: challenges and prospects for commercialization. In: Biohydrogen production: sustainability of current technology and future perspective. Springer, New Delhi, pp 253–267

Silva SS, Chandel AK (2014) Biofuels in Brazil: fundamental aspects, recent developments, and future perspectives. Springer Verlag, Heidelberg, Germany

Stoeglehner G, Narodoslawsky M (2009) How sustainable are biofuels? Answers and further questions arising from an ecological footprint perspective. Bioresour Technol 100 (16):3825–3830

Szabó G, Fazekas I, Szabó S, Szabó G, Buday T, Paládi M, Krisztián K, Kerényi A (2014) The carbon footprint of a biogas power plant. Environ Eng Manag J 13(11):2867–2874

Thangavelu SK, Ahmed AS, Ani FN (2016) Review on bioethanol as alternative fuel for spark ignition engines. Renew Sust Energ Rev 56:820–835

Torres-Jimenez E, Svoljšak-Jerman M, Gregorc A, Lisec I, Dorado MP, Kegl B (2009) Physical and chemical properties of ethanol– biodiesel blends for diesel engines. Energy Fuel 24 (3):2002–2009

Vargaftik, N. B. (1975). Handbook of physical properties of liquids and gases-pure substances and mixtures

Verma P, Sharma MP (2016) Review of process parameters for biodiesel production from different feedstocks. Renew Sust Energ Rev 62:1063–1071

Wang M, Wang Z, Gong X, Guo Z (2014) The intensification technologies to water electrolysis for hydrogen production - a review. Renew Sust Energ Rev 29:573–588

Weber CL, Matthews HS (2008) Quantifying the global and distributional aspects of American household carbon footprint. Ecol Econ 66(2–3):379–391

Wei P, Cheng LH, Zhang L, Xu XH, Chen HL, Gao CJ (2014) A review of membrane technology for bioethanol production. Renew Sust Energ Rev 30:388–400

Weiland P (2010) Biogas production: current state and perspectives. Appl Microbiol Biotechnol 85 (4):849–860

Whiting A, Azapagic A (2014) Life cycle environmental impacts of generating electricity and heat from biogas produced by anaerobic digestion. Energy 70:181–193

Wright LA, Kemp S, Williams I (2011) 'Carbon footprinting': towards a universally accepted definition. Carbon management 2(1):61–72

Wu CZ, Yin XL, Yuan ZH, Zhou ZQ, Zhuang XS (2010) The development of bioenergy technology in China. Energy 35(11):4445–4450

Yousuf A, Khan MR, Pirozzi D, Ab Wahid Z, Atnaw SM (2017) Economic and market value of biogas technology. In: Waste biomass management - a holistic approach. Springer, Cham, pp 137–158

Zachow CR (2000) Biogás. Monografia de Graduação, Ijuí-RS, Brazil

Part II
Improving Biogas Production: Progress, Challenges and Perspectives

Potential Feedstock for Sustainable Biogas Production and its Supply Chain Management

Richa Singh, Meenu Hans, Sachin Kumar, and Yogender Kumar Yadav

Abstract Biogas is a potential alternative to the nonrenewable energy sources due to its renewable and carbon-neutral nature; thus abundant feedstocks lead to convenient generation/management and relatively cleaner and efficient burning, causing reduced greenhouse gas (GHG) emissions. A variety of biodegradable biomass resources, termed as anaerobic digestion (AD) feedstocks, are categorized based on their source of origin, viz., agricultural, forestry, industrial, kitchen, municipality, sewage, aquatic, etc. However, the major issues and challenges in dealing with AD feedstocks are encountered during harvesting, collection, transportation, and storage to the production site. In this chapter, the basics of feedstock types, availability, supply, and logistics for biogas production are discussed along with issues of supply chain management analysis.

Richa Singh and Meenu Hans are equal contributors as first author.

R. Singh
Biochemical Conversion Division, Sardar Swaran Singh National Institute of Bio-Energy, Kapurthala, India

I. K. Gujral Punjab Technical University, Kapurthala, India

M. Hans
Biochemical Conversion Division, Sardar Swaran Singh National Institute of Bio-Energy, Kapurthala, India

São Carlos Institute of Physics, University of São Paulo, São Paulo, Brazil

Department of Microbiology, Guru Nanak Dev University, Amritsar, Punjab, India

S. Kumar (✉)
Biochemical Conversion Division, Sardar Swaran Singh National Institute of Bio-Energy, Kapurthala, India
e-mail: sachin.kumar20@gov.in

Y. K. Yadav
Department of Renewable and Bio-Energy Engineering, College of Agriculture Engineering & Technology, CCS Haryana Agriculture University, Hisar, India

N. Balagurusamy, A. K. Chandel (eds.), *Biogas Production*,
https://doi.org/10.1007/978-3-030-58827-4_8

Keywords Anaerobic digestion · Biogas · AD feedstocks · Feedstock logistics · Supply chain

1 Introduction

Continuous population expansion, industrial development, and adverse environmental impacts associated with constant long-term use of conventional nonrenewable energy resources have raised the demand for alternative energy sources to meet the interminable requirements for livelihood (Nazrin et al. 2016). This has encouraged the research activities toward the development of renewable source-based technologies with low environmental impact. Presently, biofuels have emerged as cleaner and renewable sources of energy for the upliftment of human living standards worldwide. Likewise, it has been expected that bioenergy will contribute about 30% of the global energy requirements by 2050 (Guo et al. 2015). Correspondingly, biogas is one of the most convenient, traditional, and efficient energy carrier produced in nature via anaerobic digestion (AD). It has gained attention since the last few decades around the major parts of the world to meet the demands for heat, electricity, and transportation fuel while reducing the environmental impact.

Considering its vast potential as sustainable energy source, various countries have invested on biogas production in the last few years, for instance. Germany, being the pioneer global biogas producer with 25% installation capacity, has installed more than 8000 biogas plants based on agricultural feedstocks, whereas USA, China, and India have invested on biogas production based on cellulosic feedstocks (Achinas et al. 2017). Benefits of utilizing biogas for heating, cooking, and electricity generation and as compressed natural gas are not only due to its environmentally sound features (carbon neutral, reduced emissions of NOx and SOx) but also due to the immense availability of its feedstocks (Hans and Kumar 2019).

AD is one of the excellent choices for sustainable management of enormously available lignocellulosic biomass (LCB) by converting its organic matter into biogas (main product) and digestate as biofertilizer (by-product) (Mussoline et al. 2013). Various biomass sources such as agriculture sector, industrial sector, and domestic/municipality could be used as AD feedstocks provided that they contain a substantial amount of organic matter. However, certain other characteristics of biomass such as inherent dispersion and highly variable and unstable existence make its utilization a bit challenging (Roni et al. 2016). Technologies need to be developed that could provide a sustainable, secure, and affordable supply of quality feedstocks for biogas production at an industrial scale. Consistent supply of feedstocks to achieve nationwide goals based on cost, quality, quantity, availability, and accessibility at any given time could determine the maximum amount of biogas generation (Al Seadi et al. 2013).

AD feedstock supply chain or logistics system includes a variety of operations, viz., harvesting, collection, preprocessing, handling, transportation, and storage. The logistics system acquires a cost for every operation included while influencing the quality of feedstock. This chapter sums up the improvisation taken up for each operation of the logistics system and aims to minimize the cost, improve the quality, and enhance the access of feedstocks to regulate the biogas plants sustainably and efficiently.

2 Anaerobic Digestion (AD) Feedstocks

Biomass consists of biodegradable organic matter, which makes it suitable for biogas production, frequently named as AD or biogas feedstocks. AD feedstocks' composition could derive the potential of biogas production, which can broadly be categorized into two groups, viz., lignocellulosic and non-lignocellulosic feedstocks (IEA 2016). Moreover, lignocellulosic and non-lignocellulosic feedstocks can be obtained from various sources as shown in Fig. 1. There are few major characteristics (may be regional or seasonal) of any feedstock, which make it eligible for AD such as its composition (cellulose, hemicellulose, lignin, protein, fat, sugars, etc.), its proximate analysis (total solids, TS; volatile solids, VS), carbon-to-nitrogen (C/N) ratio, and biodegradability (Steffen et al. 1998).

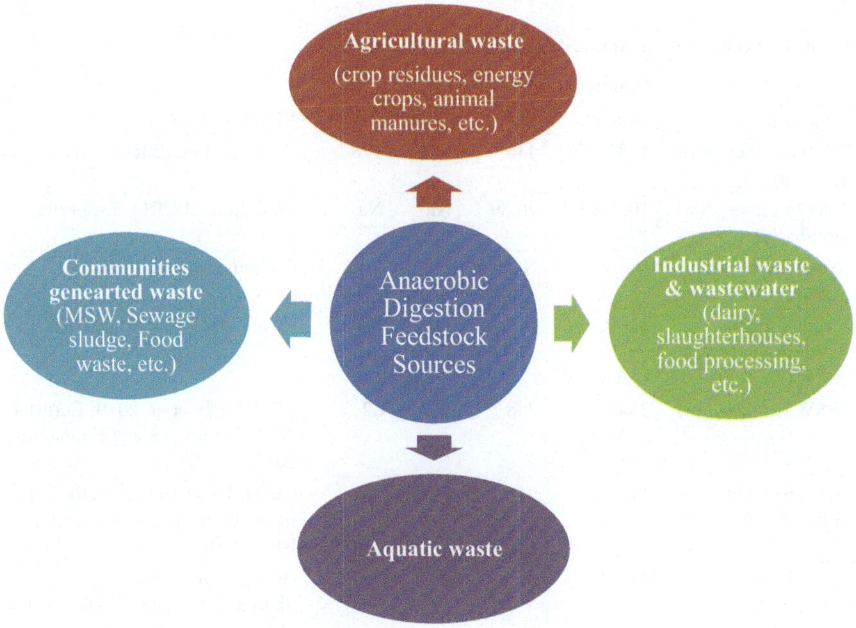

Fig. 1 AD feedstock sources from different waste-generating sectors

2.1 Lignocellulose-Based Feedstocks

Lignocellulosic wastes are promising feedstocks for biogas production, even though the complex and recalcitrant structure (made of strongly linked cellulose, hemicellulose, and lignin) create economic and technical barriers for the operation of AD process. The bioprocess efficiency of lignocelluloses is related to the pretreatment performance, which mainly aims to fasten the process and increases the biogas yield (Patinvoh et al. 2017). Pretreatment can be a simple particle size reduction or complex operation such as thermal, thermochemical, and biological, aiming to remove or redistribute the recalcitrant lignin molecule in order to facilitate accessibility of carbohydrates, fats, and protein moiety of biomass to anaerobic microorganisms (Hernandez and Jimenez 2018). There are several sources of lignocellulosic feedstocks such as animal wastes, agricultural wastes (crop residues, grass clippings, energy crops), forestry residues and wastes from agro-industries, etc. Table 1 represents the availability of the potential lignocellulosic feedstocks.

2.1.1 Crop Residues

Crop residues are the unutilized or leftover agricultural materials in the cultivated lands after harvesting. The benefit of crop residues in comparison to other biomass is that the separate land space is not required because they are a part of edible portion of crops and thus grow altogether. Apart from crop residues, leftovers and garden

Table 1 Availability of AD feedstocks in different continents

	Continents				
Availability	America	Europe	Asia	Africa	References
Crop residues (million ton)	1237	445	1793	178	Bentsen et al. (2014)
Energy crops (ton/ha/yr)	10.9–17.5	26.89	Na	Na	(Wright et al. 2011; Lasorella et al. 2011)
Wood processing waste (million ton)	Na	Na	Na	8.1	Gadonneix et al. (2010)
Animal manure (million ton of wet weight)	1100	Na	Na	Na	Zhang and Schroder (2014)
MSW (million ton)	254	248	280	62	(Gadonneix et al. 2010; Eurostat 2017; Hoornweg and Bhada-Tata 2012)
Sewage sludge (million ton)	17.8	9.0	7.52	1.0	(Asian Development Bank 2012; Indah Water Konsortium Sdn. Bhd 2010)
Algal biomass (million ton)	45–85	50	30	0.13	(Barry et al. 2016; Skarka 2012; FAO 2018; FAO 2013)

Na data not available

Table 2 Composition and methane yield of different lignocellulosic residues

Residue	C:H:L	TS %	VS %	Methane yield (m^3/kg-VS)	References
Silage maize	16:11:38	30.8	94.1	0.259	Hutňan (2016)
Grass silage	31:29:10	50	92	0.344–0.383	(Cadavid Rodríguez and Bolaños Valencia 2016; Sawasdee and Pisutpaisal 2014)
Paddy straw	38:23:13	93	80	0.202	(Singh and Kumar 2019; Dinuccio et al. 2010)
Wheat straw	33:22:19	93.1	76.8	0.282	(Mancini et al. 2018; Yong et al. 2015)
Corn Stover	39:26.6:19	86	94.3	0.296	(Lizasoain et al. 2017; Liew et al. 2012)
Sugarcane bagasse	42:22:18	94	97	0.122–0.236	(Kumari and Das 2015; Janke et al. 2015)
Coffee pulp	31:11:23	55	91	0.131	(Battista et al. 2016; Ulsido and Li 2016)
Pulp and paper sludge	na	24.2	77	0.432	Kamali et al. (2016)
Forestry residues	42: na: 44	50	64	0.214	Teghammar et al. (2014)
Banana stalks (sun-dried)	56:8:18	92	83	0.236	Kalia et al. (2000)
Chicken manure	12:20:2	40	75	0.309[a]	(FNR 2016; Paterson 2015)
Cattle manure	27:12:13	25	76	0.236[a]	(FNR 2016; Paterson 2015)

C:H:L cellulose/hemicellulose/lignin, *TS* total solids, *VS* volatile solids, *na* data not available
Methane estimated as 55% of the reported biogas yield values
[a]Methane estimated as 60% of reported biogas yield values

wastes remaining on the field could also be utilized as feedstocks for biogas plants situated nearby or in the farm land along with utilization of digestate/effluent in the same land for farming. The estimated and projected current theoretical potential of crop residues is 116 and 140–170 EJ/yr., respectively, by 2100 depending on the production circumstances (Daioglou et al. 2016). However, Table 2 represents the reported methane yield along with various crop residues with different compositions.

2.1.2 Energy Crops

Crops that are solely grown for the production of bioenergy purpose not for food, and cultivated only once in the season, could be known as energy crops. They are the herbaceous agricultural wastes, which include plants with little woody or nonwoody tissues like grasses and legumes (miscanthus, sorghum, switchgrass, etc.). Herbaceous energy crops have the tendency of fast growth, which contribute to generate large amount of biomass in short span of time (Al Seadi et al. 2013). These crops are preferably cultivated in the countries with very high energy costs and adequate agricultural land with suitable climatic conditions. Haberl et al. (2010) have shown in their studies that global energy crops have a potential of 81 EJ $yr.^{-1}$ and have huge possibilities to be used as AD feedstock for biogas/bioenergy generation.

2.1.3 Wood Energy Crops/Wood Processing Waste

Wood energy crops are woody in nature but include fast-growing trees like eucalyptus, poplar, etc. The advantages of these types of energy crops are that they require less input during cultivation compared to annual crops. Besides, wood processing waste mainly includes residues obtained from wood-based industries such as sawdust and wood shavings. However, these kind of woody biomass could be utilized as AD feedstocks in the form of uniformly cut pieces (Al Seadi et al. 2013).

2.1.4 Grass Clippings/Garden Waste

Clippings of grasses could be obtained from yard trimmings, lawn and landscape management, etc. Few species of grasses, which consist of high amounts of carbohydrates and fibers, could be utilized in the form of silage as a feedstock for AD (Dussadee et al. 2016). However, the theoretical methane yield of grass clippings varied in the range of 500 m^3/kg-VS to 700 m^3/kg-VS determined by an elemental analysis via the modified Buswell equation (Oldenburg et al. 2011).

2.1.5 Pulp and Paper Industry Waste

These kinds of industrial wastes include woody as well as nonwoody materials which contribute in the production of pulp and paper (P&P) as waste. P&P industries generate huge amounts of solid wastes as well as wastewater. However, in order to be used as AD feedstocks, C/N ratio needs to be maintained due to low nitrogen content in P&P wastes by incorporation of a suitable nitrogen-rich co-substrate (Kamali et al. 2016).

2.1.6 Animal Manure

Animal manure is the ancient AD feedstock; for instance, cattle dung is the traditionally used feedstock for biogas production at a large scale (e.g., gobar gas plant in

India) in most of the countries under ambient atmospheric conditions. Animal manure constitutes dung along with urine which is highly useful as buffering agent during the AD process. Animal wastes can be obtained from any large livestock farming facility. Generally, one livestock unit (LU), taken as aggregation of livestock of different ages and species, is equal to 500 kg of live weight composed of 1 cow, 250 laying hens, or 6 smashed pigs (Steffenet al. 1998). Typically, animal manures are used as co-substrates, which enhance the organic matter as well as the inherent microbial flora along with balancing the C/N ratio, leading to higher biogas yield, and stabilization of two types of wastes simultaneously (Achinas et al. 2017).

2.2 *Non-lignocellulosic Feedstocks*

Non-lignocellulosic feedstocks are defined as biomass, which are composed of mainly proteins, lipids, saccharides, inorganics, etc. Compared to lignocellulosic biomass, non-lignocellulosic biomasses are generated worldwide in huge quantity. AD is one of the key conversion routes to stabilize these types of wastes from an economic and environmental point of view (Li and Jiang 2017). Further detailed discussion on each type of non-lignocellulosic biomass is given in the subsequent sections along with compositional analysis and methane yields in Table 3.

2.2.1 Municipal Solid Waste

MSW is collected by municipality from a variety of sources such as residential, commercial, industrial, construction, etc. and disposed of at the waste disposal site.

Table 3 Composition and methane yields of different non-lignocellulosic feedstocks

Residue	TS %	VS%	Methane yield (m^3/kg-VS)	Reference
Potato effluent	4	90	0.611	FNR (2016); Hung et al. (2006)
Palm oil mill effluent	3.1	86	0.562	Sri Rahayu et al. (2015)
Slaughtering waste	15	80	0.700	Patinvoh et al. (2017)
Vinasse	1	90	0.246	Janke et al. (2015)
Coffee wastewater	2.3–5.5	52–76	na	Syarief et al. 2012
Food waste	20	92	0.260	Yong et al. (2015)
MSW (biodegradable)	20	92	0.386[a]	The World Bank (2018); FNR (2016)
Algal biomass	20	93	0.450[a]	Murphy et al. (2015)

na data not available
Methane estimated as 55% of the reported biogas yield values
[a]Methane estimated as 60% of reported biogas yield values

MSW possesses a high potential for biogas generation via AD (Hilkiah Igoni et al. 2008). Feedstocks such as discarded papers, newspapers, cardboards, food packaging material, and food/fruit/vegetable/kitchen waste are the attractive sources of biodegradable waste. These wastes are generally in either a solid or a semisolid form (Negi et al. 2018). However, all MSW could not be used for AD due to their nonbiodegradable character. Therefore, generally organic fraction of MSW is considered as AD feedstock (OMSW) than entire MSW. Based on about 60 and 40% of organic matter and moisture content, respectively, theoretical biogas yield using MSW as feedstock is estimated as 150 kg/tone (Scarlat et al. 2015). MSW needs to be pre- and post-treated for digestion during the process due to its complex composition (Iacovidou et al. 2013).

2.2.2 Sewage Sludge

Sewage sludge is generated as a by-product from wastewater treatment of municipality or industries in semisolid form of material. This type of waste is high in chemical oxygen demand (COD), which makes it one of the prominent biomass for AD. Typically, sewage sludge is reported to have a low C/N ratio which varies between 6.0 and 16.0 due to the presence of high nitrogen content (Iranzo et al. 2004). It could be used as co-substrate with other biomasses to maintain the C/N ratio of the AD process for uninterrupted progress. Sewage sludge like animal manure is generally used as inoculum (source of microorganisms) in AD (Komatsu and Kudo 2007).

2.2.3 Industrial Wastes and Wastewater

Industrial wastes to be used as AD feedstock include only the organic wastes mostly generated from food processing industries or agro-industries; for example, dry solids or effluents generated from various sources which are rich in proteins and sugars could be potential feedstock for AD. Further, there are industries which generate waste in the form of oils and fats in urban and rural areas having high potential for AD (Aftab et al. 2014).

2.2.4 Aquatic Waste

Algae are the biomass sources which exist naturally in oceans as seaweed or on marginal land as microalgae with significant growth rates. Abundant availability of algae encourages its utilization as AD feedstock for biogas production and paves the way toward third-generation biofuels (biogas here) production. The compositional characterization of algal biomass revealed that they consist of large amount of proteins and lipids, which make them high-potential substrates for AD. Some of algae are found to have doubling time of 24 h and a negligible or low lignin content, which allows its easier digestion than that of lignocellulosic biomass without the need of pretreatment (Montingelli et al. 2015). Furthermore, water hyacinth is one of the most attractive aquatic biomasses for biogas production with a potential of about 12.1 L/kg biomass due to its fast growing, nontoxic features and availability in major parts of the world (Kunatsa and Mufundirwa 2013).

3 Challenges Encountered by Biogas Producers

All over the world, the availability and compositional characteristics of diverse biomass feedstocks have shown the potential for biogas production; however, due to some constraints in handling of these feedstocks from the fields or place of generation to the biogas production site/plant, the whole system could not be considered as sustainable and economical at an industrial level. Cost of the feedstock depends on various factors such as biomass type, location, yield/production, weather, harvesting systems, collection methods, preprocessing, storage, and transportation distance (Roni et al. 2016). However, the foremost challenges are associated with collection and distribution of the feedstocks, which further affect the realizable potential. These constraints involved in collection and distribution are summarized below:

(i) *Constraints of equipment:* Collection of feedstocks is highly different in developed and developing countries. Generally, in developed countries the harvesting and collection processes of feedstocks are integrated, mechanized, and highly efficient but may also require sufficient investment on associated sensitive and expensive equipment (Karlsson et al. 2014).

(ii) *Harvesting methods and practices:* Lignocellulose-based feedstocks such as agricultural wastes (paddy straw, wheat straw, corn stover, etc.) are practiced to be burnt after harvest or ploughed back into the fields in order to improve soil quality or suppress the weed growth. These practices are the main cause of environmental pollution and wastage of a potential feedstock for bioenergy production (Wendt et al. 2018).

(iii) *Physical constraints:* The scattered location of feedstocks affects the harvest and collection, which makes the supply process inefficient and uneconomical. The following factors should be taken into account especially for the logistics of large-scale biogas plants:

a. The low energy density, high volume characteristics, and high moisture content of biomass compared to fossil fuels make the transportation often expensive in terms of cost as well as energy requirements. Whilst, Searcy et al. (2007) reported that their transportation by trucks is limited to the maximum of 100 km, which bound their supply to a limited region. Therefore, preprocessing the biomass into bales, bundles, chips, etc. and moisture reduction are recommended for elevating the energy density and for decreasing the transportation costs for process efficiency (Berglund 2006).

b. Seasonal availability largely affects the supply chain of feedstocks for biogas plants. Likewise, the crop residues and perennial energy crops have relatively short period of harvesting season, but their availability is required all around the year for continuous operation of biogas plants. Therefore, storage of these types of feedstocks is essential at the site along with suitable advanced pretreatment methods for maintaining the physical properties.
c. Collection, segregation, and storage methods of non-lignocellulosic feedstocks such as MSW, sewage sludge, and industrial wastes represent the physical constraints to the biogas producers (Bing et al. 2015). All these types of feedstocks are collected in mixed form, i.e., biodegradable (organic matter) along with nonbiodegradable (metals, plastics, hazardous matter, etc.) material. Therefore, the segregation of biodegradable material from nonbiodegradable portion of waste is an important step, which raises the capital cost.

In order to attain sustainable bioenergy/biogas from AD feedstocks, proper understanding of supply chain along with all limitations of logistics is required. This can leads to deploy large-scale biogas plants and trading.

4 Feedstock Supply Chain/Logistics System

A feedstock supply system, named as supply chain or logistics system, is an organization of different operations required to convert biomass into usable product. In general, the term feedstock logistics system involves harvesting or collection, preprocessing, handling, transportation, and storage of the feedstock to the particular production site (AD plant) to be used as raw material (Fig. 2). The logistics system is categorized by the characteristics of feedstock related to energy content, density, composition, seasonality, and regional characteristics. The elaborated flow chart of the logistics system of AD feedstock is presented in Fig. 3.

4.1 Feedstock Harvesting and Collection

Harvesting and collection operations are the prior steps in any feedstock organizing logistics, which cover all the practices from cutting to removal of feedstock from the production site to collection site (Ren et al. 2015). Harvesting is the act of gathering a

Fig. 2 Major components of a feedstock logistics system

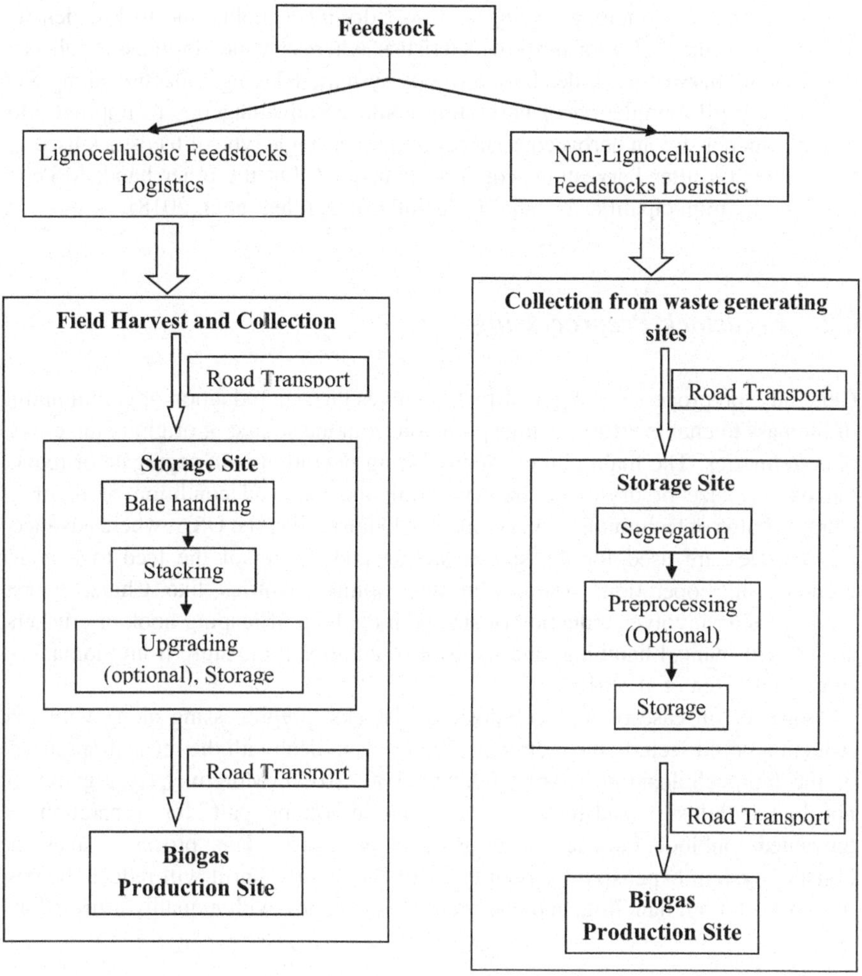

Fig. 3 Elaborated flow chart of feedstock logistics system

reaped (cut) crop from field and is highly dependent on feedstock type such as single-pass harvest system (involves single type of feedstock) or multi-pass harvest system (involves multiple types of feedstocks). For example, lignocellulosic feedstock (corn stover, switchgrass, wheat straw, etc.) harvest system consists of a combine, rake, baler, windrower, and forage chopper, while a woody feedstock harvest system includes a feller buncher, skidder, and chipper (Langeveld and Peterson 2018). However, the harvesting time of certain crops can vary from its usual period due to environmental and crop conditions.

Collection includes shifting of harvested biomass to a field side stack or deck, centralized location. Potential collection equipment includes road-siders, loaders, skidders, and cable systems (Langeveld and Peterson 2018). For an herbaceous biomass, collection includes baling and post drying in fields due to low-density, high-volume, and high-moisture characteristics, whereas woody biomass is collected directly after harvesting. Indeed, for a woody system, delaying collection along with field drying till completion of harvesting could be advantageous to improve efficiency, whereas for an herbaceous supply system, a single type of feedstock used for baled biomass after harvesting would be beneficial. On the other hand, MSW is collected by municipalities or other local authorities (Negi et al. 2018).

4.2 Feedstock Preprocessing

Generally, preprocessing of a feedstock involves the size reduction or comminuting of biomass to enable effortless transportation from the source of origin to the biogas plant/refineries. The main aim of preprocessing operation is to segregate or reduce the particle size or drying of biomass from the received condition of biomass. Current status of technology is developed in countries like the USA, where advanced technologies are used for the size reduction and drying of the feedstock under preprocessing operation, whereas in developing countries like China, typical preprocessing involves reduction of size with the help of reaping hook or crushers, subsequent manual handling, and removal of additional moisture from biomass by sun drying (Ren et al. 2015).

However, in case of lignocellulosic feedstocks, preprocessing along with size reduction could include the pretreatment process to deal with the recalcitrant nature of the lignocellulosic structure (Achinas et al. 2017). Moreover, segregation of non-lignocellulosic feedstocks is also an important part for separation of unwanted/nonbiodegradable waste (Bing et al. 2015). This process makes the logistics system expensive for both types of feedstocks, but it will reduce the cost of transportation, handling, and end product conversion with a quality affirmation.

4.3 Feedstock Handling and Transportation

Transportation of feedstocks to the biogas plant site requires a well-managed transport infrastructure, which includes proper road accessibility (Miao et al. 2012). The unplanned transportation of feedstocks from the harvesting point to the site causes deviation from road accessibility such as due to heavily loaded vehicles. This will bring the negative impact on overall logistics. Moreover, the selection of vehicles for transportation plays an important role and specifically depends on the characteristics of feedstocks. For instance, feedstocks in liquid and solid forms will be transported by different kind of vehicles such as vacuum tankers, dustbin lorries, or trailers, respectively (Searcy et al. 2007). On the other hand, large size vehicles are most preferable for transportation of feedstocks considering the minimum consumption of energy and sustainable economy. However, heavy vehicles can cause major soil devastation of agricultural fields during collection of harvested feedstocks. This soil damage can be avoided by using the recommended low-tier pressure vehicles as possible. Roni et al. (2014) stated that transportation of feedstocks for long distances could be economical, if public transportations such as railways and waterways will be used. This will enhance the transportation capacity and reduce the road congestion usually faced by the biogas refineries.

4.4 Feedstock Storage

Storage of feedstocks is an essential part of the supply chain due to the absence of feedstock availability in all seasons over the year. Feedstock storage provides a continuous supply of required feedstock in the biogas digesters (Al Seadi et al. 2013). The form of feedstock, i.e., liquid, solid, or semisolid, determines the choice of storage facilities such as silos bunker, tanks, etc. Generally, silos bunker are used for solid types of feedstocks, which have a capacity to store for more than a year, whereas tanks are used for liquid types of feedstock (sludge, slurries) which have a capacity to store for a few days.

The storage facility dimensions are measured by the available quantity of feedstock to be delivered and daily amount to be fed into the biogas digesters. The location of the storage facilities can be placed at the biogas plant site itself or decentralized near to the plant (Kenney et al. 2013). Decentralized storage of feedstock has the advantage to facilitate availability of feedstocks for those biogas plants where long-distance transportation is involved.

5 How to Improve AD Feedstock Supply Further?

It has been observed that the present AD feedstock supply chain system needs to incorporate the feedstock location of local as well as remote regions. It will aid in reducing the risk of feedstock supply and enhancing the trading market (Atashbar et al. 2016). Further, advance technologies are also an essential part of the AD feedstock supply system improvement such as equipment for harvesting and collection. There are few strategies, which could help in improving the overall feedstock supply chain system economically as follows.

5.1 Integration of Mechanized/Equipped Harvesting

Harvesting is the very first step for lignocellulose-based feedstocks, which is commonly performed at labor-based harvesting system. According to the case study of Ren et al. (2015), the estimated harvesting investment at labor-based logistics accounts \$11.61/ton and \$9.88/dry metric ton for corn stover and sweet sorghum stalk, respectively. However, when harvesting of these feedstocks is mechanized, the cost decreases to \$3.94/ton and \$1.1/ton, respectively. The above case shows that if the harvesting of the supply chain system is integrated by well-equipped machines, it will definitely reduce the cost due to only one-time investment on the procurement and preprocessing of the feedstock for size reduction.

5.2 *Mixed Feedstock Strategy*

The mixing of different feedstocks is advantageous to achieve the balanced AD parameters, viz., C/N ratio, pH, volatile solids, etc. This type of mixing usually termed as co-digestion in AD process especially, when agricultural wastes like straw, energy crops, grass clippings, forest waste, animal manure, etc. are mixed with each other to balance the operational parameters, gives a way to reduce the access cost and enhance the availability. Roni et al. (2016) have used mixing of various feedstocks altogether, which is termed as co-digestion. They used three feedstocks, namely, switchgrass, wheat straw, and corn stover, and supplied to a biorefinery of 800,000 ton/year capacity. The feedstock access cost of \$45/ton as compared to the \$70/ton cost of corn stover alone. Therefore, mixed feedstock strategy in the above case accounts to an annual saving of \$20 million.

Similarly, MSW, sewage sludge, and food waste are such feedstocks, which have highly degradable organic matter, and if mixed with any lignocellulosic biomass could enhance the biogas production by reducing the lag phase of AD process. Hence, mixing of multiple feedstocks in AD plays a vital role in reducing the access costs (grower payment) and volume of different wastes simultaneously.

5.3 *Incorporation of Geographical Information Systems (GIS) for Site Selection*

The selection of site before establishing a biogas plant is the most essential part of the process. The selected site must include feedstock resource regions, accessible roads near the farms or waste-generating sites, and the distance measurements for transportation (Alam et al. 2012). In this respect, geographic information system (GIS) is quite helpful, providing data in an organized way based on the geographical positions on the earth's surface through a computer system.

The case study reported by Epp et al. (2008) shows that biogas plant site selection should be possibly at short distance because for long distances, transportation will bring uneconomical impact on the logistics. Studies also reported about the economically sustainable average distance of transportation as 5 km for low energy content per volume feedstocks such as slurries or sludge. On the other hand, for energy crops more than 15 km is uneconomical. Moreover, the benefit of location of biogas plants based on agricultural feedstocks near to the animal farms is that the liquid animal slurries can be directly supplied to the plant.

5.4 Inclusion of Government Policies and Support

The governmental policies and support are required for biogas industries at every level from the producers to the end users. If we analyze the supply chain of biogas industries at primary level, the farmers are the producers of LCB, whereas non-lignocellulosic biomass is usually collected by municipality personnel. Therefore, both farmers and municipality personnel should know the management skills of the generated waste at respective levels. Awareness about the upcoming policies of government will be beneficial if available at a common platform (Martin 2015). The governmental support in the form of funds should be essential for the growth of research and development organizations.

6 Future Prospects

Biogas as a cleaner and renewable source of energy has gained attention and has been recognized as a future sustainable energy source. An ideal biogas industry should deal with proper logistics system which involves networking of producers to the end users (Redman 2010). This could be possible only by social awareness to distinguish the waste generated at their house or society as biodegradable and nonbiodegradable. In this direction, governmental support and awareness policies should be implemented. Moreover, the digestate generated from the AD processing of feedstocks could be used as fertilizer, which generates revenue (Plana and Noche 2016). Therefore, considering it's economical and agricultural benefits, maintenance of digestate from storage to the handling and transportation may also be prioritized along with AD feedstocks.

Lastly, the advanced technologies are needed to enhance AD feedstock logistics system more efficiently and sustainably in the future for biogas biorefineries.

7 Conclusions

Biogas is considered as an excellent bioenergy carrier almost all around the world. Along with traditional feedstocks (animal manure), other biomass such as lignocellulosic as well as non-lignocellulosic wastes have been explored to be utilized as potential AD feedstocks. The major components of a feedstock supply chain or logistics system include harvesting or collection, preprocessing, handling, transportation, and storage. Besides large availability and compositional and management benefits of these wastes, a variety of challenges are associated with sustainable feedstock supply chain or logistics system. The top most constraints are linked to collection and distribution of feedstock, which are related to equipment, harvesting methods, practices, etc. AD feedstock supply could be further improved by adapting integration of mechanized/equipped harvesting, mixed feedstock strategy, geographical information systems (GIS) for site selection, government policies and support, and society awareness about categorization and management of wastes and utilizing AD digestate for efficient biomass conversion system to biogas, future energy security, and sustainability.

Acknowledgments Authors gratefully acknowledge the Ministry of New and Renewable Energy, Government of India, for providing funds to carry out the research activities. RS and MH are thankful to Sardar Swaran Singh National Institute of Bio-Energy (SSS-NIBE), Kapurthala, India, for providing Senior Research Fellowship. RS acknowledges I.K. Gujral Punjab Technical University, Kapurthala, India, for her Ph.D. registration (Reg. No. 1422003). MH also acknowledges Guru Nanak Dev University, Amritsar, India, for her Ph.D. registration [Reg. No. 2007.KJ/A.519].

References

Achinas S, Achinas V, Euverink GJW (2017) A technological overview of biogas production from biowaste. Engineering 3:299–307. https://doi.org/10.1016/J.ENG.2017.03.002

Aftab T, Iqbal K, Iqbal J et al (2014) Production of biogas from an agro-industrial waste and its characteristics. J Sci Res 6:347–357. https://doi.org/10.3329/jsr.v6i2.17320

Al Seadi T, Rutz D, Janssen R, Drosg B (2013) Biomass resources for biogas production

Alam B, Pulkki R, Shahi C (2012) Road network optimization model for supplying Woody biomass feedstock for energy production in northwestern Ontario. Open For Sci J:1–14

Asian Development Bank (2012) Promoting beneficial sewage sludge utilization in the people' s republic of China

Atashbar NZ, Labadie N, Prins C (2016) Modeling and optimization of biomass supply chains: a review and a critical look. IFAC-PapersOnLine 49:604–615. https://doi.org/10.1016/j.ifacol.2016.07.742

Barry A, Wolfe A, English C et al (2016) National Algal Biofuels Technology Review. U.S. Department of Energy

Battista F, Fino D, Mancini G (2016) Optimization of biogas production from coffee production waste. Bioresour Technol 200:884–890. https://doi.org/10.1016/j.biortech.2015.11.020

Bentsen NS, Felby C, Jellesmark B (2014) Agricultural residue production and potentials for energy and materials services. Prog Energy Combust Sci 40:59–73. https://doi.org/10.1016/j.pecs.2013.09.003

Berglund M (2006) Biogas production from a systems analytical perspective. Lund university

Bing X, Bloemhof JM, Wong CY et al (2015) Research challenges in municipal solid waste logistics management. Waste Manag 48:584–592. https://doi.org/10.1016/j.wasman.2015.11.025

Cadavid Rodríguez LS, Bolaños Valencia IV (2016) Grass from public green spaces an alternative source of renewable energy in tropical countries. Rev Ion 29:109–116. https://doi.org/10.18273/revion.v29n1-2016009

Daioglou V, Stehfest E, Wicke B et al (2016) Projections of the availability and cost of residues from agriculture and forestry. GCB Bioenergy 8:456–470. https://doi.org/10.1111/gcbb.12285

Dinuccio E, Balsari P, Gioelli F, Menardo S (2010) Bioresource technology evaluation of the biogas productivity potential of some Italian agro-industrial biomasses. Bioresour Technol 101:3780–3783. https://doi.org/10.1016/j.biortech.2009.12.113

Dussadee N, Unpaprom Y, Ramaraj R (2016) Grass silage for biogas production. Adv Silage Prod Util. https://doi.org/10.5772/64961

Epp C, Rutz D, Kottner M, Finsterwalder T (2008) Guidelines for selecting suitable sites for biogas plants. Romania:1–18

Eurostat (2017) Municipal waste statistics. Europe. Source Eurostat (online data code: env_wasmun)

FAO (2013) The global aquaculture production statistics for the year 2011

FAO 2018 (2018) The state of world fisheries and aquaculture

FNR (2016) Leitfaden - Biogas. Gulzow Fachagentur Nachwachsende Rohstoffe

Gadonneix P, de Castro FB, de Medeiros NF, Drouin RD (2010) 2010 survey of energy resources. United Kingdom, London

Guo M, Song W, Buhain J (2015) Bioenergy and biofuels: History, status, and perspective. Renew Sustain Energy Rev 42:712–725. https://doi.org/10.1016/j.rser.2014.10.013

Haberl H, Beringer T, Bhattacharya SC et al (2010) The global technical potential of bio-energy in 2050 considering sustainability constraints. Curr Opin Environ Sustain 2:394–403. https://doi.org/10.1016/j.cosust.2010.10.007

Hans M, Kumar S (2019) Biohythane production in two-stage anaerobic digestion system. Int J Hydrog Energy 44:17363–17380. https://doi.org/10.1016/j.ijhydene.2018.10.022

Hernandez SC, Jimenez LD (2018) The potential for biogas production from agriculture wastes in Mexico. INTECH:15–36. https://doi.org/10.5772/intechopen.7545735

Hilkiah Igoni A, Ayotamuno MJ, Eze CL et al (2008) Designs of anaerobic digesters for producing biogas from municipal solid-waste. Appl Energy 85:430–438. https://doi.org/10.1016/j.apenergy.2007.07.013

Hoornweg D, Bhada-Tata P (2012) What s waste: a global review of solid waste management. Washington, DC: World Bank. Available from: http://hdl.handle.net/10986/17388 (accessed on 18 July 2019)

Hung Y-T, Salman H, Lo H, Awad A (2006) Potato Wastewater Treatment

Hutňan M (2016) Maize silage as substrate for biogas production

Iacovidou E, Vlachopoulou M, Mallapaty S et al (2013) Anaerobic digestion in municipal solid waste management: part of an integrated, holistic and sustainable solution. Waste Manag 33:1035–1036. https://doi.org/10.1016/j.wasman.2013.03.010

IEA (2016) Mobilising sustainable supply chains – biogas cases

Indah Water Konsortium Sdn.Bhd (2010) Sustainability Report 2010

Iranzo M, Cañizares JV, Roca-Perez L et al (2004) Characteristics of rice straw and sewage sludge as composting materials in Valencia (Spain). Bioresour Technol 95:107–112. https://doi.org/10.1016/j.biortech.2004.01.013

Janke L, Leite A, Nikolausz M et al (2015) Biogas production from sugarcane waste: assessment on kinetic challenges for process designing. Int J Mol Sci 16:20685–20703. https://doi.org/10.3390/ijms160920685

Kalia VC, Sonakya V, Raizada N (2000) Anaerobic digestion of banana stem waste. Bioresour Technol 73:191–193. https://doi.org/10.1016/S0960-8524(99)00172-8

Kamali M, Gameiro T, Costa MEV, Capela I (2016) Anaerobic digestion of pulp and paper mill wastes - an overview of the developments and improvement opportunities. Chem Eng J 298:162–182. https://doi.org/10.1016/j.cej.2016.03.119

Karlsson A, Björn A, Yekta SS, Svensson BH (2014) Improvement of the biogas production process

Kenney KL, Smith WA, Gresham GL, Westover TL (2013) Understanding biomass feedstock variability. Biofuels 4:111–127. https://doi.org/10.4155/bfs.12.83

Komatsu T, Kudo K (2007) Anaerobic codigestion of sewage sludge and rice straw. .... Manag Public 16:495–501

Kumari S, Das D (2015) Improvement of gaseous energy recovery from sugarcane bagasse by dark fermentation followed by biomethanation process. Bioresour Technol 194:354–363. https://doi.org/10.1016/j.biortech.2015.07.038

Kunatsa T, Mufundirwa A (2013) Biogas production from water hyacinth case of Lake Chivero-Zimbabwe a review. Int J Recent Technol Eng 2:135–142

Langeveld JWA, Peterson EC (2018) Feedstocks for Biogas Production : Biogas and Electricity Generation Potentials. 122–135. doi:https://doi.org/10.1007/978-3-319-77335-3

Lasorella MV, Monti A, Alexopoulou E et al (2011) Yield comparison between switchgrass and miscanthus based on multi-year side by side comparison in Europe. Conf Proc 19th Eur Biomass Conf Exhib:729–733. https://doi.org/10.2478/s11756-009-0109-4.6

Li DC, Jiang H (2017) The thermochemical conversion of non-lignocellulosic biomass to form biochar: a review on characterizations and mechanism elucidation. Bioresour Technol 246:57–68. https://doi.org/10.1016/j.biortech.2017.07.029

Liew LN, Shi J, Li Y (2012) Methane production from solid-state anaerobic digestion of lignocellulosic biomass. Biomass Bioenergy 46:125–132. https://doi.org/10.1016/j.biombioe.2012.09.014

Lizasoain J, Trulea A, Gittinger J et al (2017) Corn Stover for biogas production: effect of steam explosion pretreatment on the gas yields and on the biodegradation kinetics of the primary structural compounds. Bioresour Technol 244:949–956. https://doi.org/10.1016/j.biortech.2017.08.042

Mancini G, Papirio S, Lens PNL, Esposito G (2018) Increased biogas production from wheat straw by chemical pretreatments. Renew Energy 119:608–614. https://doi.org/10.1016/j.renene.2017.12.045

Martin M (2015) Potential of biogas expansion in Sweden: identifying the gap between potential studies and producer perspectives. Biofuels 6:233–240. https://doi.org/10.1080/17597269.2015.1090769

Miao Z, Shastri Y, Grift TE et al (2012) Lignocellulosic biomass feedstock transportation alternatives, logistics, equipment configurations, and modeling. Biofuels Bioprod Biorefin 6:351–362. https://doi.org/10.1002/bbb

Montingelli ME, Tedesco S, Olabi AG (2015) Biogas production from algal biomass: a review. Renew Sust Energ Rev 43:961–972. https://doi.org/10.1016/j.rser.2014.11.052

Murphy JD, Drosg B, Allen E, et al (2015) A perspective on algal biogas A perspective on algal biogas IEA Bioenergy A perspective on algal biogas

Mussoline W, Esposito G, Giordano A, Lens P (2013) The anaerobic digestion of rice straw: a review. Crit Rev Environ Sci Technol 43:895–915. https://doi.org/10.1080/10643389.2011.627018

Nazrin M, Shiun J, Lip W, Hashim H (2016) Optimisation and targeting of supply-demand of biogas system through gas system cascade analysis (GASCA) framework. J Clean Prod. https://doi.org/10.1016/j.jclepro.2016.06.057

Negi S, Dhar H, Hussain A, Kumar S (2018) Biomethanation potential for co-digestion of municipal solid waste and rice straw: a batch study. Bioresour Technol 254:139–144. https://doi.org/10.1016/j.biortech.2018.01.070

Oldenburg S, Westphal L, Körner I (2011) Energy recovery of grass biomass. WIT Trans Ecol Environ:383–395

Paterson M (2015) Implementation guide for small-scale biogas plants. Farmers handbook:1–122

Patinvoh RJ, Osadolor OA, Chandolias K et al (2017) Innovative pretreatment strategies for biogas production. Bioresour Technol 224:13–24. https://doi.org/10.1016/j.biortech.2016.11.083

Plana PV, Noche B (2016) A review of the current digestate distribution models: storage and transport. Waste Manag Environ VIII 1:345–357. https://doi.org/10.2495/wm160311

Redman G (2010) A detailed economic assessment of anaerobic digestion technology and its suitability to UK farming and waste systems

Ren L, Cafferty K, Roni M et al (2015) Analyzing and comparing biomass feedstock supply systems in China: corn Stover and sweet sorghum case studies. Energies 8:5577–5597. https://doi.org/10.3390/en8065577

Roni MS, Cafferty KG, Hess JR, et al (2016) Lignocellulosic crop supply chains (eg, Miscanthus, switchgrass, reed canary grass, rye, giant reed, etc.). Elsevier Ltd

Roni S, Eksioglu SD, Searcy E, Jacobson JJ (2014) Estimating the variable cost for high-volume and long-haul transportation of densified biomass and biofuel. Transp Res PART D 29:40–55. https://doi.org/10.1016/j.trd.2014.04.003

Sawasdee V, Pisutpaisal N (2014) Feasibility of biogas production from napier grass. Energy Procedia 61:1229–1233. https://doi.org/10.1016/j.egypro.2014.11.1064

Scarlat N, Motola V, Dallemand JF et al (2015) Evaluation of energy potential of municipal solid waste from African urban areas. Renew Sust Energ Rev 50:1269–1286. https://doi.org/10.1016/j.rser.2015.05.067

Searcy E, Flynn P, Ghafoori E, Kumar A (2007) The relative cost of biomass energy transport. Appl Biochem Biotechnol 137–140:639–652. https://doi.org/10.1007/s12010-007-9085-8

Singh R, Kumar S (2019) A review on biomethane potential of paddy straw and diverse prospects to enhance its biodigestibility. J Clean Prod 217:295–307. https://doi.org/10.1016/j.jclepro.2019.01.207

Skarka J (2012) Microalgae biomass potential in Europe

Sri Rahayu A, Karsiwulan D, Yuwono H, et al (2015) Handbook POME to biogas. Project development in Indonesia, Washington, Winrock International. 100

Steffen R, Szolar O, Braun R (1998) Feedstocks for anaerobic digestion. Inst Agrobiotechnology Tulin, Univ Agric Sci Vienna:1–29

Syarief R, Novita E, Noor E, Mulato SRI (2012) Smallholder coffee processing design using wet, vol 7, pp 93–102

Teghammar A, Forgács G, Sárvári Horváth I, Taherzadeh MJ (2014) Techno-economic study of NMMO pretreatment and biogas production from forest residues. Appl Energy 116:125–133. https://doi.org/10.1016/j.apenergy.2013.11.053

The World Bank (2018) Solid Waste Management-Urban Development. Accessed on 12th June, 2018. http://www.worldbank.org/en/topic/urbandevelopment/brief/solid-waste-management

Ulsido MD, Li M (2016) Solid waste management practices in wet coffee processing industries of Gidabo watershed, Ethiopia. Waste Manag Res 34:638–645. https://doi.org/10.1177/0734242X16644519

Wendt LM, Smith WA, Hartley DS et al (2018) Techno-economic assessment of a chopped feedstock logistics supply chain for corn Stover. Front Energy Res 6:1–14. https://doi.org/10.3389/fenrg.2018.00090

Wright LL, Perlack RD, Turhollow AF, Eaton LM (2011) Switchgrass production in the USA. IEA Bioenergy:1–18

Yong Z, Dong Y, Zhang X, Tan T (2015) Anaerobic co-digestion of food waste and straw for biogas production. Renew Energy 78:527–530. https://doi.org/10.1016/j.renene.2015.01.033

Zhang H, Schroder J (2014) Animal manure production and utilization in the US. In: Applied Manure and Nutrient Chemistry for Sustainable Agriculture and Environmen. Springer Science +Business Media Dordrecht 2014, Oklahoma, p 22

Potentials and Challenges of Micro- and Macroalgae as Feedstock for Biogas Production

Adriana López-Villanueva, Sanjuanita María Del Sol Ugalde-Hurtado, Inty Omar Hernández-De Lira, and Miriam Paulina Luevános-Escareño

Abstract Nowadays the use and the search of different sources of energy non-derived from fossil fuels are increasing. Biogas represents the most prominent bioenergy technology worldwide; it can be used as fuel by combustion or for the generation of electrical energy. It is estimated that global biogas production in 2017 was 1.33 EJ. This renewable technology can be produced for a while range of sources, in which micro- and macroalgae represent an attractive substrate because their typical composition possesses high-energy compounds, such as starch as well as an abundant amount of lipids and protein. In addition, micro- and macroalgae are capable of rapidly growing in non-arable land area that are suitable for food production. However, there are challenges that must be faced in order to get the correct performance of an anaerobic digester, such as recalcitrant compounds, low ratio of C/N, and inhibition for the generation of fatty acids or ammonia content. The aim of this chapter is to explain potential and challenges of the use of micro- and macroalgae as feedstock for biogas production.

Keywords Microalgae · Macroalgae · Biogas · Anaerobic digestion · Bioenergy

A. López-Villanueva
Laboratorio de Biorremediación, Facultad de Ciencias Biológicas, Universidad Autónoma de Coahuila, Torreón, Coahuila, Mexico

S. M. D. S. Ugalde-Hurtado · M. P. Luevános-Escareño (✉)
Laboratorio de Bioprocesos, Facultad de Ciencias Biológicas, Universidad Autónoma de Coahuila, Torreón, Coahuila, Mexico
e-mail: miriam_luevanos@uadec.edu.mx

I. O. H.-D. Lira
Laboratorio de Biorremediación, Facultad de Ciencias Biologicas, Universidad Autonoma de Coahuila, Torreón, Coahuila, Mexico

Facultad de Ingeniería Mecánica y Eléctrica, Universidad Autónoma de Coahuila, Torreón, Coahuila, Mexico
e-mail: inty.hernandez@uadec.edu.mx

N. Balagurusamy, A. K. Chandel (eds.), *Biogas Production*,
https://doi.org/10.1007/978-3-030-58827-4_9

1 Introduction

Algae biomasses represent an important potential in fuel production and value-added products. It is estimated that the theoretical yield of biomass from microalgae is about 280 ton ha^{-1} $yr.^{-1}$ (Khan et al. 2018). Algae-derived bioenergy suggest a great opportunity in contributing to world energy security and in mitigating the environmental impacts associated with fossil fuels. The ability of algae to grow in low-quality water or wastewater and the limitation of the use of agriculture land in its production highlight the idea to establish algae cultivation as bioenergetic biomass. Furthermore, the increasing requirements for biomass-based transportation fuels and electricity generation by biomethane through anaerobic digestion (AD) indicate that algae can help to address these challenges. Also, the energy recovery in the form of methane from algal biomass using AD process has become an attractive and feasible approach. Typically, the methane yield of algal biomass has been reported in a ratio from 0.024 to 0.6 L CH_4 g^{-1} VS, depending on the algal species and experimental conditions (Roberts et al. 2016). In addition, Allen et al. (2015) have anticipated that the gross energy from AD of algae is equivalent to 365 GJ ha^{-1} $yr.^{-1}$, when *Saccharina latissima* was cultivated and used as substrate in AD process. This represents a higher amount of energy compared with liquid biofuel systems such as ethanol from sugarcane (135 GJ ha^{-1} $yr.^{-1}$) and biodiesel from palm oil (120 GJ ha^{-1} $yr.^{-1}$) (Milledge et al. 2019). Albeit, despite their clear potential, there are no evidence of commercial-scale amounts of techno-economic and viable-commercial fuels from algae sources. On the other hand, several technical challenges on the AD process of algae have been described previously, such as the low concentration of biodegradable substrate, recalcitrant components, cell wall degradability, low C/N ratio, effects from salinity concentration, and ammonia inhibition. The Low total solid (TS) content in their growth medium ranges from 0.05–0.075% to 0.3–0.4% dry matter (dm) for open pond and closed systems, respectively (Fasaei et al. 2018). Meanwhile, (Santos-Ballardo et al. (2016) indicated that microalgae typically have a C/N ratio of 3–10. Finally, high salt concentration ($\geq$10 $g{\cdot}L^{-1}$) might inhibit AD, because of osmotic pressure and dehydration of microbial community. A recent study found a salt concentration of 15% dm in unwashed *Sargassum muticum* (Milledge and Harvey 2016b). This chapter describes the potential of biogas as energy source from either micro- or macroalgae, the current production technology, current contribution to the global energy needs, challenges of algae as AD substrate, and at the same time, highlighting the areas of improvement in algae biogas production technology.

2 Anaerobic Digestion and Biogas Production from Algae

2.1 Algae Composition

Algae are photosynthetic organisms that grow in a range of aquatic habitats, such as ponds, rivers, lakes, oceans, and wastewater plants. Algae are generally classified as red algae (Rhodophyta), brown algae (Phaeophyta), and green algae (Chlorophyta) (Khan et al. 2018); also algae can be categorized by size into macroalgae usually known as seaweed and microalgae, microscopic single-cell organisms (Milledge and Harvey 2016a). The relevance of algal cellular composition for biofuels production is due to its lipid content 50% dw (Juneja et al. 2013). Another important component is carbohydrates (mainly starch), which in dry weight represent a range from 20% to 40% of total cell mass (Hu 2007). Meanwhile, after the utilization of lipids and starch fractions for biofuel production, the residual crude algae can constitute up to 60% dw of proteins (Becker 2007). The success of biogas production with a high methane yield depends on several factors, such as substrate composition, reactor design, the environmental gradients (temperature), the dry matter content on the feedstock, and the way substrate is fed (Montañez-Hernández et al. 2018). However, the relative composition of carbohydrates, proteins, and lipids present in algal biomasses have a precise impact in methane yield production. Morales-Polo et al. (2018) established that methane yield from lipids is higher than either carbohydrates or protein. Concurrently, the proportion of different components of algae changes over species, cultivation techniques, and environmental gradients, which creates challenges in AD process of algae (Ward et al. 2014). In addition, the microbial community structure involve during AD is vulnerable to the main component of the feedstock (García-Lozano et al. 2019). The main composition in terms of lipids, carbohydrates, and proteins of macroalgae is shown in Table 1 and for microalgae in Table 2. Commonly, microalgae are characterized by higher contents of lipids and proteins but lower carbohydrate content compared to macroalgae. As shown in Table 1, carbohydrate composition in macroalgae depends mostly in their category: red, green, or brown. Starch, cellulose, agarose, and alginates are the main carbohydrates in macroalgae species (Monlau et al. 2014). In contrast, microalgae strains present mainly starch as a source of carbohydrates; they are the main constituent of inner cell wall. The outer cell wall is composed for polysaccharides such as agar, alginate, and pectin (Chen et al. 2013).

2.2 Biomethane Potential of Algal Biomasses

AD is considered the most major bioenergy technology across the world; it is a profitable option that provides a sustainable solution to treat complex organic matter and reduce greenhouse gases emission while producing clean energy, enhancing fertilizer potency, and decreasing contamination. AD is a process usually divided in

Table 1 Macroalgae composition of lipids, proteins, and carbohydrates expressed in % of dry matter (Monlau et al. 2014)

%	Green algae			Red algae			Brown algae			
	Codium fragile	*Enteromorpha linza*	*Ulva lactuca*	*Gelidium amansii*	*Porphyra tenera*	*Gracilaria verrucosa*	*Laminaria japonica*	*Ecklonia stolonifera*	*Saccharina japonica*	*Sargassum fulvellum*
Lipids	1.8	1.8	6.2	0–3.1	4.4	3.2	1.8–2.4	2.4	0.5	0.5
Proteins	10.9	31.6	20.6	16.3	38.7	15.6	9.4–14.8	13.6	19.9	10.6
Carbohydrates	32.3	37.4	54.3	61–67.3	35.9	33.5	51.9–60	48.6	44.5	66

Table 2 Microalgae composition of lipids, proteins, and carbohydrates expressed in % of dry matter (Roy and Pal 2015)

%	*Spirulina platensis*	*Chlorella* sp.	*Scenedesmus* sp.	*Dunaliella* sp.	*Synechococcus* sp.	*Euglena* sp.	*Prymnesium* sp.	*Anabaena* sp.	*Spirulina maxima*
Lipids	4–9	14–22	12–14	6–8	11	14–20	22–38	4–7	6–7
Proteins	50–65	51–28	50–56	49–57	63	39–61	28–45	48	60–71
Carbohydrates	8–14	12–17	10–52	4–32	15	14–18	25–33	22–38	13–16

four steps: hydrolysis, fermentation, acetogenesis, and methanogenesis. In the first phase, complex polymers are hydrolyzed into soluble monomers, which then are converted to volatile fatty acids (VFAs), alcohols, hydrogen (H_2), and CO_2 by fermentative bacteria. In the next step, VFAs with a carbon chain longer than two are transformed to acetic acid, H_2, and CO_2 by acetogenic bacteria. Finally, methanogens produce methane utilizing as substrates the products from acetogenesis (Alvarado et al. 2014). Global biogas generation increased from 0.28 EJ in 2000 to 1.28 EJ in 2014, with a global volume of 59 billion m^3 biogas (Scarlat et al. 2018). The total production of electricity derived from biogas in 2007 represented 87,932 GWh, showing the importance of this technology for energy security production (IRENA 2019). Due to the algae biomass composition, it can be used in AD process in different manners: directly, after cell wall disruption, after lipid extraction, and algae-derived biochar (Fig. 1). The first studies of anaerobic digestion of algae biomass were in the 1950s. The first article in this area evaluated the biomethane potential (BMP) of two microalgae species, *Chlorella* and *Scenedesmus*, obtaining 170–320 ml CH_4 g VS^{-1} (Golueke et al. 1957).

In the recent years, several studies were carried out to determine the BMP from algae substrates (Allen et al. 2015; Gonzalez-Fernandez et al. 2018; Zhu et al. 2018). Furthermore, another important number of strategies have been developed to increase methane yield of algae (Du et al. 2020; He et al. 2016; Keymer et al. 2013). Klassen et al. (2017) prove the feasibility of *Chlamydomonas reinhardtii* as mono-substrate, derived from nitrogen-limited growth conditions. The results demonstrated a methane yield of 462 ml CH_4 g VS^{-1}. Markou et al. (2020) compared the residual biomass of microalgae *Chlorella vulgaris* after the extraction of chlorophyll, protein, and lipid fractions; the residual biomass showed a methane yield in the range of 207–209 ml CH_4 g VS^{-1}. Calicioglu and Demirer (2016) also evaluated the

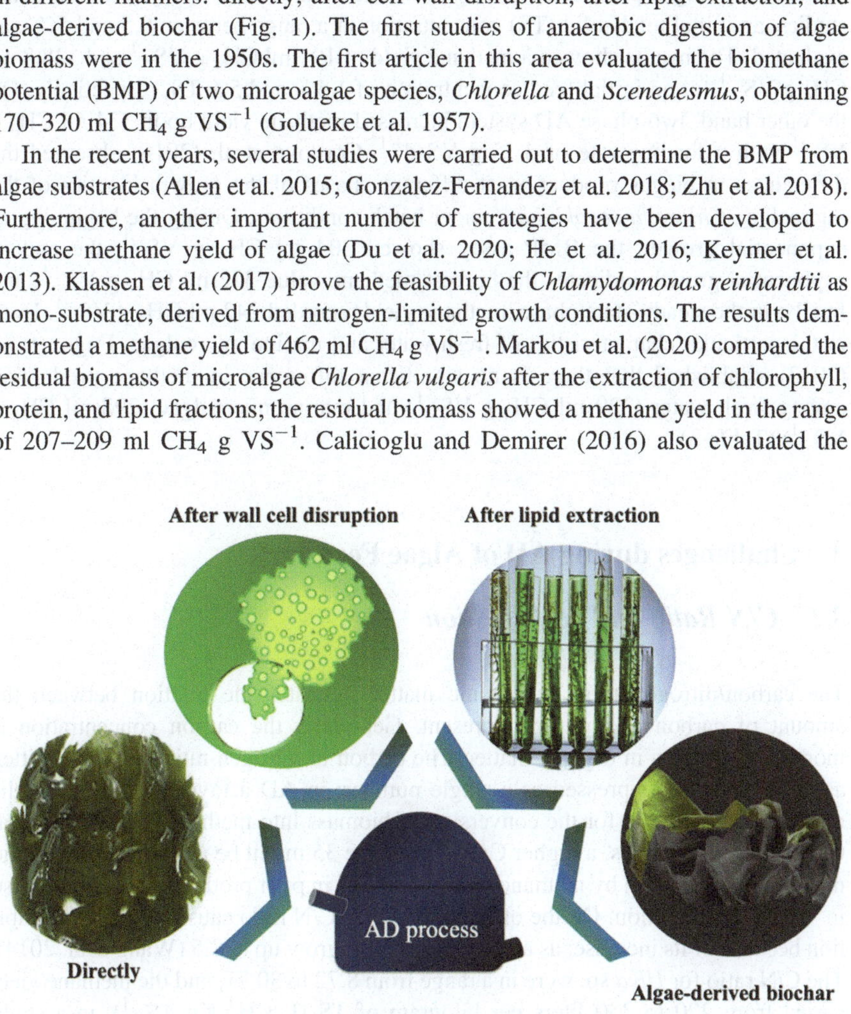

Fig. 1 Ways in which algae can be incorporate into AD scheme

biogas performance of *Chlorella vulgaris*. The microalgal slurry obtained from the effluent of a photobioreactor for municipal wastewater treatment was subjected to BMP assay. Crude algae slurry accounted an amount of methane of 249 ml CH_4 g VS^{-1}. Nonetheless, after application of heat, autoclave, and thermochemical pretreatment, the methane yield improved as levels as 408, 398, and 258 ml CH_4 g VS^{-1} for each pretreatment, respectively. As mentioned previously, microbial structure in the inoculum plays an important role during AD of algae. A previous study demonstrates that ammonia-tolerant inoculum provides better performance in terms of methane yield (415 ml CH_4 g VS^{-1}) of AD process of *Chlorella vulgaris*, which is characterized for its high nitrogen content. (Mahdy et al. 2017). Varol and Ugurlu (2016) tested the utilization of blue-green algae *Spirulina platensis* as substrate in biogas production under batch and semicontinuous systems. In batch studies, an initial total solid (TS) concentration in a range between 0.5 and 5% was evaluated. The results showed a methane yield of 191 ml CH_4 g VS^{-1} and 138.8 ml CH_4 g VS^{-1} for an initial feed concentration of 1.5% and 5% TS, respectively. On the other hand, two-phase AD system increased methane yield upon 334 ml CH_4 g VS^{-1} with a loading rate of 3.75 g VS d^{-1}. Córdova et al. (2018) observed the differences in biogas production at different phases of the growth kinetics of the microalgae *Chlorella sorokiniana* under batch conditions. During the beginning of exponential growth, the BMP assay showed 304 ml CH_4 g VS^{-1}, the end of exponential growth indicates the highest methane value 322 ml CH_4 g VS^{-1}, and finally, in the steady-state phase, methane yield recorded 242 ml CH_4 g VS^{-1}. Even in the particular characteristics of fresh water and marine microalgae, Frigon et al. (2013) established that there is no significant difference in methane production among fresh water (329 ml CH_4 g VS^{-1}) and marine microalgae (298 ml CH_4 g VS^{-1}) (Table 3).

3 Challenges during AD of Algae Feedstock

3.1 C/N Ratio and co-Digestion

The carbon/nitrogen ratio of organic matter indicates the relation between the amount of carbon and nitrogen present. Generally, the carbon concentration is more than nitrogen in organic matter. The carbon to nitrogen ratio which is written as C/N is usually expressed as a single number; in AD a favorable value for the microbes requirement for the conversion of biomass into methane is ~30 (Talavera-Caro et al. 2020). Thus, a higher C/N ratio above 35 might be an indication of rapid nitrogen consumption by methanogens and results in poor production of biogas, also in VFAs accumulation. On the other hand, lower C/N ratio causes ammonia inhibition because of its increase; as a result the pH can grow up to 8.5 (Wang et al. 2014). The C/N ratio for *Ulva* sp. were in a range from 8.72 to 30.71, and the methane yield varied from 220 to 330 liters per kilogram of TS (L CH^4 Kg TS^{-1}), in a study reported by Wise et al. (1979). The C/N ratio for *Saccorhiza polyschides* and

Table 3 Biomethane production from anaerobic digestion of microalgae biomass

Microalgae species	Fresh or marine	Methane yield	Loading rate	Reference
Chlamydomonas reinhardtii	Fresh	462 ml g^{-1} VS	761 g L^{-1} VS	(Klassen et al. 2017)
Chlorella vulgaris	Fresh	361 ml g^{-1} VS	200 g kg^{-1} VS	(Frigon et al. 2013)
Scenedesmus sp.	Fresh	258 ml g^{-1} VS	234 g kg^{-1} VS	(Frigon et al. 2013)
Micractinium sp.	Fresh	360 ml g^{-1} VS	215 g kg^{-1} VS	(Frigon et al. 2013)
Scenedesmus dimorphus	Fresh	397 ml g^{-1} VS	246 g kg^{-1} VS	(Frigon et al. 2013)
Spirulina maxima	Fresh	330 ml g^{-1} VS	22.5 g L^{-1} VS	(Varel et al. 1988)
Chlorella kessleri	Fresh	217 ml g^{-1} VS	2 g L^{-1} TS	(Mussgnug et al. 2010)
Euglena gracilis	Fresh	325 ml g^{-1} VS	2 g L^{-1} TS	(Mussgnug et al. 2010)
Scenedesmus obliquus	Fresh	178 ml g^{-1} VS	2 g L^{-1} TS	(Mussgnug et al. 2010)
Chlorella sorokiniana	Fresh	212 ml g^{-1} VS	–	(Zabed et al. 2019)
Dunaliella sp.	Marine	440 ml g^{-1} VS	0.9 g L^{-1} TS	(Polakovicová et al. 2012)
Palmaria palmata	Marine	279 ml g^{-1} VS	738 g kg^{-1} VS	(Jard et al. 2013)
Saccharina latissimi	Marine	209 ml g^{-1} VS	564 g kg^{-1} VS	(Jard et al. 2013)
Sargassum muticum	Marine	130 ml g^{-1} VS	634 g kg^{-1} VS	(Jard et al. 2013)
Ulva lactuca	Marine	241 ml g^{-1} VS	821 g kg^{-1} VS	(Jard et al. 2013)
Isochrysis sp.	Marine	408 ml g^{-1} VS	305 g kg^{-1} VS	(Frigon et al. 2013)
Thalassiosira weissflogii	Marine	265 ml g^{-1} VS	133 g kg^{-1} VS	(Frigon et al. 2013)
Dunaliella tertiolecta	Marine	286 ml g^{-1} VS	5 g L^{-1} TS	(Lakaniemi et al. 2011)
Nannochloropsis oculata	Marine	204 ml g^{-1} VS	–	(Buxy et al. 2013)

Saccorhiza latissimi is 23.2 and 24, respectively. However, their BMP yield indicate a great difference, 263 ml CH_4 g VS^{-1} for *S. polyschides* and 341,263 ml CH_4 g VS^{-1} for *S. latissimi* (Allen et al. 2015). Anjaneyulu et al. (1989) tested *Sargassum tenerrimum*; the C/N ratio found was 18, which produced levels at 132 L CH^4 Kg TS^{-1}. The ptimization of C/N kinetics by increasing its ratio from 8 to 24 in a brown seaweed (*Fucus serratus*), by solvent extraction, performed a methane yield increasing of 70%. (Tedesco and Daniels 2018). Albeit, no positive correlation between methane yield and methane potential was shown in the anaerobic digestion of *Dunaliella*, *Nostoc*, and *Scenedesmus*, which appears with low C/N ratio (below 12) (Frigon et al. 2013). These results may suggest that the effect of C/N ratio depends on algal species rather than AD conditions.

Anerobic co-digestion is the simultaneous digestion of two of more substrates intendants to increase methane yield and overcome other limitations of mono-digestion process (Xie et al. 2016). Mono-digestion process using algae often faces of ammonium inhibition due to the high nitrogen content in the substrate. Therefore, co-digestion aid to reach the optimal C/N ratio (25–30) thus is a process suitable for the use of algae biomass (Mata-Alvarez et al. 2011). The co-digestion of *Spirulina platensis* with food waste and sludge improved methane yield by 37.5% and 10.3%, respectively, compared to the different materials digested individually (Du et al. 2019). *Chlorella vulgaris* co-digested with potato processing waste increased biogas yields in the range of 22–47% (Zhang et al. 2019). Herrmann et al. (2016) evaluated the co-digestion of *Arthrospira platensis* with carbon-rich co-substrates (barley straw, beet silage, and brown seaweed). Nevertheless, only the co-digestion of *A. platensis* with beet silage in a proportion of 45% and 55%, respectively, resulted in a favorable methane yield of 361 ml CH_4 g VS^{-1}. (Romagnoli et al. (2019) performed an experimental co-digestion to evaluate the effect of wheat straw and straw pellets over Baltic seaweed (*Cladophora* sp. and *Ulva intestinalis*). In terms of BMP through a synergy index (α), important increase has been seen for the co-digestion of finely treated straw pellet in both seaweed (*Cladophora* sp. $\alpha = 1.007$; *Ulva intestinalis* $\alpha = 1.083$).

4 Conclusions

Biogas is one of the most promising biofuels, and algae biomass is an attractive feedstock for energy recovery in the form of methane. The widespread studies' efforts to make algae substrates affordable for AD provide much experience and comprehension of the overall challenges and complexity of many algae species in its role as biofuel feedstock. A great amount of algal species has been tested for biomethane potential using anaerobic digestion process. However, the potential of micro- and macroalgae in methane production could increase its promising technology if the scientific community can integrate wastewater treatment and bioenergy production by algae being cultivated in the waste effluents. Even though this process is not new, several technical improvements might be applied for increasing methane yields, and look forward to the implementation at industrial level. The main improvements to scale up AD process of algae must be related with the development of novel pretreatments and new bioreactor design, increase the low biomass loading during AD, and remove undesirable components and biodegradability of recalcitrant materials. Finally, economical and environmental impacts must be assessed for each scenario in order to determine the true commercial value of biogas production through algae feedstocks.

References

Allen E, Wall DM, Herrmann C, Xia A, Murphy JD (2015) What is the gross energy yield of third generation gaseous biofuel sourced from seaweed? Energy 81:352–360. https://doi.org/10.1016/j.energy.2014.12.048

Alvarado A, Montañez-Hernández LE, Palacio-Molina SL, Oropeza-Navarro R, Luévanos-Escareño MP, Balagurusamy N (2014) Microbial trophic interactions and mcrA gene expression in monitoring of anaerobic digesters. Front Microbiol 5. https://doi.org/10.3389/fmicb.2014.00597

Anjaneyulu K, Tarwadi SJ, Mehta DJ (1989) Anaerobic digestion of seaweed for biogas: a kinetic evaluation. J Chem Technol Biotechnol 45(1):5–14. https://doi.org/10.1002/jctb.280450103

Becker EW (2007) Micro-algae as a source of protein. Biotechnol Adv 25(2):207–210. https://doi.org/10.1016/j.biotechadv.2006.11.002

Buxy S, Diltz R, Pullammanappallil P (2013) Biogasification of marine algae Nannochloropsis Oculata, Materials Challenges in Alternative and Renewable Energy II (pp. 59–67). Wiley. https://doi.org/10.1002/9781118585160.ch6

Calicioglu O, Demirer GN (2016) Biogas production from waste microalgal biomass obtained from nutrient removal of domestic wastewater. Waste and Biomass Valorization 7(6):1397–1408. https://doi.org/10.1007/s12649-016-9546-9

Chen C-Y, Zhao X-Q, Yen H-W, Ho S-H, Cheng C-L, Lee D-J, Bai F-W, Chang J-S (2013) Microalgae-based carbohydrates for biofuel production. Biochem Eng J 78:1–10. https://doi.org/10.1016/j.bej.2013.03.006

Córdova O, Ruiz-Filippi G, Fermoso FG, Chamy R (2018) Influence of growth kinetics of microalgal cultures on biogas production. Renew Energy 122:455–459. https://doi.org/10.1016/j.renene.2018.01.125

Du X, Tao Y, Li H, Liu Y, Feng K (2019) Synergistic methane production from the anaerobic co-digestion of Spirulina platensis with food waste and sewage sludge at high solid concentrations. Renew Energy 142:55–61. https://doi.org/10.1016/j.renene.2019.04.062

Du X, Tao Y, Liu Y, Li H (2020) Stimulating methane production from microalgae by alkaline pretreatment and co-digestion with sludge. Environ Technol 41(12):1546–1553. https://doi.org/10.1080/09593330.2018.1540665

Fasaei F, Bitter JH, Slegers PM, van Boxtel AJB (2018) Techno-economic evaluation of microalgae harvesting and dewatering systems. Algal Res 31:347–362. https://doi.org/10.1016/j.algal.2017.11.038

Frigon J-C, Matteau-Lebrun F, Hamani Abdou R, McGinn PJ, O'Leary SJB, Guiot SR (2013) Screening microalgae strains for their productivity in methane following anaerobic digestion. Appl Energy 108:100–107. https://doi.org/10.1016/j.apenergy.2013.02.051

García-Lozano M, Hernández-De Lira IO, Huber DH, Balagurusamy N (2019) Spatial variations of bacterial communities of an anaerobic lagoon-type biodigester fed with dairy manure. PRO 7 (7):408. https://doi.org/10.3390/pr7070408

Golueke CG, Oswald WJ, Gotaas HB (1957) Anaerobic digestion of algae. Appl Microbiol 5 (1):47–55

Gonzalez-Fernandez C, Barreiro-Vescovo S, de Godos I, Fernandez M, Zouhayr A, Ballesteros M (2018) Biochemical methane potential of microalgae biomass using different microbial inocula. Biotechnol Biofuels 11. https://doi.org/10.1186/s13068-018-1188-7

He S, Fan X, Katukuri NR, Yuan X, Wang F, Guo R-B (2016) Enhanced methane production from microalgal biomass by anaerobic bio-pretreatment. Bioresour Technol 204:145–151. https://doi.org/10.1016/j.biortech.2015.12.073

Herrmann C, Kalita N, Wall D, Xia A, Murphy JD (2016) Optimised biogas production from microalgae through co-digestion with carbon-rich co-substrates. Bioresour Technol 214:328–337. https://doi.org/10.1016/j.biortech.2016.04.119

Hu Q (2007) Environmental effects on cell composition, En *Handbook of Microalgal Culture* (pp. 83–94). Wiley, Ltd. https://doi.org/10.1002/9780470995280.ch5

IRENA (2019) Renewable energy statistics 2019. The International Renewable Energy Agency, Abu Dhabi

Jard G, Marfaing H, Carrère H, Delgenes JP, Steyer JP, Dumas C (2013) French Brittany macroalgae screening: composition and methane potential for potential alternative sources of energy and products. Bioresour Technol 144:492–498. https://doi.org/10.1016/j.biortech.2013.06.114

Juneja A, Ceballos RM, Murthy GS (2013) Effects of environmental factors and nutrient availability on the biochemical composition of algae for biofuels production: a review. Energies 6 (9):4607–4638. https://doi.org/10.3390/en6094607

Keymer P, Ruffell I, Pratt S, Lant P (2013) High pressure thermal hydrolysis as pre-treatment to increase the methane yield during anaerobic digestion of microalgae. Bioresour Technol 131:128–133. https://doi.org/10.1016/j.biortech.2012.12.125

Khan MI, Shin JH, Kim JD (2018) The promising future of microalgae: current status, challenges, and optimization of a sustainable and renewable industry for biofuels, feed, and other products. Microb Cell Factories 17. https://doi.org/10.1186/s12934-018-0879-x

Klassen V, Blifernez-Klassen O, Wibberg D, Winkler A, Kalinowski J, Posten C, Kruse O (2017) Highly efficient methane generation from untreated microalgae biomass. Biotechnol Biofuels 10 (1):186. https://doi.org/10.1186/s13068-017-0871-4

Lakaniemi A-M, Hulatt CJ, Thomas DN, Tuovinen OH, Puhakka JA (2011) Biogenic hydrogen and methane production from Chlorella vulgaris and Dunaliella tertiolectabiomass. Biotechnol Biofuels 4(1):34. https://doi.org/10.1186/1754-6834-4-34

Mahdy A, Fotidis IA, Mancini E, Ballesteros M, González-Fernández C, Angelidaki I (2017) Ammonia tolerant inocula provide a good base for anaerobic digestion of microalgae in third generation biogas process. Bioresour Technol 225:272–278. https://doi.org/10.1016/j.biortech.2016.11.086

Markou G, Ilkiv B, Brulé M, Antonopoulos D, Chakalis L, Arapoglou D, Chatzipavlidis I (2020) Methane production through anaerobic digestion of residual microalgal biomass after the extraction of valuable compounds. Biomass Conversion and Biorefinery. https://doi.org/10.1007/s13399-020-00703-3

Mata-Alvarez J, Dosta J, Macé S, Astals S (2011) Codigestion of solid wastes: a review of its uses and perspectives including modeling. Crit Rev Biotechnol 31(2):99–111. https://doi.org/10.3109/07388551.2010.525496

Milledge JJ, Harvey PJ (2016a) Potential process 'hurdles' in the use of macroalgae as feedstock for biofuel production in the British Isles. J Chem Technol Biotechnol 91(8):2221–2234. https://doi.org/10.1002/jctb.5003

Milledge JJ, Harvey PJ (2016b) Ensilage and anaerobic digestion of Sargassum muticum. J Appl Phycol 28(5):3021–3030. https://doi.org/10.1007/s10811-016-0804-9

Milledge JJ, Nielsen BV, Maneein S, Harvey PJ (2019) A brief review of anaerobic digestion of algae for bioenergy. Energies 12(6):1166. https://doi.org/10.3390/en12061166

Monlau F, Sambusiti C, Barakat A, Quéméneur M, Trably E, Steyer J-P, Carrère H (2014) Do furanic and phenolic compounds of lignocellulosic and algae biomass hydrolyzate inhibit anaerobic mixed cultures? A comprehensive review. Biotechnol Adv 32(5):934–951. https://doi.org/10.1016/j.biotechadv.2014.04.007

Montañez-Hernández LE, Hernández-De Lira IO, Rafael-Galindo G, de Lourdes Froto Madariaga M, Balagurusamy N (2018) Sustainable production of biogas from renewable sources: global overview, scale up opportunities and potential market trends. In: Singh OV, Chandel AK (eds) *Sustainable Biotechnology-Enzymatic Resources of Renewable Energy* (pp. 325–354). Springer International Publishing, Cham. https://doi.org/10.1007/978-3-319-95480-6_13

Morales-Polo C, Cledera-Castro MDM, Moratilla Soria BY (2018) Reviewing the anaerobic digestion of food waste: from waste generation and anaerobic process to its perspectives. Appl Sci 8(10):1804. https://doi.org/10.3390/app8101804

Mussgnug JH, Klassen V, Schlüter A, Kruse O (2010) Microalgae as substrates for fermentative biogas production in a combined biorefinery concept. J Biotechnol 150(1):51–56. https://doi.org/10.1016/j.jbiotec.2010.07.030

Polakovicová G, Kušnír P, Nagyová S, Mikulec J (2012) Process integration of algae production and anaerobic digestion. Chem Eng Trans 29:1129–1134. https://doi.org/10.3303/CET1229189

Roberts KP, Heaven S, Banks CJ (2016) Comparative testing of energy yields from micro-algal biomass cultures processed via anaerobic digestion. Renew Energy 87:744–753. https://doi.org/10.1016/j.renene.2015.11.009

Romagnoli F, Dorella M, Gruduls A, Collotta M, Tomasoni G (2019) Anaerobic co-digestion of Baltic seaweeds with wheat straw and straw pellets: synergetic effects on biomethane yield and kinetic biodegradability constant. Energy Procedia 158:854–860. https://doi.org/10.1016/j.egypro.2019.01.221

Roy SS, Pal R (2015) Microalgae in aquaculture: a review with special references to nutritional value and fish dietetics. Proceedings of the Zoological Society 68(1):1–8. https://doi.org/10.1007/s12595-013-0089-9

Santos-Ballardo DU, Rossi S, Reyes-Moreno C, Valdez-Ortiz A (2016) Microalgae potential as a biogas source: current status, restraints and future trends. Rev Environ Sci Biotechnol 15 (2):243–264. https://doi.org/10.1007/s11157-016-9392-z

Scarlat N, Dallemand J-F, Fahl F (2018) Biogas: developments and perspectives in Europe. Renew Energy 129:457–472. https://doi.org/10.1016/j.renene.2018.03.006

Talavera-Caro AG, Sánchez-Muñoz MA, de Lira Hernandez IO, Montañez-Hernández LE, Hernández-Almanza AY, Morlett-Chávez JA, Esparza-Perusquia M, de Las M, Balagurusamy N (2020) Proteomics of Lignocellulosic substrates bioconversion in anaerobic digesters to increase carbon recovery as methane. In: Zakaria ZA, Boopathy R, Dib JR (eds) Valorisation of agro-industrial residues–volume I: biological approaches. Springer International Publishing, pp 81–110. https://doi.org/10.1007/978-3-030-39137-9_4

Tedesco S, Daniels S (2018) Optimisation of biogas generation from brown seaweed residues: compositional and geographical parameters affecting the viability of a biorefinery concept. Appl Energy 228:712–723. https://doi.org/10.1016/j.apenergy.2018.06.120

Varel VH, Chen TH, Hashimoto AG (1988) Thermophilic and mesophilic methane production from anaerobic degradation of the cyanobacterium Spirulina maxima. Resour Conserv Recycl 1 (1):19–26. https://doi.org/10.1016/0921-3449(88)90004-3

Varol A, Ugurlu A (2016) Biogas production from microalgae (Spirulina platensis) in a two stage anaerobic system. Waste and Biomass Valorization 7(1):193–200. https://doi.org/10.1007/s12649-015-9442-8

Wang X, Lu X, Li F, Yang G (2014) Effects of temperature and carbon-nitrogen (C/N) ratio on the performance of anaerobic co-digestion of dairy manure, chicken manure and Rice straw: focusing on ammonia inhibition. PLoS One 9(5). https://doi.org/10.1371/journal.pone.0097265

Ward AJ, Lewis DM, Green FB (2014) Anaerobic digestion of algae biomass: a review. Algal Res 5:204–214. https://doi.org/10.1016/j.algal.2014.02.001

Wise DL, Augenstein DC, Ryther JH (1979) Methane fermentation of aquatic biomass. Resour Recover Conserv 4(3):217–237. https://doi.org/10.1016/0304-3967(79)90002-7

Xie S, Hai FI, Zhan X, Guo W, Ngo HH, Price WE, Nghiem LD (2016) Anaerobic co-digestion: a critical review of mathematical modelling for performance optimization. Bioresour Technol 222:498–512. https://doi.org/10.1016/j.biortech.2016.10.015

Zabed HM, Qi X, Yun J, Zhang H (2019) Anaerobic digestion of microalgae biomass for methane production. In: Alam A, Wang Z (eds) *Microalgae Biotechnology for Development of Biofuel and Wastewater Treatment* (pp. 397–421). Springer, Cham. https://doi.org/10.1007/978-981-13-2264-8_16

Zhang Y, Caldwell GS, Zealand AM, Sallis PJ (2019) Anaerobic co-digestion of microalgae Chlorella vulgaris and potato processing waste: effect of mixing ratio, waste type and substrate to inoculum ratio. Biochem Eng J 143:91–100. https://doi.org/10.1016/j.bej.2018.12.021

Zhu L, Li Z, Hiltunen E (2018) Theoretical assessment of biomethane production from algal residues after biodiesel production. WIREs Energy and Environment 7(1):e273. https://doi.org/10.1002/wene.273

The Realm of Microorganisms in Biogas Production: Microbial Diversity, Functional Role, Community Interactions, and Monitoring the Status of Biogas Plant

Alicia Guadalupe Talavera-Caro, Inty Omar Hernández-De Lira, Efraín Reyes Cruz, María Alejandra Sánchez-Muñoz, and Nagamani Balagurusamy

Abstract Anaerobic digestion is being considered as a sustainable technology to treat organic wastes to reduce contamination and emission of greenhouse gasses and at the same time produce energy in the form of methane. The microbiological process of AD represents the most challenged step during biogas production due to microbial complexity. At the time, at least 11 microbial groups have been described. These populations have been shown unique metabolism and an interspecies interaction because of the limited amount of energy available for growth. The microbial community structure is considered as the core in the success of AD method. Furthermore, to expand AD technology in order to approach an economically feasible process under the concept of biorefinery and not only the advances on engineering processes, the design of new biogas digesters and tools for real-time monitoring for AD are the keys for a successful implementation of this process. In addition, the classification of the microbial community structure and the understanding of the metabolic networks play a crucial role for its development. In this chapter, different aspects of the microbiology of AD of full-scale biogas digesters are discussed with specific focus on the presence of different microbial groups, their activity, and interactions.

Keywords Anaerobic digestion · Microbial communities · Biogas · Microbial networks · Biogas reactors

---

A. G. Talavera-Caro · I. O. H.-D. Lira (✉) · E. R. Cruz · M. A. Sánchez-Muñoz · N. Balagurusamy
Laboratorio de Biorremediación, Facultad de Ciencias Biologicas, Universidad Autonoma de Coahuila, Torreón, Coahuila, Mexico
e-mail: omar.hernandez@uadec.edu.mx

N. Balagurusamy, A. K. Chandel (eds.), *Biogas Production*,
https://doi.org/10.1007/978-3-030-58827-4_10

1 Introduction

Anaerobic digestion (AD) represents the most prominent worldwide technology to convert organic wastes, such as livestock manures, municipal solid waste, municipal and industrial wastewaters, and agro-industrial residues, into biogas due to an engineered and biochemical process, which involves a series of operational parameters, such as organic loading rate, and the interactions of at least eleven microbial groups (Alvarado et al. 2014). The importance of AD is not only because of its significance in waste management but also because AD offers carbon recovery in the form of methane, which demonstrates to be a sustainable manner to produce clean energy as electricity and heat and as vehicle fuel. Notwithstanding the advances on the engineering processes, the design of new biogas digesters, and tools for real-time monitoring for AD, the microbiology aspect always poses challenges. Microbial community composition analyses, in biogas digesters of several substrates, have been widely reported. However, due to their complexity, these populations have been shown unique metabolism and an interspecies interaction which have not been yet precisely characterized (García-Lozano et al. 2019). This process is still contemplated as the core in the success of AD method. In addition, the classification of the microbial community structure and the understanding of the metabolic networks are crucial to expand the implementation of AD technology in order to achieve an economically feasible process. Moreover, the purpose of this process is presently exploring to include the generation of value-added products, under the concept of biorefinery, not only the energy generation and nutrient recovery (Schnürer 2016). This chapter describes several aspects of the microbiology of AD, including the presence of different microbial groups, their activity and interactions, and the consequent response of the different operational parameters in a full-scale biogas digesters are discussed.

2 Metabolism of Anaerobic Digestion Process

2.1 Anaerobic Digestion: Functional Role

Anaerobic digestion is a chain of interconnected biological reactions in which at least 11 groups of microorganisms, belonging to domain bacteria and archaea, interact in numerous associations where the organic matter, as carbohydrates, proteins, lipids, or more complex compounds, is transformed into biogas (containing ~65% CH_4, 35% CO_2, and trace amounts of H_2S, NH_3, and H_2) and anaerobic biomass. Besides bioenergy production in the form of methane, AD presents several advantages, such as lesser biomass sludge production in comparison to aerobic treatment technologies, elimination of pathogens, the digestate produced is an improved fertilizer, and the reduction of greenhouses gasses (GHG) emissions.

Usually AD is conceptually divided into three or four stages, hydrolysis and/or fermentation, acetogenesis, and methanogenesis. The performance of these processes is carried out by the combined action of hydrolytic-fermentative bacteria, syntrophic acetogenic bacteria, and methanogenic archaea. During the first stage, insoluble and complex polymers (carbohydrates, lipids, proteins, etc.) are hydrolyzed and converted into simple and soluble products (sugars, long-chain fatty acids, glycerol, amino acids, etc.), which are catabolized by fermentative bacteria into alcohol, fatty acids, hydrogen, and carbon dioxide. Subsequent steps involve the oxidation of such alcohols and fatty acids by syntrophic acetogens, forming acetate, H_2, and CO_2. Finally, during methanogenesis, acetate and other methyl-containing C1 compounds are reduced to methane by aceticlastic and methylotrophic methanogens, and CO_2 is reduced by H_2-oxidizing methanogens (Nagamani and Ramasamy 1999).

2.2 *Hydrolysis*

This first step of AD is considered as the rate limiting performed by the microbial decomposition of organic matter (proteins, lipids, and polysaccharides) into soluble small molecules by extracellular enzymes of facultative and obligate anaerobic bacteria (Cazier et al. 2015; Boontian 2014). Substrates are cleaved enzymatically, mainly by the amylases, cellulases, proteases, and lipases excreted by microorganisms (Bajpai 2017). Interaction networks from domains help us to understand the substrate conversion process (Shaw et al. 2017). Usually AD is greater than 10^{16} cells/mL which involves saccharolytic bacteria (~10^8 cells/mL), proteolytic bacteria (~10^6 cells/mL), and lipolytic bacteria (~10^5 cells/mL) (Amani et al. 2010). Proportions of the enzymes excreted from these bacteria and the optimum operation of a biogas plant will depend on the substrate and its degradation characteristics (Weinrich and Nelles 2015). Mostly, substrates employed to start up this stage are wastes like animal waste and lipid-rich wastes from oil industry, pulp-paper processing, wastewaters, animal fat, agricultural waste, or energy crops, which show different microbial communities according to degradation demand (Montañez-Hernández et al. 2018; Tabatabaei et al. 2010; Appels et al. 2011).

2.2.1 Polysaccharide Hydrolysis

Lignocellulosic biomass is mainly found in biodigesters and consists of cellulose (30–56%), hemicellulose (10–27%), and lignin (3–30%). It is worth to mention that lignin is a recalcitrant compound that can limit the hydrolysis rate for biogas production (Sawatdeenarunat et al. 2016; Venkiteshwaran et al. 2016). At present, two types of polysaccharide hydrolysis systems are known: multienzymatic complex systems, called cellulosomes, and free enzymatic systems (Felix and Ljungdahl, 1993). Anaerobic microorganisms produce cellulosomes, fixed on the bacterial cell wall, which bind to the substrate for its hydrolysis. Aerobic microorganisms degrade

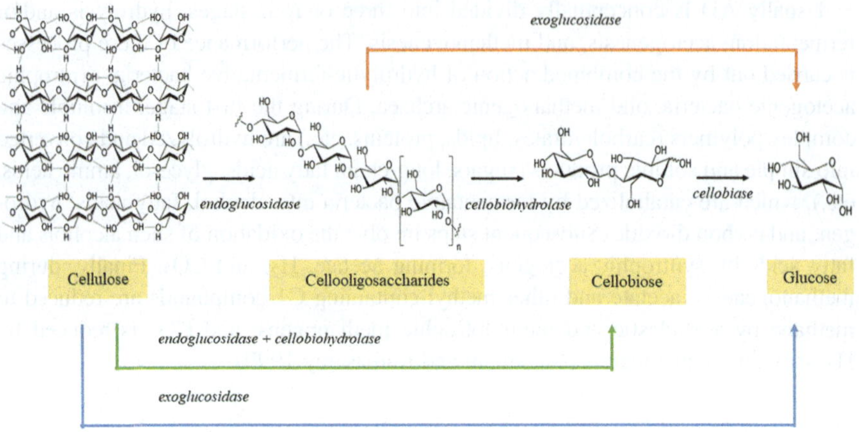

Fig. 1 Schematic representation of enzymatic hydrolysis of cellulose

cellulose, by secreting a set of enzymes, viz., endoglucanase, exoglucanase (cellobiohydrolase), and cellobiases (Fig. 1). Meanwhile, the hydrolysis of starch is performed by a mixture of amylases as α-amylase, α-xylosidase, and β-xylosidase able to hydrolyze amylose and amylopectin (Zhu et al. 2016).

2.2.2 Protein Hydrolysis

As well as carbohydrates and lipids, protein constitutes a major percentage of the organic load in anaerobic sludges and wastewaters. Wastewater and sewage from food processing industries as abattoir, dairy, fish, and vegetables comprise around 40% of protein (Barnett et al. 1994; Ramsay and Pullammanappallil 2001). Proteins are natural nitrogen-rich polymers that are mainly composed of amino acids linked by peptide bonds. Nitrogen provides an essential element for the synthesis of amino acids, protein, and nucleic acids and acts as a strong base when it is converted to ammonia. Physiologically, proteases are released to the extracellular media to cleavage proteins into its constituents, peptides and free amino acids, which are subsequently metabolized to VFAs, CO_2, H_2, NH_4^+, and S_2^-. Proteases are classified principally based on their site of action in two major groups: exoproteases (carboxipeptidases or aminopeptidases) and endoproteases. Further classification is based on the functional group (serine, cysteine, aspartate, or metallo) and optimal pH (acidic, neutral, or alkaline) (Schaechter 2009).

2.2.3 Lipid Hydrolysis

Lipid is the term used to describe fat, oil, and grease contained mostly in wastewater stream and other sources. Lipids are considered as excellent substrates for anaerobic digestion and co-digestion due to the higher methane yield obtained when compared to proteins or carbohydrates (Yang et al. 2016). Most of lipids in wastes are present as triacylglycerides, a glycerol ester with three long-chain fatty acids (LCFA). During hydrolysis of triacylglycerides, glycerol and LCFA (typically 14 to 24 carbon atoms) are produced by extracellular lipases in order to increase lipid solubility. These enzymes are excreted by acidogenic bacteria, and the further conversion of the hydrolysis products takes place inside the bacterial cells.

2.3 *Acidogenesis*

The second stage from AD is fermentation, also called acidogenesis, where monomers will be further decomposed by fermentative bacteria into short-chain fatty acids or volatile fatty acids (VFAs). Generally, acetate, butyrate, and propionate (most prevalent VFAs), lactate, valerate, pyruvate, formic acids, CO_2, and/or hydrogen are present as by-products of this stage (Chen et al. 2017; Mani et al. 2016; Ren et al. 2018). During AD, acidogenesis is the quickest step producing precursors of methane. Three main types of fermentation are known: ethanol/acetic acid-type, butyric acid-type, and propionic acid-type. These pathways are determinant to achieve a high performance of methane production, where the major products are butyric and acetic acid (70–90%) (Chen et al. 2015). The performance of the fermentation stage is one of the most attractive strategies for biogas production enhancement in AD process goals, especially on organic wastes (Lu et al. 2018).

2.3.1 Carbohydrates Fermentation

In the absence of methanogens, the major products of sugar fermentation by anaerobic bacteria are acetate, ethanol, H_2, and CO_2. When H_2-utilizing bacteria are active, acetate production is increased. Formerly, for most of microorganisms, fermentation of glucose occurs by the glycolytic pathway, producing pyruvate, or by-products of pyruvate (Fig. 2). Glucose can be fermented to lactate by homofermentative bacteria to lactate or to multiple end products as acetate, formate, butyrate, propionate, valerate, and CO_2, by heterofermentative bacteria. Usually, these microorganisms produce CO_2 and H_2 with the concomitant production of formate, acetate, lactate, and succinate. Commonly, heterofermentative bacteria include *Lactobacillus*, *Microbacterium*, and *Leuconostoc*. The main product of clostridia, eubacteria, fusobacteria, and butivibrios is butyrate, acetate, CO_2, and H_2, while *Clostridium* species can ferment those end products plus others, as acetone. Other anaerobic bacteria, as *Propionibacterium* species, ferment glucose to form CO_2, propionate, acetate, and succinate. Propionate is produced by the partial reversal of Krebs cycle reactions and implies a CO_2 fixation by pyruvate (the Wood-Werkman reaction) which forms oxaloacetate. Subsequently, oxaloacetate is reduced in three steps and then decarboxylated to propionate. In another three-carbon pathway, propionate is formed by a lactyl-SCoA intermediate.

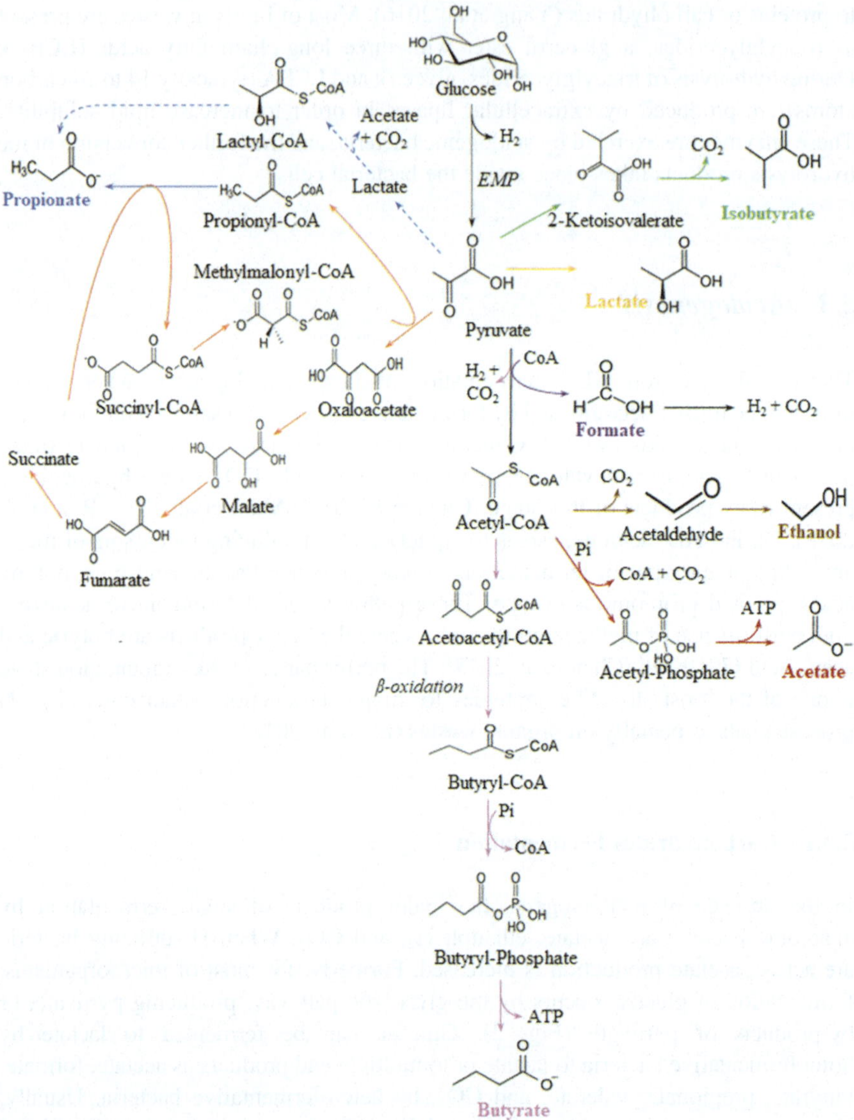

Fig. 2 Fermentative pathways occurring in anaerobic digestion for bacteria, and the major end products formed from glucose. EMP: Embden-Meyer Pathway. Orange line: Reversal Krebs cycle. Blue dotted line: Acrylate pathway

2.3.2 Amino Acid Fermentation

Amino acid can be fermented anaerobically by two principal ways: a pair of amino acids can be decomposed through the Stickland reaction, or one single amino acid can be degraded by H_2-utilizing bacteria. The end products of fermentation include short-chain and branched-chain organic acids, NH_3, CO_2, and small amounts of H_2 and sulfur-containing compounds (Ramsay and Pullammanappallil 2001). The Stickland reaction implies one amino acid, used as electron donor, while another amino acid acts as an electron acceptor. This reaction produces 0.5 mole of ATP per mole of amino acid transformed, and their utilization may be linked to the oxidative deamination step and/or the decarboxylation step (Andreesen et al. 1989). The alternative pathway to the Stickland reaction proceeds when hydrogen partial pressure is sufficient low, releasing hydrogen as electrons (Schnürer 2016). It is worth to mention that the oxidative deamination reactions are endergonic under standard conditions. Thus, the reaction cannot proceed unless the reducing equivalents produced are taken up via interspecies hydrogen transfer by methanogens, sulfate reducers, or acetogens and by another amino acid in the Stickland reaction or in the reduction of acetate to butyrate (Örlygsson et al. 1995).

2.3.3 Glycerol Fermentation

As mentioned earlier, most of glycerol present in biodigesters is a product of lipid hydrolysis plus LCFA. Glycerol is a source of carbon and energy, and its uptake can occur by active or passive transport (Holst et al. 2000). Anaerobic fermentation of glycerol can be carried out by a reductive or an oxidative pathway (Biebl et al. 1999).

The reductive pathway leads to 1,3-propanediol production by means of glycerol dehydration. The oxidative pathway leads to glycerol dehydrogenation to produce phosphoenolpyruvate which can in turn be converted to propionate by several decarboxylations, or it can be converted to pyruvate. Thus, pyruvate can be then be fermented in simpler compounds, depending of the microorganism and the environmental conditions, such as 2,3-butanediol, lactate, butyrate, n-butanol, ethanol, acetate, formate, hydrogen, and carbon dioxide (Siles et al. 2009).

2.4 *Acetogenesis and Syntrophy*

Obligate anaerobic bacteria that synthesize acetyl-CoA by the reductive acetyl-CoA or Wood-Ljungdahl pathway, for energy and cell carbon obtaining from CO_2, are called acetogens; acetogenic bacteria that produce acetate as sole end product are called homoacetogen (Drake 1994). The pathway consists in the reduction of 2 moles of CO_2 to 1 mole of acetate by 8 protons (Hattori 2008). In addition, the participant key enzyme for the pathway is the acetyl-CoA synthase (ACS) (Müller and Frerichs 2013). Thereby, the aforementioned statements separate acetogens from those microbial groups that produce acetate as an end product of fermentation (Schuchmann and Müller 2016). Acetogens versatility is demonstrated by their wide variety of useful substrates, i.e., sugars, $CO_2 + H_2$, C_1 compounds, dicarboxylic acids, and alcohols (Müller and Frerichs 2013). In addition, electron acceptors such as nitrate, nitrite, thiosulfate, and fumarate can be used by acetogens. However, repression of acetyl-CoA pathway takes place (Müller and Frerichs 2013). In anaerobic digestion, acetogenic bacteria contribute in the formation of acetate as a precursor for CH_4 production by aceticlastic methanogenesis. Therefore, their presence in anaerobic digesters benefits the process.

There are other bacteria that are in the presence of hydrogen-scavenger microorganisms, such as hydrogenotrophic methanogens, which act in syntrophic relationship to obtain energy. Syntrophic acetate oxidation (SAO) process consists in the oxidation of acetate, by syntrophic acetate-oxidizing bacteria (SAOB), to produce H_2 and CO_2, available substrates for hydrogenotrophic methanogens to form CH_4 (Sun et al. 2014). It is believed that acetate oxidation is carried out by the reversible reactions of the Wood-Ljungdahl pathway (Müller and Frerichs 2013). The process of transferring reducing equivalents (such as H_2) from bacteria to archaea is called interspecies electron transfer (Stams and Plugge 2009). Oxidation of acetate is a highly endergonic reaction under standard conditions ('$\Delta G^{0\prime} = +104.6$ kJ/mol) (Hattori 2008). Therefore, it requires low H_2 partial pressure (<10 Pa) that can be obtained by the activity of hydrogenotrophic methanogens (Schink 1997). Coupling of both pathways results in an overall exergonic reaction ('$\Delta G^{0\prime} = -31.0$ kJ/mol). However, this small energy released is shared by both microorganisms, explaining their slow growth rate (Hattori 2008).

2.5 *Methanogenesis*

Methanogenesis is the final stage of anaerobic digestion, where the biological formation of methane is performed by methanogens, an obligate anaerobic archaeon. Methanogens use three main substrates to obtain energy. The first type of substrate is CO_2; most of the methanogens are capable to reduce CO_2 to methane by electrons from H_2, but also other electron donors such as formate, secondary alcohols such as 2-propanol, 2-butanol, and even ethanol might be used by methanogens. The oxidation of these last compounds occurs partially generating ketones and acetate. The second substrate is compounds that contain methyl groups, such as methanol and amines. The last group corresponds to acetate.

Methane biosynthesis occurs through two main pathways known as hydrogenotrophic or CO_2-reduction and aceticlastic. In the CO_2-reduction pathway, formate (Hungate et al. 1970; Archer and Harris 1986) or H_2 is oxidized, and CO_2 is reduced to CH_4, whereas in the aceticlastic pathway, acetate is cleaved with the carbonyl group oxidized to CO_2 and the methyl group reduced to CH_4 (Ferry 2011). Although both routes differ in terms of reactions and enzymes, the last step that corresponds to the production of methane and the formation of heterodisulfide is common in both pathways. The reduction of CO_2 to CH_4 reaction sequence starts with a two-electron reduction of CO_2 and methanofuran (MFR) to formyl-MFR where the formyl group is bound to the amino group of the coenzyme. The formyl group is then transferred to the N^5 of tetrahydromethanopterin (H_4MPT); the formyl-H_4MPT thus generated cyclizes to the methenyl-H_4MPT, which is reduced in two steps to the methyl-H_4MPT. Finally, the methyl group is transferred to the thiol group of coenzyme M. The methyl thioether formed is reduced to CH_4 in the final step of the pathway (Hedderich and Whitman 2013).

In methanogenesis from methanol, the methyl group enters the C1 pathway at the level of coenzyme M and is reduced to methane. The electrons for this reduction are obtained from the oxidation of an additional methyl group to CO_2 using the reverse of the steps of the reductive C1 pathway (Hedderich and Whitman 2013). During growth on acetate, the methyl (C-2) carbon of acetate is reduced to methane using electrons obtained from the oxidation of the carboxyl (C-1) carbon of acetate. In this metabolism, the methyl group enters the C1 pathway at the level of methyl-H_4MPT (Hedderich and Whitman 2013).

Around 70% of the methane synthesized by methanogens in a full-scale biogas plant comes from the acetoclastic pathway, while the remaining percentage comes from the CO_2-reduction. However, phylogenetic studies recognize the CO_2-reduction pathway as the oldest since some of the specific enzymes for this pathway are not distributed in other microorganisms. In contrast, the enzymes required for the acetoclastic pathway are also found in some acetogenic and fermentative bacteria, suggesting that the appearance of this pathway occurred much later than hydrogenotrophic pathway (Bapteste et al. 2005).

3 Microbial Composition in Full-Scale Biogas Digesters

3.1 Hydrolytic Bacteria

The implementation of polysaccharides, as substrate for carbon source, is generally used on biogas plants. Mostly, lignocellulosic-rich substrates are feedstock in high-capacity bioreactor and present as energy crops, agricultural residues, animal manure, and food waste as a sustainable source (Koch et al. 2010; Ziganshin et al. 2013). Microorganism presence found in biogas plants for the hydrolytic anaerobic process varies on the type of reactor as is shown in Table 1. However, mostly the phyla of *Proteobacteria* (within *Deltaproteobacteria*, *Gammaproteobacteria*, *Betaproteobacteria*, *Alphaproteobacteria* classes) are present in the initial phase where *Clostridiales* (*Clostridium*, *Ruminococcus*, *Butyrivibrio*, *Acetivibrio*, and *Eubacterium*), *Thermoanaerobacterales* (*Caldicellulosiruptor*), *Fibrobacteres* (*Fibrobacter*), *Spirochaetales* (*Spirochaeta*), *Tissierellia* (*Anaerococcus*) orders are involved and some archaea started to appear, as well as in the hydrolytic stage (Cirne et al. 2007; Manyi-Loh et al. 2013; Narihiro and Sekiguchi 2007). A study reported by Tian et al. (2017) observed the order of *Bacteroidales* accounted for the 30% of the total prokaryote population; in this order of microorganisms, the family of *Marinilabiaceae* accounted the 85% of the order. Thus, *Bacteroidales* were predicted as the microorganisms able to degrade biopolymers, including xylan, and also reported to degrade chitin in anaerobic conditions. Nonetheless, *Bacteroidales* was identified by their abundance and its role in anaerobic digestion of cellulose and hemicellulose.

Table 1 Microorganisms predominant identified in anaerobic digestion from different types of reactors, with relative order abundance (%) and OTUs of species within the orders

Order	Species related to order	OTUs	Order abundance %	Type of reactor	References
Erysipelotrichales	*Erysipelotrichaceae*	NA	<0.1–2%	Wet fermentation BGP CSTR	(Sundberg et al. 2013)
Clostridiales	– *Pelotomaculum isophthalicicum* – *Ruminococcus* – *Acetivibrio cellulolyticus* – *clostridium thermocellum* – *Pelotomaculum thermopropionicum* – *Pelotomaculum isophthalicicum* – *Heliobacillus mobilis*	42 NA NA NA 110 270 1078	15–18%	– Wet and dry fermentation BGP – UASB – Continuous anaerobic digester – Full-scale thermophilic and mesophilic – CSTR	(Sundberg et al. 2013; Stolze et al. 2015; Lee et al. 2012; Sun et al.2015; Doloman et al. 2017)
Spirochaetales	– *Treponema sp.* – *Spirochaeta cellobiosphila* – *Spirochaeta halophila*	NA 6 36	>16%	– Wet and dry fermentation BGP – Full-scale thermophilic and mesophilic	(Stolze et al. 2015; Lee et al. 2012; Sundberg et al. 2013)
Bacteroidales	*Prevotella sp.*	NA	08–44%	– Wet fermentation BGP – CSTR – UASB	Ziganshin et al. 2013; Sun et al. 2015; Doloman et al. 2017)
Cytophagales	*Fulvivirga kasyanovii* *Sediminitomix flava*	11 62	NA	Full-scale thermophilic and mesophilic	(Lee et al. 2012)
Burkholderiales	*Curvibacter delicatus* *Acidovorax defluvii*	46 280	NA	Full-scale thermophilic and mesophilic	(Lee et al. 2012)

Thermoanaerobacterales	*Caldanaerobius fijiensis*	68	NA	Full-scale thermophilic and mesophilic	(Lee et al. 2012)
Synergistales	*Anaerobaculum mobile* *Treponema primitia*	453 396	1.9%	Full-scale thermophilic and mesophilic UASB	(Lee et al. 2012; Doloman et al. 2017)

UASB upflow anaerobic sludge blanket reactor, *CSTR* continuous stirred-tank reactor, *BGP* biogas plant

Generally, the main microorganisms in anaerobic digesters involved in *protein hydrolysis* are from the order *Bacteroidales*, *Clostridiales*, *Fusobacteriales*, *Selenomonadales*, and *Lactobacillales* (Amani et al. 2010). *Clostridiales* and *Bacteroidales* are recognized as the main contributors in polymer hydrolysis and fermentation steps. In biogas plants (BGPs), these versatile orders are capable of hydrolyzing a wide range of substrates, including carbohydrates, lipids, and proteins. Previous metagenomic studies in BGPs have demonstrated its dominance ranging from 15 to 84% of the total microorganisms (Schlüter et al. 2008; Sundberg et al. 2013). More recently, its prevalence is shown in mesophilic and thermophilic biogas plants fed with lignocellulosic wastes (agricultural) and animal manure (Table 2).

Nonetheless, other works have acknowledged groups as *Spirochaetales* and *Bacillales* participating in protein degradation specifically of maize silage either with pig or chicken manure (Ortseifen et al. 2016; Stolze et al. 2015) and *Candidatus Cloacamonas* as the main protein degrader phylum from a BGP treating dairy manure (Li et al. 2014). Particularly wastes from food or ethanol fermentation have shown an increased abundance of specific groups of microorganisms. *Thermotogales* (10.4%) dominated a BGP treating food waste wastewater, while *Coprothermobacterales* (68.2%) showed a marked dominance in a mesophilic farm-scale digester treating brewery and swine wastes (Cho et al. 2017; Lee et al. 2016).

As seen in Table 2, serine and metalloproteases seem to have an important role generally in anaerobic bacteria. *Clostridiales*, *Bacteroidales*, and *Coprothermobacterales* encode for both enzymes, while other groups of microorganisms synthetize mostly for serine proteases. Both proteases are ubiquitously found in prokaryotes, and its mechanism depends on the active site which include a nucleophilic serine amino acid (serine proteases) or generally requires zinc or cobalt (metalloproteases) (Hedstrom 2002; Rawlings and Barrett 1995). However, little is known about the role of proteases in anaerobic digestion. Hence, a deeper insight is still necessary in order to known which specific proteases are participating in AD and to understand the dynamic changes within the community treating specific kind of substrates.

On the other hand, when treating lipids, *Firmicutes* and *Bacteroidetes* are crucial for the performance of the biodigester (Salama et al. 2019). Syntrophic β-oxidizing bacteria from microbial consortium as *Proteobacteria* and *Syntrophomonas* sp. from *Clostridiales* order have been reported as FOG degraders. Moreover, within the *Proteobacteria* phylum, *Rheinheimera* sp. and *Bacillus* sp. can digest FOG under anaerobic conditions and decrease LCFA deposition (Klaucans and Sams 2018). Studies on reactor treating palmitate and oleate revealed a predominance of *Clostridiaceae* and *Syntrophomonadaceae* from *Clostridiales* (Alves et al. 2009).

It is well-known that long-chain fatty acid (LCFA) oxidation on AD is performed through the path of β-oxidation, where coenzyme A is utilized for LCFAs conversion into acetate and hydrogen (Rasit et al. 2015). Studies have reported that FOG biodegradability has high potential biogas production on methane yielding (~1200 L CH_4/kg VS) on full-scale wastewater treatment plants (WWTPs) (Shen et al. 2015). Lipid degradation is critical for the effective degradation of food waste to produce biogas; also lipids are considered as a good substrate to produce renewable energy at an industrial level (Ziels et al. 2016). A study by He et al. (2018) presented an

Table 2 Principal orders involved in protein hydrolysis identified in biogas plants

Order	Biogas plant	Substrate	Methane yield	Functional role	Abundance	References
Clostridiales	Full-scale mesophilic digester (34–35 °C)	Low substrate loading of sewage sludge	7000 m^3/d with a methane content of 61.26%	It is the most abundant order in anaerobic digesters, encodes principally metalloproteases	7.03%	(Świątczak et al. 2017)
	Industrial-scale biogas plants (37–38 °C)	CD1: Slaughterhouse waste CD2: Thin stillage CD4: Grass, wheat based stillage	CD1, 319 ± 24 CD2, 307 ± 54 CD4, 348 ± 24 (methane potential)		CD1, 4.3% CD2, 3.8% CD4, 56.3%	(Sun et al. 2016)
	Thermophilic pilot plant digester (56 °C)	Poultry litter	57.7 ± 2.2% methane of the total biogas		46%	(Smith et al. 2014)
Bacteroidales	Mesophilic pilot-BGP (36–38 °C)	Above ground biomass of Jerusalem artichoke	50–60% methane content	*Bacteroidales* secretes carboxypeptidases, metalloproteases, serine proteases, and ATP-dependent	ND	(Ciccoli et al. 2018)
	Thermophilic BGP (54 °C)	Maize silage, barley, cattle manure, and pig manure	0.62 $m^3{}_{biogas}$ $kg^{-1}{}_{VS}$ d^{-1} with a methane content of 53-54%.		4.5% of bacterial domain	(Maus et al. 2016)
Coprothermobacterales	Mesophilic farm-scale digester (35 °C) during a period of starvation	Sludge from brewery and swine wastewater and sewage waste	0.1–0.3 m^3/m^3/d with a methane content from 20-60%	Encoding mostly metalloproteases and serine proteases	68.2%	(Cho et al. 2017)
Bacillales	Mesophilic BGP3 (40 °C)	Maize silage and pig manure	528.5 l/kg oDM	Encoding for serine proteases	0.07 FPKM (fragment per kilobase)	(Ortseifen et al. 2016)
Thermotogales	Thermophilic full-scale anaerobic digester (58.5 °C)	Food waste-recycling wastewater (34% of protein)	ND	Encodes serine proteases	10.4%	(Lee et al. 2016)

(continued)

Table 2 (continued)

Order	Biogas plant	Substrate	Methane yield	Functional role	Abundance	References
Spirochaetales	Mesophilic BGP (40 °C)	Dry fermentation (DF): Maize silage, green rye, chicken manure. Wet fermentation (WF): Maize silage, pig manure	DF, 350.5 l/kg oDM WF, 417.8 l/kg oDM	From genus *Treponema* encodes serine proteases	DF, 0.5% WF, 1.7%	(Stolze et al. 2015)
C. Cloacamonas	Mesophilic mixed plug-flow loop reactor (37 °C)	Dairy manure	60% of methane	Encodes all the machinery for protein degradation including proteases and peptidases	20–27%	(Li et al. 2014)

TAN Total ammonia nitrogen, *ND* Not determined
*Genus. #Phylum.

organic loading rate for stable biogas production of 0.5–1.5 g VS^{-1} $days^{-1}$ using cooking oil skimmed from food waste as the only carbon source, where *Anaerovibrio* (lipid hydrolysis bacteria) hydrolyze triglycerides to produce glycerol and fatty acids. This increased from 9.3 to 40% in a relative high concentration of lipids with the highest value of 2.0 g VS L^{-1} $days^{-1}$, while the genus of *Syntrophomonas* increased to ~29%, playing significant roles in the mesophilic anaerobic digestion.

3.2 Fermentative Bacteria

Biogas reactors have been tested in different manners during monosacharides fermentation, sucha as ADM1 model, with lactate suggesting that *Clostridiales* is a butyrate-producing bacterium predominantly, and other microorganisms were found *Propionibacteriales* synthetizing propionate, *Lactobacillales* (*Carnobacterium* sp.), a lactic acid bacteria, and *Synergistales* (*Lactivibrio alcoholicus*) a lactate-degrading bacteria (Satpathy et al. 2016). On thermophilic biogas plants, the order of *Petrogales*, *Defluviitoga tunisiensis*, and *Desulfotomaculum australicum* are described as lactic acid degraders, also contained acidogenic/acetogenic bacteria belonging to the *Clostridiales*, *Tissierellales*, and *Bacillales* orders (Table 3) (Maus et al. 2016).

When a high rate of *amino acid fermentation* occurs, high amounts of NH_3 and ammonium (NH_4^+) are produced, mostly when treating a proteinaceous-rich feedstock as animal wastes as slaughterhouse waste, dairy manure, animal manure, and aquaculture sludge and wastes from food industry and households. In AD, high concentrations of NH_3 are toxic to some microorganisms inhibiting cytosolic enzymes, as well as NH_4^+ which can be intracellular accumulated modifying the

Table 3 Main orders involved in monosaccharides and other fermentative substrates in a thermophilic biogas plant

Order	Strain	Substrate	Acid formed
Clostridiales	▪ *Clostridium kluyveri* DSM-555. ▪ *Clostridium cochlearium* JCM 1396. ▪ *Sporanaerobacter acetigenes* DSM-13106. ▪ *Desulfotomaculum guttoideum* JCM-11016.	▪ Succinate ▪ Glucose ▪ Glucose ▪ Ethanol	▪ Acetic acid. ▪ Acetic, butyric and propionic acid. ▪ Acetic acid. ▪ Acetic acid.
Petrotogales	▪ *Dendrosporobacter quercicolus.* ▪ *Selenomonas bovis* WG.	▪ Lactic acid ▪ Glycerol	▪ Acetic and propionic acid. ▪ Lactic, propionic acids, and succinate.
Bacillales	▪ *Bacillus thermoamylovorans* BHK67. ▪ *Soehngeria saccharolytica* DSM-12858.	▪ Glucose ▪ Lactic acid	▪ Acetic and propionic acid. ▪ Acetic acid.

pH and K^+ concentration causing process instability. Hence, an overproduction of ammonia can inhibit the whole process of AD due to that protein hydrolysis is faster than carbohydrate or lipid hydrolysis (Andreesen et al. 1989).

Although several studies have demonstrated ammonia-tolerant bacteria population by high methane yields, the fraction of NH_3 relative to the total (NH_3 + NH_4^+)-nitrogen (TAN) should be monitored (Hansen et al. 1993). TAN concentration of 0.68 g L^{-1} does not affect the methanogenic activity at mesophilic conditions. However, a range between 1.5 and 3 g L^{-1} of TAN is inhibitory, and a TAN concentration $>$ 4 g L^{-1} fully inhibits AD (Angelidaki and Ahring 1993; Hansen et al. 1993).

Carbohydrate-fermenting bacteria usually degrade proteins in a process energetically favorable. Many studies have shown proteolytic bacteria from the genus *Clostridia*, which also play an important role in amino acid fermentation (de Vladar 2012). In fact, *Clostridia* species only carry out Stickland reaction using all amino acids and producing δ-aminovalerate, α-aminobutyrate, or γ-aminobutyrate as intermediates in the fermentation (Mead 1971). As shown in Table 4 several orders have been grouped as including the order *Clostridiales*. However, other groups as *Synergistales*, *Thermotogales*, and *Thermoanaerobacterales* have been found in biogas plant treating agricultural wastes, food waste wastewater, and sewage sludge, and they have been recognized to degrade several amino acids to produce propionate and/or acetate (Lee et al. 2016; Maus et al. 2016; Świątczak et al. 2017).

Nonetheless, a phylum lately recognized as protein degrader and amino acid fermenter is *Candidatus Cloacamonas acidaminovorans* belonging to WWE1 candidate division, which encode all the machinery for protein degradation and derive most of the carbon and energy from amino acid fermentation (Pelletier et al. 2008). *C. Cloacamonas* has been found in great abundance in mesophilic BGPs, mainly digesting agricultural wastes and animal manure (Stolze et al. 2015, 2015; Sun et al. 2016). However, this phylum was more abundant (28.6%) in a mesophilic-thermophilic lagoon-type reactor treating pig manure and several wastes (Pampillón-González et al. 2017). In spite of proteinaceous feedstock are usually no recommended for biogas production considering the increased risk of inhibition by ammonium (Kragl and Aivasidis 2005), several studies had led to reach an adaptation of the microbial community to protein-rich biomass which can be appropriate to sustainable biogas production (Kovács et al. 2015, 2013).

Anaerobic digestion of *glycerol* as sole source or in co-digestion with other organic materials has been widely explored (Viana et al. 2012). However, both ways showed clear limitations mainly associated (1) to the presence of toxic compounds as LCFAs and inorganic salts of chloride and sulfates and (2) to the high chemical oxygen demand of glycerol. Despite of such disadvantages, microbial communities are able to adapt to high salinity, achieving promising methane potentials in anaerobic reactors treating only glycerol. Various works have shown that methane potential values are near to the theoretical methane production potential for glycerol (0.426m^3 CH_4/kg glycerol), making glycerol a challenge (Kolesárová et al. 2011; Siles et al. 2009; Yang et al. 2008).

Table 4 Principal orders involved in amino acid fermentation identified in biogas plants to reach an adaptation of the microbial community to protein-rich biomass which can be appropriate to sustainable biogas production (Kovács et al. 2015; Kovács et al. 2013)

Order	Biogas Plant	Substrate	Methane yield	Tan (g/l)	Functional role	Abundance	References
Clostridiales	Sewage treatment plants (STP, 37–40 °C)	STP03: Municipal and textile wastes STP07: Municipal, food, brewing and slaughterhouse wastes STP08: Municipal waste	STP03, 0.20 STP07, 0.80 STP08, 0.28 (m^3 biogas/m^3 fermenter volume/day)	ND	Only carry out Stickland reactions. All species use proline as main substrate forming acetate as end product	0.015–0.16% (OBP54 class *Clostridia*)	(Buettner and Noll 2018)
	Thermophilic BGP (54 °C)	Maize silage, barley, cattle manure, and pig manure	0.62 m^3_{biogas} kg^{-1} vs d^{-1}. Methane content, 54%	ND		36.5%	(Maus et al. 2016)
Synergistales	Full-scale mesophilic digester (34–35 °C)	Low substrate loading of sewage sludge	7000 m^3/d with a methane content of 61.26%	ND	*Synergistales* able to degrade most of the amino acids	4.61%	(Świątczak et al. 2017)
Thermotogales and Thermoanaerobacterales	Full-scale anaerobic	Food waste-recycling	ND	2–4.3	*Thermoanaerobacterales* produce acetate and propionate while	*Thermoanaerobacterales*, 7.9%	(Lee et al. 2016)

(continued)

Table 4 (continued)

	digester (58.5 °C)	wastewater (34% of protein)			*Thermotogales* produce acetate. All from a mix of L-alanine, L -serine, L-threonine, L-cysteine, L- glutamate, and L-methionine		
	Thermophilic BGP (54 °C)	Maize silage, barley, cattle manure, and pig manure	0.62 m^3_{biogas} kg^{-1}vs d^{-1}. Methane content, 54%	ND		*Thermotogales*, 7.1% *Thermoanaerobacterales*, 0.22%	(Maus et al. 2016)
Candidatus Cloacamonas	Mesophilic-thermophilic (35–55 °C) AD reactor	Pig manure and wastes	318.6–543.9 m^3/day with 67% of CH_4	0.64	*Cloacimonetes* degrade only proline, alanine, aspartate, glutamate, lysine, and asparagine, as source of energy and generate CO_2 and H_2 as end products Other groups: *Marinimicrobia* and *Fusobacteriales* capable of ferment glutamate by five different pathways	*Cloacimonetes* 28.6%	(Pampillón-González et al. 2017)
	Industrial-scale biogas plants (37–38 °C)	CD3: Slaughter-house waste	347 ± 15 (methane potential)	CD3, 2.6		*Cloacimonetes*, 18% *Marinimicrobia*, 8.2%	(Sun et al. 2016)
	Mesophilic BGPs (2 and 3, 40 °C)	B2: Maize silage, grass, poultry/pig/cattle manure B3: Maize silage, pig manure	B2, 336.68 l/kg oDM B3, 276.93. l/kg oDM	B2, 2.32 B3, 3.15		*Cloacimonetes*, B2, 8% B3, 8–12% *Fusobacteriales*, B3, 15–18%	(Stolze et al. 2016)

	Mesophilic BGP (40 °C)	Dry fermentation (DF): Maize silage, green rye, chicken manure. Wet fermentation (WF): Maize silage, pig manure	DF, 350.5 l/kg oDM WF, 417.8 l/kg oDM	DF, 2.25WF, 2.85		*Cloacimonetes*, DF, 0.01% *Synergistales* WF, 0.4% DF, 0.1%	(Stolze et al. 2015)

TAN Total ammonia nitrogen, *ND* Not determined
*Phylum

Many microorganisms are able to metabolize glycerol in aerobic conditions; nevertheless, few are able to do it anaerobically. Species from the order *Enterobacteriales*, *Clostridiales*, *Lactobacillales*, *Bacillales*, and *Burkholderiales* have been reported to ferment glycerol in 1,3-propanediol and ethanol (Varrone et al. 2013; Yazdani and Gonzalez 2007; Zhou et al. 2017). More recently, sludge from brewery and glycerol used to methane production in a shock loading consortia acclimation showed that species from order *Thermotogales*, *Lactobacillales*, and *Clostridiales* were strongly dependent on the glycerol feeding system (Vásquez and Nakasaki 2016). Microbial dynamicity on glycerol fermentation has been also evaluated in anaerobic reactors overloaded with lipids, demonstrating a predominant order of *Selenomodales*, *Lactobacillales*, *Clostridiales*, and *Bacteroidales* (De Francisci et al. 2015). Lately, only one work has analyzed the enrichment of ammonia oxidation bacteria as *Candidatus Brocadia caroliniensis* from a full-scale process treating anaerobic digester effluent with the addition of glycerol. This worked attributed greatly the order *Brocadiales*, a partial transformation capability of glycerol (Park et al. 2017).

3.3 Acetogens and Syntrophic Acetate Oxidizers

Acetogens are mainly found in three phyla *Firmicutes*, *Acidobacteria*, and *Spirochaetes*. Nevertheless, most of them are inside of the first phylum and belong to *Clostridia* class (Scherer et al. 2018), as it can be observed in Table 5. In the study of St-Pierre and Wright (2014), the mentioned phyla were found in three full-scale digesters fed with cow manure as the main substrate. *Firmicutes* phylum was the most diverse and predominant in all digesters, the same occurred with *Clostridia* class. In contrast, the presence of *Negativicutes*, another class were acetogens can be found, was almost null (0.1% from *Firmicutes* reads). Interestingly, the dominant pathway for methane production affected *Clostridia* presence, showing less abundance when hydrogenotrophic methanogenesis prevail.

In an anaerobic digester fed with excess activated sludge, *Clostridium*, *Eubacterium*, *Thermoanaerobacter*, *Moorella* (all from *Clostridia* class) and *Treponema* (from *Spirochaetia* class) were the dominant acetogenic genus, but just the first two were among the top 50 in abundance. Furthermore, the prevalence of genes involved in the Wood-Ljungdahl pathway (i.e., acetate kinase and phosphate acetyltransferase) confirmed the constant formation of acetate and its role as precursor for CH_4 production; this latter was observed by the higher abundance of *Methanosaeta* (26.2% from total reads of methanogens) and *Methanosarcina* (12.8%) genera over hydrogenotrophic methanogens (*Methanospirillum*, 13.1%; *Methanoculleus*, 11.1%; *Methanoregula*, 7.6%) (Guo et al. 2015). A similar outcome was reported by Zhang et al. (2009) after the implementation of a Focused-Pulse treatment in a WWTP for biosolids removal enhancement. Here, microbial populations suffered a shift that caused the loss of hydrogenotrophic methanogens dominance against aceticlastic methanogenesis. In addition, an acetogenic group

Table 5 Taxonomical groups belonging to acetogenic bacteria from *Firmicutes* (1), *Acidobacteria* (2), and *Spirochaetes* (3) phylum (Müller and Frerichs 2013; Schink 1994; Ragsdale and Pierce 2008; complemented with NCBI taxonomic information)

Class	Order	Family	Genera and species
1.Clostridia	Clostridiales	*Clostridiaceae*	*Caloramator fervidus* *clostridium aceticum* *Oxobacter pfennigii* *Natronincola histidinovorans* *Tindallia californiensis*
		Eubacteriaceae	*Acetobacterium woodii* *Eubacterium limosum*
		Lachnospiraceae	*Acetitomaculum ruminis* *Marvinbryantia formatexigens* *Syntrophococcus sucromutans*
		Peptostreptococcaceae	*Acetoanaerobium noterae*
	Halanaerobiales	*Halobacteroidaceae*	*Acetohalobium arabaticum* *Natroniella acetigena*
	Thermoanaerobacterales	*Thermoanaerobacteraceae*	*Moorella glycerini* *Moorella thermoacetica* *Thermacetogenium phaeum* *Thermoanaerobacter kivui*
Negativicutes	*Selenomonadales*	*Sporomusaceae*	*Sporomusa ovata*
2.Holophagae	*Holophagales*	*Holophagaceae*	*Holophaga foetida*
3.Spirochaetia	*Spirochaetales*	*Spirochaetaceae*	*Treponema primitia*

called *Treponema primitia* was favored by the shift and increased its abundance from 7.1 to 11.5% (from *Spirochaetes* reads) even it is phylum reads decreased (18.8 to 13.2%), supporting acetate production.

As we have seen in digesters with cow manure as substrate, acetogens can also be found in reactors treating poultry or pig manure. Furthermore, their presence is not limited by awkward conditions such as the predominance of hydrogenotrophic methanogens. For example, in a pilot-scale digester exclusively fed with poultry manure, *Firmicutes* dominated bacterial abundance with 76%. Within it, *Clostridia* was composed of *Clostridiales* (64%) and *Thermoanaerobacterales* (11%). Two OTUs in *Clostridia* probably belonged to the last-mentioned order because they had a close similarity to *Moorella glycerini* and *Moorella thermoacetica*. Furthermore,

Table 6 Identified SAOB in *Firmicutes* (1) and *Thermotogae* (2) phylum (Hattori 2008; Maus et al. 2016; Ruiz-Sánchez et al. 2018; Westerholm et al. 2016; complemented with NCBI taxonomic information)

Class	Order	Family	Genera and species
1.Tissierellia	*Tissierellales*	*Tissierellaceae*	*Clostridium ultunense*
Clostridia	*Thermoanaerobacterales*	*Thermoanaerobacteraceae*	*Thermacetogenium phaeum*
		Thermoanaerobacterales family III. *Incertae sedis*	*Syntrophaceticus schinkii*
2.Thermotogae	*Thermotogales*	*Thermotogaceae*	*Thermotoga lettingae* *Thermotoga maritima*

1.6% of total OTUs abundance belonged to *Negativicutes* class and possibly to acetogenic microorganisms (Smith et al. 2014). However, aceticlastic methanogenesis did not prevail in the reactor, explaining the limited abundance of acetogens. The same prevalence of hydrogenotrophic methanogens was observed in another pilot-scale digester reported by Liu et al. (2009); pig manure was managed in this case. From microbial analysis, the abundant presence of phylum *Firmicutes* and *Spirochaetes* was observed with 47.2 and 13.2%, respectively. The former contained *Clostridia* class and most of its OTUs belonged to *Clostridiaceae* family, a striking source of acetogens, and the latter contained *Treponema* genus but not *T. primitia* species. More species related to homoacetogens were found, including *M. glycerini* and *Sporobacter termitidis*, but just comprised the 0.5 and 1% of total OTUs abundance.

Due to the prevalence of acetogens in *Clostridia* class, a common taxonomic classification for microorganisms participating in hydrolysis and acidogenesis phase, it is complicated to ensure which of them are present in biodigesters. Furthermore, direct production of acetate by fermentation can produce more confusion. Therefore, combination of metagenomic studies with the analysis of specific genes present in Wood-Ljungdahl pathway can be a useful method to present a clearer image of microbial species involved in this phase of the process. In addition, metatranscriptomic and metaproteomic studies can enhance this purpose.

Species belonging to SAOB are *Clostridium ultunense*, *Thermacetogenium phaeum*, and *Thermotoga lettingae*, the only mesophilic microorganism of the group is *C. ultunense*, and the rest are thermophilic. In addition, it has demonstrated ammonium resistance. Furthermore, the first two microorganisms have shown the ability to produce acetate with H_2 and CO_2 as substrates (Hattori 2008). Therefore, SAOB strains can belong to acetogenic bacteria. More examples of these microorganisms are given in Table 6.

The predominance of SAO for CH_4 production in biodigesters requires to overcome acetogenic bacteria and aceticlastic methanogens; this can be reached by their inhibition. It is known that both groups are susceptible to high ammonia concentrations (Chen et al. 2014); on the other hand, SAOB and hydrogenotrophic methanogens are more resistant to this compound (Ruiz-Sánchez et al. 2018). Therefore, it is feasible that in anaerobic digestion of nitrogen-rich compounds, which present constant release of large ammonia-nitrogen amounts, such as animal manures, slaughterhouse, and food wastes, supremacy of CH_4 production by SAO will be observed (Chen et al. 2014; Ruiz-Sánchez et al. 2018). Hence, according to ammonia content, microbial populations in biodigesters will differ (Ruiz-Sánchez et al. 2018).

A scenario that favors SAO occurs in thermophilic digesters, which are present in high metabolic rates, large OLR management, greater CH_4 production, and lower HRT (Zinder 1990) compared with mesophilic digesters. In a thermophilic biogas plant, *Tissierella* class (*Firmicutes* phylum) and a species confirmed as SAOB, *Tepidanaerobacter* and *Syntrophaceticus* (*Thermoanaerobacteraceae* family), were found. As it was expected, hydrogenotrophic methanogens (*Methanomicrobiales* and *Methanobacteriales* orders) dominate the digester (Maus et al. 2016). Due to that the identified SAOB species are very scarce, this grants the possibility that unknown taxonomic groups belong to them. In the mentioned biogas plant, *Defluviitoga tunisiensis* abundance overpassed by far other microorganisms. Albeit, this genus is not identified as a SAOB; it probably participates in syntrophy with hydrogenotrophic methanogens in the digestion of biomass (Maus et al. 2016). Stolze et al. (2016) suggested the same idea and confirmed strain's ability to produce H_2 in a thermophilic digester dominated by hydrogenotrophic methanogens.

Another investigation that strengths the taxonomic diversity of SAOB was the one done by Ruiz-Sánchez et al. (2018). Their investigation of microbial diversity in mesophilic full-scale CSTR fed with pig manure and agricultural wastes at high ammonia levels (6–7 g TAN/L) did not find known SAOB species. Instead, genera *Longilinea* and *Alloprevotella* from *Chloroflexi* and *Bacteroidetes* phylum, respectively, predominated along with hydrogenotrophic methanogens (*Methanoculleus* and *Methanomassiliicoccus*). Furthermore, reactors developed well, CH_4 content in biogas ranged between 66 and 74%, and it was positively correlated with TAN concentration (Ruiz-Sánchez et al. 2018).

Sun et al. (2014) investigated specifically the presence of SAOB and the dominant methanogenic pathway in 13 well-functioning biogas plants and three thermophilic, and the rest were mesophilic. All thermophilic and seven mesophilic SAO coupled with hydrogenotrophic methanogens prevailed. In contrast, the rest three mesophilic digesters were dominated by aceticlastic methanogenesis. Interestingly, SAO process was observed in co-digestion plants, while one substrate biogas plant was managed by aceticlastic methanogenesis. In addition, higher free ammonia levels were present in co-digestion plants compared with single substrate plants; the former ranged values between 0.16 and 0.82 g/L and the latter 0.03–0.09 g/L of free ammonia. The results from metagenomic reads demonstrated that in all digesters, *Syntrophaceticus schinkii* was present; clearly, in digesters dominated by SAO, their abundance was higher. Another representative microorganism was *Tepidanobacter acetotoxidans*; however, it wasn't found where aceticlastic methanogenesis predominated. The mesophilic *Clostridium ultunense* was limited to the digesters with high ammonia content, and *Thermacetogenium phaeum* showed the same behavior but in thermophilic conditions. Dominant methanogenic orders that accompanied SAOB were *Methanomicrobiales* and *Methanobacteriales* with more than 80% of methanogenic reads, overpassing the percent reached by aceticlastic methanogens in their digesters. The first order demonstrated high abundance in all digesters, and the second was preferentially found in thermophilic digesters (Sun et al. 2014).

It is generally thought that hydrogenotrophic methanogens contributes for a little fraction of the total methane produced. However, in digesters dominated by SAO, these methanogens dominate and can work under stressful conditions in a stable way. Therefore, the development of knowledge about the network involving this syntrophic relationship is an outstanding topic for a better understanding in process efficiency. Finally, Table 7 shows the microorganisms that have been identified in anaerobic full-scale reactors involved in acetate oxidation.

3.4 Methanogens

Methanogenesis is an antique process carried by methanogenic archaea which belongs to the phylum *Euryarchaeota*. These microorganisms are distributed around the planet, and they are the main source of methane emissions to the atmosphere. Up this date, seven taxonomic orders are acknowledged, each one grouping members with unique features. The methanogens' division included the orders *Methanobacteriales*, *Methanomicrobiales*, *Methanocellales*, *Methanopyrales*, *Methanococcales*, *Methanosarcinales*, and *Methanoplasmatales* (Alvarado et al. 2014). Albeit, in anaerobic biogas digesters, only *Methanobacteriales*, *Methanomicrobiales*, and *Methanosarcinales* group members were recognized (Alvarado et al. 2014). In biogas reactors, the amount of profitable methanogenesis is perhaps the most prominent indicator of a good performance and efficient.

It is well documented that 70% of methane production in biodigesters is carried out by the acetoclastic pathway, meanwhile the other 30% corresponding to the CO_2-reduction pathway. Members of *Methanobacteriales* and *Methanomicrobiales* utilize the hydrogenotrophic pathway. Hydrogen is commonly used as electron donor in this case, but some species also use formate and alcohols. On the other hand, *Methanosarcinales* are the most diverse in terms of metabolism. Acetate, hydrogen, format, ethanol, isopropanol, and methylated compounds can be metabolized by members from this order (Kendall and Boone 2006). Microbial community structure of archaeal communities has been evaluated in recent publications on full-scale biodigesters (Cheon et al. 2008; Werner et al. 2011; Regueiro et al. 2012; Sundberg et al. 2013; Li et al. 2015; Abendroth et al. 2015). Studies that describe the archaeal population in full-scale mesophilic biodigesters feed with dairy manure indicates a major prevalence of *Methanosarcina thermophila* with an abundance of 98.5% (St-Pierre and Wright 2013). In addition, Li et al. (2014) evaluated a mixed plug-flow loop reactor; the results indicate a high proportion (86%) assigned to the

Table 7 Presence of possible acetogenic and syntrophic acetate-oxidizing bacteria in anaerobic full-scale reactors

Substrate	Reactor type	HRT (D)	T (°C)	CH_4 yield (m^3/d)	kWh (month)	Phylum	%	Taxonomic classification	%	References
Acetogens										
Activated sludge	WWTP	5	35	1500	N/A	*Firmicutes*	12.5	*Clostridia*[C] *Negativicutes*[C]	72.5 3.32	(Guo et al. 2015)
Cattle manure, whey silage	Plug-flow digester	21	37.8	N/A	$1.29x10^5$	*Firmicutes*	52	*Clostridia*[C] *Negativicutes*[C]	44.4 0.1	(St-Pierre and Wright 2014)
Cattle manure, waste of ice cream factory	Plug-flow digester	25-27	38.3	N/A	$9.49x10^4$	*Firmicutes*	34.1	*Clostridia*[C] *Negativicutes*[C]	25.7 0.1	(St-Pierre and Wright 2014)
Cattle manure, oil waste	Plug-flow digester	30	36.1	N/A	$1.42x10^5$	*Firmicutes*	49.5	*Clostridia*[C] *Negativicutes*[C]	25.7 0.1	(St-Pierre and Wright 2014)
Primary sludge, waste, and activated sludge	WWTP	30-35	35-38	N/A	$6.39x10^5$	*Spirochaetes*	13.8	*Treponema primitia*[S]	11.5	(Rittmann et al. 2008; Zhang et al. 2009)
Poultry manure	Pilot-scale CSTR	29.5	56.7	7.088	N/A	*Firmicutes*	76	*Clostridia*[C] *Clostridiales*[O]*Thermoanaerobacterales*[O]	52 64 11	(Bombardiere et al. 2007; Smith et al. 2014.)
Pig manure	Pilot-scale digester	N/A	N/A	150	N/A	*Firmicutes* *Spirochaetes*	47.2 13.2	*Clostridia*[C] *Moorella glycerini*[S]	45.5 0.5	(Liu et al. 2009)

(continued)

Table 7 (continued)

Substrate	Reactor type	HRT (D)	T (°C)	CH_4 yield (m^3/d)	kWh (month)	Phylum	%	Taxonomic classification	%	References
								Sporobacter termitidis[S] *Treponema*[G]	1 4.2	
Syntrophic acetate oxidizers										
Maize and barley silage, cattle and pig manure	Thermophilic biogas plant (three digesters)	19.8	54	394.2	$5.13x10^3$	*Firmicutes* *Thermotogae*	36.5 7.1	*Tepidoanaerobacter*[G] *Tissierella*[G] *Defluviitoga*[G]	0.41 0.02 5.53	(Maus et al. 2016)
Poultry manure	Pilot-scale CSTR	29.5	56.7	7.088	N/A	*Thermotoga*	N/A	*Defluviiotoga teniensis*[C]	3	(Bombardiere et al. 2007; Smith et al. 2014)

HRT hydraulic retention time, *WWTP* wastewater treatment plant; *N/A* not available; [C]Class; [O]Order; [G]Genus; [S]Species system. Methanogenic communities possess a specialized metabolism, features that make them more likely to be inhibited genus *Methanosaeta* sp. Apart from that, a lagoon-type biodigester feed with pig waste showed a relative abundance of 52% and 42% of hydrogenotrophic *Methanospirillum* and acetoclastic *Methanosaeta*, respectively (Pampillón-González et al. 2017). Sundeberg et al. (2013) carried out a study in which 21 full-scale biogas digesters were evaluated. The microbial diversity indicates a prevalence of acetoclastic *Methanosaeta* sp. across all sewage sludge digesters. Meanwhile co-digestion at mesophilic conditions reactors was dominated by hydrogenotrophic *Methanoculleus* sp. and *Methanobrevibacter* sp. In addition, reactors operated under thermophilic conditions were dominated by *Methanobacterium* sp. Kirkegaard et al. (2017) evaluated 32 full-scale anaerobic digester systems fed with activated sludge. The report found that *Methanosaeta* sp. genus dominates the mesophilic reactors, and genus *Methanothermobacter* sp. was more abundant in thermophilic conditions over the 6-year period of the study.

4 Conclusions

The complexity of microbial diversity, their functional role, and its community interactions in a specialized environment such as biogas digester denote how particular is the phenomena behind AD process. Several microbial groups belonged to the phylum of *Firmicutes*, *Proteobacteria*, and *Bacteroidetes* indicate that AD have stronger relationship between community structure and its function rather than its environment. On the other hand, methanogens seem to have more heterogeneity across full-scale biogas digesters and might substrate and operational conditions be the main factors that affect methanogen populations. In this regard, hydrogenotrophic methanogens show a high relative abundance in more biodigesters.

The multifunctionally of this process and the recent advances in next-generation sequencing technology will allow a best understanding of the microbial populations and their responses to environmental gradients during the digestion course. Furthermore, a comprehensive analysis of the microbial populations in full-scale anaerobic digesters allows to create economical strategies to improve bioenergy production in form of methane.

References

Abendroth C, Vilanova C, Günther T, Luschnig O, Porcar M (2015) Eubacteria and archaea communities in seven mesophile anaerobic digester plants in Germany. Biotechnol Biofuels 8:87. https://doi.org/10.1186/s13068-015-0271-6

Alvarado A, Montañez-Hernández LE, Palacio-Molina SL, Oropeza Navarro R, Luévanos-Escareño MP, Balagurusamy N (2014) Microbial trophic interactions and mcrA gene expression in monitoring of anaerobic digesters. Front Microbiol 5:1–14. https://doi.org/10.3389/fmicb.2014.00597

Alves MM, Pereira MA, Sousa DZ, Cavaleiro AJ, Picavet M, Smidt H, Stams AJM (2009) Waste lipids to energy: how to optimize methane production from long-chain fatty acids (LCFA). Microb Biotechnol 2(5):538–550. https://doi.org/10.1111/j.1751-7915.2009.00100.x

Amani T, Nosrati M, Sreekrishnan TR (2010) Anaerobic digestion from the viewpoint of microbiological, chemical, and operational aspects — a review. Environ Rev 18(NA):255–278. https://doi.org/10.1139/A10-011

Andreesen J R, Bahl H, Gottschalk G (1989) Introduction to the Physiology and Biochemistry of the Genus Clostridium. En N. P. Minton & D. J. Clarke (Eds.), Clostridia (pp. 27-62). https://doi.org/10.1007/978-1-4757-9718-3_2

Angelidaki I, Ahring BK (1993) Thermophilic anaerobic digestion of livestock waste: the effect of ammonia. Appl Microbiol Biotechnol 38(4):560–564

Appels L, Lauwers J, Degrève J, Helsen L, Lievens H, Willems K, Van Impe J, Dewil R (2011) Anaerobic digestion in global bio-energy production: potential and research challenges. Renew Sust Energ Rev

Archer DB, Harris JE (1986) Methanogenic bacteria and methane production in various habitats. Soc Appl Bacteriol Symp Ser 13:185–223

Bajpai P (2017) "Basics of anaerobic digestion process". SpringerBriefs in applied sciences and technology, 7-12. https://doi.org/10.1007/978-981-10-4130-3_2

Bapteste É, Brochier C, Boucher Y (2005) Higher-level classification of the Archaea: evolution of methanogenesis and methanogens. Archaea 1:353–363

Barnett JW, Kerridge GJ, Russell JM (1994) Effluent treatment systems for the dairy industry. Australas Biotechnol 4:26–26

Biebl H, Menzel K, Zeng AP, Deckwer WD (1999) Microbial production of 1,3-propanediol. Appl Microbiol Biotechnol 52(3):289–297

Bombardiere J, Espinosa-Solares T, Domaschko M, Chatfield M (2007) Thermophilic anaerobic digester performance under different feed-loading frequency. Appl Biochem Biotechnol 137–140(1–12):765–775

Boontian N (2014) Conditions of the anaerobic digestion of biomass. International Journal of Biological, Biomolecular, Agricultural, Food and Biotechnological Engineering 8 (9):1036–1114

Buettner C, Noll M (2018) Differences in microbial key players in anaerobic degradation between biogas and sewage treatment plants. Int Biodeterior Biodegradation 133:124–132

Cazier EA, Trably E, Steyer JP, Escudie R (2015) Biomass hydrolysis inhibition at high partial pressure in solid-state anaerobic digestion. Bioresour Technol 190:106–113. https://doi.org/10.1016/j.biortech.2015.04.055

Chen Y, Liu H, Zheng X, Wang X, Wu J (2017) New method for enhancement of bioenergy production from municipal organic wastes via regulation of anaerobic fermentation process. Appl Energy 196:190–198. https://doi.org/10.1016/j.apenergy.2017.01.100

Chen JL, Ortiz R, Steele TWJ, Stuckey DC (2014) Toxicants inhibiting anaerobic digestion: a review. Biotechnol Adv 32(8):1523–1534

Chen X, Yuan H, Zou D, Liu Y, Zhu B, Chufo A, Jaffar M, Li X (2015) Improving biomethane yield by controlling fermentation type of acidogenic phase in two-phase anaerobic co-digestion of food waste and rice straw. Chem Eng J 273:254–260. https://doi.org/10.1016/j.cej.2015.03.067

Cheon J, Hidaka T, Mori S, Koshikawa H, Tsuno H (2008) Applicability of random cloning method to analyze microbial community in full-scale anaerobic digesters. J Biosci Bioeng 106:134–140. https://doi.org/10.1263/jbb.106.134

Cho K, Shin SG, Kim W, Lee J, Lee C, Hwang S (2017) Microbial community shifts in a farm-scale anaerobic digester treating swine waste: correlations between bacteria communities associated with hydrogenotrophic methanogens and environmental conditions. Sci Total Environ 601:167–176

Ciccoli R, Sperandei M, Petrazzuolo F, Broglia M, Chiarini L, Correnti A et al (2018) Anaerobic digestion of the above ground biomass of Jerusalem artichoke in a pilot plant: impact of the preservation method on the biogas yield and microbial community. Biomass Bioenergy 108:190–197

Cirne DG, Lehtomäki A, Björnsson L, Blackall LL (2007) Hydrolysis and microbial community analyses in two-stage anaerobic digestion of energy crops. J Appl Microbiol 103(3):516–527. https://doi.org/10.1111/j.1365-2672.2006.03270.x

De Francisci D, Kougias PG, Treu L, Campanaro S, Angelidaki I (2015) Microbial diversity and dynamicity of biogas reactors due to radical changes of feedstock composition. Bioresour Technol 176:56–64

de Vladar HP (2012) Amino acid fermentation at the origin of the genetic code. Biol Direct 7:6. https://doi.org/10.1186/1745-6150-7-6

Doloman A, Soboh Y, Walters AJ, Sims RC, Miller CD (2017) Qualitative analysis of microbial dynamics during anaerobic digestion of microalgal biomass in a UASB reactor. Int J Microbiol 2017:1–12. https://doi.org/10.1155/2017/5291283

Drake HL (1994) Acetogenesis. (H. L. Drake, Ed.), Acetogenesis. Boston, MA: Springer US

Felix CR, Ljungdahl LG (1993) The cellulosome: the exocellular organelle of clostridium. Annu Rev Microbiol 47(1):791–819. https://doi.org/10.1146/annurev.mi.47.100193.004043

Ferry JG (2011) Fundamentals of methanogenic pathways that are key to the biomethanation of complex biomass. Curr Opin Biotechnol 22:351–357. https://doi.org/10.1016/j.copbio.2011.04.011

García-Lozano M, Hernández-De Lira IO, Huber DH, Balagurusamy N (2019) Spatial variations of bacterial communities of an anaerobic lagoon-type biodigester fed with dairy manure. PRO 7:408. https://doi.org/10.3390/pr7070408

Guo J, Peng Y, Ni B-J, Han X, Fan L, Yuan Z (2015) Dissecting microbial community structure and methane-producing pathways of a full-scale anaerobic reactor digesting activated sludge from wastewater treatment by metagenomic sequencing. Microb Cell Factories 14(1):33

Hansen RC, Keener HM, Marugg C, Dick WA, Hoitink HAJ (1993) Composting of poultry manure. Design, Environmental, Microbiological and Utilization Aspects. Renaissance Publications, Ohio, Science and Engineering of Composting, pp 131–153

Hattori S (2008) Syntrophic acetate-oxidizing microbes in Methanogenic environments. Microbes Environ 23(2):118–127

He J, Wang X, Yin X, Li Q, Li X, Zhang Y, Deng Y (2018) Insights into biomethane production and microbial community succession during semi-continuous anaerobic digestion of waste cooking oil under different organic loading rates. AMB Expr 8:92. https://doi.org/10.1186/s13568-018-0623-2

Hedderich R, Whitman WB (2013) Physiology and biochemistry of the methane-producing Archaea. In: Rosenberg E, DeLong EF, Lory S, Stackebrandt E, Thompson F (eds) The prokaryotes: prokaryotic physiology and biochemistry (pp. 635–662). Springer. https://doi.org/10.1007/978-3-642-30141-4_81

Hedstrom L (2002) Serine protease mechanism and specificity. Chem Rev 102(12):4501–4524

Holst B, Lunde C, Lages F, Oliveira R, Lucas C, Kielland-Brandt MC (2000) GUP1 and its close homologue GUP2, encoding multimembrane-spanning proteins involved in active glycerol uptake in Saccharomyces cerevisiae. Mol Microbiol 37(1):108–124

Hungate RE, Smith W, Bauchop T, Yu I, Rabinowitz JC (1970) Formate as an intermediate in the bovine rumen fermentation. J Bacteriol 102:389–397

Kendall MM, Boone DR (2006) The order Methanosarcinales. In: Dworkin M, Falkow S, Rosenberg E, Schleifer K-H, Stackebrandt E (eds) The prokaryotes: volume 3: Archaea. Bacteria: Firmicutes, Actinomycetes (pp. 244-256). Springer. https://doi.org/10.1007/0-387-30743-5_12

Kirkegaard RH, McIlroy SJ, Kristensen JM, Nierychlo M, Karst SM, Dueholm MS, Albertsen M, Nielsen PH (2017) The impact of immigration on microbial community composition in full-scale anaerobic digesters. Sci Rep 7:1–11. https://doi.org/10.1038/s41598-017-09303-0

Klaucans E, Sams K (2018) Problems with fat, oil, and grease (FOG) in food industry wastewaters and recovered FOG recycling methods using anaerobic co-digestion: a short review. Key Eng Mater 762:61–68. https://doi.org/10.4028/www.scientific.net/kem.762.6

Koch K, Lübken M, Gehring T, Wichern M, Horn H (2010) Biogas from grass silage – measurements and modeling with ADM1. Bioresour Technol 101:8158–8165. https://doi.org/10.1016/j.biortech.2010.06.009

Kolesárová N, Hutňan M, Špalková V, Kuffa R, Bodík I (2011) Anaerobic treatment of biodiesel by-products in a pilot scale reactor. Chem Pap 65(4):447–453

Kovács E, Wirth R, Maróti G, Bagi Z, Nagy K, Minárovits J et al (2015) Augmented biogas production from protein-rich substrates and associated metagenomic changes. Bioresour Technol 178:254–261

Kovács E, Wirth R, Maróti G, Bagi Z, Rákhely G, Kovács KL (2013) Biogas production from protein-rich biomass: fed-batch anaerobic fermentation of casein and of pig blood and associated changes in microbial community composition. PLoS One 8(10). https://doi.org/10.1371/journal.pone.0077265

Kragl U, Aivasidis A (eds) (2005) Technology transfer in biotechnology: from lab to industry to production. Springer, Berlin

Lee J, Han G, Shin SG, Koo T, Cho K, Kim W, Hwang S (2016) Seasonal monitoring of bacteria and archaea in a full-scale thermophilic anaerobic digester treating food waste-recycling wastewater: correlations between microbial community characteristics and process variables. Chem Eng J 300:291–299

Lee S-H, Kang H-J, Lee YH, Lee TJ, Han K, Choi Y, Park H-D (2012) Monitoring bacterial community structure and variability in time scale in full-scale anaerobic digesters. J Environ Monit 14(7):1893. https://doi.org/10.1039/c2em10958a

Li Y-F, Chen P-H, Yu Z (2014) Spatial and temporal variations of microbial community in a mixed plug-flow loop reactor fed with dairy manure. Microb Biotechnol 7:332–346. https://doi.org/10.1111/1751-7915.12125

Li J, Rui J, Yao M, Zhang S, Yan X, Wang Y, Yan Z, Li X (2015) Substrate type and free ammonia determine bacterial community structure in full-scale Mesophilic anaerobic digesters treating cattle or swine manure. Front Microbiol 6. https://doi.org/10.3389/fmicb.2015.01337

Liu FH, Wang SB, Zhang JS, Zhang J, Yan X, Zhou HK et al (2009) The structure of the bacterial and archaeal community in a biogas digester as revealed by denaturing gradient gel electrophoresis and 16S rDNA sequencing analysis. J Appl Microbiol 106(3):952–966

Lu X, Wang H, Ma F, Zhao G, Wang S (2018) Improved process performance of the acidification phase in a two-stage anaerobic digestion of complex organic waste: effects of an iron oxide-zeolite additive. Bioresour Technol 262:169–176. https://doi.org/10.1016/j.biortech.2018.04.052

Mani S, Sundaram J, Das KC (2016) Process simulation and modeling: Anerobic digestion of complex organic matter. Biomass Bioenergy 93:158–167. https://doi.org/10.1016/j.biombioe.2016.07.018

Manyi-Loh CE, Mamphweli S, Meyer EL, Okoh AI, Makaka G, Simon M (2013) Microbial anaerobic digestion (bio-digesters) as an approach to the decontamination of animal wastes in pollution control and the generation of renewable energy. Int J Environ Res Public Health 10 (9):4390–4417. https://doi.org/10.3390/ijerph10094390

Maus I, Koeck DE, Cibis KG, Hahnke S et al (2016) Unraveling the microbiome of a thermophilic biogas plant by metagenome and metatranscriptome analysis complemented by characterization of bacterial and archaeal isolates. Biotechnol Biofuels 9(1):171. https://doi.org/10.1186/s13068-016-0581-3

Mead GC (1971) The amino acid-fermenting clostridia. J Gen Microbiol 67(1):47–56. https://doi.org/10.1099/00221287-67-1-47

Montañez-Hernández LE, Lira H-D, Rafael-Galindo G, Froto Madariaga L, Balagurusamy N (2018) Sustainable production of biogas from renewable sources: global overview. Scale Up

Opportunities and Potential Market Trends Sustainable Biotechnology-Enzymatic Resources of Renewable Energy:325–354. https://doi.org/10.1007/978-3-319-95480-6_13

Müller V, Frerichs J (2013) Acetogenic Bacteria. In eLS (pp. 1–9). Wiley, Chichester, UK

Nagamani B, Ramasamy K (1999) Biogas production technology: an Indian perspective. Curr Sci 77:44–55

Narihiro T, Sekiguchi Y (2007) Microbial communities in anaerobic digestion processes for waste and wastewater treatment: a microbiological update. Curr Opin Biotechnol 18(3):273–278. https://doi.org/10.1016/j.copbio.2007.04.003

Örlygsson J, Houwen FP, Svensson BH (1995) Thermophilic anaerobic amino acid degradation: deamination rates and end-product formation. Appl Microbiol Biotechnol 43(2):235–241. https://doi.org/10.1007/BF00172818

Ortseifen V, Stolze Y, Maus I, Sczyrba A, Bremges A, Albaum SP et al (2016) An integrated metagenome and-proteome analysis of the microbial community residing in a biogas production plant. J Biotechnol 231:268–279

Pampillón-González L, Ortiz-Cornejo NL, Luna-Guido M, Dendooven L, Navarro-Noya YE (2017) Archaeal and bacterial community structure in an anaerobic digestion reactor (lagoon type) used for biogas production at a pig farm. Microbiol Physiol 27(5):306–317. https://doi.org/10.1159/000479108

Park H, Brotto AC, van Loosdrecht MC, Chandran K (2017) Discovery and metagenomic analysis of an anammox bacterial enrichment related to Candidatus "Brocadia caroliniensis" in a full-scale glycerol-fed nitritation-denitritation separate centrate treatment process. Water Res 111:265–273

Pelletier E, Kreimeyer A, Bocs S, Rouy Z, Gyapay G, Chouari R et al (2008) "Candidatus Cloacamonas Acidaminovorans": genome sequence reconstruction provides a first glimpse of a new bacterial division. J Bacteriol 190(7):2572–2579. https://doi.org/10.1128/JB.01248-07

Ragsdale SW, Pierce E (2008) Acetogenesis and the wood–Ljungdahl pathway of CO2 fixation. Biochimica et Biophysica Acta (BBA) - Proteins and Proteomics 1784(12):1873–1898

Ramsay IR, Pullammanappallil PC (2001) Protein degradation during anaerobic wastewater treatment: derivation of stoichiometry. Biodegradation 12(4):247–256. https://doi.org/10.1023/A:1013116728817

Rasit N, Idris A, Harun R, Wan Ab Karim Ghani WA (2015) Effects of lipids inhibition on biogas production of anaerobic digestion from oily effluents and sludges: an overview. Renew Sust Energ Rev 45:351–358. https://doi.org/10.1016/j.rser.2015.01.066

Rawlings ND, Barrett AJ (1995) Evolutionary families of metallopeptidases. En Proteolytic Enzymes: Aspartic and Metallo Peptidases: Vol. 248. Methods in Enzymology (pp. 183-228). https://doi.org/10.1016/0076-6879(95)48015-3

Regueiro L, Veiga P, Figueroa M, Alonso-Gutierrez J, Stams AJM, Lema JM, Carballa M (2012) Relationship between microbial activity and microbial community structure in six full-scale anaerobic digesters. Microbiol Res 167:581–589. https://doi.org/10.1016/j.micres.2012.06.002

Ren Y, Yu M, Wu C, Wang Q, Gao M, Huang Q, Liu Y (2018) A comprehensive review on food waste anaerobic digestion: research updates and tendencies. Bioresour Technol 247:1069–1076. https://doi.org/10.1016/j.biortech.2017.09.109

Rittmann BE, Lee HS, Zhang H, Alder J, Banaszak JE, Lopez R (2008) Full-scale application of focused-pulsed pre-treatment for improving biosolids digestion and conversion to methane. Water Sci Technol 58(10):1895–1901

Ruiz-Sánchez J, Campanaro S, Guivernau M, Fernández B, Prenafeta-Boldú FX (2018) Effect of ammonia on the active microbiome and metagenome from stable full-scale digesters. Bioresour Technol 250(October 2017):513–522

Salama E, Saha S, Kurade MB, Dev S, Chang SW, Jeon B-H (2019) Recent trends in anaerobic co-digestion: fat, oil, and grease (FOG) for enhanced biomethanation. Prog Energy Combust Sci 70:22–42. https://doi.org/10.1016/j.pecs.2018.08.002

Satpathy P, Biernacki P, Uhlenhut F, Cypionka H, Steinigeweg S (2016) Modelling anaerobic digestion in a biogas reactor: ADM1 model development with lactate as an intermediate (part I). J Environ Sci Health A 51(14):1216–1225. https://doi.org/10.1080/10934529.2016.1212558

Sawatdeenarunat C, Nguyen D, Surendra KC, Shrestha S, Rajendran K, Oechsner H et al (2016) Anaerobic biorefinery: current status, challenges, and opportunities. Bioresour Technol 215:304–313. https://doi.org/10.1016/j.biortech.2016.03.074

Schaechter M (2009) Encyclopedia of microbiology (third edition). Elsevier Science

Scherer P, Klocke M, Pühler A, Off S, Schlüter A, Maus I, Hassa J (2018) Metagenome, metatranscriptome, and metaproteome approaches unraveled compositions and functional relationships of microbial communities residing in biogas plants. Appl Microbiol Biotechnol 102 (12):5045–5063

Schink B (1994) Diversity, ecology, and isolation of Acetogenic bacteria. In: Drake HL (ed) Acetogenesis. Chapman & Hall Microbiology Series (physiology / ecology / molecular biology / biotechnology). Springer, Boston, MA

Schink B (1997) Energetics of syntrophic cooperation in methanogenic degradation. Microbiology and Molecular Biology Reviews : MMBR 61(2):262–280

Schlüter, A., Bekel, T., Diaz, N. N., Dondrup, M., Eichenlaub, R., Gartemann, K.-H., … Goesmann, A. (2008). The metagenome of a biogas-producing microbial community of a production-scale biogas plant fermenter analysed by the 454-pyrosequencing technology

Schnürer A (2016) Biogas production: microbiology and technology. In: Hatti-Kaul R, Mamo G, Mattiasson B (eds) Anaerobes in biotechnology (pp. 195-234). Springer International Publishing. https://doi.org/10.1007/10_2016_5

Schuchmann K, Müller V (2016) Energetics and application of heterotrophy in Acetogenic bacteria. Appl Environ Microbiol 82(14):4056–4069

Shaw GT-W, Liu A-C, Weng C-Y, Chou C-Y, Wang D (2017) Inferring microbial interactions in thermophilic and mesophilic anaerobic digestion of hog waste. PLoS One 12(7):e0181395. https://doi.org/10.1371/journal.pone.0181395

Shen Y, Linville JL, Urgun-Demirtas M, Mintz MM, Synder SW (2015) An overview of biogas production and utilization at full-scale wastewater treatment plants (WWTPs) in the United States: challenges and opportunities towars energy-neutral WWTPs. Renew Sust Energ Rev 50:346–362. https://doi.org/10.1016/j.rser.2015.04.129

Siles JA, Martín M, Chica AF, Martín A (2009) Anaerobic digestion of glycerol derived from biodiesel manufacturing. Bioresour Technol 100(23):5609–5615. https://doi.org/10.1016/j.biortech.2009.06.017

Smith AM, Sharma D, Lappin-Scott H, Burton S, Huber DH (2014) Microbial community structure of a pilot-scale thermophilic anaerobic digester treating poultry litter. Appl Microbiol Biotechnol 98(5):2321–2334

Stams AJM, Plugge CM (2009) Electron transfer in syntrophic communities of anaerobic bacteria and archaea. Nat Rev Microbiol 7(8):568–577

Stolze Y, Bremges A, Rumming M, Henke C, Maus I, Pühler A et al (2016) Identification and genome reconstruction of abundant distinct taxa in microbiomes from one thermophilic and three mesophilic production-scale biogas plants. Biotechnol Biofuels 9(1):1–18

Stolze Y, Zakrzewski M, Maus I, Eikmeyer F, Jaenicke S et al (2015) Comparative metagenomics of biogas producing microbial communities from production-scale biogas plants operating under wet or dry fermentation conditions. Biotechnol Biofuels 8(1):14. https://doi.org/10.1186/s13068-014-0193-8

St-Pierre B, Wright A-DG (2013) Metagenomic analysis of methanogen populations in three full-scale mesophilic anaerobic manure digesters operated on dairy farms in Vermont, USA. Bioresour Technol 138:277–284. https://doi.org/10.1016/j.biortech.2013.03.188

St-Pierre B, Wright ADG (2014) Comparative metagenomic analysis of bacterial populations in three full-scale mesophilic anaerobic manure digesters. Appl Microbiol Biotechnol 98 (6):2709–2717

Sun L, Liu T, Müller B, Schnürer A (2016) The microbial community structure in industrial biogas plants influences the degradation rate of straw and cellulose in batch tests. Biotechnol Biofuels 9 (1):128

Sun L, Müller B, Westerholm M, Schnürer A (2014) Syntrophic acetate oxidation in industrial CSTR biogas digesters. J Biotechnol 171(1):39–44

Sun L, Pope PB, Ejisink VGH, Schnürer A (2015) Characterization of microbial community structure during continuous anaerobic digestion of straw and cow manure. Microb Biotechnol 8(5):815–827. https://doi.org/10.1111/1751-7915.12298

Sundberg C, Al-Soud WA, Larsson M, Alm E, Yekta SS, Svensson BH, Sørensen SJ, Karlsson A (2013) 454 pyrosequencing analyses of bacterial and archaeal richness in 21 full-scale biogas digesters. FEMS Microbiol Ecol 85(3):612–626. https://doi.org/10.1111/1574-6941.12148

Świątczak P, Cydzik A, Rusanowska P (2017) Microbiota of anaerobic digesters in a full-scale wastewater treatment plant. Archives of Environmental Protection 43(3):53–60

Tabatabaei M, Rahim RA, Abdullah N, Wright A-DG, Shirai Y, Sakai K, Sylaiman A, Hassan MA (2010) Importance of the methanogenic archaea populations in anaerobic wastewater treatments. Process Biochem 45(8):1214–1225. https://doi.org/10.1016/j.procbio.2010.05.017

Tian J-H, Pourcher A-M, Bureau C, Peu P (2017) Cellulose accessibility and microbial community in solid state anaerobic digestion of rape straw. Bioresour Technol 223:192–201. https://doi.org/10.1016/j.biortech.2016.10.009

Varrone C, Rosa S, Fiocchetti F, Giussani B, Izzo G, Massini G et al (2013) Enrichment of activated sludge for enhanced hydrogen production from crude glycerol. Int J Hydrog Energy 38 (3):1319–1331

Vásquez J, Nakasaki K (2016) Effects of shock loading versus stepwise acclimation on microbial consortia during the anaerobic digestion of glycerol. Biomass Bioenergy 86:129–135

Venkiteshwaran K, Bocher B, Maki J, Zitomer D (2016) Relating anaerobic digestion microbial community and process function. Microbiology Insights 8(S2):37–44. https://doi.org/10.4137/MBI.S33593

Viana M, Freitas A, Leitão R, Pinto G, Santaella S (2012) Anaerobic digestion of crude glycerol: a review. Environ Technol 1:81–92. https://doi.org/10.1080/09593330.2012.692723

Weinrich S, Nelles M (2015) Critical comparison of different model structures for the applied simulation of the anaerobic digestion of agricultural energy crops. Bioresour Technol 178:306–312. https://doi.org/10.1016/j.biortech.2014.10.138

Werner JJ, Knights D, Garcia ML, Scalfone NB, Smith S, Yarasheski K, Cummings TA, Beers AR, Knight R, Angenent LT (2011) "bacterial community structures are unique and resilient in full-scale bioenergy systems". Proceedings of the National Academy of Sciences of the United States of America, 108, 4158-4163. https://doi.org/10.1073/pnas.1015676108

Westerholm M, Moestedt J, Schnürer A (2016) Biogas production through syntrophic acetate oxidation and deliberate operating strategies for improved digester performance. Appl Energy 179:124–135. https://doi.org/10.1016/j.apenergy.2016.06.061

Yang Y, Tsukahara K, Sawayama S (2008) Biodegradation and methane production from glycerol-containing synthetic wastes with fixed-bed bioreactor under mesophilic and thermophilic anaerobic conditions. Process Biochem 43(4):362–367

Yang Z-H, Xu R, Zheng Y, Chen T, Zhao L-J, Li M (2016) Characterization of extracellular polymeric substances and microbial diversity in anaerobic co-digestion reactor treated sewage sludge with fat, oil, grease. Bio/Technology 212:164–173. https://doi.org/10.1016/j.biortech.2016.04.046

Yazdani SS, Gonzalez R (2007) Anaerobic fermentation of glycerol: a path to economic viability for the biofuels industry. Curr Opin Biotechnol 18(3):213–219

Zhang H, Banaszak JE, Parameswaran P, Alder J, Krajmalnik-Brown R, Rittmann BE (2009) Focused-pulsed sludge pre-treatment increases the bacterial diversity and relative abundance of acetoclastic methanogens in a full-scale anaerobic digester. Water Res 43(18):4517–4526

Zhou J-J, Shen J-T, Jiang L-L, Sun Y-Q, Mu Y, Xiu Z-L (2017) Selection and characterization of an anaerobic microbial consortium with high adaptation to crude glycerol for 1, 3-propanediol production. Appl Microbiol Biotechnol 101(15):5985–5996

Zhu N, Yang J, Ji L, Liu J, Yang Y, Yuan H (2016) Metagenomic and metaproteomic analyses of a corn Stover-adapted microbial consortium EMSD5 reveal its taxonomic and enzymatic basis for degrading lignocellulose. Biotechnol Biofuels 9(243):1–23. https://doi.org/10.1186/s13068-016-0658-z

Ziels RM, Karlsson A, Beck DAC, Ejlertsson J, Yekta SS, Bjorn A, Stensel HD, Svensson BH (2016) Microbial community adaptation influences long chain fatty acid conversion during anaerobic co digestion of fats, oils, and grease with municipal sludge. Water Res 103:372–382. https://doi.org/10.1016/j.watres.2016.07.043

Ziganshin AM, Liebetrau J, Pröter J, Kleinsteuber S (2013) Microbial community structure and dynamics during anaerobic digestion of various agricultural waste materials. Appl Microbiol Biotechnol 97(11):5161–5174. https://doi.org/10.1007/s00253-013-4867-0

Zinder SH (1990) Conversion of acetic acid to methane by thermophiles. FEMS Microbiol Lett 75 (2):125–137

Metabolic Engineering: A Tool to Increase the Methane Yield and Efficiency of Anaerobic Digestion Process

María José Castro Alonso, Mariel Rubí Macías Franco, Armando Pacheco Valenciana, and Lilia Ernestina Montañez-Hernández

Abstract Anaerobic digestion constitutes a mature biotechnology for organic waste treatment and bioenergy production. However, the process is still subject of bad performance or even failure due to critical factors such as incorrect management and lack of knowledge about the microbial activities that occur within the process. Over the years, several approaches have been developed in order to enhance the performance of anaerobic digestion systems. These usually address plant design, type of feedstock, monitoring parameters, and characterization of functional microbial communities, among others. In this background, metabolic engineering surges as a strategy to maximize the metabolic capabilities of key microorganisms with the purpose of increasing the methane yield. Although this area has remained largely unexploited in the context of anaerobic digestion, there are few reports about genetically engineered strains aimed to improve methane production in anaerobic reactors. Also, it is worth of mention the development of methane bioconversion processes mediated by engineered microorganisms. Hence, this chapter discusses the current situation of metabolic engineering regarding methane production and its subsequent conversion to other value-added products.

Keywords Metabolic engineering · Genetic tools · Anaerobic digestion

1 Introduction

Despite the widespread use of anaerobic digestion (AD) as a biotechnological process for the treatment of organic wastes with the consequent generation of bioenergy (Ahring et al. 1992; Jain et al. 2015), its performance is still subject to failure due to several environmental and biological factors (Chen et al. 2008; Chen et al. 2014). In order to avoid inhibition of the process and achieve higher methane

M. J. C. Alonso · M. R. M. Franco · A. P. Valenciana · L. E. Montañez-Hernández (✉)
Laboratorio de Biorremediación, Facultad de Ciencias Biológicas, Universidad Autónoma de Coahuila, Torreón, Coahuila, Mexico
e-mail: liliamontanezhernandez@uadec.edu.mx

N. Balagurusamy, A. K. Chandel (eds.), *Biogas Production*,
https://doi.org/10.1007/978-3-030-58827-4_11

yields in biogas, several strategies have been described, addressing reactor design (Mumme et al. 2010; Garfí et al. 2016), system configuration (Demirel and Yenigün 2002; Schievano et al. 2014; Lindner et al. 2016), type of feedstock and their pretreatment methods (Pitk et al. 2012; Lisboa and Lansing 2013; Zheng et al. 2014), monitoring and proper control of physicochemical parameters (Boe et al. 2010; Jimenez et al. 2015; Ferguson et al. 2016; Wu et al. 2019), and elucidation and description of microbial communities and their metabolic activities (Cabezas et al. 2015; Carballa et al. 2015; Lim et al. 2018), among others (Budzianowski 2016; Xia et al. 2016; Zema 2017).

Recently, metabolic engineering of microorganisms has increased its importance in the field of biotechnology due to the growing demand for renewable fuels and chemicals (Hollinshead et al. 2014; Merlin Christy et al. 2014; Zeldes et al. 2015; Bilal et al. 2018). Therefore, it is possible to conceive it as a strategy to increase methane production in AD by genetically altering the metabolism of key microorganisms in the process. To achieve this purpose, successful genetic tools and protocols must be available. These include broad choice of plasmids along with selection and counterselection markers, genome editing systems, DNA delivery systems, and procedures for promoter, codon, and ribosome binding optimization (Hollinshead et al. 2014).

In this regard, of all microorganisms involved in AD, acetogenic bacteria are the microbial group in which most research and progress have been made concerning strain engineering (Straub et al. 2014; Cho et al. 2015; Nybo et al. 2015; Bengelsdorf et al. 2016). However, these efforts have been directed toward the improvement of native products yields such as butanol, ethanol, 2,3-butanediol, acetate, and butyrate or for the production of non-native compounds such as acetone and isopropanol (Schiel-Bengelsdorf and Dürre 2012; Humphreys and Minton 2018).

Unfortunately, for the rest of microbial groups involved in AD, the lack or limitation of genetic tools, as consequence of the poor understanding of the metabolism, represents the first limitation for metabolic engineering. Furthermore, the effective performance of anaerobic systems is highly influenced by the metabolic cooperation between the microorganisms rather than the individual activity of a certain group (De Vrieze and Verstraete 2016). This complexity might make difficult the success of genetically modified microorganisms inside the process and may interfere in verifying that the methane production obtained is linked to the genetic modification introduced. Such is the case of bioaugmentation, an approach to enhance methane production in AD by adding specific microorganisms to the process, which until this day has shown inconsistent results regarding its efficacy (Nzila 2017).

However, these drawbacks can be overcome by increasing the knowledge about the composition of microbial groups, their metabolic capacities, and the nexus they establish, a task that has shown to progress in the last years (De Vrieze and Verstraete 2016; Rotaru et al. 2014a, b; Treu et al. 2016). Moreover, co-culture engineering has emerged as an option to improve productivity by relieving metabolic stress imposed when only one strain is employed (Zhang and Wang 2016). In this way, it is possible to design and construct polycultures in which each member can contribute by expressing different genes and performing specific pathways. Although co-culture engineering is still in its early stages of development and to this date only well-known metabolically modified microorganisms such as *Escherichia coli* and *Saccharomyces cerevisiae* have been employed, it represents a potential opportunity for the success of engineered microorganisms in AD process.

In this chapter, the current knowledge about metabolic engineering in the context of AD is presented and critically compared with the prevalent trends employed for improved methane yields. Moreover, bioconversion processes using methane as a feedstock are also discussed since they have heavily relayed on engineered strains.

2 Metabolic Engineering in Hydrolysis

The first step for the conversion of organic matter to methane AD is hydrolysis, where complex compounds of the substrates are reduced to monomeric products by the activity of extracellular enzymes excreted by fermentative bacteria (Oh et al. 2018; Yu et al. 2008; Salama et al. 2019). The products obtained are then transformed into alcohols, fatty acids, CO_2, and H_2, which will be consumed in the following steps of anaerobic digestion.

The microbial communities that intervene in the hydrolysis provide enzymes that help the degradation of specific compounds in such a way that the synergy between enzyme-substrate will determine the efficiency of the hydrolysis rate. Otherwise, this process can become null. Hydrolytic bacteria are present in different environments and feedstock; nevertheless the most studied sources of these microorganisms for AD are rumen, animal manure, and anaerobic sludge (Shrestha et al. 2017). Hydrolytic bacteria belonging to phyla of *Firmicutes* have been reported with high cellulolytic activity (Cirne et al. 2006), as shown in Table 1. Similarly, for the degradation of other polymers such as proteins, lipids, and other carbohydrates, the predominant phyla involved in this process are *Proteobacteria* and *Firmicutes* (Zhao et al. 2017).

Microbial hydrolysis is considered a limiting step in AD when high organic load rates are used (Li et al. 2018). In addition to the microbial groups present in AD, the composition of substrate and the particle size and temperature are other predominant

Table 1 Hydrolytic bacteria involved in lignocellulosic hydrolysis (Azman et al. 2015)

Phyla	Genus	Phyla	Genus
Firmicutes	*Clostridium* *Ruminococcus* *Caldicellulosiruptor* *Caldanaerobacter* *Butyrivibrio* *Acetivibrio* *Halocella* *Eubacterium*	*Bacteroidetes*	
		Fibrobacteres	
		Spirochaete	*Spirochaeta*
		Thermotogae	*Fervidobacterium* *Thermotoga*

factors that influence the velocity rate of hydrolysis (Ge et al. 2011). Therefore the time in which the substrate is hydrolyzed greatly impacts the progress of the rest of the steps. A slow hydrolysis rate indicates the scarce availability of carbon, while a fast hydrolysis rate propitiates the generation and accumulation of volatile fatty acids (VFA), which can cause inhibition of the process in short periods (10–12 days) (Cirne et al. 2006; Tian et al. 2017; Zhao et al. 2019). In this regard, substrates such as lignocellulosic biomass present great challenges for its conversion to methane by AD, due to its recalcitrant composition and lignin content (Cirne et al. 2006; Tsavkelova et al. 2018).

Several approaches have been developed in order to overcome the limitations of microbial hydrolysis rate for recalcitrant substrates like lignocellulosic biomass (Alfenore and Molina-Jouve 2016). The most common one is the application of a pretreatment step, which consists in disrupting the structure of cellulose, hemicellulose, and lignin through chemical, physical, or biological methods (Zheng et al. 2014). Although many of them have demonstrated their effectiveness in decreasing particle size and increasing methane production, their major drawbacks are the high costs associated with their implementation and the environmental impacts caused by the use of harsh chemicals in some methods. Another strategy employed to achieve high hydrolysis rate is bioaugmentation, which relays in the addition of specific cellulolytic microorganisms to AD systems treating lignocellulosic feedstock (Costa et al. 2012; Martin-Ryals et al. 2015; Ecem Öner et al. 2018). However, as previously stated, the effects of bioaugmentation over AD performance are still not clear (De Vrieze and Verstraete 2016).

In this scenery, genetic engineering surges as an attractive alternative to enhance microbial degradation of complex polymers (Merlin Christy et al. 2014). However, this approach has been commonly dedicated to create metabolic modifications in microorganisms such as *S. cerevisiae* and *E. coli* for the improvement of biofuel production, mainly bioethanol (den Haan et al. 2015; Bilal et al. 2018). Since the hydrolysis step of AD is a complex process carried out by the activity and cooperation of microbial consortia, the application of genetic modifications as a tool to improve hydrolytic activity might be impractical, compared with processes mediated by single microorganisms (Azman et al. 2015).

Nonetheless, efforts have been made to create strains capable of degrade efficiently lignocellulosic biomass. Such is the case of the study done by (Guedon et al. 2002), where they genetically modified *Clostridium cellulolyticum* for its adaptation to high amounts of carbon. It has been established that excessive production of pyruvate causes inhibition of energy metabolism and bacterial growth (Desvaux et al. 2000). Therefore, in order to overcome inhibition by high concentrations of cellulose, a heterologous expression of pyruvate decarboxylase and alcohol dehydrogenase from *Zymomonas mobilis* was achieved in *C. cellulolyticum*. The engineered strain reached higher cellulose consumption (150%) over the wild-type strain, with increased excretion of acetate, which is a key compound for the next steps of AD.

Although there are not directly involved with metabolic engineering, genetically modified enzymes also constitute a viable option to enhance hydrolysis. Several enzymes with interesting features such as thermostability and resistance to acids and alkalis have been reported (Manisha 2017). However, they are not further discussed in this section since enzymatic addition for improved hydrolysis is considered within pretreatment methods.

In conclusion, genetic engineering might be considered as great alternative for improving methane production in AD. Nonetheless, it has been difficult to consolidate this approach in the microorganisms involved in the hydrolysis step, despite the progress achieved for other biotechnological process such as bioethanol production. Moreover, approaches such as pretreatment step and bioaugmentation are still the first choices for the increase in hydrolysis rate and methane production. Therefore, the modification of specific microorganisms in hydrolysis step should be planned carefully and implemented according to the needs and features of the process.

3 Metabolic Engineering in Fermentation

3.1 Acidogenesis

Fermentation, or more commonly known as acidogenesis, is the fastest phase of anaerobic digestion (Deublein and Steinhauser 2011). In this stage, the products from hydrolysis are assimilated by fermentative microorganisms through their cells membranes and then transformed into fermentation products such as volatile fatty acids (VFAs), ethanol, lactate, CO_2, and H_2, among others. Although the main substrates for fermentation are monosaccharides and amino acids, long chain fatty acids, glycerol, and organic halogenates may also be metabolized (Angelidaki et al. 2011). Out of all fermentation products, VFAs are the most remarkable for AD process since they are the direct precursors for methanogenesis substrates, hence the use of the term acidogenesis (Akuzawa et al. 2011; Huang et al. 2015; Meegoda et al. 2018). They are also considered common monitoring parameters because their accumulation can lead to acidification of anaerobic reactors, resulting in process failure (Li et al. 2019). The most abundant VFAs in AD are acetic, propionic, and butyric acids, but isobutyric, valeric, and isovaleric acids may also be important intermediates (Bergman 1990).

Several bacterial groups such as *Clostridium*, *Acetobacter*, *Kluyveromyces*, *Propionibacterium*, *Moorela*, *Butyrivibrio*, and *Butyribacterium*, among others, have been recognized to produce VFAs (Bhatia and Yang 2017). These bacteria employ numerous pathways, some of which are summarized in Fig. 1.

Recently, VFAs have gained attention due to their use in textile, food, pharmaceutical, and plastic industries and their high added value (Bhatia and Yang 2017; Zhou et al. 2018). Therefore, several strategies have been developed in order to increase the microbial production of these compounds, involving improved hydrolysis rate, adjustment of process parameters, continuous removal of the acids produced, and metabolic engineering of bacterial strains (Zhou et al. 2018).

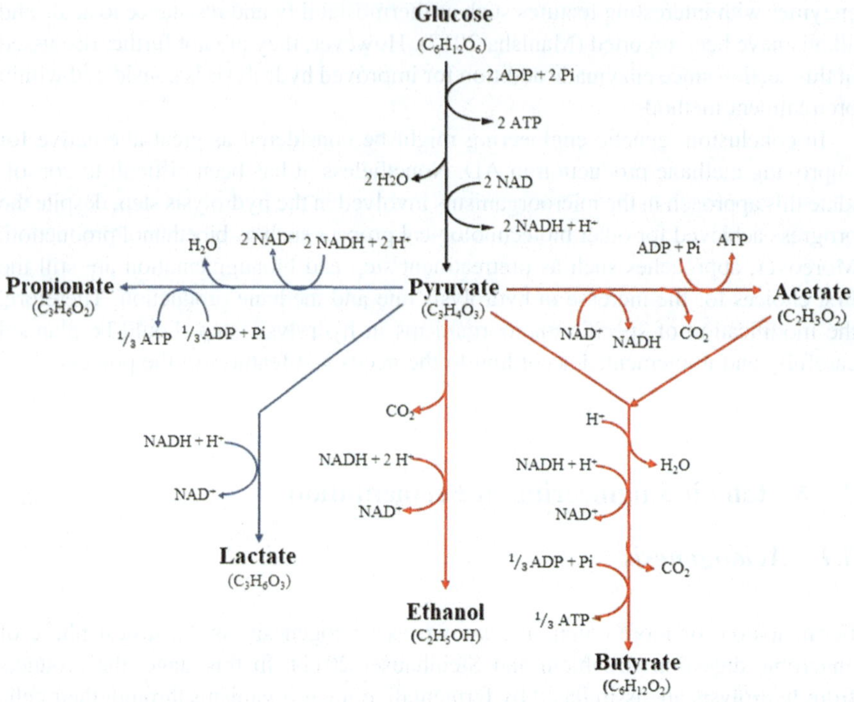

Fig. 1 Acidogenesis products from pyruvate. Marked in blue are products obtained by Embden-Meyerhof-Parnas, pentose phosphate, or Entner-Doudoroff pathways. Marked in red are products formed by Acetyl- CoA pathway

Increased propionic acid production has been reported in genetically modified *Propionibacterium* strains employing different approaches. These include the disruption of acetate kinase gene (*ack*) in *P. acidipropionici* (Suwannakham et al. 2006; Zhang and Yang 2009); the overexpression of endogenous enzymes such as propionyl-CoA:succinate CoA transferase in *P. freudenreichii* (Wang et al. 2015); and the expression of exogenous enzymes from *E. coli* and *Klebsiella pneumoniae* (e.g., phosphoenolpyruvate carboxylase, glycerol dehydrogenase, malate dehydrogenase, etc.), also in *P. freudenreichii* and in *P. jensenii* (Zhuge et al. 2013; Ammar et al. 2014; Liu et al. 2015; Liu et al. 2016). Moreover, it was demonstrated that deletion of lactate dehydrogenase also increased propionic acid production in *P. jensenii* (Liu et al. 2015).

Regarding butyric acid production, the metabolic engineering approaches have been focused in the deletion of key genes involved in acetate or solvent production pathways. For example, high butyric acid production was achieved in *Clostridium tyrobutyricum* by inactivating the genes encoding the enzymes phosphotransacetylase (*pta*) and acetate kinase (*ack*) (Zhu et al. 2005; Liu et al. 2006a). However, unexpected high acetic acid production was registered in the mutant strains but not in the wild type, indicating the possible activation of unknown pathways or activity of isoenzymes for acetic acid production (Liu et al. 2006b). Although *Clostridium acetobutylicum* has been recognized for its ability to produce solvents such as butanol, acetone, and ethanol, increased production of butyric acid has also been demonstrated by simultaneous disruption of the genes *pta*, Coenzyme-A transferase subunit B (*ctfB*), and aldehyde/alcohol dehydrogenase (*adhE1*) (Jang et al. 2013). The production and selectivity of butyric acid of *C. acetobutylicum* were further improved by knocking out the butyrate kinase I gene (*buk*), favoring the activity of butyrate kinase II (*bukII*), which presumably has higher specificity for butyryl phosphate (Jang et al. 2014). This mutant achieved butyric acid titers (32 g L^{-1}) comparable those obtained with *C. tyrobutyricum*, with also the highest selectivity reported in *Clostridium* species. Moreover, Jang et al. (2014) created a quintuple mutant strain by also knocking out the hydrogenase gene *hydA*, but in this case, the effect on butyric acid production was only significant at pH 5.

Although it is clear that important progress has been made in the development of mutant strains with high VFAs yields, the studies mentioned above are focused in the improvement of VFAs production as final products. It is not clear if the results of these reports can be applied with the purpose of increasing methane production in AD, considering that propionic and butyric acid accumulation promotes inhibition of the process. Nevertheless, resistance traits achieved in acidogenic bacteria might serve as a reference for the development of these features in other microorganisms that are very sensitive to organic acids (e.g., methanogens). For example, Guan et al. (2016) overexpressed several enzymes involved in arginine deaminase and glutamate decarboxylase systems in order to increase acid resistance in *P. jensenii*. Among the five overexpressed enzymes, glutamate decarboxylase (*gadB*) showed the best resistance to propionic acid with satisfactory yields. This study also investigated the changes in expression of related genes to the engineered pathways and the effects in amino acid distribution and metabolism with the purpose of understanding the mechanisms for acid resistance and production in the genetically modified strains.

Another area of acidogenesis that hasn't been fully explored for metabolic engineering is the fermentation of amino acids. Amino acids can be degraded by two mechanisms: the Stickland reaction and syntrophic degradation (Ramsay and Pullammanappallil 2001). The Stickland reaction is the common pathway for amino acid fermentation in AD and the most simple. This pathway usually involves the participation of two amino acids: one acts as electron donor, while the other acts as an electron acceptor, generating approximately 0.5 moles of ATP per mole of amino acid oxidized (Nisman 1954; Andreesen et al. 1989).

Several studies have identified the operons of the enzymes glycine reductase and D-proline reductase involved in the reductive branch of the Stickland reaction in *Clostridium* species (Fig. 2) (Graentzdoerffer et al. 2001; Bouillaut et al. 2013). Both enzymes are characterized for containing selenocysteine and by being induced in the presence of glycine, proline, and selenium in the medium (Kabisch et al. 1999; Jackson et al. 2006). Moreover, a regulator protein PrdR, which preferentially activates transcription of proline reductase genes, was identified in *C. difficile* (Bouillaut et al. 2013). Nevertheless, at present, there is no available information

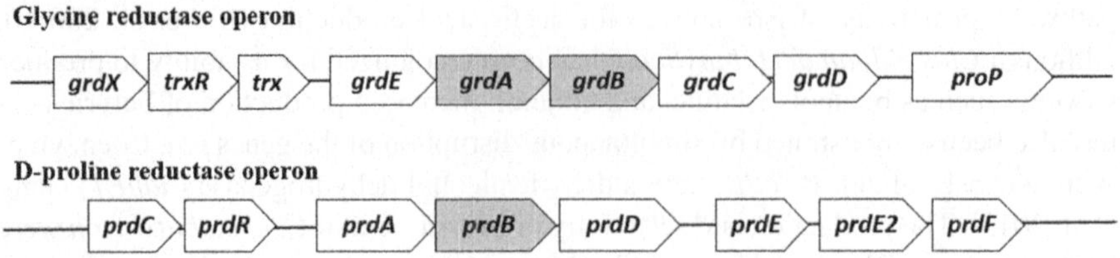

Fig. 2 Operons involved in the reductive branch of Stickland reaction in *C. difficile*. The *grdA*, *grdB*, and *prdB* subunits encode proteins that contain selenocysteine

about the regulation of Stickland reaction during AD, which limits the incursion of metabolic engineering strategies for improved amino acid fermentation in anaerobic reactors.

3.2 *Secondary Fermentations: Acetogenesis and Syntrophic Degradations*

Although acetic acid is actively produced during the acidogenesis step, it is one of the main products of secondary fermentations and a direct substrate for methanogenesis. Hence, the production of acetate from primary fermentation products is often considered a separate stage of AD. Secondary production of acetate can be achieved in two ways: (1) by the reduction of CO_2 and (2) by the syntrophic oxidation of longer VFAs, ethanol, and other substrates (Almeida et al. 2011; Gomez Camacho and Ruggeri 2018).

3.2.1 Acetogenesis

The term "acetogenesis" is often used to describe the reductive production of acetic acid from CO_2 rather than other metabolic processes with the same outcome (Drake et al. 2013). Microorganisms that perform acetogenesis employing the Wood-Ljungdahl pathway are known as acetogens. Acetogenic bacteria are highly diverse; however most of them are grouped in the phylum Firmicutes, and the most studied species belong to the genera *Clostridium*, *Acetobacterium*, *Moorella*, and *Ruminococcus*, among others (Insam et al. 2010; Drake et al. 2013). Although these microorganisms utilize a CO_2 fixation pathway, many of them can grow heterotrophically and are capable of producing other compounds besides acetic acid (Schiel-Bengelsdorf and Dürre 2012).

As stated in the introduction of this chapter, acetogens are one of the microbial groups involved in AD in which significant progress has been made regarding metabolic engineering. However, most of the research has focused in the production and process optimization of value-added compounds that aren't related to the performance of AD (Schiel-Bengelsdorf and Dürre 2012; Straub et al. 2014; Humphreys and Minton 2018). This is not necessarily a drawback since the genetic tools developed for such purposes may also be directed toward improvement of AD performance. A remarkable example is the genus *Clostridium*, which is comprised of several species of high interest in biotechnology due to their diverse metabolic capabilities such as degradation of lignocellulosic biomass and production of solvents, organic acids, biofuels, etc. Consequently, important advances have been made in the creation of genetic tools and methods (Table 2) in order to exploit the biotechnological potential of this genus (Heap et al. 2009; Mearls et al. 2015; Charubin et al. 2018; Joseph et al. 2018). As the presence and activity of several members of *Clostridium* are crucial in several stages of AD, the existence of sophisticated and efficient protocols for genetic manipulation could support a fast progress in the introduction of modified *Clostridium* strains in anaerobic systems.

Few papers have reported increased acetic acid production by engineered acetogens. For example, Straub et al. (2014) overexpressed either the four tetrahydrofolate-dependent enzymes of the methyl branch of the Wood-Ljungdahl pathway or the enzymes phosphotransacetylase and acetate kinase in *Acetobacterium woodii*. The modified strains registered increased acetate production with overall concentration of ~50 g L^{-1}, but only during autotrophic growth. In a similar fashion, a modified transformation protocol for *Clostridium ljungdahlii*, developed by Leang et al. (2013), was tested for the disruption of the genes *adhE1* and *adhE2*, encoding for bifunctional alcohol/aldehyde dehydrogenases. The genetic system successfully allowed the creation of either single or double deletion strains, which were further evaluated regarding their growth in fructose and their ethanol and acetate production. It was demonstrated that deletion of *adhE1*, but not of *adhE2*, decreased ethanol production ~6 times, while acetate production increased ~1.6 times, compared with the wild type strain.

3.2.2 Syntrophic Metabolism: Interspecies Electron Transfers

Syntrophy is defined as a mutualistic relationship between organisms, in which the exchanged intermediates must be kept at low concentrations in order to support the metabolic cooperation (Sieber et al. 2012; Kamagata 2015).

Interspecies electron transfer (IET) is the most important mechanism promoting syntrophic relations in anaerobic digestion (Morita et al. 2011; Leng et al. 2018). Therefore, the understanding of IET and syntrophic relations is essential for enhanced efficiency of methane production in biological-based engineering systems such as aerobic digesters (Gomez Camacho and Ruggeri 2018; Leng et al. 2018).

As stated before, a crucial factor in the inhibition of methanogenesis is the accumulation of short-chain fatty acids and alcohols as a result of slow metabolism of microbial communities (Zhao et al. 2015). Methane production is driven by terminal electron accepting reactions that involved tightly coupled metabolic associations (McInerney et al. 2008; Zhao et al. 2015). In these metabolic associations, fatty acid degraders transform short-chain fatty acids such as propionate and butyrate

Table 2 Genetic tools developed for manipulation of *Clostridium* species (Heap et al. 2009; Mearls et al. 2015; Charubin et al. 2018; Joseph et al. 2018)

Tools	Description
Transformation methods	Mediated by electroporation Conjugation with *E. coli*
Shuttle vectors	pMTL8000 series modular plasmids Each plasmid contains: 1. Gram + replicon: pBP1, pCB102, pCD6, and pIM13 2. Selectable marker (antibiotic resistance genes): chloramphenicol (*catP*), erythromycin (*ermB*), spectinomycin (*aad9*) and tetracycline (*tetA*) 3. Gram – replicon: ColE1 and p15a, with or without conjugal transfer function (*traJ*) 4. Application-specific component
Counterselection methods and markers	*Methods* Genetic complementation Allelic coupled exchange Toxin-antitoxin systems CRISPR/Cas9 system *Markers* Resistance to 5-fluoroorotic acid or 5-fluorouracil by deletion of orotate phosphoribosyltransferase (*pyrE*), orotidine-5-phosphate decarboxylase (*pyrF*), or uracil phosphoribosyltransferase (*upp*) genes Resistance to galactose by deletion of the galactose-1-phosphate uridylyltransferase (*galT*) gene in the *galKT* operon Resistance to 5-fluorocytosine by deletion of cytosine deaminase gene (*codA*) from *E. coli* Survival of the mutant cell in lactose by deletion of mRNA interferase (*mazF*) from *E. coli*
Reporter gene systems	*Enzymatic* Chloramphenicol acetyltransferase (*catP*) β-galactosidase (*lacZ*) from *Thermoanaerobacterium thermosulfurigens* EM1 β-glucuronidase (*gusA* or *uidA*) from *E. coli* Alkaline phosphatase (phoZ) from *Enterococcus faecalis* β-1,4-endoglucanase (*eglA*) from *Clostridium saccharobutylicum* *Bioluminescent* Luciferase (*luc* and *lucB*) from *Photinus pyralis* Luciferases (*luxAB*) from *Vibrio fischeri* *Fluorescence reporters* Green fluorescent protein (GFP) variants: YFP (yellow) and CFP (cyan) mCherryOpt FMN-based fluorescent proteins (FbFPs): PpFbFP and BsFbFP, LOV (light oxygen voltage), and improved LOV (iLOV)
Gene editing systems	*CRISPR/Cas9 system* *ClosTron system* Mobile Group II introns with retrotransposition-activated selection marker *Transposon-based systems*

(continued)

Table 2 (continued)

Tools	Description
	Tn916/Tn1545 family Mu phage and EZ-Tn5 Mariner-transposable Himar1
Promoters	*Constitutive* Phosphotransbutyrylase (*ptb*) and thiolase (*thl*) from *Clostridium acetobutylicum* and *Clostridium pasteurianum* *Inducible* Xylose promoter-repressor system from *Staphylococcus xylosus* Lactose promoter (*bgaRL*) from *Clostridium perfringens* Laminaribiose promoter from *Clostridium thermocellum* Arabinose promoter from *C. cellulolyticum* Tetracycline system from *Clostridium difficile* Anhydrotetracycline system from *C. acetobutylicum*

to acetate, CO_2, and electrons. Subsequently, methanogens accept the electrons from this metabolic process to reduce carbon dioxide to methane (Zhao et al. 2015; Cheng and Call 2016).

Metabolic interdependence between syntrophic microorganisms and methanogens is so extreme that degradation of substrates is not possible without close association between partners (Sieber et al. 2018). Several studies of anaerobic digesters have revealed that under standard conditions (P_{H2} of 1 atm, substrate and product concentrations of 1 M, pH 7 and temperature 298 K), degradation of butyrate and propionate to H_2, formate, and acetate by fermentative bacteria involves endergonic reactions. Unless low concentrations of H_2 (<10 Pa) and formate (<10 μm) are kept by hydrogenotrophic methanogens and acetotrophic methanogens, these reactions become thermodynamically favorable due to a change in redox potential from −410 mV and -420 mV to -260 mV and -290 mV for H_2 and formate, respectively (de Bok et al. 2004; Stams and Plugge 2009; Müller et al. 2010).

Over the years, interspecies H_2 transfer and interspecies formate transfer have been widely studied to determine their contribution as extracellular electron carriers in energetics and metabolism of syntrophic short-chain fatty acid oxidation with the purpose of improve stability of microbial functions in anaerobic reactors. Moreover, in the last years, other shuttle molecules and structures such as cytochromes (Qian et al. 2007), conductive pili (Summers et al. 2010), humic substances (Lovley et al. 1999), sulfur compounds (Biebl and Pfennig 1978), cysteine (Kaden et al. 2002), and carbon-based conductive particles (Dang et al. 2016) have been recognized to be involved in IET.

Nowadays, IET can be divided in two mechanisms: (1) mediated interspecies electron transfer, with H_2 and formate as the major electron carriers, and (2) direct interspecies electron transfer, with biological and non-biological conductive materials as electron carriers (McInerney et al. 2008; Cheng and Call 2016; Leng et al. 2018).

Mediated Interspecies Electron Transfer

Mediated interspecies electron transfer (MIET) also known as indirect interspecies electron transfer refers to metabolite exchange among two or more different species of microorganisms through chemical shuttle molecules, mainly H_2 and formate, as discussed above (Cheng and Call 2016; Gomez Camacho and Ruggeri 2018).

MIET has been recognized since the discovery of the key role of H_2 as electron carrier in metabolic interdependence of *S organism* and *Methanobacillus melianskii* (Bryant et al. 1967). Subsequently, other researchers found the essential function of formate as extracellular electron carrier (Zindel et al. 1988; Hattori et al. 2001; Zhang et al. 2014; Zhu et al. 2019). The first example of interspecies formate transfer without hydrogen transfer was syntrophic amino-acid metabolism between *Eubacterium acidaminophilum* and sulfate-reducing bacteria (Zindel et al. 1988). Since then, syntrophic associations have been described and analyzed over which chemical compound is predominant as electron carrier, especially among microorganisms involved in anaerobic digestion (McInerney et al. 2008; Müller et al. 2010; Sieber et al. 2012; Schink et al. 2017; Gomez Camacho and Ruggeri 2018).

However, the molecular and biochemical mechanisms involved in MIET are not completely understood. Although molecular analyzes have revealed multiple genes involved in energy-conserving metabolism for pathways leading to acetate formation from propionate and butyrate degradation (McInerney et al. 2008), the lack of information regarding this type of metabolism and the difficult cultivation of syntrophic microorganisms in laboratory conditions have restricted the inclusion of metabolic engineering in syntrophy for increased methane production. Nonetheless, it is worth of mention the advances that have been made concerning the understanding of this unique type of metabolism.

Propionate Metabolism

The oxidation of propionate by syntrophic metabolism involves two types of metabolic pathways with either formate or H_2 transfer or both at the same time. These are methylmalonyl-CoA (MMC) pathway and dismutation pathway (Müller et al. 2018).

In the MMC pathway, one ATP is generated by substrate-level phosphorylation, and three pairs of electrons are released in the form of H_2 and/or formate per one molecule of propionate degraded. The key reactions are the oxidations of succinate to fumarate ($E^{\circ\prime} = +30$ mV), malate to oxaloacetate ($E^{\circ\prime} = -176$ mV), and pyruvate to acetil-CoA and CO_2 ($E^{\circ\prime} = -470$ mV) (Liu and Lu 2018; Sieber et al. 2018). From these reactions, the oxidation of succinate to fumarate is the most energy-consuming since it requires an input of two-thirds of ATP for the generation of a proton gradient across the membrane in order to transfer electrons from succinate to H_2 (1Pa) or formate (10 μm) by means of reverse electron transfer (Van Kuijk et al. 1998; Stams and Plugge 2009; Schink and Stams 2013; Sieber et al. 2018).

The most studied microorganisms in propionate degradation through MMC pathway are *Pelotomaculum thermopropionicum* and *Syntrophobacter fumaroxidans*. *P. thermopropionicum* is the only propionate degrader with a complete set of arranged genes in their genome (mmc cluster; 10kbp in length) involved in MMC pathway (Kosaka et al. 2006, 2008; Kato et al. 2009; Müller et al. 2010). Also it has been demonstrated that *P. thermopropionicum* is unable to grow in propionate without the presence of methanogens and cannot use other electron acceptors such as sulfate, oxygen, nitrate, and metals (Kosaka et al. 2008; Kato et al. 2009; Worm et al. 2010, 2014).

Kosaka et al. (2006) showed that MMC pathway in *P. thermopropionicum* conforms the central catabolic pathway and it is linked with several peripheral pathways. They observed that catabolic pathways are regulated in response of some global cellular situations and co-culture conditions rather than available substrates since some genes of *mmc* cluster are physically linked to those for PAS-domain-containing regulator (PTH_1355). In bacteria, PAS proteins have functions related to the detection of changes in electron transport systems, and the PAS-domain-containing regulator in *P. thermopropionicum* is homologous to RocR of *Bacillus subtilis* (Calogero et al. 1994). Furthermore, Kosaka et al. (2006) found that putative σ^1 (σ54)-dependent promoter sequences were upstream the *mmc* cluster and sdh1 operon. Later, several studies have confirmed that genes of *mmc* cluster are expressed in response of co-culture conditions and specific substrates (Kato et al. 2009; Worm et al. 2011, 2014).

Many studies have showed that *P. thermopropionicum* have the ability to communicate with methanogens via flagellum, forming cell aggregates (Imachi et al. 2007; Shimoyama et al. 2009; Kato and Watanabe 2010). Genomic and proteomic analysis of co-cultures of *P. thermopropionicum* and *Methanothermobacter thermautotrophicus* demonstrated that filament cap protein (FliD) of *P. thermopropionicum* modified the genetic expression of over 50 genes of *M. thermautotrophicus*, increasing methanogenesis rate. Shimoyama et al. (2009) showed that up-regulated genes by FliD were involved in ATP synthesis and H_2 utilization during methanogenesis, which indicated that *M. thermautotrophicus* perceives signals from *P. thermopropionicum* FliD to initiate syntrophic association (Ishii et al. 2005; Kato and Watanabe 2010).

In contrast to MMC pathway, the second type of propionate metabolism, the dismutation pathway, is not well understood yet. It is only known from analysis of *Smithella propionica* that two propionate molecules are coupled to form a C6 intermediate and that this intermediate is rearranged to a 3-ketohexanoic acid intermediate before it is cleaved for its conversion to acetate and butyrate. Finally, these products are degraded via beta-oxidation. The enzymes and intermediates involved in these complex reactions have not been identified (de Bok et al. 2002, 2004; McInerney et al. 2008; Sieber et al. 2018).

Butyrate Metabolism

The most studied microorganisms involved in syntrophic degradation of short-chain fatty acids are *Syntrophomonas wolfei* (McInerney et al. 1981; Sieber et al. 2010), *Syntrophus aciditrophicus* (Jackson et al. 1999; McInerney et al. 2008),

Syntrophothermus lipocalidus (Sekiguchi et al. 2000), and *Syntrophosphora bryantii* (Dong et al. 1994). These microorganisms have demonstrated high degree of syntrophic specialization employing MIET, although some of them can also grow in pure cultures with unsaturated fatty acids such as crotonate as carbon source (Schink 1997; McInerney et al. 2008; Sieber et al. 2010, 2015, 2018).

Syntrophomonas wolfei is considered one of the model microorganisms for syntrophy due to its limited metabolic repertory, which makes it a syntrophic specialist. However, this feature makes difficult the incursion of genetic approaches aimed to study the molecular mechanisms involved in its syntrophic metabolism (Sieber et al. 2010). The co-culture between *S. wolfei* and *Methanospirillum hungatei* is the most studied system for syntrophic butyrate oxidation (McInerney et al. 1981; Müller et al. 2010; Sieber et al. 2010, 2015). β-oxidation is the pathway employed by *S. wolfei*, and multiple copies of the enzymes involved in β-oxidation are found in its genome, indicating possible differential expression under different environmental conditions (Sieber et al. 2010). Overall, one mole of ATP is produced per mole of butyrate, by means of substrate-level phosphorylation (Worm et al. 2010). However, over two-thirds of the ATP formed is spent in reverse electron transport mechanisms, with the subsequent production of either H_2 or formate (Schmidt et al. 2013; Crable et al. 2016; Sieber et al. 2018).

Indeed, most of the research done in *S. wolfei* growing syntrophically in butyrate has focused in the identification of the elements involved in reverse electron transfer mechanisms (Sieber et al. 2012, 2018; Müller et al. 2018). Proteomic analyses performed in syntrophic butyrate degradation co-cultures of *S. wolfei* and *M. hungatei* have indicated that an electron transfer flavoprotein (ETF) and an iron-sulfur oxidoreductase are implicated in the transference of electrons from butyryl-CoA to outer hydrogenases or formate dehydrogenases via menaquinone, through a proton gradient (Schmidt et al. 2013; Sieber et al. 2018). In the chemiosmotic scheme described in Crable et al. (2016), two protons from outside the membrane are collected during the reduction of menaquinone by the iron-sulfur oxidoreductase and then released inside the cell when menaquinol is oxidized by an outer hydrogenase or formate dehydrogenase. The hydrolysis of ATP by the ATP synthase, the inward movement of protons already described, and their consumption during H_2 and formate production drive this reverse electron transfer mechanism.

Finally, in contrast to syntrophic propionate degradation, communication mechanisms between *S. wolfei* and its methanogen partner haven't been reported. The presence of flagella has been observed in cultures of *S. wolfei*, and genes encoding flagellar proteins have been identified in its genome (Sieber et al. 2010). However, it is unknown if flagellar proteins are involved in cell-to-cell communication, as in the case of *P. thermopropionicum* (Shimoyama et al. 2009; Kato and Watanabe 2010). On the other hand, Sieber et al. (2015) detected 83 proteins unique to a butyrate degrading *S. wolfei* in co-culture with *M. hungatei*. Seven of them corresponded to putative transcriptional regulators, which might be involved in the perception of the environmental conditions and the physiological responses to them during syntrophic growth.

Direct Interspecies Electron Transfer

In 2010, Summers et al. demonstrated another type of mechanism involved in IET, direct interspecies electron transfer (DIET). In their study, biological electrical connections, such as e-pili and the multiheme c-type cytochrome OmcS, rather than interspecies hydrogen transfer favored the electron exchange between *Geobacter metallireducens* and *Geobacter sulfurreducens*, during growth in co-culture metabolizing ethanol. Since then, two other DIET mechanisms such as electron transfer through conductive materials and electron exchange through transport proteins in outer cell membranes have also been proposed (Lovley 2017).

Direct evidence of DIET related to AD is detailed in the reports of Rotaru et al. (2014a, b) and Holmes et al. (2018, 2017), in which it was found that co-cultures of *G. metallireducens* with *Methanotrix harundinacea* or *Methanosarcina barkeri* formed biological electrical connections through conductive pili and c-type cytochromes, during ethanol conversion to methane. The stoichiometric ethanol consumption and methane production of the co-cultures, the formation of microbial aggregates in the medium, the inability of co-cultures containing PilA-defficient *G. metallireducens* strains to metabolize ethanol and produce methane, and the transcriptomic analysis of *M. harundicea* and *M. barkeri* demonstrated the existence of DIET in their experiments.

In contrast, the majority of the research involving complex systems has focused in the addition of conductive materials such as granular activated carbon, carbon cloth, graphene, biochar, and magnetite into anaerobic reactors with the purpose of stimulating DIET and, thus, increasing the methane production (Baek et al. 2018). The settings and outcomes of this kind of studies have been already reviewed and discussed (Barua and Dhar 2017; Baek et al. 2018; Park et al. 2018); however, the observed effects of this approach over methane production are broadly inconsistent with those observed in co-cultures and might not be related to the existence of DIET at all (Cheng and Call 2016). Therefore, it is necessary to design strategies that validate the presence of DIET in anaerobic reactors as well as its role in AD performance (Van Steendam et al. 2019).

In conclusion, as neither of the DIET strategies has been extensively studied and, therefore, knowledge about the molecular and biochemical processes as well as the thermodynamics involved in this type of IET is largely incomplete, metabolic engineering in this area remains a challenging task.

4 Metabolic Engineering in Methanogenesis

The methanogenesis pathways are known for being part of the most ancient metabolism for energy generation and carbon fixation in the Archaea domain (Borrel et al. 2016). Methanogenesis also comprises the last step of anaerobic digestion, which is the most crucial phase if methane is the main product of interest (Lee et al. 2017).

To this day, methanogenic archaea is distributed in seven taxonomic orders: *Methanopyrales*, *Methanobacteriales*, *Methanopyrales*, *Methanosarcinales*, *Methanomicrobiales*, *Methanocellales*, and *Methanomassiliicoccales* (Dridi et al. 2012; Hedderich and Whitman 2013; Sakai et al. 2008). However, this group of microorganisms is commonly divided according to the different pathways they use to obtain energy and produce methane, which in turn depend on the substrate that is consumed. In general, methanogens can utilize as substrates acetate (aceticlastic pathway), other methyl-containing C1 compounds such as methanol and methylamines (methylotrophic pathway), and CO_2 and H_2 (hydrogenotrophic pathway) (Thauer et al. 2008; Alvarado et al. 2014). Members of all orders except *Methanomassiliicoccales* can perform hydrogenotrophic methanogenesis, while only some members of *Methanosarcinales* carry out aceticlastic pathway. In contrast, methylotrophic methanogenesis is a pathway found in the orders *Methanomassiliicoccales*, *Methanomicrobiales*, and *Methanosarcinales* (Enzmann et al. 2018).

Due to the increasing demand for renewable energy sources such as methane, a great amount of research has been aimed to the improvement of the performance of anaerobic digestion reactors along with the development of real-time monitoring strategies. Since methanogenesis step is where the product of interest is formed, most of the novel approaches have been directed toward enhancing the conditions for growth and metabolic activity of methanogens, which in turn will increase methane production (Demirel and Scherer 2008; Munk et al. 2012). In this regard, genetic engineering represents a currently unexploited option that could improve methane yield by manipulating the metabolic pathways of methanogens. The incorporation or even deletion of key genes in target methanogenic species could achieve such thing. Likewise, metabolic engineering could also help to improve the structure and catalytic properties of different proteins, creating enzymes that will be more resistant and stable to a range of operational conditions. To fully implement this approach, efficient methods to isolate and culture methanogenic populations as well as techniques to clone and transfer genetic material are needed.

In methanogens, genetic techniques and tools have been developed for *Methanosarcina* and *Methanoccocus* species (Costa and Leigh 2014) as described in Table 3. In general, transformation methods in both species are mediated by liposome and polyethylene glycol (PEG) with puromycin resistance as a genetic marker (Sarmiento et al. 2011; Kohler and Metcalf 2012). However, new genetic tools need to be developed in order to make a progress on the understanding of the methanogenesis (Enzmann et al. 2018). Recently, Jennings (2018) attempted a faster cloning process for the genera *Methanosarcina* and *Methanosaeta* by optimizing a genetic cloning kit (Gibson kit), which consists in the construction of linearized plasmids from several DNA fragments in a single step (Gibson et al. 2009). Although the experiments had no positive results due to primer failure, this study demonstrates that further research on this topic still continues.

Table 3 Genetic tools and techniques developed for methanogens (Sarmiento et al. 2011; Kohler and Metcalf 2012)

	Methanosarcina species	*Methanococcus* species
Transformation methods	Mediated by liposome and polyethylene glycol (PEG)	Natural transformation. Mediated by protoplasting, polyethylene glycol (PEG) and liposome
Genetic markers	Resistance to puromycin by introduction of the *pac* gene (puromycin transacetylase) from *Streptomyces alboniger* Resistance to pseudomonic acid by mutagenesis of the *ileS12*gene (isoleucyl-tRNA synthetase) Counterselection with 8-aza-2, 6-diaminopurine (8ADP) by deletion of the *hpt* gene (hypoxanthine phosphoribosyltransferase)	Resistance to puromycin as described for *Methanosarcina* Resistance to neomycin by cloning of the genes APH30 I and APH30 II (aminoglycoside phosphotransferase) Counterselection with 6-azauracil by deletion of the *upt* gene (uracil phosphoribosyltransferase)
Reporter gene systems	Based in *uidA* gene, encoding the β-glucuronidase from *E. coli*, e.g., Plasmid pAB79	Based in *uidA* and *lacZ* genes, encoding β-glucuronidase and β-galactosidase, e.g., vector Mip1
Shuttle vectors	Contain *pac* cassette and pC2A replicon for selection and replication, respectively, in *Methanosarcina.*For *E. coli*, contain ampicillin (*bla*), chloramphenicol (*cat*) or kanamycin-resistance (*aph*) genes for selection, and pMB1ori, oriR6Kγ, or oriS for replication, e.g., Plasmid pWM321	Plasmids most frequently used (pMEV1 and pMEV2) contain *pac* cassette or genes APH30 I and APH30 II for selection, and ORFless1–2 for replication in *M. maripaludisFor E. coli*, contain ampicillin (*bla*) for selection and ColE1 for replication. Include *lacZ* gene under control of histone promoter PhmvA
Integration systems	Contain the *int* gene (host factor independent integrase) and integration site from *Streptomyces* bacteriophage ΦC31. Under control of constitutive (pmcrB) or tetracycline controlled (pmcrB[tetO1/03/04]) promoters. Might include λ attachment sites for plasmid retrofitting, e.g., Plasmid pMP44	First plasmids (Mip1 and Mip2) contain *pac* cassette for selection and *hisA* gene for homologous recombination Common plasmid piJ03 contain *pac* cassette for selection, flanked by multicloning sites
Transposon methods for mutagenesis	Modified transposon system from *mariner* transposable element *Himar1*, containing *pac* cassette and *aph* gene as selection markers in *Methanosarcina* and *E. coli*, respectively	Mudpur transposon system containing *pac* cassette as selection marker System based in Tn5 transposable element with *pac* cassette and *aph* gene as selection markers in *M. maripaludis* and *E. coli*, respectively

4.1 Genome-Scale Metabolic Reconstruction of Methanogens

Before trying to implement metabolic engineering approaches to create new strategies for increasing methane production, a better understanding about the metabolism and gene expression in methanogens is needed. In this regard, genome-scale metabolic reconstruction (GSMR) is a scheme that has helped to provide valuable information for metabolic engineering purposes. The goal of GSMR studies is to discover and understand metabolic scenarios of an organism through biochemical knowledge that has been already compiled and computational analysis. However, this tool could be prone to a lot of errors in model predictions because reconstructions heavily relay on genomes that have been automatically annotated. Therefore, most of the results generated from these analysis need to be corrected by manually curating the reconstruction. Additionally, the model will likely require to be tested experimentally (Schellenberger et al. 2010; Richards et al. 2016).

A recent genome-scale metabolic reconstruction was conducted by Richards et al. (2016), focusing on the hydrogenotrophic metabolism of *Methanococcus maripaludis* to support future metabolic engineering research. The authors added to the reconstruction pathways for uncommon methanogenic coenzymes as well as other coenzymes for lipid and sugar metabolism and demonstrated that ferredoxin reduction by electron bifurcation is essential for growth of *M. maripaludis*. In this case, the energy-converting hydrogenase (Eha) functions as an alternative for the reduction of ferredoxin. This enzyme, however, requires the presence of a sodium ion gradient that limits ATP synthesis and, therefore, growth. This conclusion was very similar to the one obtained by the work of Lie et al. (2012). They previously had also proposed a cyclical pathway that involved electron bifurcation, which nowadays is a better representation of hydrogenotrophic methanogenesis performed by methanogens that doesn't have cytochromes.

It is worth to mention that in the reconstruction of Richards et al. (2016), it was highlighted the fact that *M. maripaludis* can't grow in the presence of acetate as energy source despite data supporting assimilation of acetate as carbon source. The model didn't show thermodynamic constrains for growth on acetate; however the authors hypothesized that the anapletoric role of hydrogenases like Eha and Ehb would prevent aceticlastic pathway in *M. maripaludis*, as these hydrogenases are highly important for the oxidation of ferredoxin in *Methanosarcina barkeri*, an aceticlastic methanogen. Finally, another important modification they tested was the substitution of a sulfate transporter and sulfate reduction pathway for one that utilizes H_2S, in order to comprehend the mechanisms involved in sulfur assimilation by *M. maripaludis*.

Other genome-scale metabolic reconstructions have been done employing different methanogens, and their findings are summarized in Table 4.

Despite the valuable information that GSMR studies have provided using only methanogens, Bizukojc et al. (2010) took a step further and applied genome reconstruction on a syntrophic association between *Clostridium butyricum* and *Methanosarcina mazei*, in order to characterize the metabolic products generated

Table 4 Genome-scale metabolic reconstructions of different methanogens and their proposed proteins and genes involved in methanogenesis and cellular growth

Microorganism	Proteins and genes of interest included in the model	Function	Reference
Methanosarcina acetivorans C2A	Coenzymes and cofactors of methanogenesis	Acetoclastic, methylotrophic and hydrogenotrophic pathway	(Benedict et al. 2012)
	Heterodisulfide reductase complexes (HdrABC and HdrDE)	Growth in methylated compounds	
	Rnf complex and F_{420}-ferredoxin oxidoreductase reaction	Regeneration of F_{420}	
	Methenyl-tetrahydrosarcinopterin cyclohydrolase (Mch)	Essential for growth on acetate	
	$F_{420}H_2$ dehydrogenase complex (Fpo)	Necessary for growth on acetate and for regeneration F_{420}	
	Exclusion of F_{420} reducing hydrogenase (Frh) and methanophenazine-dependent hydrogenase (Vht)	Inactive and non-functional hydrogenases	
	Exclusion of pentose phosphate pathway	No genes are found encoding the oxidative pathway	
	Addition of a ribulose-5-phosphate pathway from *Methanocaldococcus jannaschii*	Synthesis of ribulose-5-phospate with generation of CH_2S	
	Addition of phosphotransacetylase (Pta) and acetate kinase (Ack)	Ability to grow on CO	
Methanosarcina barkeri fusaro	Heterodisulfide reductase complex HdrDE	Translocation of two H^+ across the membrane per CH_4 produced	(Gonnerman et al. 2013)
	F_{420} reducing hydrogenase (Frh)	Production of hydrogen	
	Methanophenazine-dependent hydrogenase (Vht)	Production of two H^+ from H_2	
	Addition of cofactor 5-hydroxybenzimidazolycolbamide (5HBC)	Cofactor for methyl transfer reactions	
	N5-methyl-tetrahydrosarcinapterin:coenzyme M methyltransferase (Mtr)	Translocation of two Na^+	
	Na+/H+ antiporter	Translocation of two Na^+	
	Energy converting hydrogenase (Ech)	Translocation of two H^+	

(continued)

Table 4 (continued)

Microorganism	Proteins and genes of interest included in the model	Function	Reference
	Addition the S-adenosyl-L-methionine	Heme synthesis	
	Addition of dichlorination mechanism	Dichlorination	
	Addition of a ribose-5-phosphate pathway	Synthesis pathway	
	Addition a new coenzyme M pathway	Biosynthesis pathway	
Methanococcus maripaludis S2	Addition of synthesis pathways for proline, methionine, glycine, histidine, and cysteine	Biomass precursors	(Goyal et al. 2014)
	Addition of synthesis pathways for coenzyme B, coenzyme M, FAD, and H_4MPT	Cofactors	
	Addition of synthesis pathways for folic acid, riboflavin and cobalamin	Vitamins	
	Addition of histidinol-phosphatase (Hisj)	Synthesis of histidine	
	Carbon monoxide dehydrogenase/acetyl-CoA synthase (CODH/ACS)	Synthesis of acetyl-CoA	
	Formate dehydrogenases (*fdhA* and *fdhB*)	Generation of $F_{420}H_2$	
	Expression of *nif* and *glnA* genes	N_2 fixation, unique in hydrogenotrophic methanogenesis	
	Alanine as a N and C source	Increases Methane Evolution Rate (MER)	
	Nitrogenase	Increase of MER	
	5,10-methylen – H_4MPT reductase	Increase of MER	
	High activity of coenzyme F_{420} hydrogenase with reduction of CO_2	Increase of MER	
	Deletion of adenylate kinase (adkA), acetate CoA ligase (acd), and malate dehydrogenase (mdh)	Increase of MER	
	Methylene dehydrogenase and hmd	Increase of methane production	

from biodiesel degradation. In the first place, *C. butyricum* is known to produce 1,3-propanediol (PDO) and other by-products like acetate, formate, butyrate, and even methanol from glycerol. Because these last minor products are inhibitory for the growth and production of PDO, the authors proposed their consumption by *M. mazei* due to its ability to produce methane principally from methanol and acetate. In the first place, they identified two pathways in *C. butyricum*. The catabolic pathway involved the transformation of glycerol to PDO and other by-products, while the anabolic pathway comprised formation of amino acids. On the other hand, the reconstruction of *M. mazei* omitted the intermediate formyl-MFR (methanofuran) in the hydrogenotrophic pathway for simplicity reasons. Since the main goal was to increase the consumption of *C. butyricum* by-products by *M. mazei*, the authors tested several sceneries to achieve that. They found out that methanogenic activity was stimulated by the presence of methanol in the medium, resulting in a faster depletion of acetate and formate, which was further favored by no hydrogen production by C. *butyricum*. The results obtained in this model have demonstrated that it is possible to employ GSMR as a tool to predict the behavior of a syntrophic culture under different sceneries, generating information that will facilitate the interpretation of further experiments and will aid in the understanding of the nexus that methanogens established with other microorganisms.

Finally, a recent genome analysis reported the comparison of four methanogen species belonging to three different classes (Gilmore et al. 2017). The methanogen species selected were *Methanobacterium bryantii* and *Methanosphaera cuniculi* as Class I methanogens, *Methanocorpusculum parvum* as a member of Class II, and *Methanosarcina spelaei* as representative of Class III. The authors focused in the prediction of the proteins of the membrane and their relation toward energy conservation, where they found a high number of genes related to energy production and conversion; coenzyme transport and metabolism; and translation, ribosomal structure, and biogenesis. Table 5 summarizes the remaining results of their comparisons.

In conclusion, the results of the reconstructions discussed in the previous paragraphs greatly contribute to the elucidation of metabolic and biochemical pathways of methanogens. Moreover, these studies present the additional advantage that the constructed models can be updated as more experimental information becomes available. The knowledge obtained from these studies will certainly not only serve as a device to identify specific traits for metabolic engineering of methanogens; it will also help to predict the behavior of such modified methanogens in the desired environments with the purpose of increase biological methane production (Benedict et al. 2012; Richards et al. 2016).

4.2 *Transcriptomic Approaches in Methanogenesis*

Transcriptomic analysis of an organism can help to identify the transcription of several genes under certain conditions and, consequently, provide information about their specific functions in metabolism (Gruninger et al. 2018). Transcriptional studies of methanogens have become relevant in the context of AD because their results have improved the understanding not only of genome regulation for methane production under different operational conditions; they have also allowed to identify

Table 5 Comparison of important features from four methanogens, deduced by their metabolic reconstructions (Gilmore et al. 2017)

	M. bryantii	*M. cuniculi*	*M. spelaei*	*M. parvum*
Energy conservation mechanism	Electron bifurcation	Electron bifurcation	Chemiosmotic coupling	Chemiosmotic coupling
Genes involved on energy conservation	Energy converting hydrogenases (Eha, Ehb)	Non-F420-reducing hydrogenase (Mvh) Heterodisulfide reductase complex (Hdr) Energy converting hydrogenase (Ehb)	Complex methanophenazine reducing hydrogenase/heterodisulfide reductase (Vho/Hdr) Energy converting hydrogenase (Ech) Formylmethanofuran dehydrogenase complex (Fmd)	Formylmethanofuran dehydrogenase complex (Fwd) Energy-conserving hydrogenase (Eha) Heterodisulfide reductase complex (HdrABC) F420-reducing hydrogenase (FrhABG)
Substrate for growth	Formate and isopropanol/isobutanol	Methanol and CO_2/H_2	Methanol, methylamines and acetate	H_2/CO_2, formate and isopropanol
Genes for substrate uptake	Formate dehydrogenase Alcohol-dehydrogenase NADP-dependent F_{420} reductase	Corrinoid protein, Methyltransferase corrinoid activation protein Methanol methyltransferase	Methylamine transferases Acetate kinase Phosphotransacetylase	Formate dehydrogenase

changes in gene expression when methanogens establish interactions with other microorganisms.

Kato et al. (2014) studied the effects of ammonia inhibition over *Methanosaeta thermophila* strain TP (now *Methanothrix thermophila*) and compared them with the activity of a syntrophic acetate oxidation co-culture comprising *Methanothermobacter thermautotrophicus* strain TM and *Thermacetogenium phaeum* strain PB. Their results showed that methane production of *M. thermophila* was inhibited at 100 mM of NH_4Cl, while methane production of the syntrophic co-culture was barely affected at such concentration. This outcome was further confirmed by analyzing the methanogenic activity of tri-cultures and the number of 16S rRNA transcripts of *M. thermophila*, which indicated that levels of $NH_4Cl > 50$ mM favored the switch from aceticlastic methanogenesis to syntrophic acetate degradation. Likewise, transcriptome analysis of *M. thermophila* cultures growing with 100 mM NH_4Cl and without NH_4Cl revealed that of 1671 genes, 308 were up-regulated and 342 down-regulated due to exposure to ammonia, indicating possible damaging processes such as protein denaturation, oxidative stress, and intracellular cation imbalance.

It has been highlighted that transcripts related to methanogenesis are very abundant in methanogens. However, not all genes involved in these pathways are expressed constitutively; instead, many of them are regulated by the presence or absence of specific substrates in the environment (Browne and Cadillo-Quiroz 2013). The regulation of the methanogenesis pathways has been extensively studied in *Methanosarcina* species, since many members can produce methane employing any of the three pathways described. For example, several transcriptional regulators have been identified in *M. acetivorans* for expression of methanol methyltransferases and putative methyl-sulfide methyltransferases (Bose and Metcalf 2008; Bose et al. 2009). Moreover, a global regulator, MreA, which is involved in the regulation of acetoclastic and methylotrophic pathways as well as other regulator families, has been also characterized in *M. acetivorans* with homologs in *M. mazei* and *M. barkeri* (Reichlen et al. 2012).

The interactions between methanogens and bacteria can also affect the transcription of genes related to methanogenesis. *Methanotrix harundinacea*, a traditionally aceticlastic methanogen, was shown to highly express the genes for reduction of CO_2 to methane during co-culture with *Geobacter metallireducens* as a consequence of direct interspecies electron transfer (DIET) (Rotaru et al. 2014a, b). This event has been further confirmed in microbial aggregates from laboratory digesters treating brewery wastewater and in samples of rice paddy soils (Shrestha et al. 2014; Holmes et al. 2017), which indicates that CO_2 reduction induced by DIET in *Methanotrix* species plays an important role in methane production in several environments.

Finally, for the study of consortia in different environments, transcriptional studies of the *mcrA* gene, encoding for the alpha subunit of the key enzyme methyl-coenzyme M reductase (MCR), constitute a common analysis that has been largely associated to the presence of methanogenic archaea and their methane production rate (Freitag and Prosser 2009; Freitag et al. 2010; Alvarado et al. 2014). Consequently, its current role heavily leans toward monitoring of methanogenic activity, especially in anaerobic reactors. However, the discovery that MCR is also required for anaerobic oxidation of methane (Scheller et al. 2010) and the presence of *mcr*-like genes in other archaeal lineages (Evans et al. 2019) have not only challenged the function of *mcrA* as a molecular marker for methanogenesis; it has also opened the opportunities to manipulate the metabolism of methanogens for other purposes than methane production (Soo et al. 2016).

4.3 New Techniques in Genome Edition of Methanogens: Cas9 System

Besides the common genetic tools employed for the creation of mutant and recombinant strains, new genetic techniques, such as RNA Interference, CRISPRs, or TALENs, offer new opportunities to edit microbial metabolism and therefore elucidate metabolic pathways in a fast and efficient way. Nonetheless, these techniques had never been tested to modify the members of the methanogenic consortium present in AD process, until now (Hollinshead et al. 2014).

Nayak and Metcalf (2017) published the first study in which a Cas9 protein editing system was employed to edit the genome of *Methanosarcina acetivorans*. The system consisted in a series of plasmids containing endonuclease Cas9 from *Streptococcus pyogenes*, chimeric single-guide (sg) RNAs targeting several genes and repair templates for homology-dependent repair (HDR) since *M. acetivorans* lacks a nonhomologous end joining (NHEJ) repair pathway. The genome-editing technique was very successful in terms of capability, effectiveness, and velocity, recording frequencies of edited strains of about ~20% compared to transformation efficiency. The authors also demonstrated that it was possible to introduce simultaneous mutations by co-expression of several sgRNAs and insert large fragments of DNA (up to 3 kbp), as long as proper templates were introduced along with the genome editing tool. In this regard, the system was furthermore tested for NHEJ repair by introducing genes encoding for this pathway from a related methanogen, *Methanocella paludicula*, in conjunction with the genome editing machinery. Although transformation efficiencies were low in this case, the system was able to introduce deletions into the ssuC gene, at regions with microhomology.

The tools developed by this study will certainly provide opportunities to construct mutant strains in a shorter time, facilitating gene expression studies, molecule interaction analyses, and protein characterization not only in methanogens but also in other members of Archaea domain. It is expected that with the increase of efficient genetic tools, metabolic engineering will become a viable option for improvement of methanogenic activity within anaerobic reactors.

4.4 Metabolic Engineering for Methane Production and Synthesis of Value-Added Products in Methanogens

Although most of the metabolic engineering in methanogenic archaea has been aimed to the understanding of their unique metabolism, the genetic strategies developed for these microorganisms can also be applied to enhance the methanogenic pathways and, therefore, the methane production in anaerobic reactors. In the case of the hydrogenotrophic methanogenesis, metabolic engineering could be employed to expand the range of substrates utilized by this type of methanogens, besides H_2 and CO_2. However, it is required that new substrates in this pathway be able to forcefully reduce coenzyme F_{420}, the flavin group of heterodisulfide reductase complex (Hdr), and Ferredoxin (Costa and Leigh 2014). Similarly, methylotrophic methanogenesis also offer possibilities for the engineering since methanogens carrying out this pathway can already use a wide range of carbon substrates like acetate or a variety of methylated compounds (e.g., methanol, methylamines, dimethyl sulfide, among others) (Ferry 2011).

The engineering of a methanogen capable of using complex substrates for methanogenesis has already been demonstrated. Lessner et al. (2010) expressed an esterase (MekB) from *Pseudomonas veronii* in *Methanosarcina acetivorans* CA2 and then evaluated its growth in either methyl acetate (MeAc) or methyl propionate (MePr) esters. Their results showed that cultures of *M. acetivorans* CA2 containing MekB consumed ~97% of both esters, achieving generation times of 10.7 ± 0.2 h and 18.5 ± 0.2 h in MeAc and MePr, respectively. The methanol produced by the hydrolysis of the methyl esters was further metabolized through the methylotrophic pathway for methane production, while acetate and propionate accumulated in the medium. However, after methanol was consumed, acetate was also metabolized by *M. acetivorans*. These results demonstrate that it is possible to express bacterial enzymes in methanogenic archaea that improve their metabolism; nevertheless, the application of engineered methanogens in anaerobic reactors still requires additional considerations and further optimization in order to become a possible strategy for enhancing the performance of methane-producing anaerobic reactors.

Furthermore, methanogens have also been considered as attractive microorganisms for the production of chemicals due to several advantages in which are included the use of cheaper substrates such as CO_2 and their energy efficient processes. In this regard, Lyu et al. (2016) cloned and transformed geraniol synthase from *Ocimum basilicum* into *Methanococcus maripaludis* and examined geraniol production under different growth conditions. The highest yield of 4.6 mg of geraniol per g of dry weight was obtained by autotrophic growth, employing minimal formate medium without puromycin. In their proposed metabolic model, the authors suggested that active heterotrophic growth and presence of puromycin in the medium decreased geraniol production due to utilization of intermediates of the mevalonate pathway for biomass and lipid biosynthesis and for detoxification processes, respectively. They also suggested further optimization strategies, which included the overexpression of rate-limiting enzymes in the mevalonate pathway, the increase of soluble fraction of geraniol synthase, and the overall rearrangement of carbon flux in the cell.

Lastly, metabolically engineered microorganisms in AD, including methanogens, have recently gained attention not for a potential higher production of methane but also for the production of valuable products such as acetate or electricity from C1 substrates (Wood 2017). In this regard, the interest for methane has switched from end product to feedstock, as it will describe in the following section. Nevertheless, it is possible that the new significance of methane as substrate will boost the development of genetic tools and techniques that ultimately benefit its production in anaerobic systems.

5 Monitoring of Anaerobic Reactors: Can Metabolic Engineering Enhance Control over AD Processes?

Conventional monitoring of anaerobic reactors for biogas production has traditionally relied on the values of operational and biochemical parameters such as pH, temperature, organic load rate (OLR), volatile suspended solids (VSS), concentration of volatile fatty acids (VFAs), and biogas composition, among others (Boe et al. 2010; Jimenez et al. 2015; Ferguson et al. 2016; Wu et al. 2019). However, with the development of molecular techniques, the presence and metabolic activity of microbial populations have also become an important factor that determines good performance and high methane production (Bozan et al. 2017; Lim et al. 2018). Therefore, in well-established anaerobic systems, environmental parameters as well as microbial composition are employed in conjunction not only to monitor fluctuations in biogas production; their relationship has been also evaluated for prediction of process imbalance and failure.

For analysis of microbial populations, several molecular techniques are available, and the election of a specific one depends of what aspects of the microbial communities are desired to know. For example, sequencing of genomic DNA is usually employed for establishing which microorganisms are present in the reactor at a certain time, while methods like denaturing gradient gel electrophoresis (DGGE) and terminal-restriction fragment length polymorphism (T-RFLP) are used to monitor changes in microbial groups at different times and under different operational conditions. Quantification of the microorganisms can be achieved by quantitative PCR (qPCR) or fluorescent in situ hybridization (FISH), and the metabolic function of a sole member or of a group of microorganisms can be assigned by stable isotope probing (SIP) and the meta-omics (metagenomics, metatranscriptomics, metaproteomics, and metabolomics) (Vanwonterghem et al. 2014; Cabezas et al. 2015; Treu et al. 2016).

Despite the great number of molecular analysis and their related indicators, their successfully application as regular monitoring tools like operational parameters have not been possible due to the complexity of the procedures and data interpretation, the variation of system configurations in anaerobic processes, and the retrospective management in which most analysis are carried out in anaerobic digestion (Carballa et al. 2015; De Vrieze and Verstraete 2016). To overcome these limitations, it has been suggested that monitoring of anaerobic reactions should be carried out in a proactive way, in which prediction and anticipation are the key elements. In this regard, microbial analysis should be performed routinely, optimizing the acquisition, processing and interpretation of the data, and defining the most suitable indicators for good performance, imbalance, and failure of the process at the early stages (Koch et al. 2014; Carballa et al. 2015). Such way of monitoring is still in process to be accomplished.

It could be proposed that metabolic engineering of the most representative microorganisms in anaerobic digestion could improve the monitoring of anaerobic reactors by creating reliable indicators, which would be easy to implement and interpret. However, this area has remained largely unexploited due to several constrains such as:

1. Most of the microorganisms described in anaerobic digestion have not been isolated in laboratory or are difficult to culture.
2. The understanding of major metabolic pathways, including energy conservation pathways, is still incomplete.
3. There are few protocols and tools for genetic manipulation for even the most characterized microorganisms.
4. The complicated nexus that bacterial and archaeal microorganisms create between each other might affect the desired signal.

To this date, novel monitoring approaches have focused in the parameters and techniques described in previous paragraphs (Narihiro and Sekiguchi 2011; Alvarado et al. 2014; Ziels et al. 2015; De Vrieze et al. 2016; Wu et al. 2019), which are more concentrated in describing microbial composition and metabolic activity of the members. Other attempts of creating monitoring tools combining engineering and microorganisms are microbial biosensors, microbial fuel cells (MFC), and microbial electrolysis cells (MEC) (Liu et al. 2011; Chen et al. 2015; Yang et al. 2015; Cerrillo et al. 2016; Jin et al. 2016; Kretzschmar et al. 2016; Sweeney et al. 2018; Yu et al. 2018; Sun et al. 2019). However, all the methods discussed employ native microorganisms or biofilms without further modifications.

As few reports of genetic modifications of microorganisms aimed to enhance the performance of anaerobic reactors for methane production are available (Guedon et al. 2002; Lessner et al. 2010), it could be deduced that there is still a long way before metabolic engineering could be applied in the development of monitoring of anaerobic digestion. Hopefully, with the increasing knowledge about the microorganisms, genomes, and pathways that govern the process, a future where such monitoring tools are accessible is not impossible to conceive.

6 Metabolic Engineering in Methane Bioconversion Processes

Traditionally, the biogas produced in AD systems has been employed directly as a heat fuel for cooking and other domestic activities, or it has been destined to combined heat and power (CPH) units for electricity generation. Nevertheless, in recent years, there has been an increasing interest in transforming biogas into other biofuels (e.g., compressed natural gas, hydrogen, and gasoline) or into value-added products (e.g., biomass, pigments, and lactate) (Yang et al. 2014; Strong et al. 2015; Gür 2016; Wang et al. 2017). While physical and chemical processes dominate methane conversion into biofuels, there is still a growing interest in the production of compounds by biological strategies, employing wild type or genetically modified microorganisms.

Several microbial groups can oxidize methane through aerobic and anaerobic pathways, which is a feature that promotes them as the best candidates for the development of methane bioconversion processes. The most studied group for this purpose is methanotrophic bacteria, which are mainly distributed in the phyla *Proteobacteria*, *Verrucomicrobia*, and candidate phylum NC10 (Dedysh and Knief 2018). These microorganisms are known for employing the enzyme methane monooxygenase (MMO) for the first step oxidation of methane to methanol, which can be found in the cells in two forms: soluble or particulate (Khmelenina et al. 2018).

It has been previously demonstrated that methanotrophs can be employed for the production of compounds such as methanol, ectoine, single cell protein, lactate, and biopolymers (Pieja et al. 2017), using natural gas (CH_4) or biogas as feedstock (Table 6). Of all these products, single cell protein is perhaps the most successful one in terms of commercialization and scale-up, and several trademarks such as Uniprotein® (https://www.unibio.dk) and FeedKind® (http://www.calysta.com) are available as animal feed or additives. However, current methane bioconversion processes employing native methanotrophic bacteria still present several issues, which have restricted their commercial potential. These limitations include technical aspects such as high costs and risks associated with the utilization of two flammable gasses (CH_4 and O_2) as substrates, the need to optimize process configuration due to the low oxygen diffusion rate, the requirement of upgraded biogas or natural gas without impurities, and the addition of downstream treatments for purification of the products (Kalyuzhnaya et al. 2015; Pieja et al. 2017). Moreover, methanotrophs are difficult to isolate and present slow growth rates, and their conversion efficiencies are low for industrial considerations, especially at the initial activation of methane (Haynes and Gonzalez 2014; Liao et al. 2016).

In response to the previous disadvantages, metabolic engineering might provide alternatives to overcome the metabolic constrains associated with low yields. Likewise, genetic modifications in methanotrophs can expand the spectrum of bioproducts by deleting or reducing the interference of competitive pathways (Kalyuzhnaya et al. 2015; Henard and Guarnieri 2018).

Table 6 Products synthetized by methanotrophic bacteria

Product	Genetically engineered	Strain/culture conditions	Maximum production(*yield*)	Reference
Poly-3-hydroxybutyrate (PHB)	No	*Methylocystis parvus* OBBP Under nitrogen limitation CH_4:O_2 mixture of 1:1	68% cell dry weight at 170–180 h *(0.67 g PHB* g^{-1} *CH_4)*	(Asenjo and Suk 1986)
	No	Methanotrophic mixture dominated by *Methylocystis sp.* GB 25 DSMZ 7674 Under nitrogen limitationCH_4: 20%; O_2: 15% of air saturation	51.3% cell dry weight at 24 h(*0.55 g PHB* g^{-1} *CH_4)*	(Wendlandt et al. 2001)
	No	*Methylosinus trichosporium* OB3b *Methylocystis* SC2 *Methylocystis* 42/22 Under nitrogen limitation CH_4:O_2 mixture of 1:1	38 ± 4% cell dry weight 30 ± 13% cell dry weight 25 ± 7% cell dry weight All at 24 h	(Pieja et al. 2011)
	No	*Methylosinus trichosporium* OB3b *Methylocystis parvus* OBBP Under nitrogen limitation CH_4:O_2 mixture of 1:1	29% cell dry weight *(1.13 ± 0.02 g PHB* g^{-1} *CH_4)* 60% cell dry weight *(0.88 ± 0.12 g PHB* g^{-1} *CH_4)* All at 22 h	(Rostkowski et al. 2013)
Polyhydroxybutyrate-co-hydroxyvalerate (PHB-co-HV)	No	Methanotrophic enrichment dominated by genus *Methylocystis* Under nitrogen limitation CH_4:O_2 mixture of 1:1.5 With 100 mg/L of sodium valerate	44 ± 2% cell dry weight at 48 h *(0.36 mg PHB-co-HV* mg^{-1} *CH_4)*	(Myung et al. 2015)
	No	*Methylocystis sp.* WRRC1 Under nitrogen limitation CH_4:O_2 mixture of 1:1 With 0.34% of sodium valerate and free copper medium	78% cell dry weight at 72 h *(0.67 g PHB-co-HV* g^{-1} *CH_4)*	(Cal et al. 2016)

(continued)

Table 6 (continued)

Product	Genetically engineered	Strain/culture conditions	Maximum production(*yield*)	Reference
Methanol	No	*Methylosinus trichosporium* OB3b In medium containing 400 mmol/L of phosphate and 10 mmol/L of $MgCl_2$ as inhibitors for MEDH, and 20 mmol/L of sodium formate CH_4:O_2 mixture of 1:1	1.13 g L^{-1} at 40 h, with conversion rate of 64%	(Duan et al. 2011)
	No	Isolated and tolerant to H_2S *Methylocaldum gracile* SAD2 No MEDH inhibitor Biogas: air mixture of 1:2 CH_4 and H_2S content in biogas of 70.6% and < 50 ppm, respectively	343 mg L^{-1} at 48 h without H2S added. Conversion rate of 34% 276 mg L^{-1} at 36 h with 500 ppm of H2S added. Conversion rate of 30%	(Zhang et al. 2016)
	No	Isolated *Methylocaldum sp.* 14B In medium containing 50 mM of phosphate as MEDH inhibitor, and 80 mM of formate Biogas: air mixture of 1:2.5 CH_4 content in biogas of 69.8%	0.43 ± 0.00 g L^{-1} at 48 h, with conversion rate of 25.5 ± 1.1%	(Sheets et al. 2016)
Single cell protein	No	Enrichment containing *Methylomonas* sp. Natural gas: air mixture of 60:40	69.3% of bacterial dry weight	(Yazdian et al. 2005)
	No	Bacterial meal culture dominated by *Methylococcus capsulatus* Grow in natural gas	70.26 g per 100 g of dry matter*	(Øverland et al. 2010)
Ectoine	No	Methylobacter alcaliphilus 20Z CH_4: air mixture 1:1 In a medium containing 6% NaCl and 30 mM NH_4CI	1021 nmol mg^{-1} dry biomass	(Khmelenina et al. 2000)
	No	*Methylomicrobium alcaliphilum* 20Z In medium containing 6% NaCl and 0.05 mM Cu^2	66.9 ± 4.2 mg g^{-1} of biomass	(Cantera et al. 2016)

		$^{+}$ with agitation at 600 rpm. CH_4 headspace concentration of 4%		
	No	*Methylomicrobium alcaliphilum* 20Z In medium containing 6% NaCl and 25 mM Cu^{2+} with agitation at 300 rpm. CH_4 headspace concentration of 4%	37.4 ± 3.8 mg g^{-1} of biomass (intracellular) 12.9 ± 0.7 mg L^{-1} (extracellular)	(Cantera et al. 2017)
Lipids for biodiesel	No	*Methylomicrobium buryatense* CH_4: air mixture of 1:4 With agitation at 500 rpm in a 5 L CSTR reactor	Fatty acids content of 9.5 ± 0.8% of cell dry weight. Reported efficiency of lipid conversion to hydrocarbons of 99%	(Dong et al. 2017)
Succinic acid	Yes	*Methylococcus capsulatus* Bath Overexpression of malate dehydrogenase and isocitrate lyase of *E. coli*, in medium containing 30 μg ml^{-1} of kanamycin Growth in methane mixture (95% CH_4; 5% CO2)	~0.105 g L^{-1} of succinic acid at 72 h.	(Subbian 2017)
Isobutanol	Yes	*Methylococcus capsulatus* Bath With constitutive expression of ketoacid decarboxylase from *M. capsulatus* and alcohol dehydrogenase from *Saccharomyces cerevisiae* In medium containing 15 g ml^{-1} kanamycin and ~ 2 g L^{-1} of 2-ketoisovalerate	~ 0.22 g L^{-1} of isobutanol at 72 h	(Coleman et al. 2016)
Isoprene	Yes	Methylococcus capsulatus Bath With inducible expression of isoprene synthase from *Pueraria montana* Unspecified CH_4: air mixture	~ 10 mg L^{-1}	(Leonard et al. 2016)
Carotenoids	Yes	*Methylomonas* sp. 16a With transposon carrying a carotenoid biosynthetic pathway gene cluster (*crtEiYIB*) from strain DC404, isolated from soil and carotenoid genes (*crtWZ*) from *Paracoccus* sp. N81106 CH_4: air mixture 1:3	~ 2.0 mg of total carotenoids g^{-1} of dry cell weight	(Sharpe et al. 2007)

(continued)

Table 6 (continued)

Product	Genetically engineered	Strain/culture conditions	Maximum production(*yield*)	Reference
	Yes	*Methylomonas* sp. 16a With two copies of the gene cluster *crtWZEidiYIB* integrated into the chromosome. The gene cluster consisted of genes from *Paracoccus* sp. N81106, environmental strains DC404 and DC413, and *Brevundimonas vesicularis* strain DC263 CH_4: air mixture 1:3	1.5 ± 0.5 mg of astaxanthin g^{-1} of dry cell weight	(Ye et al. 2007)
Lactate	Yes	*Methylococcus capsulatus* Bath *Methylosinus trichosporium* OB3b *Methylomicrobium buryatense* 5G Expression of lactate dehydrogenase from several organisms under constitutive and inducible promoters CH_4: air mixture of 1:1	>1000 μM of lactic acid at 72 h	(Saville et al. 2016)
	Yes	*Methylomicrobium buryatense* Overexpression of L-lactate dehydrogenase from *Lactobacillus helveticus*, induced by 2 μg mL^{-1} of anhydrotetracycline CH_4: air mixture of 1:4	0.808 ± 0.343 g lactate L^{-1} at 96 h (*0.05 g lactate* g^{-1} CH_4)	(Henard et al. 2016)
	Yes	*Methylomicrobium alcaliphilum* 20ZR Pyruvate dehydrogenase deficient mutant Content of biogas in air of 33% (20% CH_4 & 13% CO_2)	~2.5 g lactate g^{-1} of dry cell weight at 96 h Specific productivity of 0.027 g lactate g^{-1} of dry cell weight h^{-1}	(Henard et al. 2018)

Although limited, there are genetic tools available for methanotrophs. Most of them involve the utilization of broad-host range plasmids from the incompatibility groups P and Q, as well as pBBR-based plasmids for the expression of homologous or heterologous genes (Kalyuzhnaya et al. 2015; Lee et al. 2016; Henard and Guarnieri 2018). Also, these tools can also be employed for chromosomal insertions and gene knockouts, incorporating antibiotic and/or counterselection markers (Puri et al. 2015; Ishikawa et al. 2018). To this date, the most efficient method for the delivery of the plasmids is conjugation; however, few electroporation methods have been developed recently (Crombie and Murrell 2011; Yan et al. 2016). In this way, the range of products that could be obtained through aerobic oxidation of methane have recently expanded to include biofuels such as ethanol, isobutanol, and 1,4-butanediol or soluble metabolites like succinate, lactic acid, isoprene, and carotenoids, among others (Table 6) (Strong et al. 2015).

Due to the still incomplete knowledge about the kinetic and genetic features of methanotrophic bacteria, only few species such as *Methylococcus capsulatus* Bath, *M. trichosporium* OB3b, and *Methylocystis* spp. have been considered for metabolic engineering purposes and industrial processes (Kalyuzhnaya et al. 2015). However, there is no doubt that with the continuous advancements in the biochemistry, genetics, and physiology of methanotrophic bacteria, more species will become candidates for the development of methane bioconversion industries.

In contrast to the progress made in aerobic methane oxidation, anaerobic oxidation of methane (AOM) has remained largely unexploited due to slow growth rates, the lack of an isolated pure strain, and the uncertainty of the pathways that this group of microorganisms use for their growth (Cui et al. 2015; McGlynn 2017). Nevertheless, anaerobic methane conversion presents several advantages over aerobic processes such as higher conversion efficiencies, the possibility of directing the reducing power of the cell to the production of reduced products, and omission of oxygen supply (Bennett et al. 2018).

Recently, anaerobic methanotrophic archaea (ANME) have gained special attention because they also encode the key enzyme for methanogenesis, methyl-coenzyme M reductase (MCR), which is able to catalyze the reverse reaction (Eq. 1) of that observed in methanogenesis (Knittel et al. 2018; Evans et al. 2019).

$$CH_4 + CoM - S - S - CoB \rightarrow CH_3 - SCoM + SH - CoB \tag{1}$$

This feature, also observed in methanogenic archaea such as *Methanothermobacter marburgensis* and *Methanosarcina acetivorans*, has prompted the interest in reverse methanogenesis as a central pathway for the production of chemicals (Moran et al. 2005; Moran et al. 2007; Scheller et al. 2010).

Indeed, Soo et al. (2016) genetically engineered *M. acetivorans* to grow on methane as a sole carbon source, using iron as electron acceptor. To achieve this, MCR from ANME-1 populations from the Black Sea was cloned and transformed into *M. acetivorans*. The authors reported methane consumptions of 109 ± 12 μmol and 143 ± 16 μmol employing low and high inoculum size, respectively, of the engineered methanogen. Moreover, an extracellular acetate concentration of

10.3 ± 0.8 mM was measured in cultures expressing MCR from ANME-1, which was ~1.7 times higher than the one recorded for cultures expressing the empty vector.

According to the results from analysis of incorporation into acetate of methane and bicarbonate labeled with ^{13}C, genome-scale metabolic modeling and RNA sequencing, Soo et al. proposed a reverse aceticlastic pathway as the one operating in the engineered *M. acetivorans* strain, following the stoichiometry (Eq. 2):

$$4CH_4 + 2HCO_3^- + 8Fe^{3+} \rightarrow 3CH_3COO^- + 8Fe^{2+} + 9H^+ \qquad (2)$$

This achievement has opened the possibilities of exploiting the strain and its reverse pathway for the synthesis of different products from acetate, mainly biofuels. In fact, the same group tried to further engineer their *M. acetivorans* strain for butanol production, employing a pathway previously described in *Escherichia coli* (McAnulty et al. 2017). The pathway included the enzymes crotonase (Crt), 3-hydroxybutyryl-CoA dehydrogenase (Hbd), and aldehyde/alcohol dehydrogenase (AdhE2) from *Clostridium acetobutylicum*; the trans-enoyl-CoA reductase (Ter) from *Treponema denticola*; and the acetyl-CoA acetyltransferase (AtoB) from *E. coli* (Shen et al. 2011).

With the exception of AtoB from *E. coli*, all enzymes were cloned into *M. acetivorans*, along with MCR from ANME-1 (Soo et al. 2016) and the native acetyl-CoA acetyltransferase (MA_4042) from the same methanogen. However, instead of butanol, L-lactate was produced from methane, reaching an extracellular yield of 0.105 ± 0.009 mol lactate per mol methane, which was ~10 times higher than the one obtained through aerobic methane oxidation (Table 6) (Henard et al. 2016; McAnulty et al. 2017). The authors hypothesized that the production of L-Lactate might have been due to a lactate dehydrogenase activity of Hbd of *C. acetobutylicum*. Although in vitro essays using purified Hbd failed to demonstrate such activity, the fact that lactate was produced only in *M. acetivorans* containing plasmids with Hbd indicated the role of enzyme in lactate production. It was further theorized that post-translational modifications were probably responsible for the lactate dehydrogenase activity of Hbd observed in engineered *M. acetivorans*.

There is no doubt that the works of Soo et al. (2016) and McAnulty et al. (2017) from the Thomas Wood Research Group have demonstrated that is possible to produce high-value products such as acetate and lactate from methane through reverse methanogenesis. Further research surely will be focused on increasing the molecular tools for metabolic engineering and the understanding of the pathways involved in anaerobic methane bioconversion processes in order to optimize them for the competitive market.

Finally, the reverse methanogenesis pathway constructed in *M. acetivorans* not only has been applied for the production of chemicals of interest; it also has been tested in a microbial fuel cell (MFC) for electricity production with methane as substrate, in combination with an adapted sludge and *Geobacter sulfurreducens* (McAnulty et al. 2017). The constructed MEC consumed 260 ± 40 μmol of methane and yielded a Coulombic efficiency of 90 ± 10%, which was associated with maximum power generation and current density of 168 ± 9 mW m^{-2} and 273 ± 7 mA m^{-2}, respectively. However, these values were only reached when all three biological components (air-adapted *M. acetivorans* expressing MCR from ANME-1, methane-consuming sludge and *G. sulfurreducens*) were present in the MFC.

The biological scheme proposed by the authors to explain the electricity generation in the MEC consists in the initial methane oxidation carried out by air-adapted and engineered *M. acetivorans*. Then, the acetate produced in the reverse methanogenesis pathway (Soo et al. 2016) is further oxidized to CO_2 by *G. sulfurreducens* and *Paracoccus* spp. (the dominating microorganism in the adapted sludge), generating even more electrons. Finally, the excess of electrons is channeled to the anode by humic acids from sludge and membrane cytochromes from the microorganisms, acting as electron shuttles.

While production of bio-compounds and other commodities from methane seems a promising technology, there are still important issues to overcome such as low yields, lack of optimal microbial strains, limited metabolic engineering tools, rigorous process conditions, and high investment costs. A recent approach to overcome the limitations associated with native methane oxidizers is the construction of synthetic methylotrophic pathways in well-known industrial microorganisms (Whitaker et al. 2015). However, this strategy has directed its attention toward the optimization of growth and consumption of methanol instead of methane, since the latter can be converted into the former by chemical methods, which might exhibit higher conversion efficiencies and production rates than most biological processes (Bennett et al. 2018). Nevertheless, as stated by Clomburg et al. (2017), biological-based industries relying on sustainable feedstocks like methane could still take advantage of their smaller-scale processes by exploiting the continuous knowledge and improvement that will be generated through time. In either regard, there is no doubt that metabolic engineering along with efficient process design will be key elements to promote the development of mature methane bioconversion industries that will be able to contribute to the global demand for chemicals.

7 Conclusion

Important advances have been made in the area of metabolic engineering for fuel production; however, most of the progress has been focused on mono- and co-cultures with microorganisms in which genomic tools and extensive metabolic knowledge are available. Although some microbial groups involved in AD have been subject of metabolic engineering with different purposes, their successful application for the improvement of methane production has been not tested. The complexity of the system along with the lack of knowledge about the metabolic nexus are significant challenges to overcome if engineered strains are introduced into the process, as previously demonstrated with bioaugmentation strategies.

Nevertheless, metabolic engineering still stands as an attractive option for enhancing AD performance. For example, the construction and study of mutant strains, whether they are finally introduced to the process or not, will help to increase the understanding of the microbial metabolism and, thus, to create approaches that will allow proper control and monitoring of anaerobic reactors. The real scope of metabolic engineering has not been reached.

References

Ahring BK, Angelidaki I, Johansen K (1992) Anaerobic Treatment Of Manure Together With Industrial Waste. Water Sci Technol 25:311–318

Akuzawa M, Hori T, Haruta S et al (2011) Distinctive Responses of Metabolically Active Microbiota to Acidification in a Thermophilic Anaerobic Digester. Microb Ecol 61:595–605. https://doi.org/10.1007/s00248-010-9788-1

Alfenore S, Molina-Jouve C (2016) Current Status And Future Prospects Of Conversion Of Lignocellulosic Resources To Biofuels Using Yeasts And Bacteria. Process Biochem 51:1747–1756. https://doi.org/10.1016/j.procbio.2016.07.028

Almeida A, Nafarrete-Rivera E, Alvarado A et al (2011) Expresión Genética En La Digestión Anaerobia: Un Paso Adelante En La Comprensión De Las Interacciones Tróficas De Esta Biotecnología. Acta Quím Mex 3:14–34

Alvarado A, Montañez-Hernández LE, Palacio-Molina SL et al (2014) Microbial Trophic Interactions And *mcrA* Gene Expression In Monitoring Of Anaerobic Digesters. Front Microbiol 5. https://doi.org/10.3389/fmicb.2014.00597

Ammar EM, Jin Y, Wang Z, Yang S-T (2014) Metabolic Engineering Of *Propionibacterium freudenreichii*: Effect Of Expressing Phosphoenolpyruvate Carboxylase On Propionic Acid Production. Appl Microbiol Biotechnol 98:7761–7772. https://doi.org/10.1007/s00253-014-5836-y

Andreesen JR, Bahl H, Gottschalk G (1989) Introduction to the Physiology and Biochemistry of the Genus *Clostridium*. In: Minton NP, Clarke DJ (eds) Clostridia. Springer, Boston, MA, pp 27–62

Angelidaki I, Karakashev D, Batstone DJ et al (2011) Biomethanation and Its Potential. In: Methods in Enzymology. Elsevier, pp 327–351

Asenjo JA, Suk JS (1986) Microbial Conversion Of Methane Into poly-β-hydroxybutyrate (PHB): Growth And Intracellular Product Accumulation In A Type II Methanotroph. J Ferment Technol 64:271–278. https://doi.org/10.1016/0385-6380(86)90118-4

Azman S, Khadem AF, van LJB et al (2015) Presence and Role of Anaerobic Hydrolytic Microbes in Conversion of Lignocellulosic Biomass for Biogas Production. Crit Rev Environ Sci Technol 45:2523–2564. https://doi.org/10.1080/10643389.2015.1053727

Baek G, Kim J, Kim J, Lee C (2018) Role and Potential of Direct Interspecies Electron Transfer in Anaerobic Digestion. Energies 11:107. https://doi.org/10.3390/en11010107

Barua S, Dhar BR (2017) Advances Towards Understanding And Engineering Direct Interspecies Electron Transfer In Anaerobic Digestion. Bioresour Technol 244:698–707. https://doi.org/10.1016/j.biortech.2017.08.023

Benedict MN, Gonnerman MC, Metcalf WW, Price ND (2012) Genome-Scale Metabolic Reconstruction and Hypothesis Testing in the Methanogenic Archaeon *Methanosarcina acetivorans* C2A. J Bacteriol 194:855–865. https://doi.org/10.1128/JB.06040-11

Bengelsdorf FR, Poehlein A, Linder S et al (2016) Industrial Acetogenic Biocatalysts: A Comparative Metabolic and Genomic Analysis. Front Microbiol 7. https://doi.org/10.3389/fmicb.2016.01036

Bennett RK, Steinberg LM, Chen W, Papoutsakis ET (2018) Engineering The Bioconversion Of Methane And Methanol To Fuels And Chemicals In Native And Synthetic Methylotrophs. Curr Opin Biotechnol 50:81–93. https://doi.org/10.1016/j.copbio.2017.11.010

Bergman EN (1990) Energy Contributions Of Volatile Fatty Acids From The Gastrointestinal Tract In Various Species. Physiol Rev 70:567–590. https://doi.org/10.1152/physrev.1990.70.2.567

Bhatia SK, Yang Y-H (2017) Microbial Production Of Volatile Fatty Acids: Current Status And Future Perspectives. Rev Environ Sci Biotechnol 16:327–345. https://doi.org/10.1007/s11157-017-9431-4

Biebl H, Pfennig N (1978) Growth Yields Of Green Sulfur Bacteria In Mixed Cultures With Sulfur And Sulfate Reducing Bacteria. Arch Microbiol 117:9–16. https://doi.org/10.1007/BF00689344

Bilal M, Iqbal HMN, Hu H et al (2018) Metabolic Engineering And Enzyme-Mediated Processing: A Biotechnological Venture Towards Biofuel Production – A review. Renew Sust Energ Rev 82:436–447. https://doi.org/10.1016/j.rser.2017.09.070

Bizukojc M, Dietz D, Sun J, Zeng A-P (2010) Metabolic Modelling Of Syntrophic-Like Growth Of A 1,3-propanediol Producer, *Clostridium butyricum*, And A Methanogenic Archeon, *Methanosarcina mazei*, Under Anaerobic Conditions. Bioprocess Biosyst Eng 33:507–523. https://doi.org/10.1007/s00449-009-0359-0

Boe K, Batstone DJ, Steyer J-P, Angelidaki I (2010) State Indicators For Monitoring The Anaerobic Digestion Process. Water Res 44:5973–5980. https://doi.org/10.1016/j.watres.2010.07.043

Borrel G, Adam PS, Gribaldo S (2016) Methanogenesis And The Wood–Ljungdahl Pathway: An Ancient, Versatile, And Fragile Association. Genome Biol Evol 8:1706–1711. https://doi.org/10.1093/gbe/evw114

Bose A, Kulkarni G, Metcalf WW (2009) Regulation Of Putative Methyl-Sulphide Methyltransferases In *Methanosarcina acetivorans* C2A. Mol Microbiol 74:227–238. https://doi.org/10.1111/j.1365-2958.2009.06864.x

Bose A, Metcalf WW (2008) Distinct Regulators Control The Expression Of Methanol Methyltransferase Isozymes In *Methanosarcina acetivorans* C2A. Mol Microbiol 67:649–661. https://doi.org/10.1111/j.1365-2958.2007.06075.x

Bouillaut L, Self WT, Sonenshein AL (2013) Proline-Dependent Regulation Of *Clostridium difficile* Stickland Metabolism. J Bacteriol 195:844–854. https://doi.org/10.1128/JB.01492-12

Bozan M, Akyol Ç, Ince O et al (2017) Application Of Next-Generation Sequencing Methods For Microbial Monitoring Of Anaerobic Digestion Of Lignocellulosic Biomass. Appl Microbiol Biotechnol 101:6849–6864. https://doi.org/10.1007/s00253-017-8438-7

Browne PD, Cadillo-Quiroz H (2013) Contribution of Transcriptomics To Systems-Level Understanding Of Methanogenic *Archaea*. Archaea 2013:1–11. https://doi.org/10.1155/2013/586369

Bryant MP, Wolin EA, Wolin MJ, Wolfe RS (1967) *Methanobacillus omelianskii*, A Symbiotic Association Of Two Species Of Bacteria. Arch Für Mikrobiol 59:20–31. https://doi.org/10.1007/BF00406313

Budzianowski WM (2016) A Review Of Potential Innovations For Production, Conditioning And Utilization Of Biogas With Multiple-Criteria Assessment. Renew Sust Energ Rev 54:1148–1171. https://doi.org/10.1016/j.rser.2015.10.054

Cabezas A, de Araujo JC, Callejas C et al (2015) How To Use Molecular Biology Tools For The Study Of The Anaerobic Digestion Process? Rev Environ Sci Biotechnol 14:555–593. https://doi.org/10.1007/s11157-015-9380-8

Cal AJ, Sikkema WD, Ponce MI et al (2016) Methanotrophic Production Of Polyhydroxybutyrate-Co-Hydroxyvalerate With High Hydroxyvalerate Content. Int J Biol Macromol 87:302–307. https://doi.org/10.1016/j.ijbiomac.2016.02.056

Calogero S, Gardan R, Glaser P et al (1994) RocR, A Novel Regulatory Protein Controlling Arginine Utilization In *Bacillus subtilis*, Belongs To The NtrC/NifA Family Of Transcriptional Activators. J Bacteriol 176:1234–1241. https://doi.org/10.1128/jb.176.5.1234-1241.1994

Cantera S, Lebrero R, Rodríguez E et al (2017) Continuous Abatement Of Methane Coupled With Ectoine Production By *Methylomicrobium alcaliphilum* 20Z In Stirred Tank Reactors: A Step

Further Towards Greenhouse Gas Biorefineries. J Clean Prod 152:134–141. https://doi.org/10.1016/j.jclepro.2017.03.123

Cantera S, Lebrero R, Sadornil L et al (2016) Valorization Of CH_4 Emissions Into High-Added-Value Products: Assessing The Production Of Ectoine Coupled With CH_4 Abatement. J Environ Manag 182:160–165. https://doi.org/10.1016/j.jenvman.2016.07.064

Carballa M, Regueiro L, Lema JM (2015) Microbial Management Of Anaerobic Digestion: Exploiting The Microbiome-Functionality Nexus. Curr Opin Biotechnol 33:103–111. https://doi.org/10.1016/j.copbio.2015.01.008

Cerrillo M, Viñas M, Bonmatí A (2016) Overcoming Organic And Nitrogen Overload In Thermophilic Anaerobic Digestion Of Pig Slurry By Coupling A Microbial Electrolysis Cell. Bioresour Technol 216:362–372. https://doi.org/10.1016/j.biortech.2016.05.085

Charubin K, Bennett RK, Fast AG, Papoutsakis ET (2018) Engineering *Clostridium* Organisms As Microbial Cell-Factories: Challenges & Opportunities. Metab Eng 50:173–191. https://doi.org/10.1016/j.ymben.2018.07.012

Chen Y, Cheng JJ, Creamer KS (2008) Inhibition Of Anaerobic Digestion Process: A review. Bioresour Technol 99:4044–4064

Chen JL, Ortiz R, Steele TWJ, Stuckey DC (2014) Toxicants Inhibiting Anaerobic Digestion: A review. Biotechnol Adv 32:1523–1534. https://doi.org/10.1016/j.biotechadv.2014.10.005

Chen JL, Steele TWJ, Stuckey DC (2015) Modeling And Application Of A Rapid Fluorescence-Based Assay For Biotoxicity In Anaerobic Digestion. Environ Sci Technol 49:13463–13471. https://doi.org/10.1021/acs.est.5b03050

Cheng Q, Call DF (2016) Hardwiring Microbes Via Direct Interspecies Electron Transfer: Mechanisms And Applications. Environ Sci Process Impacts 18:968–980. https://doi.org/10.1039/C6EM00219F

Cho C, Choi SY, Luo ZW, Lee SY (2015) Recent Advances In Microbial Production Of Fuels And Chemicals Using Tools And Strategies Of Systems Metabolic Engineering. Biotechnol Adv 33:1455–1466. https://doi.org/10.1016/j.biotechadv.2014.11.006

Cirne DG, Björnsson L, Alves M, Mattiasson B (2006) Effects Of Bioaugmentation By An Anaerobic Lipolytic Bacterium On Anaerobic Digestion Of Lipid-Rich Waste. J Chem Technol Biotechnol 81:1745–1752. https://doi.org/10.1002/jctb.1597

Clomburg JM, Crumbley AM, Gonzalez R (2017) Industrial Biomanufacturing: The Future Of Chemical Production. Science 355:aag0804. https://doi.org/10.1126/science.aag0804

Coleman WJ, Vidanes GM, Cottarel G, et al (2016) Biological production of multi-carbon compounds from methane

Costa JC, Barbosa SG, Sousa DZ (2012) Effects of pre-treatment and bioaugmentation strategies on the anaerobic digestion of chicken feathers. Bioresour Technol 120:114–119. https://doi.org/10.1016/j.biortech.2012.06.047

Costa KC, Leigh JA (2014) Metabolic versatility in methanogens. Curr Opin Biotechnol 29:70–75. https://doi.org/10.1016/j.copbio.2014.02.012

Crable BR, Sieber JR, Mao X et al (2016) Membrane Complexes of *Syntrophomonas wolfei* Involved in Syntrophic Butyrate Degradation and Hydrogen Formation. Front Microbiol 7. https://doi.org/10.3389/fmicb.2016.01795

Crombie A, Murrell JC (2011) Development of a System for Genetic Manipulation of the Facultative Methanotroph *Methylocella silvestris* BL2. In: Methods in Enzymology. Elsevier, pp 119–133

Cui M, Ma A, Qi H et al (2015) Anaerobic oxidation of methane: an "active" microbial process. Microbiology Open 4:1–11. https://doi.org/10.1002/mbo3.232

Dang Y, Holmes DE, Zhao Z et al (2016) Enhancing anaerobic digestion of complex organic waste with carbon-based conductive materials. Bioresour Technol 220:516–522. https://doi.org/10.1016/j.biortech.2016.08.114

De Bok FAM, Luijten MLGC, Stams AJM (2002) Biochemical evidence for formate transfer in syntrophic propionate-oxidizing cocultures of *Syntrophobacter fumaroxidans* and

Methanospirillum hungatei. Appl Environ Microbiol 68:4247–4252. https://doi.org/10.1128/aem.68.9.4247-4252.2002

De Bok FAM, Plugge CM, Stams AJM (2004) Interspecies electron transfer in methanogenic propionate degrading consortia. Water Res 38:1368–1375. https://doi.org/10.1016/j.watres.2003.11.028

De Vrieze J, Regueiro L, Props R et al (2016) Presence does not imply activity: DNA and RNA patterns differ in response to salt perturbation in anaerobic digestion. Biotechnol Biofuels 9. https://doi.org/10.1186/s13068-016-0652-5

De Vrieze J, Verstraete W (2016) Perspectives for microbial community composition in anaerobic digestion: from abundance and activity to connectivity: Microbial connectivity in anaerobic digestion. Environ Microbiol 18:2797–2809. https://doi.org/10.1111/1462-2920.13437

Dedysh SN, Knief C (2018) Diversity and Phylogeny of Described Aerobic Methanotrophs. In: Kalyuzhnaya MG, Xing X-H (eds) Methane Biocatalysis: Paving the Way to Sustainability. Springer International Publishing, Cham, pp 17–42

Demirel B, Scherer P (2008) The roles of acetotrophic and hydrogenotrophic methanogens during anaerobic conversion of biomass to methane: a review. Rev Environ Sci Biotechnol 7:173–190. https://doi.org/10.1007/s11157-008-9131-1

Demirel B, Yenigün O (2002) Two-phase anaerobic digestion processes: a review: Two-phase anaerobic digestion processes. J Chem Technol Biotechnol 77:743–755. https://doi.org/10.1002/jctb.630

Den Haan R, van Rensburg E, Rose SH et al (2015) Progress and challenges in the engineering of non-cellulolytic microorganisms for consolidated bioprocessing. Curr Opin Biotechnol 33:32–38. https://doi.org/10.1016/j.copbio.2014.10.003

Desvaux M, Guedon E, Petitdemange H (2000) Cellulose Catabolism by *Clostridium cellulolyticum* Growing in Batch Culture on Defined Medium. Appl Environ Microbiol 66:2461–2470. https://doi.org/10.1128/AEM.66.6.2461-2470.2000

Deublein D, Steinhauser A (2011) Biogas from Waste and Renewable Resources: An Introduction. Wiley

Dong X, Cheng G, Stams AJM (1994) Butyrate oxidation by *Syntrophospora bryantii* in co-culture with different methanogens and in pure culture with pentenoate as electron acceptor. Appl Microbiol Biotechnol 42:647–652. https://doi.org/10.1007/BF00173934

Dong T, Fei Q, Genelot M et al (2017) A novel integrated biorefinery process for diesel fuel blendstock production using lipids from the methanotroph, *Methylomicrobium buryatense*. Energy Convers Manag 140:62–70. https://doi.org/10.1016/j.enconman.2017.02.075

Drake H, Küsel K, Matthies C (2013) Acetogenic Prokaryotes. In: Rosenberg E, DeLong E, Lory S et al (eds) The Prokaryotes. Springer Berlin Heidelberg, pp 3–60

Dridi B, Fardeau M-L, Ollivier B et al (2012) *Methanomassiliicoccus luminyensis* gen. *nov*., sp. *nov*., a methanogenic archaeon isolated from human faeces. Int J Syst Evol Microbiol 62:1902–1907. https://doi.org/10.1099/ijs.0.033712-0

Duan C, Luo M, Xing X (2011) High-rate conversion of methane to methanol by *Methylosinus trichosporium* OB3b. Bioresour Technol 102:7349–7353. https://doi.org/10.1016/j.biortech.2011.04.096

Ecem Öner B, Akyol Ç, Bozan M et al (2018) Bioaugmentation with *Clostridium thermocellum* to enhance the anaerobic biodegradation of lignocellulosic agricultural residues. Bioresour Technol 249:620–625. https://doi.org/10.1016/j.biortech.2017.10.040

Enzmann F, Mayer F, Rother M, Holtmann D (2018) Methanogens: biochemical background and biotechnological applications. AMB Express 8. https://doi.org/10.1186/s13568-017-0531-x

Evans PN, Boyd JA, Leu AO et al (2019) An evolving view of methane metabolism in the *Archaea*. Nat Rev Microbiol 17:219–232

Ferguson RMW, Coulon F, Villa R (2016) Organic loading rate: A promising microbial management tool in anaerobic digestion. Water Res 100:348–356. https://doi.org/10.1016/j.watres.2016.05.009

Ferry JG (2011) Fundamentals of methanogenic pathways that are key to the biomethanation of complex biomass. Curr Opin Biotechnol 22:351–357. https://doi.org/10.1016/j.copbio.2011.04.011

Freitag TE, Prosser JI (2009) Correlation of Methane Production and Functional Gene Transcriptional Activity in a Peat Soil. Appl Environ Microbiol 75:6679–6687. https://doi.org/10.1128/AEM.01021-09

Freitag TE, Toet S, Ineson P, Prosser JI (2010) Links between methane flux and transcriptional activities of methanogens and methane oxidizers in a blanket peat bog: Methane flux and transcriptional activities in a peat bog. FEMS Microbiol Ecol no–no. https://doi.org/10.1111/j.1574-6941.2010.00871.x

Garfí M, Martí-Herrero J, Garwood A, Ferrer I (2016) Household anaerobic digesters for biogas production in Latin America: A review. Renew Sust Energ Rev 60:599–614. https://doi.org/10.1016/j.rser.2016.01.071

Ge H, Jensen PD, Batstone DJ (2011) Temperature phased anaerobic digestion increases apparent hydrolysis rate for waste activated sludge. Water Res 45:1597–1606. https://doi.org/10.1016/j.watres.2010.11.042

Gibson DG, Young L, Chuang R-Y et al (2009) Enzymatic assembly of DNA molecules up to several hundred kilobases. Nat Methods 6:343–345. https://doi.org/10.1038/nmeth.1318

Gilmore SP, Henske JK, Sexton JA et al (2017) Genomic analysis of methanogenic archaea reveals a shift towards energy conservation. BMC Genomics 18. https://doi.org/10.1186/s12864-017-4036-4

Gomez Camacho CE, Ruggeri B (2018) Syntrophic Microorganisms Interactions in Anaerobic Digestion (AD): a Critical Review in the Light of Increase Energy Production. 64:391–396. doi: https://doi.org/10.3303/CET1864066

Gonnerman MC, Benedict MN, Feist AM et al (2013) Genomically and biochemically accurate metabolic reconstruction of *Methanosarcina barkeri* Fusaro, iMG746. Biotechnol J 8:1070–1079. https://doi.org/10.1002/biot.201200266

Goyal N, Widiastuti H, Karimi IA, Zhou Z (2014) A genome-scale metabolic model of *Methanococcus maripaludis* S2 for CO_2 capture and conversion to methane. Mol BioSyst 10:1043–1054. https://doi.org/10.1039/C3MB70421A

Graentzdoerffer A, Pich A, Andreesen JR (2001) Molecular analysis of the *grd* operon coding for genes of the glycine reductase and of the thioredoxin system from *Clostridium sticklandii*. Arch Microbiol 175:8–18. https://doi.org/10.1007/s002030000232

Gruninger RJ, Nguyen TTM, Reid ID et al (2018) Application of Transcriptomics to Compare the Carbohydrate Active Enzymes That Are Expressed by Diverse Genera of Anaerobic Fungi to Degrade Plant Cell Wall Carbohydrates. Front Microbiol 9. https://doi.org/10.3389/fmicb.2018.01581

Guan N, Li J, Shin H-D et al (2016) Metabolic engineering of acid resistance elements to improve acid resistance and propionic acid production of *Propionibacterium jensenii*. Biotechnol Bioeng 113:1294–1304. https://doi.org/10.1002/bit.25902

Guedon E, Desvaux M, Petitdemange H (2002) Improvement of Cellulolytic Properties of *Clostridium cellulolyticum* by Metabolic Engineering. Appl Environ Microbiol 68:53–58. https://doi.org/10.1128/AEM.68.1.53-58.2002

Gür TM (2016) Comprehensive review of methane conversion in solid oxide fuel cells: Prospects for efficient electricity generation from natural gas. Prog Energy Combust Sci 54:1–64. https://doi.org/10.1016/j.pecs.2015.10.004

Hattori S, Luo H, Shoun H, Kamagata Y (2001) Involvement of formate as an interspecies electron carrier in a syntrophic acetate-oxidizing anaerobic microorganism in coculture with methanogens. J Biosci Bioeng 91:294–298

Haynes CA, Gonzalez R (2014) Rethinking biological activation of methane and conversion to liquid fuels. Nat Chem Biol 10:331–339. https://doi.org/10.1038/nchembio.1509

Heap JT, Pennington OJ, Cartman ST, Minton NP (2009) A modular system for *Clostridium* shuttle plasmids. J Microbiol Methods 78:79–85. https://doi.org/10.1016/j.mimet.2009.05.004

Hedderich R, Whitman WB (2013) Physiology and Biochemistry of the Methane-Producing *Archaea*. In: DeLong EF, Lory S et al (eds) Rosenberg E. The Prokaryotes. Springer Berlin Heidelberg, Berlin, Heidelberg, pp 635–662

Henard CA, Franklin TG, Youhenna B et al (2018) Biogas Biocatalysis: Methanotrophic Bacterial Cultivation, Metabolite Profiling, and Bioconversion to Lactic Acid. Front Microbiol 9. https://doi.org/10.3389/fmicb.2018.02610

Henard CA, Guarnieri MT (2018) Metabolic Engineering of Methanotrophic Bacteria for Industrial Biomanufacturing. In: Kalyuzhnaya MG, Xing X-H (eds) Methane Biocatalysis: Paving the Way to Sustainability. Springer International Publishing, Cham, pp 117–132

Henard CA, Smith H, Dowe N et al (2016) Bioconversion of methane to lactate by an obligate methanotrophic bacterium. Sci Rep 6. https://doi.org/10.1038/srep21585

Hollinshead W, He L, Tang YJ (2014) Biofuel production: an odyssey from metabolic engineering to fermentation scale-up. Front Microbiol 5. https://doi.org/10.3389/fmicb.2014.00344

Holmes DE, Rotaru A-E, Ueki T et al (2018) Electron and Proton Flux for Carbon Dioxide Reduction in *Methanosarcina barkeri* During Direct Interspecies Electron Transfer. Front Microbiol 9. https://doi.org/10.3389/fmicb.2018.03109

Holmes DE, Shrestha PM, Walker DJF et al (2017) Metatranscriptomic Evidence for Direct Interspecies Electron Transfer between *Geobacter* and *Methanothrix* Species in Methanogenic Rice Paddy Soils. Appl Environ Microbiol 83. https://doi.org/10.1128/AEM.00223-17

Huang W, Wang Z, Zhou Y, Ng WJ (2015) The role of hydrogenotrophic methanogens in an acidogenic reactor. Chemosphere 140:40–46. https://doi.org/10.1016/j.chemosphere.2014.10.047

Humphreys CM, Minton NP (2018) Advances in metabolic engineering in the microbial production of fuels and chemicals from C1 gas. Curr Opin Biotechnol 50:174–181. https://doi.org/10.1016/j.copbio.2017.12.023

Imachi H, Sakai S, Ohashi A et al (2007) *Pelotomaculum propionicicum* sp. *nov*., an anaerobic, mesophilic, obligately syntrophic, propionate-oxidizing bacterium. Int J Syst Evol Microbiol 57:1487–1492. https://doi.org/10.1099/ijs.0.64925-0

Insam H, Franke-Whittle I, Goberna M (2010) Microbes in Aerobic and Anaerobic Waste Treatment. In: Insam H, Franke-Whittle I, Goberna M (eds) Microbes at Work. Springer, Berlin Heidelberg, pp 1–34

Ishii S, Kosaka T, Hori K et al (2005) Coaggregation facilitates interspecies hydrogen transfer between *Pelotomaculum thermopropionicum* and *Methanothermobacter thermautotrophicus*. Appl Environ Microbiol 71:7838–7845. https://doi.org/10.1128/AEM.71.12.7838-7845.2005

Ishikawa M, Yokoe S, Kato S, Hori K (2018) Efficient Counterselection for *Methylococcus capsulatus* (Bath) by Using a Mutated *pheS* Gene. Appl Environ Microbiol 84. https://doi.org/10.1128/AEM.01875-18

Jackson BE, Bhupathiraju VK, Tanner RS et al (1999) *Syntrophus aciditrophicus* sp. *nov*., a new anaerobic bacterium that degrades fatty acids and benzoate in syntrophic association with hydrogen-using microorganisms. Arch Microbiol 171:107–114

Jackson S, Calos M, Myers A, Self WT (2006) Analysis of Proline Reduction in the Nosocomial Pathogen Clostridium difficile. J Bacteriol 188:8487–8495. https://doi.org/10.1128/JB.01370-06

Jain S, Jain S, Wolf IT et al (2015) A comprehensive review on operating parameters and different pretreatment methodologies for anaerobic digestion of municipal solid waste. Renew Sust Energ Rev 52:142–154. https://doi.org/10.1016/j.rser.2015.07.091

Jang Y-S, Im JA, Choi SY et al (2014) Metabolic engineering of *Clostridium acetobutylicum* for butyric acid production with high butyric acid selectivity. Metab Eng 23:165–174. https://doi.org/10.1016/j.ymben.2014.03.004

Jang Y-S, Woo HM, Im JA et al (2013) Metabolic engineering of *Clostridium acetobutylicum* for enhanced production of butyric acid. Appl Microbiol Biotechnol 97:9355–9363. https://doi.org/10.1007/s00253-013-5161-x

Jennings M (2018) Optimizing Genetic Manipulation of Methanogens through Faster Cloning Techniques

Jimenez J, Latrille E, Harmand J et al (2015) Instrumentation and control of anaerobic digestion processes: a review and some research challenges. Rev Environ Sci Biotechnol 14:615–648. https://doi.org/10.1007/s11157-015-9382-6

Jin X, Angelidaki I, Zhang Y (2016) Microbial Electrochemical Monitoring of Volatile Fatty Acids during Anaerobic Digestion. Environ Sci Technol 50:4422–4429. https://doi.org/10.1021/acs.est.5b05267

Joseph RC, Kim NM, Sandoval NR (2018) Recent Developments of the Synthetic Biology Toolkit for *Clostridium*. Front Microbiol 9. https://doi.org/10.3389/fmicb.2018.00154

Kabisch UC, Gräntzdörffer A, Schierhorn A et al (1999) Identification of D-proline reductase from *Clostridium sticklandii* as a selenoenzyme and indications for a catalytically active pyruvoyl group derived from a cysteine residue by cleavage of a proprotein. J Biol Chem 274:8445–8454. https://doi.org/10.1074/jbc.274.13.8445

Kaden J, S Galushko A, Schink B (2002) Cysteine-mediated electron transfer in syntrophic acetate oxidation by cocultures of *Geobacter sulfurreducens* and *Wolinella succinogenes*. Arch Microbiol 178:53–58. https://doi.org/10.1007/s00203-002-0425-3

Kalyuzhnaya MG, Puri AW, Lidstrom ME (2015) Metabolic engineering in methanotrophic bacteria. Metab Eng 29:142–152. https://doi.org/10.1016/j.ymben.2015.03.010

Kamagata Y (2015) Syntrophy in Anaerobic Digestion. In: Anaerobic Biotechnology. Imperial College Press, pp 13–30

Kato S, Kosaka T, Watanabe K (2009) Substrate-dependent transcriptomic shifts in *Pelotomaculum thermopropionicum* grown in syntrophic co-culture with *Methanothermobacter thermautotrophicus*. Microb Biotechnol 2:575–584. https://doi.org/10.1111/j.1751-7915.2009.00102.x

Kato S, Sasaki K, Watanabe K et al (2014) Physiological and Transcriptomic Analyses of the Thermophilic, Aceticlastic Methanogen *Methanosaeta thermophila* Responding to Ammonia Stress. Microbes Environ 29:162–167. https://doi.org/10.1264/jsme2.ME14021

Kato S, Watanabe K (2010) Ecological and evolutionary interactions in syntrophic methanogenic consortia. Microbes Environ 25:145–151. https://doi.org/10.1264/jsme2.me10122

Khmelenina VN, Colin Murrell J, Smith TJ, Trotsenko YA (2018) Physiology and Biochemistry of the Aerobic Methanotrophs. In: Rojo F (ed) Aerobic Utilization of Hydrocarbons. Oils and Lipids. Springer International Publishing, Cham, pp 1–25

Khmelenina VN, Sakharovskii VG, Reshetnikov AS, Trotsenko YA (2000) Synthesis of osmoprotectants by halophilic and alkaliphilic methanotrophs. Microbiology 69:381–386

Knittel K, Wegener G, Boetius A (2018) Anaerobic Methane Oxidizers. In: McGenity TJ (ed) Microbial Communities Utilizing Hydrocarbons and Lipids: Members. Metagenomics and Ecophysiology. Springer International Publishing, Cham, pp 1–21

Koch C, Müller S, Harms H, Harnisch F (2014) Microbiomes in bioenergy production: From analysis to management. Curr Opin Biotechnol 27:65–72. https://doi.org/10.1016/j.copbio.2013.11.006

Kohler PRA, Metcalf WW (2012) Genetic manipulation of *Methanosarcina spp*. Front Microbiol 3. https://doi.org/10.3389/fmicb.2012.00259

Kosaka T, Kato S, Shimoyama T et al (2008) The genome of *Pelotomaculum thermopropionicum* reveals niche-associated evolution in anaerobic microbiota. Genome Res 18:442–448. https://doi.org/10.1101/gr.7136508

Kosaka T, Uchiyama T, Ishii S et al (2006) Reconstruction and regulation of the central catabolic pathway in the thermophilic propionate-oxidizing syntroph *Pelotomaculum thermopropionicum*. J Bacteriol 188:202–210. https://doi.org/10.1128/JB.188.1.202-210.2006

Kretzschmar J, Rosa LFM, Zosel J et al (2016) A Microbial Biosensor Platform for Inline Quantification of Acetate in Anaerobic Digestion: Potential and Challenges. Chem Eng Technol 39:637–642. https://doi.org/10.1002/ceat.201500406

Leang C, Ueki T, Nevin KP, Lovley DR (2013) A Genetic System for *Clostridium ljungdahlii*: a Chassis for Autotrophic Production of Biocommodities and a Model Homoacetogen. Appl Environ Microbiol 79:1102–1109. https://doi.org/10.1128/AEM.02891-12

Lee OK, Hur DH, Nguyen DTN, Lee EY (2016) Metabolic engineering of methanotrophs and its application to production of chemicals and biofuels from methane: Biofuel energy consumption-Economic growth relationship: An empirical investigation of Brazil. Biofuels Bioprod Biorefin 10:848–863. https://doi.org/10.1002/bbb.1678

Lee DJ, Jegatheesan V, Ngo HH, Hallenbeck PC (2017) Current developments in biotechnology and bioengineering: biological treatment of industrial effluents. Elsevier, Amsterdam

Leng L, Yang P, Singh S et al (2018) A review on the bioenergetics of anaerobic microbial metabolism close to the thermodynamic limits and its implications for digestion applications. Bioresour Technol 247:1095–1106. https://doi.org/10.1016/j.biortech.2017.09.103

Leonard E, Minshull J, Ness JE, et al (2016) Compositions and methods for biological production of isoprene

Lessner DJ, Lhu L, Wahal CS, Ferry JG (2010) An Engineered Methanogenic Pathway Derived from the Domains Bacteria and Archaea. mBio 1: doi: https://doi.org/10.1128/mBio.00243-10

Li Y, Chen Y, Wu J (2019) Enhancement of methane production in anaerobic digestion process: A review. Appl Energy 240:120–137. https://doi.org/10.1016/j.apenergy.2019.01.243

Li W, Khalid H, Zhu Z et al (2018) Methane production through anaerobic digestion: Participation and digestion characteristics of cellulose, hemicellulose and lignin. Appl Energy 226:1219–1228. https://doi.org/10.1016/j.apenergy.2018.05.055

Liao JC, Mi L, Pontrelli S, Luo S (2016) Fuelling the future: microbial engineering for the production of sustainable biofuels. Nat Rev Microbiol 14:288–304. https://doi.org/10.1038/nrmicro.2016.32

Lie TJ, Costa KC, Lupa B et al (2012) Essential anaplerotic role for the energy-converting hydrogenase Eha in hydrogenotrophic methanogenesis. Proc Natl Acad Sci 109:15473–15478. https://doi.org/10.1073/pnas.1208779109

Lim JW, Ge T, Tong YW (2018) Monitoring of microbial communities in anaerobic digestion sludge for biogas optimisation. Waste Manag 71:334–341. https://doi.org/10.1016/j.wasman.2017.10.007

Lindner J, Zielonka S, Oechsner H, Lemmer A (2016) Is the continuous two-stage anaerobic digestion process well suited for all substrates? Bioresour Technol 200:470–476. https://doi.org/10.1016/j.biortech.2015.10.052

Lisboa MS, Lansing S (2013) Characterizing food waste substrates for co-digestion through biochemical methane potential (BMP) experiments. Waste Manag 33:2664–2669. https://doi.org/10.1016/j.wasman.2013.09.004

Liu L, Guan N, Zhu G et al (2016) Pathway engineering of *Propionibacterium jensenii* for improved production of propionic acid. Sci Rep 6. https://doi.org/10.1038/srep19963

Liu Z, Liu J, Zhang S et al (2011) Microbial fuel cell based biosensor for in situ monitoring of anaerobic digestion process. Bioresour Technol 102:10221–10229. https://doi.org/10.1016/j.biortech.2011.08.053

Liu P, Lu Y (2018) Concerted Metabolic Shifts Give New Insights Into the Syntrophic Mechanism Between Propionate-Fermenting *Pelotomaculum thermopropionicum* and Hydrogenotrophic *Methanocella conradii*. Front Microbiol 9:1551. https://doi.org/10.3389/fmicb.2018.01551

Liu X, Zhu Y, Yang S-T (2006a) Construction and characterization of ack deleted mutant of *Clostridium tyrobutyricum* for enhanced butyric acid and hydrogen production. Biotechnol Prog 22:1265–1275. https://doi.org/10.1021/bp060082g

Liu X, Zhu Y, Yang S-T (2006b) Butyric acid and hydrogen production by *Clostridium tyrobutyricum* ATCC 25755 and mutants. Enzym Microb Technol 38:521–528. https://doi.org/10.1016/j.enzmictec.2005.07.008

Liu L, Zhuge X, Shin H et al (2015) Improved Production of Propionic Acid in *Propionibacterium jensenii* via Combinational Overexpression of Glycerol Dehydrogenase and Malate

Dehydrogenase from *Klebsiella pneumoniae*. Appl Environ Microbiol 81:2256–2264. https://doi.org/10.1128/AEM.03572-14
Lovley DR (2017) Syntrophy Goes Electric: Direct Interspecies Electron Transfer. Annu Rev Microbiol 71:643–664. https://doi.org/10.1146/annurev-micro-030117-020420
Lovley DR, Fraga JL, Coates JD, Blunt-Harris EL (1999) Humics as an electron donor for anaerobic respiration. Environ Microbiol 1:89–98
Lyu Z, Jain R, Smith P et al (2016) Engineering the Autotroph *Methanococcus maripaludis* for Geraniol Production. ACS Synth Biol 5:577–581. https://doi.org/10.1021/acssynbio.5b00267
Manisha YSK (2017) Technological advances and applications of hydrolytic enzymes for valorization of lignocellulosic biomass. Bioresour Technol 245:1727–1739. https://doi.org/10.1016/j.biortech.2017.05.066
Martin-Ryals A, Schideman L, Li P et al (2015) Improving anaerobic digestion of a cellulosic waste via routine bioaugmentation with cellulolytic microorganisms. Bioresour Technol 189:62–70. https://doi.org/10.1016/j.biortech.2015.03.069
McAnulty MJ, G. Poosarla V, Kim K-Y, et al (2017) Electricity from methane by reversing methanogenesis. Nat Commun 8:15419. doi: https://doi.org/10.1038/ncomms15419
McAnulty MJ, Poosarla VG, Li J et al (2017) Metabolic engineering of *Methanosarcina acetivorans* for lactate production from methane: Anaerobic Lactate Production From Methane. Biotechnol Bioeng 114:852–861. https://doi.org/10.1002/bit.26208
McGlynn SE (2017) Energy Metabolism during Anaerobic Methane Oxidation in ANME Archaea. Microbes Environ 32:5–13. https://doi.org/10.1264/jsme2.ME16166
McInerney MJ, Bryant MP, Hespell RB, Costerton JW (1981) *Syntrophomonas wolfei* gen. *nov*. sp. *nov*., an Anaerobic, Syntrophic, Fatty Acid-Oxidizing Bacterium. Appl Environ Microbiol 41:1029–1039
McInerney MJ, Struchtemeyer CG, Sieber J et al (2008) Physiology, ecology, phylogeny, and genomics of microorganisms capable of syntrophic metabolism. Ann N Y Acad Sci 1125:58–72
Mearls EB, Olson DG, Herring CD, Lynd LR (2015) Development of a regulatable plasmid-based gene expression system for *Clostridium thermocellum*. Appl Microbiol Biotechnol 99:7589–7599. https://doi.org/10.1007/s00253-015-6610-5
Meegoda J, Li B, Patel K, Wang L (2018) A Review of the Processes, Parameters, and Optimization of Anaerobic Digestion. Int J Environ Res Public Health 15:2224. https://doi.org/10.3390/ijerph15102224
Merlin Christy P, Gopinath LR, Divya D (2014) A review on anaerobic decomposition and enhancement of biogas production through enzymes and microorganisms. Renew Sust Energ Rev 34:167–173. https://doi.org/10.1016/j.rser.2014.03.010
Moran JJ, House CH, Freeman KH, Ferry JG (2005) Trace methane oxidation studied in several Euryarchaeota under diverse conditions. Archaea 1:303–309
Moran JJ, House CH, Thomas B, Freeman KH (2007) Products of trace methane oxidation during nonmethyltrophic growth by *Methanosarcina*. J Geophys Res 112. https://doi.org/10.1029/2006JG000268
Morita M, Malvankar NS, Franks AE, et al (2011) Potential for direct interspecies electron transfer in methanogenic wastewater digester aggregates. mBio 2:e00159–e00111. doi: https://doi.org/10.1128/mBio.00159-11
Müller N, Timmers P, Plugge CM et al (2018) Syntrophy in Methanogenic Degradation. In: Hackstein JHP (ed) (Endo)symbiotic Methanogenic *Archaea*. Springer International Publishing, Cham, pp 153–192
Müller N, Worm P, Schink B et al (2010) Syntrophic butyrate and propionate oxidation processes: from genomes to reaction mechanisms: Syntrophic butyrate and propionate oxidation processes. Environ Microbiol Rep 2:489–499. https://doi.org/10.1111/j.1758-2229.2010.00147.x
Mumme J, Linke B, Tölle R (2010) Novel upflow anaerobic solid-state (UASS) reactor. Bioresour Technol 101:592–599. https://doi.org/10.1016/j.biortech.2009.08.073
Munk B, Bauer C, Gronauer A, Lebuhn M (2012) A metabolic quotient for methanogenic *Archaea*. Water Sci Technol 66:2311. https://doi.org/10.2166/wst.2012.436

Myung J, Galega WM, Van Nostrand JD et al (2015) Long-term cultivation of a stable *Methylocystis* -dominated methanotrophic enrichment enabling tailored production of poly(3-hydroxybutyrate-co-3-hydroxyvalerate). Bioresour Technol 198:811–818. https://doi.org/10.1016/j.biortech.2015.09.094

Narihiro T, Sekiguchi Y (2011) Oligonucleotide primers, probes and molecular methods for the environmental monitoring of methanogenic archaea: Probes for methanogens. Microb Biotechnol 4:585–602. https://doi.org/10.1111/j.1751-7915.2010.00239.x

Nayak DD, Metcalf WW (2017) Cas9-mediated genome editing in the methanogenic archaeon *Methanosarcina acetivorans*. Proc Natl Acad Sci 114:2976–2981. https://doi.org/10.1073/pnas.1618596114

Nisman B (1954) The stickland reaction. Bacteriol Rev 18:16

Nybo SE, Khan NE, Woolston BM, Curtis WR (2015) Metabolic engineering in chemolithoautotrophic hosts for the production of fuels and chemicals. Metab Eng 30:105–120. https://doi.org/10.1016/j.ymben.2015.04.008

Nzila A (2017) Mini review: Update on bioaugmentation in anaerobic processes for biogas production. Anaerobe 46:3–12. https://doi.org/10.1016/j.anaerobe.2016.11.007

Oh ST, Kang S-J, Azizi A (2018) Electrochemical communication in anaerobic digestion. Chem Eng J 353:878–889. https://doi.org/10.1016/j.cej.2018.07.154

Øverland M, Tauson A-H, Shearer K, Skrede A (2010) Evaluation of methane-utilising bacteria products as feed ingredients for monogastric animals. Arch Anim Nutr 64:171–189. https://doi.org/10.1080/17450391003691534

Park J-H, Kang H-J, Park K-H, Park H-D (2018) Direct interspecies electron transfer via conductive materials: A perspective for anaerobic digestion applications. Bioresour Technol 254:300–311. https://doi.org/10.1016/j.biortech.2018.01.095

Pieja AJ, Morse MC, Cal AJ (2017) Methane to bioproducts: the future of the bioeconomy? Curr Opin Chem Biol 41:123–131. https://doi.org/10.1016/j.cbpa.2017.10.024

Pieja AJ, Rostkowski KH, Criddle CS (2011) Distribution and Selection of Poly-3-Hydroxybutyrate Production Capacity in Methanotrophic Proteobacteria. Microb Ecol 62:564–573. https://doi.org/10.1007/s00248-011-9873-0

Pitk P, Kaparaju P, Vilu R (2012) Methane potential of sterilized solid slaughterhouse wastes. Bioresour Technol 116:42–46. https://doi.org/10.1016/j.biortech.2012.04.038

Puri AW, Owen S, Chu F et al (2015) Genetic Tools for the Industrially Promising Methanotroph *Methylomicrobium buryatense*. Appl Environ Microbiol 81:1775–1781. https://doi.org/10.1128/AEM.03795-14

Qian X, Reguera G, Mester T, Lovley DR (2007) Evidence that OmcB and OmpB of *Geobacter sulfurreducens* are outer membrane surface proteins. FEMS Microbiol Lett 277:21–27. https://doi.org/10.1111/j.1574-6968.2007.00915.x

Ramsay IR, Pullammanappallil PC (2001) Protein degradation during anaerobic wastewater treatment: derivation of stoichiometry. Biodegradation 12:247–256

Reichlen MJ, Vepachedu VR, Murakami KS, Ferry JG (2012) MreA Functions in the Global Regulation of Methanogenic Pathways in *Methanosarcina acetivorans*. mBio 3: doi: https://doi.org/10.1128/mBio.00189-12

Richards MA, Lie TJ, Zhang J et al (2016) Exploring Hydrogenotrophic Methanogenesis: a Genome Scale Metabolic Reconstruction of *Methanococcus maripaludis*. J Bacteriol 198:3379–3390. https://doi.org/10.1128/JB.00571-16

Rostkowski KH, Pfluger AR, Criddle CS (2013) Stoichiometry and kinetics of the PHB-producing Type II methanotrophs *Methylosinus trichosporium* OB3b and *Methylocystis parvus* OBBP. Bioresour Technol 132:71–77. https://doi.org/10.1016/j.biortech.2012.12.129

Rotaru A-E, Shrestha PM, Liu F et al (2014a) A new model for electron flow during anaerobic digestion: direct interspecies electron transfer to *Methanosaeta* for the reduction of carbon dioxide to methane. Energy Env Sci 7:408–415. https://doi.org/10.1039/C3EE42189A

Rotaru A-E, Shrestha PM, Liu F et al (2014b) Direct Interspecies Electron Transfer between *Geobacter metallireducens* and *Methanosarcina barkeri*. Appl Environ Microbiol 80:4599–4605. https://doi.org/10.1128/AEM.00895-14

Sakai S, Imachi H, Hanada S et al (2008) *Methanocella paludicola* gen. *nov*., sp. *nov*., a methane-producing archaeon, the first isolate of the lineage "Rice Cluster I", and proposal of the new archaeal order Methanocellales ord. nov. Int J Syst Evol Microbiol 58:929–936. https://doi.org/10.1099/ijs.0.65571-0

Salama E-S, Saha S, Kurade MB et al (2019) Recent trends in anaerobic co-digestion: Fat, oil, and grease (FOG) for enhanced biomethanation. Prog Energy Combust Sci 70:22–42. https://doi.org/10.1016/j.pecs.2018.08.002

Sarmiento FB, Leigh JA, Whitman WB (2011) Genetic Systems for Hydrogenotrophic Methanogens. In: Methods in Enzymology. Elsevier, pp 43–73

Saville RM, Lee S, Regitsky DD, et al (2016) Compositions and methods for biological production of lactate from c1 compounds using lactate dehydrogenase transformants

Schellenberger J, Park JO, Conrad TM, Palsson BØ (2010) BiGG: a Biochemical Genetic and Genomic knowledgebase of large scale metabolic reconstructions. BMC Bioinformatics 11:213

Scheller S, Goenrich M, Boecher R et al (2010) The key nickel enzyme of methanogenesis catalyses the anaerobic oxidation of methane. Nature 465:606–608. https://doi.org/10.1038/nature09015

Schiel-Bengelsdorf B, Dürre P (2012) Pathway engineering and synthetic biology using acetogens. FEBS Lett 586:2191–2198. https://doi.org/10.1016/j.febslet.2012.04.043

Schievano A, Tenca A, Lonati S et al (2014) Can two-stage instead of one-stage anaerobic digestion really increase energy recovery from biomass? Appl Energy 124:335–342. https://doi.org/10.1016/j.apenergy.2014.03.024

Schink B (1997) Energetics of syntrophic cooperation in methanogenic degradation. Microbiol Mol Biol Rev 61:262–280

Schink B, Montag D, Keller A, Müller N (2017) Hydrogen or formate: Alternative key players in methanogenic degradation. Environ Microbiol Rep 9:189–202. https://doi.org/10.1111/1758-2229.12524

Schink B, Stams AJM (2013) Syntrophism Among Prokaryotes. In: DeLong EF, Lory S et al (eds) Rosenberg E. Prokaryotic Communities and Ecophysiology. Springer Berlin Heidelberg, Berlin, Heidelberg, The Prokaryotes, pp 471–493

Schmidt A, Müller N, Schink B, Schleheck D (2013) A Proteomic View at the Biochemistry of Syntrophic Butyrate Oxidation in *Syntrophomonas wolfei*. PLoS One 8:e56905. https://doi.org/10.1371/journal.pone.0056905

Sekiguchi Y, Kamagata Y, Nakamura K, et al (2000) *Syntrophothermus lipocalidus* gen. *nov*., sp. *nov*., a novel thermophilic, syntrophic, fatty-acid-oxidizing anaerobe which utilizes isobutyrate. Int J Syst Evol Microbiol 50 Pt 2:771–779. doi: https://doi.org/10.1099/00207713-50-2-771

Sharpe PL, DiCosimo D, Bosak MD et al (2007) Use of Transposon Promoter-Probe Vectors in the Metabolic Engineering of the Obligate Methanotroph *Methylomonas sp*. Strain 16a for Enhanced C40 Carotenoid Synthesis. Appl Environ Microbiol 73:1721–1728. https://doi.org/10.1128/AEM.01332-06

Sheets JP, Ge X, Li Y-F et al (2016) Biological conversion of biogas to methanol using methanotrophs isolated from solid-state anaerobic digestate. Bioresour Technol 201:50–57. https://doi.org/10.1016/j.biortech.2015.11.035

Shen CR, Lan EI, Dekishima Y et al (2011) Driving Forces Enable High-Titer Anaerobic 1-Butanol Synthesis in *Escherichia coli*. Appl Environ Microbiol 77:2905–2915. https://doi.org/10.1128/AEM.03034-10

Shimoyama T, Kato S, Ishii S, Watanabe K (2009) Flagellum mediates symbiosis. Science 323:1574. https://doi.org/10.1126/science.1170086

Shrestha S, Fonoll X, Khanal SK, Raskin L (2017) Biological strategies for enhanced hydrolysis of lignocellulosic biomass during anaerobic digestion: Current status and future perspectives. Bioresour Technol 245:1245–1257. https://doi.org/10.1016/j.biortech.2017.08.089

Shrestha PM, Malvankar NS, Werner JJ et al (2014) Correlation between microbial community and granule conductivity in anaerobic bioreactors for brewery wastewater treatment. Bioresour Technol 174:306–310. https://doi.org/10.1016/j.biortech.2014.10.004

Sieber JR, Crable BR, Sheik CS et al (2015) Proteomic analysis reveals metabolic and regulatory systems involved in the syntrophic and axenic lifestyle of *Syntrophomonas wolfei*. Front Microbiol 6. https://doi.org/10.3389/fmicb.2015.00115

Sieber JR, McInerney MJ, Gunsalus RP (2012) Genomic Insights into Syntrophy: The Paradigm for Anaerobic Metabolic Cooperation. Annu Rev Microbiol 66:429–452. https://doi.org/10.1146/annurev-micro-090110-102844

Sieber JR, McInerney MJ, Müller N et al (2018) Methanogens: Syntrophic Metabolism. In: Stams AJM, Sousa D (eds) Biogenesis of Hydrocarbons. Springer International Publishing, Cham, pp 1–31

Sieber JR, Sims DR, Han C et al (2010) The genome of *Syntrophomonas wolfei*: new insights into syntrophic metabolism and biohydrogen production: *Syntrophomonas wolfei* genome. Environ Microbiol no–no. https://doi.org/10.1111/j.1462-2920.2010.02237.x

Soo VWC, McAnulty MJ, Tripathi A et al (2016) Reversing methanogenesis to capture methane for liquid biofuel precursors. Microb Cell Factories 15. https://doi.org/10.1186/s12934-015-0397-z

Stams AJM, Plugge CM (2009) Electron transfer in syntrophic communities of anaerobic bacteria and archaea. Nat Rev Microbiol 7:568–577. https://doi.org/10.1038/nrmicro2166

Straub M, Demler M, Weuster-Botz D, Dürre P (2014) Selective enhancement of autotrophic acetate production with genetically modified *Acetobacterium woodii*. J Biotechnol 178:67–72. https://doi.org/10.1016/j.jbiotec.2014.03.005

Strong PJ, Xie S, Clarke WP (2015) Methane as a Resource: Can the Methanotrophs Add Value? Environ Sci Technol 49:4001–4018. https://doi.org/10.1021/es504242n

Subbian E (2017) Production of succinic acid from organic waste or biogas or methane using recombinant methanotrophic bacterium

Summers ZM, Fogarty HE, Leang C et al (2010) Direct exchange of electrons within aggregates of an evolved syntrophic coculture of anaerobic bacteria. Science 330:1413–1415. https://doi.org/10.1126/science.1196526

Sun H, Zhang Y, Wu S et al (2019) Innovative operation of microbial fuel cell-based biosensor for selective monitoring of acetate during anaerobic digestion. Sci Total Environ 655:1439–1447. https://doi.org/10.1016/j.scitotenv.2018.11.336

Suwannakham S, Huang Y, Yang S-T (2006) Construction and characterization ofack knock-out mutants of *Propionibacterium acidipropionici* for enhanced propionic acid fermentation. Biotechnol Bioeng 94:383–395. https://doi.org/10.1002/bit.20866

Sweeney JB, Murphy CD, McDonnell K (2018) Development of a bacterial propionate-biosensor for anaerobic digestion monitoring. Enzym Microb Technol 109:51–57. https://doi.org/10.1016/j.enzmictec.2017.09.011

Thauer RK, Kaster A-K, Seedorf H et al (2008) Methanogenic archaea: ecologically relevant differences in energy conservation. Nat Rev Microbiol 6:579–591. https://doi.org/10.1038/nrmicro1931

Tian J-H, Pourcher A-M, Bureau C, Peu P (2017) Cellulose accessibility and microbial community in solid state anaerobic digestion of rape straw. Bioresour Technol 223:192–201. https://doi.org/10.1016/j.biortech.2016.10.009

Treu L, Kougias PG, Campanaro S et al (2016) Deeper insight into the structure of the anaerobic digestion microbial community; the biogas microbiome database is expanded with 157 new genomes. Bioresour Technol 216:260–266. https://doi.org/10.1016/j.biortech.2016.05.081

Tsavkelova E, Prokudina L, Egorova M et al (2018) The structure of the anaerobic thermophilic microbial community for the bioconversion of the cellulose-containing substrates into biogas. Process Biochem 66:183–196. https://doi.org/10.1016/j.procbio.2017.12.006

Van Kuijk BL, Schlösser E, Stams AJ (1998) Investigation of the fumarate metabolism of the syntrophic propionate-oxidizing bacterium strain MPOB. Arch Microbiol 169:346–352

Van Steendam C, Smets I, Skerlos S, Raskin L (2019) Improving anaerobic digestion via direct interspecies electron transfer requires development of suitable characterization methods. Curr Opin Biotechnol 57:183–190. https://doi.org/10.1016/j.copbio.2019.03.018

Vanwonterghem I, Jensen PD, Ho DP et al (2014) Linking microbial community structure, interactions and function in anaerobic digesters using new molecular techniques. Curr Opin Biotechnol 27:55–64. https://doi.org/10.1016/j.copbio.2013.11.004

Wang B, Albarracín-Suazo S, Pagán-Torres Y, Nikolla E (2017) Advances in methane conversion processes. Catal Today 285:147–158. https://doi.org/10.1016/j.cattod.2017.01.023

Wang Z, Ammar EM, Zhang A et al (2015) Engineering *Propionibacterium freudenreichii subsp. shermanii* for enhanced propionic acid fermentation: Effects of overexpressing propionyl-CoA: Succinate CoA transferase. Metab Eng 27:46–56. https://doi.org/10.1016/j.ymben.2014.10.005

Wendlandt K-D, Jechorek M, Helm J, Stottmeister U (2001) Producing poly-3-hydroxybutyrate with a high molecular mass from methane. J Biotechnol 86:127–133. https://doi.org/10.1016/S0168-1656(00)00408-9

Whitaker WB, Sandoval NR, Bennett RK et al (2015) Synthetic methylotrophy: engineering the production of biofuels and chemicals based on the biology of aerobic methanol utilization. Curr Opin Biotechnol 33:165–175. https://doi.org/10.1016/j.copbio.2015.01.007

Wood TK (2017) Metabolic manipulation of methanogens for methane machinations. Microb Biotechnol 10:9–10. https://doi.org/10.1111/1751-7915.12425

Worm P, Koehorst JJ, Visser M et al (2014) A genomic view on syntrophic versus non-syntrophic lifestyle in anaerobic fatty acid degrading communities. Biochim Biophys Acta 1837:2004–2016. https://doi.org/10.1016/j.bbabio.2014.06.005

Worm P, Müller N, Plugge CM et al (2010) Syntrophy in Methanogenic Degradation. In: Hackstein JHP (ed) (Endo)symbiotic Methanogenic Archaea. Springer, Berlin Heidelberg, Berlin, Heidelberg, pp 143–173

Worm P, Stams AJM, Cheng X, Plugge CM (2011) Growth- and substrate-dependent transcription of formate dehydrogenase and hydrogenase coding genes in *Syntrophobacter fumaroxidans* and *Methanospirillum hungatei*. Microbiol Read Engl 157:280–289. https://doi.org/10.1099/mic.0.043927-0

Wu D, Li L, Zhao X et al (2019) Anaerobic digestion: A review on process monitoring. Renew Sust Energ Rev 103:1–12. https://doi.org/10.1016/j.rser.2018.12.039

Xia A, Cheng J, Murphy JD (2016) Innovation in biological production and upgrading of methane and hydrogen for use as gaseous transport biofuel. Biotechnol Adv 34:451–472. https://doi.org/10.1016/j.biotechadv.2015.12.009

Yan X, Chu F, Puri AW et al (2016) Electroporation-Based Genetic Manipulation in Type I Methanotrophs. Appl Environ Microbiol 82:2062–2069. https://doi.org/10.1128/AEM.03724-15

Yang L, Ge X, Wan C et al (2014) Progress and perspectives in converting biogas to transportation fuels. Renew Sust Energ Rev 40:1133–1152. https://doi.org/10.1016/j.rser.2014.08.008

Yang H, Zhou M, Liu M et al (2015) Microbial fuel cells for biosensor applications. Biotechnol Lett 37:2357–2364. https://doi.org/10.1007/s10529-015-1929-7

Yazdian F, Hajizadeh S, Shojaosadati SA et al (2005) Production of single cell protein from natural gas: parameter optimization and RNA evaluation. Iran J Biotechnol 3:235–242

Ye RW, Yao H, Stead K et al (2007) Construction of the astaxanthin biosynthetic pathway in a methanotrophic bacterium *Methylomonas sp.* strain 16a. J Ind Microbiol Biotechnol 34:289–299. https://doi.org/10.1007/s10295-006-0197-x

Yu G-H, He P-J, Shao L-M, Zhu Y-S (2008) Extracellular proteins, polysaccharides and enzymes impact on sludge aerobic digestion after ultrasonic pretreatment. Water Res 42:1925–1934. https://doi.org/10.1016/j.watres.2007.11.022

Yu Z, Leng X, Zhao S et al (2018) A review on the applications of microbial electrolysis cells in anaerobic digestion. Bioresour Technol 255:340–348. https://doi.org/10.1016/j.biortech.2018.02.003

Zeldes BM, Keller MW, Loder AJ et al (2015) Extremely thermophilic microorganisms as metabolic engineering platforms for production of fuels and industrial chemicals. Front Microbiol 6. https://doi.org/10.3389/fmicb.2015.01209

Zema DA (2017) Planning the optimal site, size, and feed of biogas plants in agricultural districts. Biofuels Bioprod Biorefin 11:454–471

Zhang W, Ge X, Li Y-F et al (2016) Isolation of a methanotroph from a hydrogen sulfide-rich anaerobic digester for methanol production from biogas. Process Biochem 51:838–844. https://doi.org/10.1016/j.procbio.2016.04.003

Zhang H, Wang X (2016) Modular co-culture engineering, a new approach for metabolic engineering. Metab Eng 37:114–121. https://doi.org/10.1016/j.ymben.2016.05.007

Zhang A, Yang S-T (2009) Propionic acid production from glycerol by metabolically engineered *Propionibacterium acidipropionici*. Process Biochem 44:1346–1351. https://doi.org/10.1016/j.procbio.2009.07.013

Zhang C, Yuan Q, Lu Y (2014) Inhibitory effects of ammonia on methanogen *mcrA* transcripts in anaerobic digester sludge. FEMS Microbiol Ecol 87:368–377. https://doi.org/10.1111/1574-6941.12229

Zhao J, Gui L, Wang Q et al (2017) Aged refuse enhances anaerobic digestion of waste activated sludge. Water Res 123:724–733. https://doi.org/10.1016/j.watres.2017.07.026

Zhao X, Li L, Wu D et al (2019) Modified Anaerobic Digestion Model No. 1 for modeling methane production from food waste in batch and semi-continuous anaerobic digestions. Bioresour Technol 271:109–117. https://doi.org/10.1016/j.biortech.2018.09.091

Zhao Z, Zhang Y, Woodard TL et al (2015) Enhancing syntrophic metabolism in up-flow anaerobic sludge blanket reactors with conductive carbon materials. Bioresour Technol 191:140–145. https://doi.org/10.1016/j.biortech.2015.05.007

Zheng Y, Zhao J, Xu F, Li Y (2014) Pretreatment of lignocellulosic biomass for enhanced biogas production. Prog Energy Combust Sci 42:35–53. https://doi.org/10.1016/j.pecs.2014.01.001

Zhou M, Yan B, Wong JWC, Zhang Y (2018) Enhanced volatile fatty acids production from anaerobic fermentation of food waste: A mini-review focusing on acidogenic metabolic pathways. Bioresour Technol 248:68–78. https://doi.org/10.1016/j.biortech.2017.06.121

Zhu X, Cao Q, Chen Y et al (2019) Effects of mixing and sodium formate on thermophilic in-situ biogas upgrading by H2 addition. J Clean Prod 216:373–381. https://doi.org/10.1016/j.jclepro.2019.01.245

Zhu Y, Liu X, Yang S-T (2005) Construction and characterization of pta gene-deleted mutant of *Clostridium tyrobutyricum* for enhanced butyric acid fermentation. Biotechnol Bioeng 90:154–166. https://doi.org/10.1002/bit.20354

Zhuge X, Liu L, Shin H et al (2013) Development of a *Propionibacterium-Escherichia coli* Shuttle Vector for Metabolic Engineering of *Propionibacterium jensenii*, an Efficient Producer of Propionic Acid. Appl Environ Microbiol 79:4595–4602. https://doi.org/10.1128/AEM.00737-13

Ziels RM, Beck DAC, Martí M et al (2015) Monitoring the dynamics of syntrophic β-oxidizing bacteria during anaerobic degradation of oleic acid by quantitative PCR. FEMS Microbiol Ecol 91. https://doi.org/10.1093/femsec/fiv028

Zindel U, Freudenberg W, Rieth M et al (1988) *Eubacterium acidaminophilum* sp. nov., a versatile amino acid-degrading anaerobe producing or utilizing H_2 or formate. Arch Microbiol 150:254–266. https://doi.org/10.1007/BF00407789

Scale-Up Operations for Biogas Production: Analysis on Critical Factors Governing Large-Scale Operations

Spyridon Achinas, Sotirios Longinos, Vasileios Achinas, and Gerrit Jan Willem Euverink

Abstract Anaerobic digestion (AD) is a unique process where different microbial species decompose organic materials in the absence of oxygen and has been widely practiced in full-scale facilities all over the world. Several AD techniques have been applied to convert livestock manures, wastewaters, and solid lignocellulosic waste into biogas. Despite the progress on the engineering of AD systems, several challenges exist for the economically and environmentally efficient way to recover carbon in the form of renewable biogas fuel. The complexity of the challenges poses constraints into the understanding of the factors associated to the scale-up of the AD operations. This study aims to review the critical factors of biogas plant project development.

Keywords Anaerobic digestion · Biogas plant · Large-scale operations · Bioreactors · Sustainability

1 Introduction

A main reason for increasing interest in biogas production from AD is the necessity for displacing the limited energy resource fossil fuels that play a big role in global warming due to greenhouse gas emissions while being processed (Davis 2018; Achinas et al. 2017). Another main environmental problem is the overproduction of organic wastes from industry, households, and agriculture. Organic wastes,

S. Achinas (✉) · G. J. W. Euverink
Faculty of Science and Engineering, University of Groningen, Groningen, Netherlands
e-mail: s.achinas@rug.nl

S. Longinos
Department of Petroleum and Natural Gas Engineering, Middle East Technical University, Ankara, Turkey

V. Achinas
Solid Waste Management Agency of Attica Region, Special Inter-Grade Association, Athens, Greece

N. Balagurusamy, A. K. Chandel (eds.), *Biogas Production*,
https://doi.org/10.1007/978-3-030-58827-4_12

together with animal manure, are both alarming waste sources but can be treated by AD to form biogas. The waste product from the AD is digestate that, in return, can be used as natural fertilizer for agriculture (RedCorn et al. 2018; Sahajwalla 2018).

Energy and environmental policies prevent global warming by prioritizing sustainable waste management and European targets of renewable energy production. Both policies can indirectly solve the main environmental issues on limited fossil fuels and waste material and besides indirectly support the biogas production by AD industry (Wen-Wei and Han-Qing 2016). Biogas production from AD releases CO_2, but compared to fossil fuels, carbon atoms in biogas originate from the short carbon cycle, which refers to photosynthesis that took place much more recent. Besides, emissions of methane and nitrous oxide are reduced during the biogas production by AD so it contributes to mitigate global warming. Other positive returns from biogas production by AD are a reduction of odors and flies that are present in stored waste sources, an increase of local economic capabilities, and improved veterinary safety due to the application of digestate as fertilizer instead of untreated manure as fertilizer (Al Seadi et al. 2008; Chen 2017a).

2 System Context Biogas Production of AD

The industry in biogas production of AD has many reasons for its promising potential. The causal loop diagram (CLD) depicted in Fig. 1 indicates the factors of the general system that contribute to the worldwide interest in biogas production by AD. The interrelations in the general system are explained in the next three paragraphs. According to Sterman (2000), the positive loops reinforce (R) change, while negative loops (B) are self-correcting; they oppose disturbances. Figure 1 shows several possible loops of which two are explained in order to understand the interrelations.

he GHG-photosynthesis-vegetable biomass-AD production-net GHG loop is a balancing loop. This is explicable by the closed (short) carbon cycle that occurs during biogas production of AD, which was explained in a previous paragraph. The gas fossil fuels-GHG-global warming-water shortages-process water loop is a reinforcing loop because water shortages will cause the need for water-saving energy production solutions. Subsequently, gas production from fossil fuels is not preferred as it requires big amounts of water during processing, while biogas production from AD does not. The loops reinforce and ultimately outweigh gas production from fossil fuels if water shortages exist. The above given context can be traced back to a system that approaches benefits and describes simplified technical aspects that influences the AD process. Anaerobic digestion of waste can improve the quality of life. The produced biogas can be used for cooking, heating water, or generation of electricity for onsite use (Wang et al. 2018). It can also mitigate deforestation by using biogas instead of firewood, and it can mitigate waste accumulation by waste processing. These factors can result in reduced public health concerns, specifically

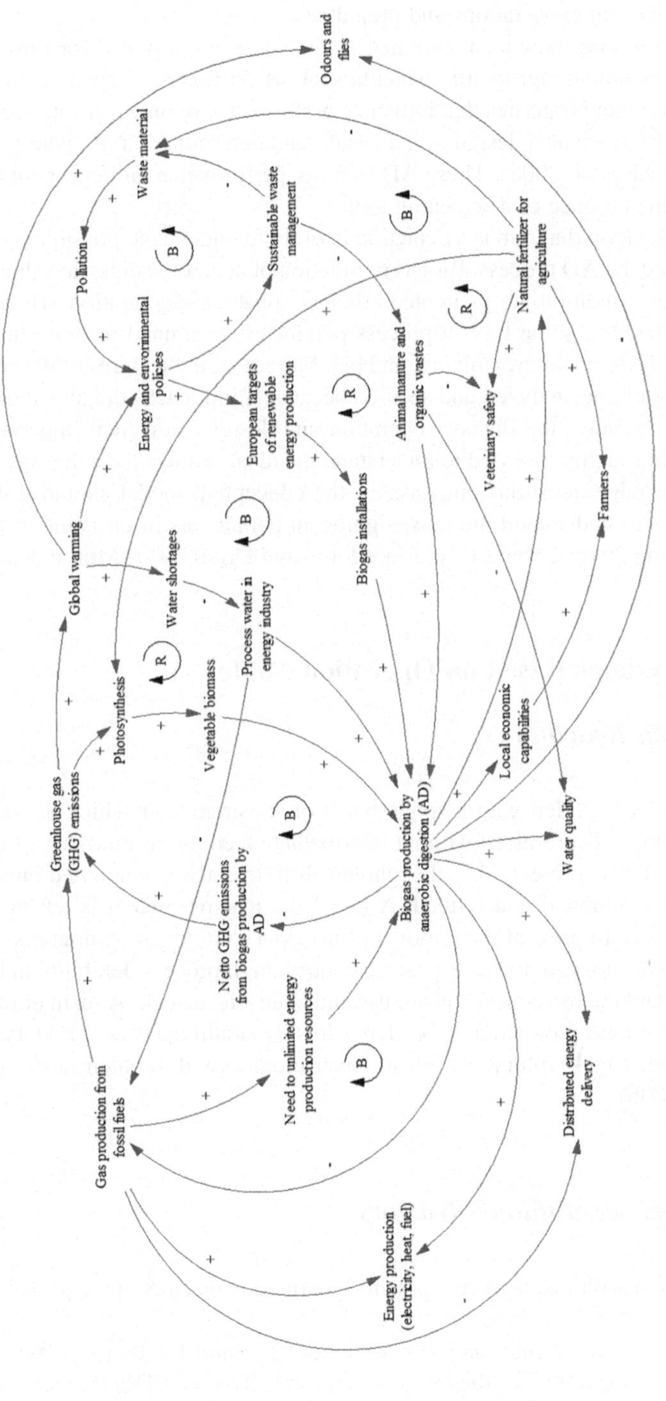

Fig. 1 Causal loop diagram on system context of biogas production from AD

children and women who are disproportionally affected by air pollution because of cultural and social expectations and prejudices.

In addition, and likewise mentioned in the CLD, the effluent contains primary nutrients that initiate agronomic benefits such as fertilizer use for improved plant growth. The main properties that influence performance of an anaerobic digester are substrate characteristics, design of a digester, and determination of operating conditions (Al Seadi et al. 2008). These AD stability performance influencers include the reason for the absence of a screening tool.

The casual loop diagram is a depiction in order to understand possible parameters that influence the AD process. Biogas production of anaerobic digestion should meet several basic conditions to guarantee efficient substrate degradation. These conditions are related to some typical process parameters examined in biotechnological processes, like temperature, mixing, and pH. Moreover, the features of the employed substrate should be analyzed and secured because it influences the abovementioned parameters. Besides, the digester operation and design is of great importance and should be thoroughly assessed to understand the ideal settings for a digester. As one of the knowledge questions emphasizes, the conceptual model should distinguish scaling sizes to understand the most significant parameters in upscaling (Wen-Wei and Han-Qing 2016; Chen 2017b; Macedonio and Drioli 2017; Moron et al. 2018).

3 AD Systems Based on Operation Mode

3.1 Batch Reactors

Batch reactors are filled with a single batch of substrate after which the AD takes place and biogas is produced without intermediate feeding or removing of liquid to or from the digester. Next, after the whole substrate batch is converted into biogas, the digester is emptied and refilled. A bit of the reactor content is left during the emptying phase to prevent the removal of necessary microbial communities. Benefits of batch reactors are the easy process control, high process flexibility in terms of cycle time, and the robustness toward dry and coarse feedstock. A main drawback is the irregular biogas production in batch reactors (Bharathiraja et al. 2016). It is stated that larger feedstock volumes result in lower biogas yield in batch mode digesters (Weiland 2006).

3.2 (Semi-)continuous Reactors

In comparison with batch reactors, (semi-)continuous reactors are applied in industrial scale.

The most common and simple reactor configuration for biogas plants are the CSTR type. Almost 90% of the existing biogas plants are CSTRs (Sanjay and Vijay

Table 1 Advantages and disadvantages of batch and CSTR operation modes in anaerobic digesters (Bharathiraja et al. 2016; Weiland 2006; Sanjay and Vijay 2012)

Mode	Advantages	Disadvantages
Batch	• Easy process control: no mixing, stirring, and pumping required • High flexibility in terms of cycle time • Low input in terms of process and mechanical demands • Robustness toward dry and coarse feedstocks • Low capital costs	• Lower and irregular biogas yield • Channeling and clogging • Large volume
CSTR	• Simplicity in design and operation • Seldom technical failure • Low setup time • Intensified process so more biogas production per time unit • Low investment and operation costs	• High HRT • Rapid acidification due to VFA accumulation • Foaming and scum formation

2012). The feedstock in CSTRs is continuously fed into the reactor, and there is a constant production of biogas. A continuous stirred digester can be horizontal, vertical, or multiple tank systems. Depending on the mixing conditions, continuous stirred digesters can be completely mixed or plug flow. The low investment and operating costs corresponding to the simple reactor design of CSTRs are a main advantage. Other benefits of CSRTs are seldom technical failure, ease of operation, less setup time, intensified processes so more production per time unit, and increased process control. Disadvantages are the high residence time and the possibility of foaming and scum formation (Weiland 2006).

Several operating modes exist in reactor industry of which batch and continuous reactors are assessed in the literature review. Batch reactors are filled with a single batch of substrate after which the AD takes place and biogas is produced without intermediate feeding or removing of liquid to or from the digester. CSTRs are continuously fed with feedstock so there is a consistent production of biogas. Benefits and drawbacks of both operation modes applied in AD are listed in Table 1. Moreover, extra survey was performed on both operation modes, however, these sections are assessed to be out of the scope of important literature findings.

4 AD Systems Based on Scale

4.1 Small-Scale AD Systems

Household digesters with an operating volume of 2–10 m^3 are small-scale digesters and mostly located in rural areas in Asia and other developing countries (Chen 2016; Manni et al. 2017). Currently, the most common small-scale anaerobic digesters are fixed dome digesters, floating drum digesters, and tubular digesters. All three digesters lack mechanical heating and mixing systems. The produced biogas from household digesters is mainly used for stoves and lamps. To illustrate, if a stove is used twice a day for a family of five, 1500–2400 L biogas is required, which requires manure from 130 chickens, 5 cows, or 1 pig (Bond and Templeton 2011).

Benefits of small-scale AD systems are agricultural, energy, environmental, health, and social benefits. These benefits are associated with burning a more environmental fuel and stabilizing residues, creating both a fertilizer and fuel source at the same time, better livestock management resulting in improved water quality, and reducing deforestation by preventing firewood from being a fuel source (Chen 2016). A drawback of a small-scale AD system is the relative high price, usually less than 800 EUR (€), in these rural areas. Therefore, countries like the Netherlands support the construction of these systems by governmental legislations or nonprofit companies that subsidize installation. These grants led to the installation of 700,000 small-scale biogas plants by 2015, impacting 3.5 million people (REN21 2017) (Table 2).

4.2 Large-Scale AD Systems

Large-scale digester size differs between hundreds to thousands cubic meters. Large-scale AD systems are popular in developed countries, since they require high capital investment and larger infrastructure. Therefore, Europe is a pioneer in large-scale AD systems (Davis 2018). The produced biogas is often upgraded to use as a transport fuel, but it is used for combined heat and power (CHP) in most of the cases (Achinas and Achinas 2017).

Currently, the two most popular large-scale digester operations are the farm-scale digesters and the centralized digesters. Farm-scale digesters usually have a capacity between 200 and 1200 m^3 and are generally constructed in swine or dairy farms. They tread agricultural residues from one till three farms. Germany is the current leader in farm-scale digesters as they have 9000 plants operating, while the country aims at extra implementation of 10.000–12.000 digesters (Wilkinson 2011). The centralized digesters have an even larger capacity of up to 8000 m^3. Denmark, the leading country in this technology, has only 20 centralized digesters running due to very high investment costs. They are willing to increase this amount to 30 centralized digesters. Another large-scale AD system is a wastewater treatment plant, which is well-established in the United States as it has 1250 wastewater treatment plants producing biogas (ABC 2016; Nelson et al. 2017).

Table 2 Benefits, drawbacks, and design and operation parameters of small-scale digesters (Achinas and Euverink 2019a, b; Achinas and Achinas 2017; Wilkinson 2011; ABC 2016; Nelson et al. 2017; Garfi et al. 2016; Choorit and Wisarnwan 2007; Chae et al. 2008; Mao et al. 2015; Verma 2002; Kigozi et al. 2014, 2014; Buekens 2005, 2005; Wang et al. 2014, 2014, 2006; Bowen et al. 2014; Karim et al. 2005; Burton and Turner 2003; Stroot et al. 2001; Gomez et al. 2006; Pinho et al. 2004; Olivet et al. 2005; Grady et al. 1999; Beccari et al. 1996; Gerardi 2006; Yu and Fang 2002; Schnurer and Jarvis 2010, 2010; Dussadee et al. 2017; Parawira et al. 2006; Guwy et al. 1997; Moller et al. 2004; Ahring et al. 1995; Costa et al. 2007; Rosato 2017; Pontoni et al. 2015; Yadvika et al. 2004; Braun et al. 2009; Zupancic and Grilc 2011; Deublein and Steinhausez 2008; Kayhanian 1999; Ariunbaatar et al. 2015; Shi et al. 2017; Rabii et al. 2019; Świątek et al. 2018; Solarte-Toro et al. 2018; Benato and Macor 2019; Oreggioni et al. 2017; Lindkvist et al. 2019; Baccioli et al. 2019; Florio et al. 2019; Lauer and Thrän 2018; Achinas 2014; Carlini et al. 2017; Dell'Antonia et al. 2013; Chen et al. 2018; Chiumenti et al. 2018; Fauzianto et al. 2014; Valenti and Porto 2019; Watts and Wiles 2007)

Parameter	Fixed dome digester	Floating drum digester	Tubular digester
Digester design, material	Fixed dome, bricks, and concrete (local materials)	Steel drum and concrete	Tubular, PVC, or polyethylene
Covering	–	Steel drum	Simple roof to protect the plastic top
Temperature range (°C)	Psychrophilic (<25 °C) Mesophilic (25–40 °C)	"	"
Total volume (m^3)	10–20	1.6–10	6–70
Hydraulic residence time (days)	~55	–	20–125
Lifespan (years)	~20	~15	5–10
Benefits	Cheap digester, interchangeable construction materials	Maintains a constant produced gas pressure	Ease of implementation and handling
Drawbacks	Produced gas pressure fluctuates much, no stirring and heating so in	High construction price, expensive construction, no stirring and heating	Reactor surface, no stirring and heating, short lifespan

5 Critical Factors for the Biogas Plant Operation

5.1 *Process Operation Parameters*

Anaerobic digestion should meet several conditions to guarantee efficient substrate degradation. These conditions are related to some typical process parameters examined in biotechnological processes, like temperature, mixing, pH, organic loading rate (OLR), hydraulic retention time (HRT, days), total alkalinity (TA, as equivalent mg $CaCO_3$), volatile fatty acids (VFAs, as equivalent mg acetic acid), VFAs/TA ratio (or FOS/TAC), redox potential, and ammonia (Garfi et al. 2016). Moreover, the features of the employed substrate should be analyzed because it influences the parameters mentioned above. Examples are the content of volatile solids (VS), total solids (TS), and the carbon-to-nitrogen ratio (C/N).

5.1.1 Temperature

Performance of AD and survival and growth of microbial consortia rely very much upon reactor temperature. Anaerobic digestion can operate in different temperature ranges that are classified as psychrophilic (<20 °C), mesophilic (20–45 °C), and thermophilic (45–70 °C) (Choorit and Wisarnwan 2007). The most occurring temperature ranges are either mesophilic temperatures or thermophilic temperatures. The operating temperature preferably does not change because mesophilic to thermophilic temperature switches (or vice versa) can result in immediately reduced biogas production until the involving microbes have increased in the required amount. Chae et al. (2008) found a significant reduction in biogas production while changing the temperature from 35 to 30 °C and 30 to 32 °C.

Several modern biogas plants operate at thermophilic temperatures due to several advantages such as reduced retention time, improved digestibility, effective destruction of pathogens, and consequently higher biogas yield compared to mesophilic temperatures. However, some outweighing disadvantages cause mesophilic temperature ranges being the most occurring process temperature range. The temperature of the thermophilic process initiates ammonia inhibition or the so-called toxicity of ammonia. Moreover, thermophilic AD requires increased energy demand and high investment costs, and there is an increased risk of process imbalance, e.g., acidification (Davis 2018; Mao et al. 2015) (Table 3). Thus, a decrease in temperature results in higher richness in microorganisms and better process stability. The optimal conditions for the AD will, therefore, be a combination of both ranges: thermophilic ranges for hydrolysis + acidogenesis, and mesophilic for acetogenesis and methanogenesis, which is convenient for a two-stage AD.

Table 3 performance comparison between mesophilic and thermophilic temperature range operation (Choorit and Wisarnwan 2007; Chae et al. 2008; Mao et al. 2015; Verma 2002; Kigozi et al. 2014; Buekens 2005; Wang et al. 2014; Bowen et al. 2014)

Performance characteristics	Mesophilic digestion	Thermophilic digestion
Biogas production	Low	High
Process stability	High	Low
Pathogens destruction	Low	High
Energy requirement	Low	High
HRT	High	Low
Effluent quality	High	Low
Odor production	Low	High
Investment costs	Low	High

5.1.2 Mixing

Most anaerobic digesters are equipped with an impeller to mix the reactor content. Mixing ensures efficient transfer of the organic compounds to the present microbial biomass. It also releases trapped gas bubbles and prevents sedimentation of dense material. The employed mixing method can differ significantly per digester (Karim et al. 2005). It can occur continuously or intermittent and activated for a few times per day. The energy input varies from 10 to 100 Whm^{-3}, depending on the type of impeller, the total solids in the digester, and the kind of reactor (Burton and Turner 2003). Currently, two types of mixing equipment are applied in Europe: a screw in a central tube that creates downward movement and an impeller attached to a central draught tube that creates upward movement.

The stirring intensity of the impeller is an important topic in digester optimization. A certain degree of mixing is essential for contact between the substrate and the microbes and therefore biogas production; however, excessive mixing can diminish biogas production. Low-speed stirring improved digester performance and stabilized an unstable continuously mixed digester (Stroot et al. 2001). Furthermore, low-speed stirring conditions better allow a digester to absorb the disturbance of shock loading compared to high-speed stirring conditions (Gomez et al. 2006).

The precise reason for this adverse effect is unclear, but the formation of anaerobic granules changed during intensive mixing and had a considerable influence on anaerobic digestion performance. Excessive stirring can disrupt the granule structure that reduces the oxidation rate of fatty acids, which might result in digester instability. Thus, low-speed mixing conditions can provide a suitable environment for the granular microbial communities (Pinho et al. 2004).

Hydrodynamic studies estimate the optimal stirring rate and determine if a digester vessel operates at its full mixing capacity (Olivet et al. 2005). In anaerobic digestion, hydrodynamics uses the nontoxic molecule Li^+, which concentration profiles can create a residence time distribution curve that reveals dead zones or areas that are not mixed well. It also reveals short channeling or circuiting where feedstock takes a more direct route between input and output instead of being distributed throughout the whole digester working volume.

5.1.3 pH

pH is the measure for acidity/alkalinity of a solution (a substrate mixture for the AD) and is an essential parameter for maintaining functional AD. Anaerobes are highly pH dependent, and methanogens are even influenced to a greater extent by pH (Grady et al. 1999). Beccari et al. (1996) established that methanogenesis is strongly affected by the pH with an optimum range between pH 6.8 and pH 7.2. If the pH value of the anaerobic digester is outside the optimum range, the activity of the methanogens decreases (Gerardi 2006). Usual progress of the pH value during AD is a decrease over time as a result of the accumulation of volatile fatty acids. Although methanogens do not prefer a lower pH value, acidogens do. The optimum pH of hydrolysis and acidogenesis is stated to be in between 5.5 and 6.5 (Yu and Fang 2002). An important reason for separate reactors (two-stage AD) for hydrolysis + acidogenesis and acetogenesis + methanogenesis is the variety in pH preferences of the anaerobes. Overall, pH control is a problematic and interactive process, whereas reduction of ammonia toxicity due to an increased concentration of free ammonia (FA) is another factor that prefers a stable monitored pH (Mao et al. 2015).

5.1.4 Alkalinity (Buffer Capacity)

Alkalinity is often known as buffer capacity and is a measure of the number of alkaline compounds (i.e., the equilibrium of bicarbonate ions and carbon dioxide) in the digester. The substrates influence alkalinity due to the ammonia that is being released by decomposition of protein- and amino acid-rich feedstocks. The ionized form of ammonia reacts with carbonate ions (i.e., dissolved carbon dioxide) to form ammonium bicarbonate ions (Schnurer and Jarvis 2010). Thus, the alkaline chemicals provide resistance to changes in pH as they neutralize the produced acids. The concentration of the alkaline chemicals is proportional to the buffering capacity (Dussadee et al. 2017, 2017). According to Parawira et al. (2006), it is of great importance that the buffering capacity of the digester remains high to stabilize the pH caused by fluctuations in VFA concentration. In the case of a stable pH (e.g., 7.0), the alkalinity is considered equivalent to the concentration of ammonia, bicarbonate, and hydrogen sulfide, which results in efficient AD (Parawira et al. 2006). Achinas and Euverink suggested the co-inoculation of the bioreactor and showed positive effect on the degradation of the organic matter (Achinas and Euverink 2019b). Alkalinity measures are more reliable process balance measures than pH measures because the accumulation of VFA will reduce the alkalinity significantly before the pH decreases. If the alkalinity shows low amounts (e.g., <4000 mg L^{-1} bicarbonate) the best-suited solution for increasing alkalinity will be reducing OLR. Alternatively, more rapid approaches are adding bicarbonate, carbonate salts, or strong bases to the digester (Guwy et al. 1997).

5.1.5 Volatile Fatty Acids (VFAs)

A low concentration of intermediate products like VFAs (e.g., acetate, butyrate, and propionate) indicates the stability of the AD process (Moller et al. 2004). The intermediate compounds are produced during acidogenesis and have a carbon chain of up to six atoms. Most of the time, an unstable AD will result in accumulation of VFAs which results in a drop of the pH. However, if the buffer capacity in the digester is high enough (i.e., a surplus of alkalinity), a pH drop will not occur. A similar concentration of VFA can be ideal for one type of digester but inhibitory for a different type. Therefore, the VFA concentration cannot be used as a stand-alone AD process monitoring parameter (Davis 2018).

Ahring et al. (1995) demonstrated that monitoring VFAs indicates process stability as increasing VFA can be indicative of an overload of the OLR. The reason here is that methanogens are not able to metabolize the produced acetate by acetogenic bacteria until the number of methanogenic archaea increased sufficiently. The total alkalinity to volatile fatty acids ratio (also referred to as FOS/TAC in German literature) is another indicator for the buffer capacity of a digester. If the two major groups of intermediary microorganisms (i.e., acidogens and methanogens) are active in the same physical space, the ideal VFA/TA ratio is between 0.1 and 0.5 for a stable AD (Costa et al. 2007). If the ratio exceeds 0.5, corrective action should be undertaken. A solution might be the addition of sodium bicarbonate to increase the amount of total alkalinity and stabilize the ratio. As it is assumable that VFA accumulation occurs due to feeding, the VFA/TA ratio is likely to increase (Rosato 2017).

5.1.6 Carbon/Nitrogen Ratio

Anaerobic digestion is sensitive to the C/N ratio as it represents the relationship between the quantity of contained carbon and nitrogen in organic matter, which illustrates the nutrient levels anaerobes require for growth (Kigozi et al. 2014). Methanogens use nitrogen to meet their protein demand. A high C/N ratio of above 40:1 initiates fast depleted nitrogen by microbes such that it will not react with the excess carbon in the feedstock, which reduces the biogas yield (Kigozi et al. 2014). In addition to the low protein solubilization rate, a high C/N ratio induces low FA and total ammonia to nitrogen concentrations within the anaerobic digester. Therefore, ammonia inhibition in the AD process can be avoided by optimizing and stabilizing the C/N ratio. Maintaining the C/N ratio can either be done by explicitly monitoring or by merely being aware of the entering waste types in the anaerobic digester and the relative composition of the wastes (Buekens 2005). The optimal C/N ratio is claimed to be in between 20:1 and 30:1, with a ratio of 25:1 being the most frequently used (Mao et al. 2015; Kigozi et al. 2014; Buekens 2005). Wang et al. (2014) tested C/N ratios of 15:1 and 20:1 at a mesophilic and thermophilic temperature resulting in excessive ammonia inhibition. Similarly, approximately threefold cumulative biogas yield was obtained from C/N ratios of 25:1 and 30:1 compared to 15:1.

5.1.7 Organic Loading Rate (OLR)

The organic loading rate shows the number of volatile solids fed into a digester per unit time (usually per day) under continuous feeding (Pontoni et al. 2015). The optimum OLR is hard to define because it is specific to the operating temperature of the digester and the substrate. Increased OLR will increase biogas production to a greater extent. However, the stability and productivity of the AD process can be severely disturbed as well. Too high OLR rate will exacerbate the methanogenic activity in a digester as hydrolysis and acidogenesis have more active bacteria than methanogens. This oblique activity results in VFA accumulation that eventually leads to irreversible acidification. Subsequently, the pH decreases, further hydrolysis is inhibited, and methanogens are not able to convert VFA into biogas anymore (Mao et al. 2015). The usual calculation followed for OLR is depicted in Eq. (1) (Davis 2018):

$$B_R = \frac{m * c}{V_r} \tag{1}$$

with $B_R = \mathrm{OLR}\left(\frac{kg}{d*m^3}\right), m =$ mass of substrate fed$\left(\frac{kg}{d}\right)$, $c =$ concentration organic matter (%), $V_R =$ digester volume(m^3).

5.1.8 Hydraulic Retention Time (HRT)

The hydraulic retention time is the average time spent by the fed substrate inside the digester. The HRT is valuable as it indicates the available time for the microorganism to grow in the reactor before they are washed out. Eventually, it establishes the conversion of the organic matter to biogas (Yadvika et al. 2004). The effect of an altered HRT on AD biogas yield is hard to determine because it is very substrate dependent. However, Yadvidka et al. (2004) state that shorter retention time is likely to face washout of active microorganisms, while longer retention time desires a large digester volume and hence more investment and equipment costs. On average, AD of lignocellulosic material needs an HRT of around 10 days (Braun et al. 2009). The HRT is defined as the ratio between digester volume and substrate volume fed per unit time (Eq. 2) (Mao et al. 2015).

$$\mathrm{HRT} = \frac{V_R}{V} \tag{2}$$

with HRT (d), V_R= digester volume (m^3), $V =$ volume fed per unit time $\left(\frac{m^3}{d}\right)$.

5.1.9 Redox Potential

Redox potential (i.e., reduction oxidizing potential) has been shown as a successful monitoring parameter in many AD systems due to redox reaction-catalyzed enzymes that degrade the organic materials in the anaerobic environment (Wang et al. 2006). The strictness of the anaerobic environment is well known, which is indicated by a redox potential of ≤ -200 mV (Zupancic and Grilc 2011). Preferably, the redox potential is between -300 and -330 mV for optimal AD process environment. If the redox potential becomes too low (more negative), adding oxidizing agents such as nitrates, nitrites, oxygen, or sulfates into the digester increases the redox potential

(Deublein and Steinhausez 2008). If the redox potential is initially too high, the facultative anaerobic microorganisms in the reactor consume the oxygen dissolved in the water and decrease the redox potential to the level required by important obligatorily anaerobic microorganisms (mostly methanogens).

5.1.10 Ammonia

Decomposition of nitrogenous matter, e.g., proteins and urea, leads to the formation of ammonia. It is an essential nutrient that serves as a precursor for the synthesis of proteins and enzymes required by the microorganisms in the reactor to survive. Ammonia is also used as a fertilizer for the growth of plants to generate feedstock (Kayhanian 1999). The total ammonia nitrogen is primarily composed of ammonium ion (NH_4^+) and free ammonia (NH_3) (i.e., free ammonia nitrogen (FAN)). The equilibrium of these two components mainly relies on process temperature and pH (Schnurer and Jarvis 2010). To illustrate, if the temperature or pH increases, the equilibrium steadily shift toward FAN. Furthermore, the FAN is the most toxic species of total ammonia nitrogen (TAN). FAN can penetrate a bacterial cell membrane resulting in a proton imbalance, altering intercellular pH, inhibiting specific enzyme activities, and increasing maintenance energy requirements (Ariunbaatar et al. 2015).

Too high ammonia content will lead to process inhibition, which Al Seadi et al. (Davis 2018) stated, where after they came up with a maximal ammonia concentration of 0.80 g L^{-1} to keep the process stable. However, Shi et al. (2017) reported that maximally allowable ammonia concentrations seem to depend on the substrate, inoculum, and environmental conditions, fluctuating from 53 mg L^{-1} to 1450 mg/L and 1500–1700 mg L^{-1} (Shi et al. 2017). Zupanic and Grilc (2011) also concluded that the allowable maximum is 2200 mg L^{-1}. The most well-known and established methods to reduce ammonia inhibition during AD are chemical (struvite precipitation) and physical (air stripping) methods. They were both effectively applied to wastewater treatment and sewage sludge AD that contained high ammonia concentrations.

5.2 *Microbial Ecology*

Anaerobic digestion requires an equal rate of degradation due to the sensitivity of the process. However, the dynamics of the separate microbes are complex and interactive, so equal degradation is hard to achieve (Rabii et al. 2019). Especially, disproportionate amounts of microbial groups influence the degradation stability. For example, complete degradation during hydrolysis is complex because organic compounds like fats and proteins are depolymerized into monomers within several days, whereas carbohydrates are depolymerized within a few hours. Additionally, if hydrolysis runs too fast, acid augmentation will occur, resulting in a lower pH and process failure (Świątek et al. 2018). As mentioned, the four steps during AD engage in syntrophic interrelation so to illustrate, if the growth rate of hydrolytic bacteria is low, the rates of the other three steps decrease, resulting in a lower biogas yield. Thus, the microbial population dynamics influence the stability of the degradation steps, and in return, the microbial population dynamics are affected by chemical conditions (e.g., alkalinity, VFA concentrations, TAN, TOC), operating parameters (e.g., OLR, pH, HRT, and temperature), and substrate characteristics (type of lignocellulosic biomass). Operating parameters and substrate characteristics control the chemical conditions. It is described that among the microbial groups involved in AD, methanogens are the key microbes for biogas production, which are the most sensitive to changes in operating parameters and are the rate-limiting step of the whole process.

6 Critical Factors for the Large-Scale Biogas Plant Investments

6.1 The Need for Renewable Gas Production

The population growth will create more need for reliable and stable energy – energy for homes, transportation, business, and industry (Solarte-Toro et al. 2018). In European level, the shifting to a low carbon economy remains a challenge. There is a consensus that CO_2 emissions have to be eliminated in order to move to a green economy. Plain gas has the lowest CO_2 emission among the fossil fuels. However, countries attempt to decarbonize the energy produced. Biogas production is regarded a way to shorten the carbon cycle. The carbon cycle of biogas is only as long as it took for the organic material to grow. The CO_2 absorbed by this organic material is released again upon combustion of the gas, but no additional CO_2 is emitted to the atmosphere. The shortage of renewable electricity can be offset by the exploitation of wastes.

6.2 Subsidy for Biogas Production

Besides the revenues of selling electricity and/or heat, biogas producers can also benefit from the subsidies (Benato and Macor 2019). Subsidizing renewable electricity production from biogas will reinforce the bioeconomy and the sustainable development. Subsidies and green policy scheme can realize the green value of the biogas and stimulate its large-scale production (Oreggioni et al. 2017). In practice the shortage of subsidy is much larger, since the subsidies scheme is not available in all the European countries.

6.3 Biogas Production

The green value of an end product can forecast the business project viability. It is essential to understand the economic drivers for the biogas project and provide financial and technical assurance for biogas-based project business case (Lindkvist et al. 2019). It is more expensive to produce biomethane instead of biogas due to the upgrading procedure (Baccioli et al. 2019). It may therefore be essential to use biogas directly to satisfy local power and heat demand. However, the subsidy policy for biomethane rather than biogas induces the biogas producers to upgrade the biogas in biomethane (Florio et al. 2019), in spite of the fact that CHP biogas plants may be a more optimal solution (Lauer and Thrän 2018).

6.4 Digestate

The digestate produced from the degradation of organic mass is a valuable end product. This product can be used as fertilizer in the agricultural land. Nitrogen, potassium, and phosphorus remain in the digester and are essential nutrients for plant growth. Besides the environmental benefit, the use of digestate is also economically essential. For instance, it avoids waste disposal costs. Nevertheless, there are stringent regulations with regard to manure use as soil conditioner in Europe (Achinas 2014). Digestate can contain various amount of hazardous matter. Hazardous matter can pose risks to human and animal health or can cause environmental pollution. There is a maximum amount of minerals that is allowed to be applied to the crops. In that case, farmers must often pay third parties to get rid of their excess manure, and subsequently additional operation costs arise.

6.5 Feedstock of Dependency of Producers

The availability and cost of waste streams is a key role for the investment success. The waste streams are often considered freely available to biogas producers during the biogas project execution (Carlini et al. 2017). As soon as an anaerobic digester is built, this can create an advantage to the supplier of waste streams for further bargain of the waste availability. The supplier has the option to not deliver the waste for free, and subsequently the owner can decide if it can proceed with the biogas production or not. At the start-up phase, it is therefore necessary to agree on long-term contracts for input material (Dell'Antonia et al. 2013).

These contracts contain at least the duration, guaranteed quality of the biomass, guaranteed amount of feedstock supply, and payments based on delivered quality and quantity. It is needed to include the available amount of feedstock before the size of the plant, based on the desired amount of output, is adjusted. Therefore, it is important to investigate the amount of feedstock that can be delivered by suppliers.

6.6 *Permitting Process*

In order to build a biogas plant, a permit from the local authorities is mandatory. New biogas producers must adhere to strict regulations, and thus it takes a long time to obtain a permit to build the digester. The permitting process renders the difficult expansion of biogas business projects. Every country has his own regulations for criteria, documentation, and procedure that are needed to get permission to install a biogas plant. Investors must document the conformity of the project with national legislation in order to get the permit for building a biogas plant. Topics that are discussed in the document are, for instance, exhaust emissions, impact on groundwater, protection of land, noise and odors, recycling and handling of organic wastes and manure, building safety, and work safety. Besides the building permit, several legislations are also taken into consideration during the permission process of biogas projects. This is specified in the environmental protection law, by-product instruction, nature protection, supply law, procedure law, environmental protection act, and land use planning law.

6.7 *Acceptability of Using Biomass for Energy Production*

In late study of the institution "Natuur en Milieu" (nature and environment), a resolution has been done concerning the sustainable development of green gas. The inference of this study is that even though the expression green gas proposes that gas has been generated from biological substances is a positive fact for the environment. The study suggests that sewage waste, landfills, and organic waste from residential houses are steadily sustainable sources for biogas generation (Chen et al. 2018; Chiumenti et al. 2018). The study also explains that for different organic substances, it should be contemplated if they cannot be used for influent or for food, or other petitions, with which there is a larger additional value. If this study is broadly obtained, large-scale production of biogas will not be material due to more decrease of the accessible feedstock. The utilization of biomass for energy generation does not give the highest benefits (Fauzianto et al. 2014). As general opinion, energy production is placed not in the initial benefits for biomass in this study (Valenti and Porto 2019). Pharmaceuticals must be the first choice of biomass use, following food, then for generation of various chemicals, and only after it is not appropriate for all the previous uses, it should be used for energy production. In a complete market, this distribution of biological substances to the output with the highest benefits would ipso facto occur. This procedure might, nevertheless, be perverted by subsidies. We have already known it from biofuel production, where high fundings were provided hence to augment biofuel production. This motive has an outcome, the augmentation of food prices, making the food not accessible for the unprivileged.

6.8 Technological Upscaling of AD Process

Upscaling bioreactors is relatively easy due to the possibility to run them parallel in high quantities without modifying the individual reactors. If one reactor runs at a desired steady state, its operating settings can be copied to other reactors such that the same steady state is reached in these reactors. Bioreactor upscaling, which Watts and Wiles (Watts and Wiles 2007) refer as scale-out, does not involve changes in hydrodynamics and reaction kinetics during the reactor process, while conventional bioreactor upscaling does. It involves a complicated iterative process during upscaling due to significantly changing hydrodynamics and reaction kinetics. The indicated difference between the described scale-up procedures is summarized in Fig. 2.

Basic design goals for a digester are a maximum volume production of biogas; to allow for a continuous, high, and sustainable organic loading rate; and to minimize reactor volume (Garfi et al. 2016). The digester size is based on the available amount of organic wastes, and the digester design preferably considers the construction practicalities of both mixing and heat loss (Achinas and Euverink 2016). To illustrate, square and rectangular underground digesters are easier to build, however, mixing will be suboptimal as flow will be stagnant in right-angled corners, resulting in a buildup of refractory compounds that will reduce the effective operating volume of the digester over time. The outlined situation might lead to process failure leading to extra downtime and maintenance. Besides, the heat loss (e.g., due to surrounding

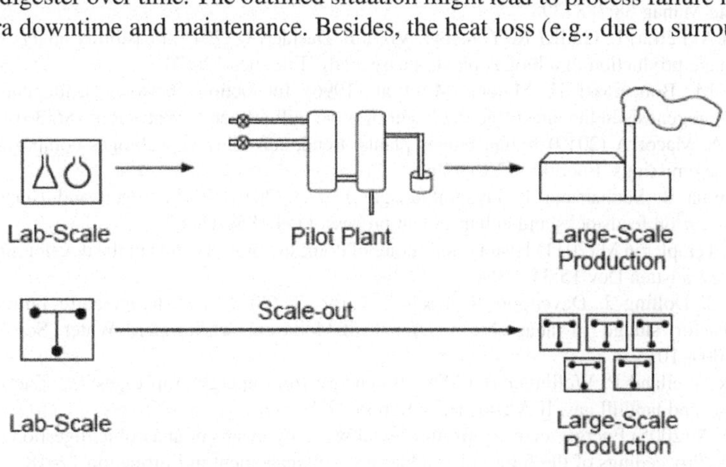

Fig. 2 Comparison between upscaling conventional bioreactors (top) and mini/micro bioreactors (bottom) (Watts and Wiles 2007)

climate conditions) influences decisions on the digester shape, material, location, and operating mode. A wide spectrum of operating modes have been applied since 1859 such as batch digesters, continuously stirred tank digesters, plug-flow digesters, and sludge bed digesters. Additionally, digesters can operate in either a one-stage or multiple-stage digestion, depending upon the scale of operation and feeding characteristics (Dussadee et al. 2017).

References

ABC (2016) Operational biogas systems in the U.S. American Biogas Council

Achinas S (2014) Material flow analysis (MFA) for waste management in olive oil industries sector in South Europe. J Sustain Dev 5(1):29–23

Achinas S, Achinas V (2017) Biogas combustion: an introductory briefing. In: Vico A, Artemio N (eds) Biogas: production, applications and global developments. Nova Science Publishers, New York, pp 179–193

Achinas S, Euverink GJW (2016) Consolidated briefing of biochemical ethanol production from lignocellulosic biomass. Electron J Biotechnol 23:44–53

Achinas S, Euverink GJW (2019a) Feasibility study of biogas production from hardly degradable material in co-inoculated bioreactor. Energies 12:1040

Achinas S, Euverink GJW (2019b) Effect of combined inoculation on biogas production from hardly degradable material. Energies 12:217

Achinas S, Achinas V, Euverink GJW (2017) A technological overview of biogas production from biowaste. Engineering 3(3):299–307

Ahring B, Sandberg M, Angelidaki I (1995) Volatile fatty-acids as indicators of process imbalance in anaerobic digesters. Appl Microbiol Biotechnol 43:559–565

Al Seadi T, Rutz D, Prassl H, Kottner M, Finsterwalder T, Volk S, Janssen R (2008) Biogas handbook, technical report, University of Southern Denmark Esbjerg. Esbjerg, Denmark

Ariunbaatar J, Scotto Di Perta E, Panico A, Frunzo L, Esposito G, Lens PNL, Pirozzi F (2015) Effect of ammoniacal nitrogen on one-stage and two-stage anaerobic digestion of food waste. Waste Manag 38:388–398

Baccioli A, Ferrari L, Guiller R, Yousfi O, Vizza F, Desideri U (2019) Feasibility analysis of bio-methane production in a biogas plant: a case study. Energies 12:473

Beccari M, Bonemazzi F, Majone M et al (1996) Interactions between acidogenesis and methanogenesis in the anaerobic treatment of olive mill effluents. Water Res 186(30):183–189

Benato A, Macor A (2019) Italian biogas plants: trend, subsidies, cost, biogas composition and engine emissions. Energies 12(6):979

Bharathiraja B, Sudharsana T, Jayamuthunagai J et al (2016) Biohydrogen and Biogas – an overview on feedstocks and enhancement process. Fuel 185:810–828

Bond T, Templeton M (2011) History and future of domestic biogas plants in the developing world. Energy Sustain Dev 15:347–354

Bowen E, Dolfing J, Davenport R, Read F, Curtis T (2014) Low-temperature limitation of bioreactor sludge in anaerobic treatment of domestic wastewater. Water Sci Technol 69:1004–1013

Braun R, Weiland P, Wellinger A (2009) Bioenergy from energy crop digestion. Energy from biogas and landfill gas. IEA Bioenergy Report 37

Buekens A (2005) Energy recovery from residual waste by means of anaerobic digestion technologies. Proceedings of the future of residual waste management in Europa, pp 17–18

Burton C, Turner C (2003) Manure management treatment strategies for sustainable agriculture, 2nd edn. Silsoe Research Institute

Carlini M, Mosconi EM, Castellucci S et al (2017) An economical evaluation of anaerobic digestion plants fed with organic agro-industrial waste. Energies 10:1165

Chae K, Jang A, Yim S, Kim I (2008) The effects of digestion temperature and temperature shock on the biogas yields from the mesophilic anaerobic digestion of swine manure. Bioresour Technol 99:1–6

Chen J (2016) A bright future for sustainable development: ushered in by innovation. Engineering 2 (1):16–18

Chen JF (2017a) Green chemical engineering for a better life. Engineering 3(3):279

Chen JF (2017b) Green chemical engineering. Engineering 3(3):283–284

Chen P, Anderson E, Addy M et al (2018) Breakthrough technologies for the biorefining of organic solid and liquid wastes. Engineering 4(4):574–580

Chiumenti A, Boscaro D, da Borso F, Sartori L, Pezzuolo A (2018) Biogas from fresh spring and summer grass: effect of the harvesting period. Energies 11:1466

Choorit W, Wisarnwan P (2007) Effect of temperature on the anaerobic digestion of palm oil mill effluent. Electron J Biotechnol 10:376–385

Costa R, Tavares C, Cossich E (2007) Stabilization of swine wastes by anaerobic digestion. Environ Technol 28:1145–1151

Davis LA (2018) The shale oil and gas revolution. Engineering 4(4):438–439

Dell'Antonia D, Cividino SRS, Carlino A et al (2013) Development perspectives for biogas production from agricultural waste in Friuli Venezia Giulia (Nord-East of Italy). J Agr Eng 44:569–572

Deublein D, Steinhausez A (2008) Biogas from waste and renewable resources. Weinheim

Dussadee N, Ramaraj R, Cheunbarn T (2017) Biotechnological application of sustainable biogas production through dry anaerobic digestion of Napier grass. Biotech 7:47

Fauzianto R, Achinas S, Achinas V (2014) Sustainable biomass production for energy strategies in Asia: a review for legal and social aspects. J Agric Sustain 7(1):1–26

Florio C, Fiorentino G, Corcelli F, Ulgiati S, Dumontet S, Güsewell J, Eltrop L (2019) A life cycle assessment of biomethane production from waste feedstock through different upgrading technologies. Energies 12:718

Garfi M, Marti-Herrero J, Garwood A, Ferrer I (2016) Household anaerobic digesters for biogas production in Latin America: a review. Renew Sust Energ Rev 60:599–614

Gerardi M (2006) Wastewater bacteria. Wiley-Interscience, New Jersey

Gomez X, Cuetos M, Cara J, Moran A, Garcia A (2006) Anaerobic co-digestion of primary sludge and the fruit and vegetable fraction of the municipal solid wastes – conditions for mixing and evaluation of the organic loading rate. Renew Energy 31:2017–2024

Grady LC, Daigger G, Lim H (1999) Biological wastewater treatment, 2nd edn. CRC Press

Guwy A, Hawkes F, Wilcox S, Hawkes D (1997) Neural network and on-off control of bicarbonate alkalinity in a fluidized-bed anaerobic digester. Water Res 31:2019–2025

Karim K, Hoffmann R, Klasson K, Al-Dahnan M (2005) Anaerobic digestion of animal waste: effect on mode of mixing. Water Res 39:3597–3606

Kayhanian M (1999) Ammonia inhibition in high-solids biogasification: an overview and practical solutions. Environ Technol 20:355–365

Kigozi R, Muzenda E, Aboyade A (2014) Biogas technology: current trends, opportunities and challenges. University of Johannesburg, pp 1–8

Lauer M, Thrän D (2018) Flexible biogas in future energy systems—sleeping beauty for a cheaper power generation. Energies 11(4):761

Lindkvist E, Karlsson M, Ivner J (2019) Systems analysis of biogas production—part I research design. Energies 12(5):926

Macedonio F, Drioli E (2017) Membrane engineering for green process engineering. Engineering 3 (3):290–298

Manni M, Coccia V, Cavalaglio G et al (2017) Best practices for recovering rural abandoned towers through the installation of small-scale biogas plants. Energies 10:1224

Mao C, Feng Y, Wang X et al (2015) Review on research achievements of biogas from anaerobic digestion. Renew Sustain Energy Rev 45:540–555

Moller H, Sommer S, Ahring B (2004) Biological degradation and greenhouse gas emissions during pre-storage of liquid animal manure. J Environ Qual 33:27–36

Moron M, Morales L, Mora C et al (2018) OMW spillage control tool based on tracking p-Coumaric acid degradation by HPLC. Environ Technol 40:1–16

Nelson MJ, Nakhla G, Zhu J (2017) Fluidized-bed bioreactor applications for biological wastewater treatment: a review of research and developments. Engineering 3(3):330–342

Olivet D, Valls J, Gordillo M, Freixo A, Sanchez A (2005) Application of residence time distribution technique to the study of the hydrodynamic behavior of a full-scale wastewater treatment plant plug-flow bioreactor. J Chem Technol Biotechnol 80:425–432

Oreggioni GD, Lesh Gowreesunker B, Tassou SA et al (2017) Potential for energy production from farm wastes using anaerobic digestion in the UK: an economic comparison of different size plants. Energies 10:1396

Parawira W, Murto M, Zvauya R, Mattiasson B (2006) Comparison of the performance of a UASB reactor and an anaerobic packed-bed reactor when treating potato waste leachate. Renew Energy 31:893–903

Pinho S, Ratusznei S, Rodrigues J, Foresti E, Zaiat M (2004) Influence of the agitation rate on the treatment of partially soluble wastewater in anaerobic sequencing batch biofilm reactor. Water Res 38:4117–4124

Pontoni L, Panico A, Salzano E, Frunzo L, Iodice P, Pirozzi F (2015) Innovative parameters to control the efficiency of anaerobic digestion process. Chem Eng Trans 43:2089–2094

Rabii A, Aldin S, Dahman Y et al (2019) A review on anaerobic co-digestion with a focus on the microbial populations and the effect of multi-stage digester configuration. Energies 12(6):1106

RedCorn R, Fatemi S, Engelberth AS (2018) Comparing end-use potential for industrial food-waste sources. Engineering 4:371–380

REN21 (2017) Renewables 2017: global status report. Secretariat, Paris

Rosato M (2017) Redimensioning the importance of the VFA/TA (FOS/TAC) method for the optimization of biogas plants, pp 1–5

Sahajwalla V (2018) Green processes: transforming waste into valuable resources. Engineering 4 (3):309–310

Sanjay P, Vijay V (2012) Biogas production, upgradation and slurry management. Narosa Publication, New Delhi, pp 116–121

Schnurer A, Jarvis A (2010) Microbiological handbook for biogas plants. Swedish Waste Management 207

Shi X, Lin J, Zuo J, Li P, Li X, Guo X (2017) Effects of free ammonia on volatile fatty acid accumulation and process performance in the anaerobic digestion of two typical bio-wastes. J Environ Sci 55:49–57

Solarte-Toro JC, Chacón-Pérez Y, Cardona-Alzate CA (2018) Evaluation of biogas and syngas as energy vectors for heat and power generation using lignocellulosic biomass as raw material. Electron J Biotechnol 33:52–62

Sterman J (2000) Business dynamics; System thinking and modeling for a complex world, Chapter 5

Stroot P, McMahon K, Mackie R, Rashkin L (2001) Anaerobic co-digestion of municipal solid waste and biosolids under various mixing conditions – I. Digester performance. Water Res 35:1804–1816

Świątek M, Lewicki A, Szymanowska D, Kubiak P (2018) The effect of introduction of chicken manure on the biodiversity and performance of an anaerobic digester. Electron J Biotechnol 37:25–33

Valenti F, Porto SMC (2019) Net electricity and heat generated by reusing Mediterranean agro-industrial by-products. Energies 12:470

Verma S (2002) Anaerobic digestion of biodegradable organics in municipal solid wastes. MSc Dissertation. Columbia University

Wang L, Zhou Q, Li F (2006) Avoiding propionic acid accumulation in the anaerobic process for biohydrogen production. Biomass Bioenergy 30:177–182

Wang X, Lu X, Li F, Yang G (2014) Effects of temperature and carbon-nitrogen (C/N) ratio on the performance of anaerobic co-digestion of dairy manure, chicken manure and rice straw: focusing on ammonia inhibition. PLoS One 9:e97265

Wang J, Wang H, Fan Y (2018) Techno-economic challenges of fuel cell commercialization. Engineering 4(3):352–360

Watts P, Wiles C (2007) Micro reactors: a new tool for the synthetic chemist. Org Biomol Chem 5:727–732

Weiland P (2006) Biomass digestion in agriculture: a successful pathway for the energy production and waste treatment in Germany. Eng Life Sci 6(3):302–309

Wen-Wei L, Han-Qing Y (2016) Advances in energy-producing anaerobic biotechnologies for municipal wastewater treatment. Engineering 2:438–446

Wilkinson K (2011) A comparison of the drivers influencing adoption of on-farm anaerobic digestion in Germany and Australia. Biomass Bioenergy 35:1613–1622

Yadvika S, Sreekrishnan T, Kohli S, Rana V (2004) Enhancement of biogas production from solid substrates using different techniques – a review. Bioresour Technol 95:1–10

Yu H, Fang H (2002) Acidogenesis of dairy wastewater at various pH levels. Water Sci Technol 45:201–206

Zupancic G, Grilc V (2011) Anaerobic treatment and biogas production from organic waste. Management of organic waste

Biogas Processing, Storage and Distribution, Transportation and Value Chain Analysis

Madhurjya Das, Mainak Chakraborty, Senthil Chinnasamy, and Balasubramanian Ramakrishnan

Abstract Biogas is a versatile renewable energy resource that has thermal, electrical and vehicular applications. The biogas systems with anaerobic digestion of diverse feedstocks or wastes not only improve the energy availability but also contribute to the preservation and protection of the environment. Optimization of biogas processing is needed as the present-day biogas systems transform only about 10% of the substrate mass into gas. Variable costs for feedstocks and operations remain a limiting factor for the successful installation and popularization of the biogas systems. The feedstocks of macro- and microalgae provide opportunities for implementing a biorefinery approach along with biogas production. The process optimization for biogas production depends significantly on the microbial community adaptability. The recently introduced microbial enrichment technologies at the laboratory-scale testing have some knowledge gaps to be filled before implementation at the large-scale operations. High sensitivities of both technological and biological aspects of the biogas system demand the knowledge- and data-driven management to ensure stable and efficient production of biogas. Biogas is expensive to store locally, necessitating the development of suitable storage systems by compression or liquefaction. Both the economic and environmental perspectives need to be considered for the creation and appreciation of the biogas value chains.

Keywords Biogas · Processing · Microbiome · Feedstocks · Algal biomass · Storage · Distribution · Transportation · Value chain analysis

M. Das · M. Chakraborty
GPS Renewables Pvt Ltd, Bengaluru, India

S. Chinnasamy
Aban Infrastructure Limited, Egmore, Chennai, India

B. Ramakrishnan (✉)
Division of Microbiology, ICAR-Indian Agricultural Research Institute, New Delhi, India
e-mail: brama@iari.res.in

N. Balagurusamy, A. K. Chandel (eds.), *Biogas Production*,
https://doi.org/10.1007/978-3-030-58827-4_13

1 Introduction

Methane, a hydrocarbon, is natural gas with an energy density of 50–55.5 MJ kg^{-1}. The biogenic methane gas is produced in landfills, from livestock activities and rice cultivation, and during biological (organic matter) waste and wastewater treatment. Methane can be captured and used as a source of energy during waste treatment. The organic waste of different origins (agricultural, industrial, animal or domestic waste) and the specific treatment processes for different feedstock determine the chemical composition and the physical characteristics of 'biogas' which is a mixture of gases including methane (55–65%) and inert carbonic gas (35–45%). The corrosive nature of biogas is due to the presence of CO_2, H_2S (100–10,000 ppm), ammonia (traces) and water vapour, and hence, it produces twice as fewer calories by combustion, compared to the fossil-derived natural gas. The biogas upgrading plants employing absorption, adsorption, membrane filtration or cryogenic separation methods can remove the CO_2 to produce biomethane, resulting in more than 95% methane, which is close to the properties of fossil-derived natural gas. The biogas as versatile renewable energy has thermal, electrical and vehicular applications. These applications can decrease the dependence on fossil fuels, which currently meet nearly 88% of the global energy demand. Methane is also a potent greenhouse gas, accounting for almost 9% of domestic greenhouse gas emissions. Methane as a greenhouse gas is more efficient at trapping solar radiation than carbon dioxide and contributes to global warming. The biogas systems have the potential to capture and use methane as energy source, which can otherwise escape into the atmosphere, and provide energy, environmental and economic benefits. The continuously increasing production of organic wastes demands the recovery of energy and recycling of nutrients. The two primary end products of anaerobic digestion are the biogas and the digestate, which can be used as plant fertilizer. Hence, the biogas systems with the energy-efficient anaerobic digestion improve not only the energy future but also contribute to the preservation of natural resources and the protection of the environment.

The form, quantity and composition of feedstock are the critical determinants of the biogas production efficiencies since the biogas plants transform only about 10% of the mass into gas. The agricultural wastes are predominantly lignocellulosic with low nitrogen content, and they require either the co-digestion or pretreatment methods (Yong et al. 2015; Rodriguez et al. 2017). Nevertheless, co-digestion or pretreatment methods are not cost-effective for the small-scale productions of biogas using agricultural wastes. The 'digester' of the biogas systems is the critical component. The life-cycle assessment analysis of three small-sized biogas systems suggested that the concrete cover slap technology is superior due to the reduction of energy dispersions and energy self-absorption, compared to bags- and balloon technologies (Collotta and Tomasoni 2017). The efficiency of anaerobic digestion of organic matter depends on bacterial and archaeal communities which mediate hydrolysis, acidogenesis, acetogenesis and methanogenesis. The organic matter is converted to volatile fatty acids and further into acetate, H_2 and CO_2, predominantly by the bacterial communities. The archaeal communities utilize acidogenic products and produce methane. The efficiency, functionality and the composition of biogas depend on the abundance and activities of these bacterial and archaeal communities. Yu et al. (2018) reported that the anaerobic digestion of rice straw at 50 °C increased the relative abundance ratio of *Methanoculleus*, *Methanosarcina* and Firmicutes, compared to that of rice straw at 39 °C. De Vrieze et al. (2018) applied the 16S rRNA (gene) sequencing as a predictive tool and reported that the active microbial community (based on the level of RNA) mirrored and predicted the functionality of the anaerobic digestion process better than the total microbial community analysis using total DNA for amplicon sequencing. The hydrogenotrophic methanogens can convert carbon dioxide to methane using hydrogen, and biomethanization by this route has the potential of carbon capture and utilization for reducing greenhouse emissions (Zabranska and Pokorna 2018). The conversion of both methane and carbon dioxide of biogas to methanol by thermochemical processing has made progress in recent times (Ghosh et al. 2019). Also, the bioconversion of methane and methanol to fuels and chemicals using methylotrophs offers new perspectives on the biological gas-to-liquid (GTL) conversion technologies (Bennett et al. 2018). More modern strategies such as the biogas storage or a biogas upgrade to biomethane for subsequent storage in a natural gas rid can achieve the demand-driven biogas supply, which can also be enhanced by changing the feeding regimes (Hahn et al. 2014; Mulat et al. 2016). High costs for feedstocks and operations of the biogas systems limit the successful installation and popularization of this technology. The government policies that aim at mitigation of greenhouse gas emissions, promotion of renewable energy sources and recycling of wastes can become the major drivers for the sustainability of biogas systems.

2 Biogas Processing

Methane gas is combustible at a boiling point of −161 °C at atmospheric pressure, with an explosive capability at 5–15% by volume. The European Biogas Association proposed that the biogas production in Europe could reach 50 Bcm year^{-1} by 2030 (Lambert 2017). The major challenge is to ensure the supply of affordable, sustainable feedstocks to the biogas systems. The feedstock for biogas production is defined as any substrate, ranging from readily degradable to highly-solid complex waste, which can be converted by anaerobic digestion to methane and carbon dioxide. Corn, soybean and sugarcane are of the first-generation feedstocks for biogas production, but they are edible and compete with the human demand for food and feed. Waste and lignocellulosic feedstocks are as the second-generation feedstocks, while the third-generation feedstocks are micro- and macro-algal biomass (Montingelli et al. 2015). The second-generation feedstocks have a strong resistance to anaerobic digestion due to the presence of higher amounts of lignin or lignocellulose, inhibitors or other molecules with the highly crystalline structure or low surface area. Even the low COD-containing industrial wastewater is used for anaerobic digestion in the high-rate reactor configurations with process control devices. The reactor configurations, the functioning of microbial communities during the anaerobic digestion and the quality of biogas are significantly influenced by the composition and quality of feedstocks. The agricultural biomass and waste account for most of the feedstocks in the current biogas systems.

2.1 Feedstocks: Energy Crops and Agricultural Residues

The plant biomass has cellulose (40–50%), hemicellulose (20–40%) and lignin in higher amounts along with smaller amounts of protein, pectin and other non-structural materials. The carbohydrates of soluble and non-structural nature and soluble cell components contribute more specific methane yield. Theoretical-specific methane yields, on average, follow the order 930 L kg volatile solid $(VS)^{-1}$ for lipids <474 L kg VS^{-1} for protein and 405 L kg VS^{-1} for carbohydrates (Herrmann and Rath 2012). Ideally, the plant biomass for biogas production should have higher amounts of degradable carbohydrates, lipids and proteins but with lower amounts of lignin and cellulose. Various plant factors such as the species and developmental stage, environmental factors such as soil fertility and moisture availability, cultivation factors such as fertilization and other management practices and processing factors such as storage, pretreatment methods and dosage influence the anaerobic digestion and the biogas production. Maize is one of the significant energy crops and the most popular feedstock for biogas production in Germany and Austria (Weiland 2010). The breeding of biogas maize varieties has led to the release of tall, late-maturing varieties with more than 30 t ha^{-1} TS yields per ha (Grieder et al. 2012; Rath et al. 2013). The specific methane yields of various biomass including those of grasses, *Miscanthus* sp., sun flower, fruits and vegetables, woody biomass, weeds and freshwater and marine biomass have been investigated (Gunaseelan 1997; Braun 2007). Vindis et al. (2012) applied a qualitative multi-attribute modelling methodology DEX, supported by the software tool DEX-i, and reported maize as the best DEX-i multicriteria evaluation appropriate, followed by sorghum, sunflower and sugar beet. The sustainability of energy crops as the biogas substrates due to the displacement of crop production for food or feed, indirect land use change (iLUC) and the food security concerns necessitate the utilization of agricultural residues (Lantz et al. 2017).

The alternatives to maize are considered to include the straw by-products from cereal production, grass from meadows and grasslands and animal manure, with the biogas energy potential of 1.2 to 2.3×10^3 PJ $year^{-1}$ and suitability for co-digestion, for the future European biogas sector (the EU28) in 2030 (Meyer et al. 2018). Breitenmoser et al. (2019) suggested that unrealistic assumptions on the quality and quantity of biowastes as one of the reasons for the failure of implementation of anaerobic digestion for the biogas production in India. The plant parts remaining on the farm after harvest or those left off after crop processing which is defined as industrial crop waste are a suitable feedstock for the dry anaerobic digestion

(Cherubin et al. 2018; Momayez et al. 2019). Crop residues of 500 Mt. are generated every year, and about 92 Mt. is burned in India. The ill-effects of crop residue burning are the emission of greenhouse gas emissions, increased levels of particulate matter, smog, the loss of biodiversity and soil fertility. The potential of rice straw and other crop residues as feedstocks for biogas generation has to be realized by the farming communities in India (Bhuvaneshwari et al. 2019). The lignocellulosic materials, the feedstock for second-generation biogas production, are highly heterogenous and recalcitrant and require pretreatment for improved accessibility and digestibility during anaerobic digestion (Wagner et al. 2018). The 'biomass recalcitrance' adds to making hydrolysis a rate-limiting process and necessitates pretreatment such as steam explosion before the anaerobic digestion of lignocellulosic materials (Mulat et al. 2018). The co-substrates such as organic waste from agriculture-related industries, biowaste from households and food waste are generally added to the biomass to increase the organic content and, consequently, higher gas yield (Achinas et al. 2017).

2.2 *Algal Biomass as Feedstocks*

Macroalgal biomass production in the marine environment offers several advantages, and this 'blue ocean strategy' has no competition with food production for land or freshwater (Hughes et al. 2012). The primary productivity rates of macroalgae (~ 1600 g C m^{-2} $year^{-1}$) are higher than those of terrestrial crops (~ 470 g C m^{-2} $year^{-1}$). Also, the macroalgal biomass contains no lignin and little cellulose, relative to those of terrestrial crops. The macroalgal biomass can provide 22 m^3 of methane per tonne on a wet weight basis. The potential to convert carbon dioxide from flue gas to biomass and the ability to grow at higher carbon dioxide concentrations make microalgal species suitable as the feedstocks for biogas production (Mudimu et al. 2014). Since each microalgal species has a characteristic cell wall and macromolecular composition, the methane yield potential is highly variable and species-specific (Mussgnug et al. 2010). Higher photosynthetic efficiencies, higher productivities and the potential to grow in saline, brackish or wastewater in non-arable lands, compared to the terrestrial crops, make algae a significant substrate for biogas production (Wiley et al. 2011). Both the macroalgae and microalgae offer the advantage of implementing a biorefinery approach along with biogas production. Nevertheless, the seasonal changes in the chemical composition, higher levels of moisture content and the presence of inhibiting chemical substances during anaerobic digestion are the limitations on the use of algal biomass as the feedstocks (Montingelli et al. 2015).

The strategies for terrestrial and marine algal growth require improvements in making growth and harvesting more efficient with the use of land, water and nutrients (Hannon et al. 2010). Seaweeds are a suitable feedstock for biogas production due to the presence of high polysaccharides (agar, alginate, carrageenan, laminaran and mannitol), but there are limitations on their cultivation and harvesting at present (Behera et al. 2015). Among the growth strategies for terrestrial alga, high-rate algal production coupled with wastewater treatment offers significant economic utility. Beal et al. (2012) quantified the energy return on investment (EROI) for the coupled system of algal production and wastewater treatment and reported that there was a second-order energy return on investment of 1.44 when algal production was coupled with wastewater treatment, without including capital, labour and other required expenses. Wannapokin et al. (2018) showed that the anaerobic co-digestion of fallen leaf leaves (*Tectona grandis*) and microalgae (*Chlorella vulgaris*) after pretreatment with 2% NaOH with a total solid (TS) ratio of 10% was better than the mono-digestion, in terms of biodegradability of TS, volatile solids (VS) and chemical oxidation demand (COD) along with biogas yield and methane potential. In addition to the optimization requirements for biomass production and harvesting, the downstream processing such as cell wall disruption and biomass solubilization is essential of anaerobic biodegradability of microalgae (Passos et al. 2014). Traditionally, the complex carbohydrates are considered to limit the microalgal digestibility. The solubilization of organic matter and subsequent higher methane production can be improved by the addition of protease. But the solubilization of protein results in the release of ammonium, an inhibition of anaerobic digestion. Hence, the protein macromolecules have a differential effect of hydrolysis and methanogenesis of anaerobic digestion of algal biomass (Magdalena et al. 2018). Harvesting of algal biomass accounts for about 20–30% of the total cost of production and necessitates a single or combination of techniques such as centrifugation, flocculation, flotation, sedimentation or filtration. The advanced micro-bio-loop, proposed by Jin and Borthwick (2016), consists of microalgae culture, de-oxygenation, anaerobic digestion and aerobic decomposition and has less than 33% environmental impacts and with a net positive energy balance of 0.0024 kWh MJ^{-1}, compared to the conventional biogas system.

2.3 *Pretreatment of Feedstocks*

The anaerobic digestion offers the advantage of utilizing different substrates as feedstocks to make biogas. The feedstocks which can float, clump or foam in the digesters and those with molecules of highly crystalline structure or low surface area require effective pretreatment. Those substrates which contain inhibitors for microbial activities also need pretreatment to make the anaerobic digestion more efficient with increased biogas yield. In the plant biomass, lignin supports the cell structure, where cellulose and hemicellulose are embedded; the lignin layers need to be broken, and the crystallinity of cellulose has to be decreased by pretreatment before anaerobic digestion. When the algal biomass or the activated sludge is used, the cell walls of microorganisms necessitate the rupturing by pretreatment.

The pretreatment types are many, based on the principles by which they function and techniques employed such as mechanical, thermal, ultrasound and electrochemical (physical principle); alkali, acid and oxidative (chemical principle);

microbiological and enzymatic (biological principle); and stream explosion, extrusion and thermochemical (combined processes). The mechanical pretreatment in the industrial scale mills involves a combination of cutting (knife milling) and grinding (hammer milling) of lignocellulosic substrates with more than 15% moisture contents (Montgomery and Bochmann 2014). While the energy demand of hammer mills is more than 2 to 5 times that of knife mills, they are relatively easy and cheap to operate (Kratky and Jirout 2011). The algal biomass may require the ball mills or colloidal mills. The biomass swelling due to the disruption of hydrogen bonds in the cellulose or lignocellulose complexes is achieved by the thermal pretreatment. The maximum temperature for pretreatment will vary depending on the nature and complexity of substrates; the energy crops can be treated at 190°°C, while the temperatures of 75–95 °C are effective for microalgal biomass (Montgomery and Bochmann 2014; Passos and Ferrer 2014). The breakdown products of xylose and lignin such as heterocyclic and phenolic compounds at high temperatures are either toxic or difficult to degrade anaerobically. Due to high costs, the alkali, acid and oxidative chemical pretreatment techniques are not preferred for biogas production at large scale. The combined processes such as a steam explosion, extrusion or thermochemical pretreatment are effective and involve a combination of mechanisms; both the nature of feedstocks and costs of process operations are critical for their successful implementation.

The biological (microbial or enzyme additions) pretreatment can be done at low temperature but is slow, relative to the physical and chemical methods. The anaerobic microbial pretreatment involves the separation of hydrolysis and acid production steps from the methane production, while the aerobic microbial pretreatment does solubilization or degradation of cellulose, hemicellulose or lignin. The dark fermentation, two-stage digestion or pre-acidification also refers to the anaerobic microbial pretreatment. The two-stage continuous digestion of household waste has been found to give an additional 21% biogas yield, with a hydraulic retention time of about 30 days (Liu et al. 2006; Morales-Polo et al. 2018). Dhouib et al. (2006) and Wagner et al. (2013) applied the fungal pretreatment (mainly white-rot fungi), for removal of phenolic toxins from wastewater or to delignify the substrates, respectively. The sewage sludge and other substrates require sanitation, ultrasound treatment or electrokinetic disintegration. Nevertheless, the large-scale application of several pretreatment techniques, including enzyme additions, depends on the effectiveness and the costs involved. The assessment of pretreatment methods employs different analytical chemistry methods such as high-performance liquid chromatography, structural carbohydrate determination and soluble chemical oxygen demand (sCOD). The assessment is critical before the implementation of pretreatment under real conditions. In addition to the chemical analysis, the biomethane potential which measures the cumulative biomethane production in a batch mode is commonly tested. However, the long-term effect of pretreatment is explained better in the continuous, at least at the pilot-scale, anaerobic digestion.

2.4 *Microbial Diversity and Functions in the Biogas Systems*

The anaerobic digestion is a complex interaction among different but interdependent microorganisms for the decomposition of organic matter under oxygen-depleted condition. The growth rate and production of microbial biomass in the anaerobic digestion are, however, lower than that of aerobic decomposition of these substrates. Biomethanation, which involves four crucial biochemical stages, is mediated by a network of complex microbial communities (Fig. 1). The substrates which are fed for the biogas processing are mainly composed of carbohydrates, proteins and lipids. These complex biomolecules are too large for the metabolic activities of a single microorganism, and they are hydrolysed to soluble sugars, peptides, amino acids and fatty acids by the extracellular enzymes produced by the microorganisms. The saccharolytic microorganisms break down sugars, while the proteolytic organisms hydrolyse proteins. The first phase of hydrolysis is followed by acidogenesis. The monomers of sugars, amino acids and fatty acids are used by the fermenting bacteria for the production of alcohols, organic acids and ammonia along with carbon dioxide, hydrogen and hydrogen sulphide during acidogenesis. In the ensuing phase of acetogenesis or dehydrogenation, alcohols and organic acids are chiefly utilized by the acetogens to produce acetic acid, hydrogen and carbon dioxide. Lastly, in the methanogenic phase, the methanogenic archaea use acetate, formate, methyl compounds, hydrogen and carbon dioxide to form methane, the major final product of biogas. Thus, the process of methane production requires the combined activity of several microbial groups with diverse metabolic capacities in all these phases (Schnurer 2016). The core microorganisms which mediate these major biochemical reactions determine the efficiency of the anaerobic digestion process. The species of about 50 bacterial phyla belonging to *Clostridium*, *Bacteroides*,

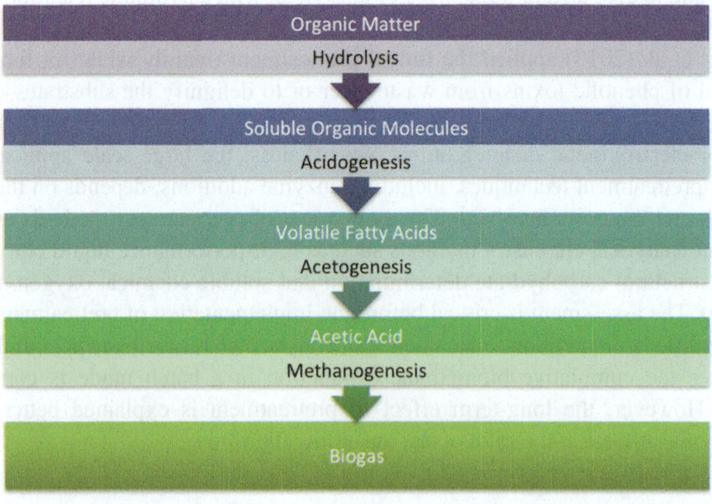

Fig. 1 Stages of biogas production

Bifidobacterium, *Butyrivibrio*, *Proteobacteria*, *Pseudomonas*, *Bacillus* and *Streptococcus* mediate hydrolysis and acidogenesis. In the beginning, *Bacteroidetes* and *Firmicutes* dominate, and then the members of *Bacteroidetes*, *Chloroflexi*, *Firmicutes* and *Proteobacteria* predominate during acidogenesis (Wang et al. 2018). The facultative anaerobic microorganisms such as *Ruminococcus*, *Paenibacillus* and *Streptococcus* can also use the monomers and produce volatile fatty acids, carbon dioxide and hydrogen, while the species of *Aminobacterium*, *Acidaminococcus* and *Desulfovibrio* produce acetic acid and hydrogen in the acidogenesis (Ziganshin et al. 2013). In general, the greater bacterial diversity and the predominance of *Firmicutes* suggest better performance of the initial reactions.

The methanogenic archaea utilize the products such as acetate, formate, carbon dioxide and hydrogen from the reactions of hydrolysis, acidogenesis and acetogenesis to produce the methane and carbon dioxide. The hydrogenotrophic methanogenic archaea oxidize formate or hydrogen and reduce carbon dioxide to methane. The aceticlastic methanogenic archaea cleave acetic acid with the methyl group reduced to methane and the carbonyl group oxidized to methane. The activities of hydrogenotrophic methanogenic archaea create a low partial pressure of hydrogen, which is critical to the oxidation of organic acids as they are endergonic under standard conditions. The hydrogenotrophic methanogenic archaea employ the Wood-Ljungdahl (CO_2 reduction or hydrogenotrophic) pathway, the most efficient pathway for energy generation, and carbon fixation (Sousa et al. 2013). Acetate is also produced through homoacetogenesis by using hydrogen and carbon dioxide from the phases of acidogenesis and acetogenesis. The aceticlastic methanogenic archaea cleaves acetate to a methyl group and CO by employing the acetotrophic (aceticlastic) pathway. Methane is produced from the methyl group, and the reducing power is obtained from the oxidation of CO. Acetate is also metabolized into hydrogen and carbon dioxide by the syntrophic acetate-oxidizing bacteria, and then the hydrogenotrophic methanogenic archaea utilize these substrates for methane production. The syntrophic bacteria can ferment formate, hydrogen and carbon dioxide, and their interactions are important for maintaining the biochemical reactions. The accumulation of volatile fatty acids leads to a higher concentration of hydrogen, but a lower concentration of hydrogen favours the formation of methane and carbon dioxide. In general, about two-thirds of methane is produced from acetate utilization by aceticlastic methanogenic archaea and bacterial syntrophic acetate oxidation, while the remaining one-third of methane is obtained by the use of hydrogen and carbon dioxide by hydrogenotrophic methanogenic archaea. Besides the hydrogenotrophic- and aceticlastic-methanogenic archaea, the methylotrophic methanogenic archaea utilize the C-1 compounds such as methanol, methyl-amines and dimethylsulphide, both as carbon and energy source. The aerobic methanotrophs or the consortia of anaerobes that reduce sulphate, nitrate, manganese or iron oxidize methane to carbon dioxide.

In the recent metagenomic and metatranscriptomic analyses of the thermophilic biogas plants, the bacterial genera of *Defluviitoga*, *Clostridium* cluster III and *Tepidanaerobacter* were found to be abundant, while the archaeal genus of *Methanoculleus* was dominant, with transcriptionally less active

Methanothermobacter (Maus et al. 2016). In the mesophilic digesters, only about 250 bacterial operational taxonomic units (OTUs) out of a total number of 5938 bacterial OTUs detected had a median relative abundance of about 70% and were considered to be the 'biogas core microbiome'. These bacterial members are the generalists, while those representing other 84% of OTUs are the specialists, found only specific anaerobic digestion plants (Calusinska et al. 2018). The bacterial and archaeal members of the microbiome are extensively surveyed with the lesser focus on fungi, protists and phages/viruses. The functional relationships among the procaryotes and eucaryotes are considered to be of cooperative in nature. But the phages or viruses may exert the regulatory effects on cell turnover (Zhang et al. 2017). The balanced biochemical processes in the anaerobic digestion are critical to improve the efficiencies of biogas production, with the rates of degradation being equal in all the phases. If the initial hydrolytic reactions are slower, the methanogenic process gets limited. Rapid acidogenesis can lead to high accumulation of volatile fatty acids, resulting in the inhibition of methanogenic archaea. The accumulation of hydrogen also influences pH and the partial pressure, limiting the mass transfer. Similarly, rapid homoacetogenesis can consume hydrogen and carbon dioxide which can cause the C-source limitation for the autotrophic methanogenic archaea. The microbial community adaptability is important for the process optimization for the biogas production (Westerholm et al. 2018). The acetate- and propionate-degrading microbial communities adapted in the thermophilic digesters showed poor resilience when the conditions were changed to mesophilic conditions. Hence, the process technologies adapted to the substrate properties are significant for an efficient degradation and biogas production.

2.5 Enhancement and Enrichment of Biogas

High sensitivities of both technological and biological aspects of the biogas system demand the knowledge- and data-driven management to ensure stable and efficient production of biogas. The decreased biogas yields are often due to disturbances and instabilities including the feedstock variability that alters the availability of nutrients, temperature variations, microbial communities with different degradation potential and accumulation of inhibitory metabolites (Theuerl et al. 2019). Hydrolysis is the rate-limited step in the anaerobic digestion, and the pretreatment methods, including the supplementation with additives, necessarily facilitate its acceleration. Integration of microbial electrolysis cell with the iron-graphite electrode led to the enhanced activities of hydrolytic enzymes and increased methane production by 22.4% (Feng et al. 2015). The application of electric hydrolysis pretreatment, based on electrophoresis, electro-osmosis and ohmic heating for lignocellulose waste, resulted in increased methane yield by 13.8% due to the disintegration of particles and microbial cell lysis (Veluchamy et al. 2018). Hydrolysis can be enhanced by the fungal pretreatment of rice straw with *Pleurotus ostreatus* and *Trichoderma reesei* (Mustafa et al. 2016). The ruminant bacteria from cattle were used as inoculum as they produced extracellular substances for the cellulolytic biofilm formation (Yue et al. 2013). The enhancement of acetic acid, an essential substrate for methanogenesis, could be done by the addition of polycylic aromatic hydrocarbon (Luo et al. 2016). The addition of ferroferric oxide could enhance the consumption of H_2 along with the production of more acetic acid by the propionic acid or butyrate acid oxidizing bacteria coupled with H_2-utilizing bacteria. Consequently, the acetogens could consume and utilize acetic acid (Yin et al. 2018). The methane production can be stimulated, and the process stability can be achieved by several additives including trace nutrients and metal oxide nanoparticles. In a two-phase anaerobic digestion plant, one of the reactors for hydrolysis, acidification and acetogenesis and another reactor for methanogenesis was found to enhance methane production (Salmoni et al. 2011).

The methane enrichment by a microbial consortium can be achieved by the in situ injection of H_2 directly into the anaerobic digester or into a separate bioreactor as ex situ injection where it is coupled with carbon dioxide (Aryal et al. 2018). H_2 can be produced on site by several methods, including the use of electrolysers, but the in situ H_2 injection of less than 1:4 CO_2:H_2 stoichiometric ratio will improve the CO_2 conversion efficiency. Improved H_2 mass transfer using a ceramic H_2 gas diffuser at 150 rpm led to the biogas upgrading up to 75% (Luo and Angelidaki 2012). The hybrid technology of coupling with ex situ and in situ injection of H_2 might require extra reactor volume (Kougias et al. 2017). The injection of H_2 could lead to the stimulation of hydrogenotrophic methanogens such as *Methanomicrobium*, *Methanoculleus* and *Methanobacterium* (Agneessens et al. 2018). In addition to the changes in the hydrogenotrophic methanogenic activity, the archaeal community structure of the anaerobic digester showed alterations, with decreased bacterial diversity (Luo and Angelidaki 2012; Bassani et al. 2015). The first commercially successful in situ H_2 injection biogas plant (BioPower2Gas), with an increased CH_4 from 50% to 75% is at Allendorf, Germany (www.biopower2gas.de) (Bailera et al. 2017). The ex situ biological methane upgrading plant has been installed at Biofos wastewater treatment plant at Avedore, Denmark (Bailera et al. 2017). The microbial enrichment technology warrants new research to fill the knowledge gaps between laboratory-scale testing and large-scale operations.

3 Biogas Storage and Distribution

The biogas is generally stored in a gas bag, water-sealed gas holder, butane or propane tanks and even commercial gas cylinders with different levels of pressure (Walsh et al. 1989; Kapdi et al. 2005; Khan et al. 2017). Since the biogas is expensive to store locally, there are large variations in their storage systems, including compression and liquefaction. But the presence of gases like CO_2, H_2S and water vapour makes it uneconomical to compress the biogas as these gases reduce the calorific value. The compression of biogas can increase the energy content, while the compressed biogas requires the storage requirements. For

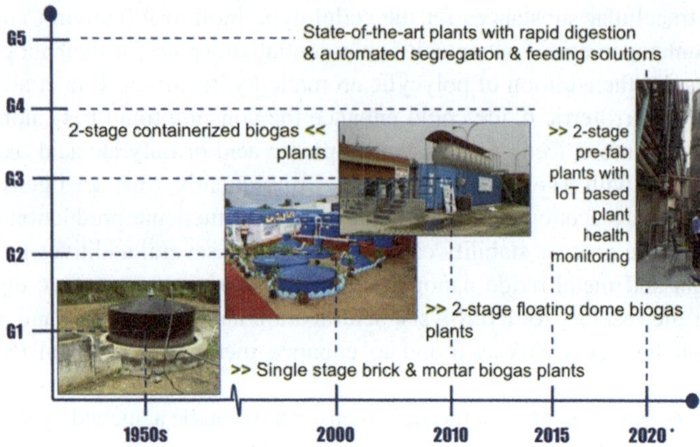

Fig. 2 Evolution of biogas plants

liquefaction of biogas, the critical temperature of about −82.5 °C and pressure of 47.5 bar are required. The renewable energy production by the biogas plants has endured its fair share of challenges along the way. The single stage brick and mortar biogas plants in the 1950s have evolved to the present-day state-of-the-art IOT-enabled 2-stage prefabricated biogas plants. In India, there are two types of biogas systems: (1) family-type small biogas systems with capacities ranging from 1 to 10 m^3 biogas per day and (2) large and industrial-scale biogas plants with a capacity above 5000 m^3 biogas per day (Mittal et al. 2018). Improvements in the technologies are rapid, and the technological evolution is categorized into four generations (G1 to G4) (Fig. 2). The future G5 plants will be completely automated, including segregation and feeding.

The traditional biogas plants (G1 to G3) have several issues related to space constraint and foul smell, safety, reliability, feedstock fluctuation and gas storage. The G4 biogas plants derive from incremental innovations to overcome these constraints. The flexible and customized layouts, including a split system, are prepared to enable different components of the plants to be placed at separate locations that are dead or non-revenue-generating spaces. Such G4 biogas plant layouts which were designed and installed by the GPS Renewables (www.greenpowersystems.co.in) are at ITC Maurya, ITC Maratha, Infosys, Bangalore and Taj Coromandel in India. Since the G1–G3 biogas plants had floating domes and balloon-based gas storage with higher chances of leakages, the end-to-end air sealed solution is provided to minimize the issues related to smell. To ensure the safety of the biogas plants of G4, fabrication is done using corrosion-resistant materials with in-built odour, scum and foaming management systems for longer plant life; the plants are fitted with multilevel gas safety design. Hence, the G4 biogas plants are safe, and the gas leakage and pressure testing procedures need to be conducted to ensure that there is no leakage of inflammable biogas. The new-generation plants with multilevel electronic safety and complete alert system, automated gas flaring system, complete remote monitoring and automated gas storage system and leakage sensing and alert mechanism for the automatic shutdown will ensure better safety. The implementation of "BioHealth Monitoring" with the proprietary auto titrator of GPS Renewables provides opportunities to ensure regular monitoring of the health of biogas plants for higher productivity with no breakdowns. The online process monitoring parameters should include not only of the metabolic process but also of the feeding substrate (Wu et al. 2019). The need for monitoring the feeding substrate is because of the sensitive nature of the core microbiome of an anaerobic digester to its changes. The challenge of fluctuation faced by Bulk Waste Generators (BWG) is addressed by having plants with a 2-stage process (Fig. 3). The first stage involves hydrolysis (as highlighted in Fig. 1), takes up the fluctuations in waste loads and acts as a buffer for 5–6 days. The second stage involves the flow of slurry from the hydrolyser (item 3 in Fig. 3) to digester (item 4 in Fig. 3), which is uniform, ensuring stability post hydrolysis.

The primary aims of biogas storage are on-site usage and before or after transportation to off-site distribution systems. Several modes of storage include low-pressure balloons, high-pressure storage cylinders, gas pipeline and low-pressure storage vessels. The neoprene balloons of G1–G3 biogas plants represent the static storage solutions of low-pressure balloons. In the high-pressure storage cylinders, the biogas after scrubbing of CO_2, H_2S and moisture are stored at a high pressure of 150–200 bar. These high-pressure storage cylinders are also referred to as the compressed biogas (CBG). Also, the purified biogas of the desired parameters can be injected into the natural gas pipeline. The biogas is also stored at a 5–6 bar in low-pressure storage vessels for more accessible transportation and distribution. The DisPred (Distributed Predigester) model (G4 biogas plants) of GPS Renewables has two units: (1) liquid composters and (2) gas generation unit (GGU). The liquid composters generate intermediate AD slurry which can be shipped to be used as an input for the gas generation unit (GGU) which has the digestor and gas purification unit. The DisPred model of GPS Renewables provides greater opportunities for creating financially viable organic waste to energy project (Fig. 4). Lindkvist and Karlsson (2018) proposed a framework for categorization, including seven categories such as substrate, organization, biogas use, digestion technology, localization, digestate and capacity. This framework of categorization will aid in knowledge sharing within and between countries. The market and commercialization of the matured biogas technologies depend on the operational cost, optimized design and type of applications (Gonzalez et al. 2017). Mittal et al. (2018) categorized the barriers to biogas technologies and found them to be different based on the scale of operations and locations such as rural areas (financial/commercial, market, social and cultural, regulatory barrier, technical and infrastructural and information barriers) and urban areas (financial/economical, market, technical and infrastructural and institutional barriers). Diouf and Miezan (2019) suggested access to financial resources by farmers in the rural areas as a critical factor to realize the potential of biogas initiative in Senegal. In general, the efforts for storage and distribution of biogas depend on the cost, design, applications, social and cultural factors and the government policies.

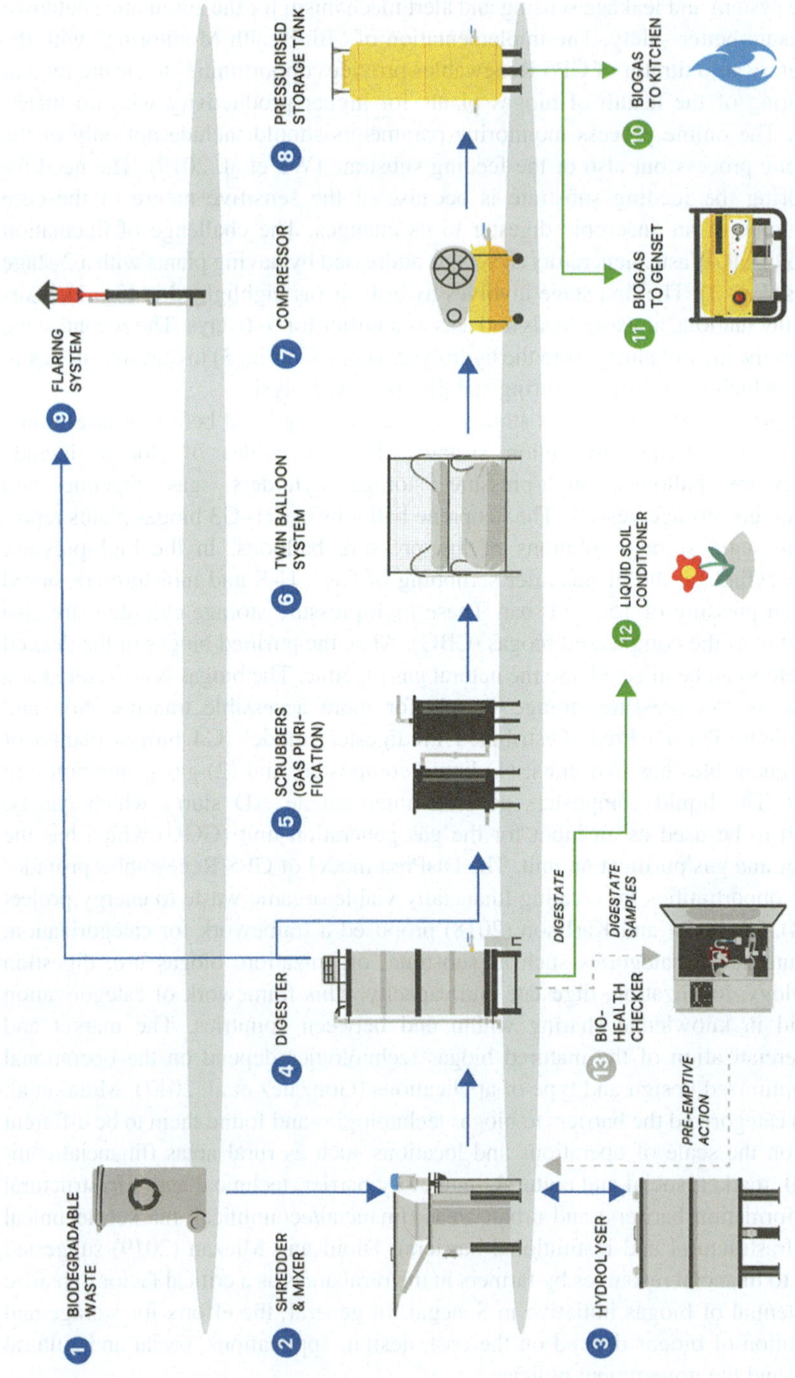

Fig. 3 Process flow of a G4 biogas plant (GPS Renewables, India)

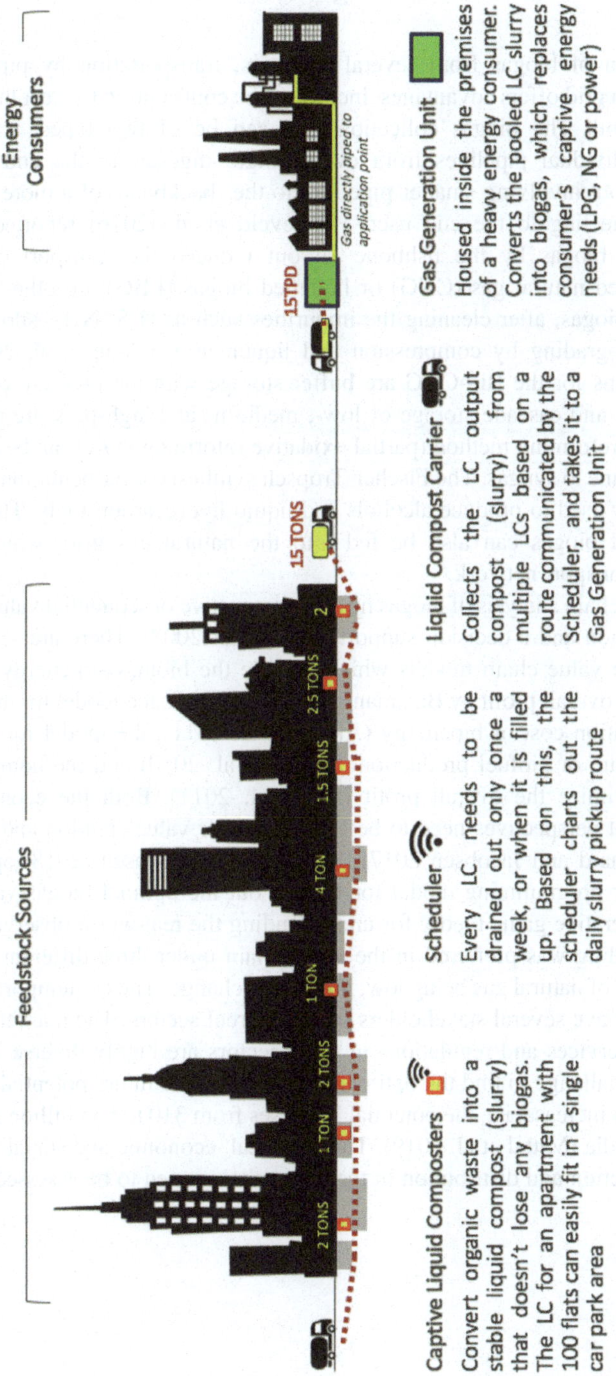

Fig. 4 Illustration of DisPred (GPS Renewables, India) in a city

4 Transportation and the Biogas Value Chain Analysis

The collection of biogas from several digesters, transportation by pipeline and delivery into a grid offers advantages, including the contribution to meet the regional energy demand. The biogas collection grid can be of two types: star lay-out involving individual pipelines from every single digester to the end user and fishbone lay-out involving smaller pipelines to the 'backbone' of a more extensive pipeline connecting to the end user. Hengeveld et al. (2016) reported that the collection of biogas by the fishbone lay-out reduced the transport costs. The bio-compressed natural gas (CNG) or liquefied biogas (LBG) are other modes of transport of biogas, after cleaning the impurities such as H_2S, NH_3, siloxanes and others and upgrading by compression and liquefaction (Yang et al. 2014). The storage systems for the Bio-CNG are buffer storage with the pressure of CNG at 20.5–25 MPa and cascade storage of low-, medium- and high-pressure reservoirs. The catalytic reforming method (partial oxidative reforming POR) can be applied to convert methane to syngas. The Fischer-Tropsch synthesis or fermentation of syngas can be further used to produce alcohols and liquid hydrocarbon fuels. The purified and upgraded biogas can also be fed into the natural gas grid, which has an established transport network.

The value chain analysis of biogas has the perspective of sequential value creation and appreciation and a decision support tool (Porter 2008). There are several proposals for the value chain models which include the biomass-to-energy model to maximize the overall profit by Balaman and Selim (2014), the model for minimizing the transportation cost of bioenergy (Aksoy et al. 2011), the model for improved economic value of biofuel production (Parker et al. 2010) and the computational model to optimize the overall profit (An et al. 2011). Both the economic and environmental perspectives need to be considered for value creation and appreciation (Skovsgaard and Jacobsen 2017). Skovsgaard and Jensen (2018) applied the mixed integer programming model for finding out the optimal biogas value chain and the cooperative game theory for understanding the real-world observation. The biogas upgrading was preferred in the value chain under three different scenarios with the price of natural gas being low, high or no change. The contemporary biogas value chains have several stakeholders from different sectors. The market values of products or services and regulations in these sectors are highly diverse, and these issues are a challenge to find the optimal value chain. The future potential of biogas is predicted to increase, i.e. the potential increases from 310 to 655 billion m^3 $year^{-1}$ by 2040 in India (Mittal et al. 2019). The technical, economic and social aspects of biogas production and distribution in the value chains need to be assessed.

5 Prospects

Biogas is an implementable renewable energy resource, at small and large scale in the developing and developed countries. This flexible energy production by anaerobic digestion offers both economic and environmental benefits including waste management, organic fertilizer production, lesser dependence on fossil fuels and mitigation of greenhouse gas emissions. The anaerobic digestion process can derive the hidden energy from different feedstocks substrates. There is a great challenge in reaching the optimal degradation efficiency by the microbial agents. The optimization of anaerobic digestion process under different scales of operation is necessary. Further research is needed to understand the network of microbial communities and their contributions to the complex process of anaerobic digestion (Xu et al. 2018). Predictions are very high on the future biogas potential at the national and global levels (Searle and Malins 2015; Mittal et al. 2019). Possible advantages of production, distribution and utilization depend on the technical-economical improvements and social benefits of the biogas value chains and the policies of the governments.

References

Achinas S, Achinas V, Euverink GJ (2017) A technological overview of biogas production from biowaste. Engineering 3(3):299–307. https://doi.org/10.1016/J.ENG.2017.03.002

Agneessens LM, Ottosen LD, Andersen M et al (2018) Parameters affecting acetate concentrations during in-situ biological hydrogen methanation. Bioresour Technol 258:33–40. https://doi.org/10.1016/j.biortech.2018.02.102

Aksoy B, Cullinan H, Webster D et al (2011) Woody biomass and mill waste utilization opportunities in Alabama: transportation cost minimization, optimum facility location, economic feasibility, and impact. Environ Prog Sustain 30:720–732. https://doi.org/10.1002/ep.10501

An H, Wilhelm WS, Searcy SW (2011) A mathematical model to design a lignocellulosic biofuel supply chain system with a case study based on a region in Central Texas. Bioresour Technol 102:7860–7870. https://doi.org/10.1016/j.biortech.2011.05.060

Aryal N, Kvist T, Ammam F et al (2018) An overview of microbial biogas enrichment. Bioresour Technol 264:359–369. https://doi.org/10.1016/j.biortech.2018.06.013

Bailera M, Lisbona P, Romeo LM et al (2017) Power to gas projects review: lab, pilot and demo plants for storing renewable energy and CO_2. Renew Sust Energ Rev 69:292–312. https://doi.org/10.1016/j.rser.2016.11.130

Balaman SY, Selim H (2014) A fuzzy multiobjective linear programming model for design and management of anaerobic digestion based bioenergy supply chains. Energy 74:928–940. https://doi.org/10.1016/j.energy.2014.07.073

Bassani I, Kougias PG, Treu L et al (2015) Biogas upgrading via hydrogenotrophic methanogenesis in two-stage continuous stirred tank reactors at mesophilic and thermophilic conditions. Environ Sci Technol 49(20):12585–12593. https://doi.org/10.1021/acs.est.5b03451

Beal CM, Hebner RE, Webber ME et al (2012) The energy return on investment for algal biocrude: results for a research production facility. Bioenergy Res 5(2):341–362. https://doi.org/10.1007/s12155-011-9128-4

Behera S, Singh R, Arora R et al (2015) Scope of algae as third generation biofuels. Front Bioeng Biotechnol 2:90. https://doi.org/10.3389/fbioe.2014.00090

Bennett RK, Steinberg LM, Chen W et al (2018) Engineering the bioconversion of methane and methanol to fuels and chemicals in native and synthetic methylotrophs. Curr Opin Biotech 50:81–93. https://doi.org/10.1016/j.copbio.2017.11.010

Bhuvaneshwari S, Hettiarachchi H, Meegoda JN (2019) Crop residue burning in India: Policy challenges and potential solutions. Int J Env Res Pub He 16(5):832. https://doi.org/10.3390/ijerph16050832

Braun R (2007) Anaerobic digestion – A multi-faceted process for energy, environmental management and rural development. In: Ranalli P (ed) Improvement of crop plants for industrial end uses. Springer, Dordrecht, pp 335–416. https://doi.org/10.1007/978-1-4020-5486-0_13

Breitenmoser L, Gross T, Huesch R et al (2019) Anaerobic digestion of biowastes in India: opportunities, challenges and research needs. J Environ Manag 236:396–412. https://doi.org/10.1016/j.jenvman.2018.12.014

Calusinska M, Goux X, Fossépré M et al (2018) A year of monitoring 20 mesophilic full-scale bioreactors reveals the existence of stable but different core microbiomes in bio-waste and wastewater anaerobic digestion systems. Biotechnol Biofuels 11(1):196. https://doi.org/10.1186/s13068-018-1195-8

Cherubin MR, Oliveira DMDS, Feigl BJ et al (2018) Crop residue harvest for bioenergy production and its implications on soil functioning and plant growth: a review. Sci Agríc (Piracicaba, Braz) 75(3):255–272. https://doi.org/10.1590/1678-992x-2016-0459

Collotta M, Tomasoni G (2017) The environmental sustainability of biogas production with small sized plant. Energy Proc 128:38–45. https://doi.org/10.1016/j.egypro.2017.09.012

De Vrieze J, Pinto AJ, Sloan WT et al (2018) The active microbial community more accurately reflects the anaerobic digestion process: 16S rRNA (gene) sequencing as a predictive tool. Microbiome 6:63. https://doi.org/10.1186/s40168-018-0449-9

Dhouib A, Ellouz M, Aloui F et al (2006) Effect of bioaugmentation of activated sludge with white-rot fungi on olive mill wastewater detoxification. Lett Appl Microbiol 42(4):405–411. https://doi.org/10.1111/j.1472-765X.2006.01858.x

Diouf B, Miezan E (2019) The Biogas initiative in developing countries, from technical potential to failure: the case study of Senegal. Renew Sust Energ Rev 101:248–254. https://doi.org/10.1016/j.rser.2018.11.011

Feng Y, Zhang Y, Chen S et al (2015) Enhanced production of methane from waste activated sludge by the combination of high-solid anaerobic digestion and microbial electrolysis cell with iron–graphite electrode. Chem Eng J 259:787–794. https://doi.org/10.1016/j.cej.2014.08.048

Ghosh S, Uday V, Giri A et al (2019) Biogas to methanol: a comparison of conversion processes involving direct carbon dioxide hydrogenation and via reverse water gas shift reaction. J Clean Prod 217:615–626. https://doi.org/10.1016/j.jclepro.2019.01.171

Gonzalez MOA, Goncalves JS, Vasconcelos RM (2017) Sustainable development: case study in the implementation of renewable energy in Brazil. J Clean Prod 142:461–475. https://doi.org/10.1016/j.jclepro.2016.10.052

Grieder C, Mittweg G, Dhillon BS et al (2012) Kinetics of methane fermentation yield in biogas reactors: genetic variation and association with chemical composition in maize. Biomass Bioenergy 37:132–141. https://doi.org/10.1016/j.biombioe.2011.12.020

Gunaseelan VN (1997) Anaerobic digestion of biomass for methane production: a review. Biomass Bioenergy 13:83–114. https://doi.org/10.1016/S0961-9534(97)00020-2

Hahn H, Ganagin W, Hartmann K et al (2014) Cost analysis of concepts for a demand oriented biogas supply for flexible power generation. Bioresour Technol 170:211–220. https://doi.org/10.1016/j.biortech.2014.07.085

Hannon M, Gimpel J, Tran M et al (2010) Biofuels from algae: challenges and potential. Biofuels 1:763–784. https://doi.org/10.4155/bfs.10.44

Hengeveld EJ, Bekkering J, van Gemert WJT et al (2016) Biogas infrastructures from farm to regional scale, prospects of biogas transport grids. Biomass Bioenergy 86:43–52. https://doi.org/10.1016/j.biombioe.2016.01.005

Herrmann A, Rath J (2012) Biogas production from maize: current state, challenges and prospects. 1. Methane yield potential. Bioenergy Res 5:1027–1042. https://doi.org/10.1007/s12155-013-9318-3

Hughes AD, Kelly MS, Black KD et al (2012) Biogas from macroalgae: is it time to revisit the idea? Biotechnol Biofuels 5:86. https://doi.org/10.1186/1754-6834-5-86

Jin Q, Borthwick AGL (2016) An advanced micro-bio-loop to produce biogas. J Clean Prod 139:1094–1097. https://doi.org/10.1016/j.jclepro.2016.08.136

Kapdi SS, Vijay VK, Rajesh SK et al (2005) Biogas scrubbing, compression and storage: perspective and prospectus in Indian context. Renew Energ 30:1195–1202. https://doi.org/10.1016/j.renene.2004.09.012

Khan IU, Othman MHD, Hashim H et al (2017) Biogas as a renewable energy fuel-A review of biogas upgrading, utilisation and storage. Energ Convers Manage 150:277–294. https://doi.org/10.1016/j.enconman.2017.08.035

Kougias PG, Treu L, Benavente DP et al (2017) *Ex-situ* biogas upgrading and enhancement in different reactor systems. Bioresour Technol 225:429–437. https://doi.org/10.1016/j.biortech.2016.11.124

Kratky L, Jirout T (2011) Biomass size reduction machines for enhancing biogas production. Chem Eng Technol 34(3):391–399. https://doi.org/10.1002/ceat.201000357

Lambert M (2017) Biogas: a significant contribution to decarbonising gas markets? The Oxford Institute for Energy Studies, Oxford. Energy Insight, 15

Lantz M, Kreuger E, Bjornsson L (2017) An economic comparison of dedicated crops vs agricultural residues as feedstock for biogas of vehicle fuel quality. AIMS Energy 5:838–863

Lindkvist E, Karlsson M (2018) Biogas production plants; existing classifications and proposed categories. J Clean Prod 174:1588–1597. https://doi.org/10.1016/j.jclepro.2017.10.317

Liu D, Liu D, Zeng RJ et al (2006) Hydrogen and methane production from household solid waste in the two-stage fermentation process. Water Res 40:2230–2236. https://doi.org/10.1016/j.watres.2006.03.029

Luo G, Angelidaki I (2012) Integrated biogas upgrading and hydrogen utilization in an anaerobic reactor containing enriched hydrogenotrophic methanogenic culture. Biotechnol Bioeng 109(11):2729–2736. https://doi.org/10.1002/bit.24557

Luo F, Devine CE, Edwards EA (2016) Cultivating microbial dark matter in benzene-degrading methanogenic consortia. Environ Microbiol 18(9):2923–2936. https://doi.org/10.1111/1462-2920.13121

Magdalena J, Ballesteros M, González-Fernandez C (2018) Efficient anaerobic digestion of microalgae biomass: proteins as a key macromolecule. Molecules 23(5):1098. https://doi.org/10.3390/molecules23051098

Maus I, Koeck DE, Cibis KG et al (2016) Unraveling the microbiome of a thermophilic biogas plant by metagenome and metatranscriptome analysis complemented by characterization of bacterial and archaeal isolates. Biotechnol Biofuels 9(1):171. https://doi.org/10.1186/s13068-016-0581-3

Meyer AKP, Ehimen EA, Holm-Nielsen JB (2018) Future European biogas: animal manure, straw and grass potentials for a sustainable European biogas production. Biomass Bioenergy 111:154–164. https://doi.org/10.1016/j.biombioe.2017.05.013

Mittal S, Ahlgren EO, Shukla PR (2018) Barriers to biogas dissemination in India: a review. Energy Policy 112:361–370. https://doi.org/10.1016/j.enpol.2017.10.027

Mittal S, Ahlgren EO, Shukla PR (2019) Future biogas resource potential in India: a bottom-up analysis. Renew Energ 141:379–389. https://doi.org/10.1016/j.renene.2019.03.133

Momayez F, Karimi K, Taherzadeh MJ (2019) Energy recovery from industrial crop wastes by dry anaerobic digestion: a review. Ind Crop Prod 129:673–687. https://doi.org/10.1016/j.indcrop.2018.12.051

Montgomery LF, Bochmann G (2014) Pretreatment of feedstock for enhanced biogas production. IEA Bioenergy. ISBN 978-1-910154-052, pp 1–24

Montingelli ME, Tedesco S, Olabi AG (2015) Biogas production from algal biomass: a review. Renew Sust Energ Rev 43:961–972. https://doi.org/10.1016/j.rser.2014.11.052

Morales-Polo C, Cledera-Castro MDM, Soria YM (2018) Reviewing the anaerobic digestion of food waste: from waste generation and anaerobic process to its perspectives. Appl Sci 8 (10):1804. https://doi.org/10.3390/app8101804

Mudimu O, Rybalka N, Bauersachs T et al (2014) Biotechnological screening of microalgal and cyanobacterial strains for biogas production and antibacterial and antifungal effects. Meta 4:373–393. https://doi.org/10.3390/metabo4020373

Mulat DG, Jacobi HF, Feilberg A et al (2016) Changing feeding regimes to demonstrate flexible biogas production: effects on process performance, microbial community structure, and methanogenesis pathways. Appl Environ Microbiol 82:438–449. https://doi.org/10.1128/AEM.02320-15

Mulat DG, Huerta SG, Kalyani D et al (2018) Enhancing methane production from lignocellulosic biomass by combined steam-explosion pretreatment and bioaugmentation with cellulolytic bacterium *Caldicellulosiruptor bescii*. Biotechnol Biofuels 11:19. https://doi.org/10.1186/s13068-018-1025-z

Mussgnug JH, Klassen V, Schuter A et al (2010) Microalgae as substrates for fermentative biogas production in a combined biorefinery concept. J Biotechnol 150:51–56. https://doi.org/10.1016/j.jbiotec.2010.07.030

Mustafa AM, Poulsen TG, Sheng K (2016) Fungal pretreatment of rice straw with *Pleurotus ostreatus* and *Trichoderma reesei* to enhance methane production under solid-state anaerobic digestion. Appl Energ 180:661–671. https://doi.org/10.1016/j.apenergy.2016.07.135

Parker N, Tittmann P, Hart Q et al (2010) Development of a biorefinery optimized biofuel supply curve for the western United States. Biomass Bioenergy 34:1597–1607. https://doi.org/10.1016/j.biombioe.2010.06.007

Passos F, Ferrer I (2014) Microalgae conversion to biogas: thermal pretreatment contribution on net energy production. Environ Sci Technol 48(12):7171–7178. https://doi.org/10.1021/es500982v

Passos F, Uggetti E, Carrère H et al (2014) Pretreatment of microalgae to improve biogas production: a review. Bioresour Technol 172:403–412. https://doi.org/10.1016/j.biortech.2014.08.114

Porter ME (2008) How competitive forces shape energy. Harv Bus Rev 57(2):137–145

Rath J, Heuwinkel H, Herrmann A (2013) Specific biogas yield of maize can be predicted by the interaction of four biochemical constituents. Bioenergy Res 6:939–952

Rodriguez C, Alaswad A, Benyounis KY et al (2017) Pretreatment techniques used in biogas production from grass. Renew Sust Energ Rev 68:1193–1204. https://doi.org/10.1016/j.rser.2016.02.022

Salmoni C, Caputo A, Bonoli M et al (2011) Enhanced methane production in a two-phase anaerobic digestion plant, after CO_2 capture and addition to organic wastes. Bioresour Technol 102:6443–6448. https://doi.org/10.1016/j.biortech.2011.03.079

Schnurer A (2016) Biogas production: microbiology and technology. In: Mamo G, Mattiasson B (eds) Anaerobes in biotechnology, advances in biochemical engineering/biotechnology, vol 156. Springer, Cham, pp 195–234. https://doi.org/10.1007/10_2016_5

Searle S, Malins C (2015) A reassessment of global bioenergy potential in 2050. GCB Bioenerg 7 (2):328–336. https://doi.org/10.1111/gcbb.12141

Skovsgaard L, Jacobsen HK (2017) Economies of scale in biogas production and the significance of flexible regulation. Energ Policy 101:77–89. https://doi.org/10.1016/j.enpol.2016.11.021

Skovsgaard L, Jensen IG (2018) Recent trends in biogas value chains explained using cooperative game theory. Energ Econ 74:503–522. https://doi.org/10.1016/j.eneco.2018.06.021

Sousa DZ, Salvador AF, Ramos J et al (2013) Activity and viability of methanogens in anaerobic digestion of unsaturated and saturated long-chain fatty acids. Appl Environ Microbiol 79 (14):4239–4245. https://doi.org/10.1128/AEM.00035-13

Theuerl S, Klang J, Prochnow A (2019) Process disturbances in agricultural biogas production—causes, mechanisms and effects on the biogas microbiome: a review. Energies 12(3):365. https://doi.org/10.3390/en12030365

Veluchamy C, Raju VW, Kalamdhad AS (2018) Electrohydrolysis pretreatment for enhanced methane production from lignocellulose waste pulp and paper mill sludge and its kinetics. Bioresour Technol 252:52–58. https://doi.org/10.1016/j.biortech.2017.12.093

Vindis P, Stajnko D, Berk P et al (2012) Evaluation of energy crops for biogas production with a combination of simulation modelling and Dex-i multicriteria method. Pol J Environ Stud 21:763–770

Wagner AO, Schwarzenauer T, Illmer P (2013) Improvement of methane generation capacity by aerobic pre-treatment of organic waste with a cellulolytic *Trichoderma viride* culture. J Environ Manag 129:357–360. https://doi.org/10.1016/j.jenvman.2013.07.030

Wagner A, Lackner N, Mutschlechner M et al (2018) Biological pretreatment strategies for second-generation lignocellulosic resources to enhance biogas production. Energies 11(7):1797. https://doi.org/10.3390/en11071797

Walsh JL, Ross CC, Smith MS et al (1989) Utilization of biogas. Biomass 20(3–4):277–290. https://doi.org/10.1016/0144-4565(89)90067-X

Wang P, Wang H, Qiu Y et al (2018) Microbial characteristics in anaerobic digestion process of food waste for methane production-a review. Bioresour Technol 248:29–36. https://doi.org/10.1016/j.biortech.2017.06.152

Wannapokin A, Ramaraj R, Whangchai K et al (2018) Potential improvement of biogas production from fallen teak leaves with co-digestion of microalgae. 3 Biotech 8(2):123. https://doi.org/10.1007/s13205-018-1084-7

Weiland P (2010) Biogas production: current state and perspectives. Appl Microbiol Biotechnol 85:849–860. https://doi.org/10.1007/s00253-009-2246-7

Westerholm M, Müller B, Singh A et al (2018) Detection of novel syntrophic acetate-oxidizing bacteria from biogas processes by continuous acetate enrichment approaches. Microbial Biotechnol 11(4):680–693. https://doi.org/10.1111/1751-7915.13035

Wiley PE, Campbell JE, McKuin B (2011) Production of biodiesel and biogas from algae: a review of process train options. Water Environ Res 83:326–338

Wu D, Li L, Zhao X et al (2019) Anaerobic digestion: a review on process monitoring. Renew Sust Energ Rev 103:1–12. https://doi.org/10.1016/j.rser.2018.12.039

Xu R, Zhang K, Liu P et al (2018) A critical review on the interaction of substrate nutrient balance and microbial community structure and function in anaerobic co-digestion. Bioresour Technol 247:1119–1127. https://doi.org/10.1016/j.biortech.2017.09.095

Yang L, Ge X, Wan C et al (2014) Progress and perspectives in converting biogas to transportation fuels. Renew Sust Energ Rev 40:1133–1152. https://doi.org/10.1016/j.rser.2014.08.008

Yin Q, He K, Echigo S et al (2018) Ferroferric oxide significantly affected production of soluble microbial products and extracellular polymeric substances in anaerobic methanogenesis reactors. Front Microbiol 9:2376. https://doi.org/10.3389/fmicb.2018.02376

Yong Z, Dong Y, Zhang X et al (2015) Anaerobic co-digestion of food waste and straw for biogas production. Renew Energ 78:527–530. https://doi.org/10.1016/j.renene.2015.01.033

Yu Q, Tian Z, Liu J et al (2018) Biogas production and microbial community dynamics during the anaerobic digestion of rice straw at 39-50 °C: a pilot study. Energy Fuel 32:5157–5163. https://doi.org/10.1021/acs.energyfuels.7b04042

Yue ZB, Li WW, Yu HQ (2013) Application of rumen microorganisms for anaerobic bioconversion of lignocellulosic biomass. Bioresour Technol 128:738–744. https://doi.org/10.1016/j.biortech.2012.11.073

Zabranska J, Pokorna D (2018) Bioconversion of carbon dioxide to methane using hydrogen and hydrogenotrophic methanogens. Biotechnol Adv 36:707–720. https://doi.org/10.1016/j.biotechadv.2017.12.003

Zhang J, Gao Q, Zhang Q et al (2017) Bacteriophage-prokaryote dynamics and interaction within anaerobic digestion processes across time and space. Microbiome 5:57. https://doi.org/10.1186/s40168-017-0272-8

Ziganshin AM, Liebetrau J, Pröter J et al (2013) Microbial community structure and dynamics during anaerobic digestion of various agricultural waste materials. Appl Microbiol Biotechnol 97(11):5161–5174. https://doi.org/10.1007/s00253-013-4867-0

Potentials and Challenges of Biogas Upgradation as Liquid Biomethane

Ram Chandra, Adya Isha, Subodh Kumar, Sameer Ahmad Khan, Paruchuri Mohan Venkata Subbarao, Virendra Kumar Vijay, Anuj Kumar Chandel, and Ved Prakash Chaudhary

Abstract Increasing human population and expansion of industries is enhancing the demand for energy sources. The conventional fossil-based energy sources are deteriorating the environment, and alternative sources are not as energy intensive as conventional fuel. Renewable fuels such as biogas seem to be a promising substitute for fossil fuel, but the incombustible fraction of carbon dioxide reduces its energy density. Many technologies have been developed to remove this incombustible fraction and other impurities of biogas. This chapter presents the technologies used for upgrading of biogas as well as natural gas and technologies to produce liquid biomethane (LBM). The chapter also focuses on the challenges with natural gas at various stage of application that could also be the challenges for the liquid biomethane (LBM) when it is utilised in the place of natural gas in the already established facilities and infrastructures.

Keywords Liquid biomethane · Biogas · Renewable fuels · Renewable energy · Membrane separations

R. Chandra (✉) · A. Isha · S. Kumar · S. A. Khan · V. K. Vijay
Centre for Rural Development and Technology, Indian Institute of Technology Delhi, New Delhi, India
e-mail: rchandra@rdat.iitd.ac.in

P. M. V. Subbarao
Department of Mechanical Engineering, Indian Institute of Technology Delhi, New Delhi, India

A. K. Chandel
Department of Biotechnology, Engineering School of Lorena (EEL), University of Sao Paulo (USP), Lorena, São Paulo, Brazil

V. P. Chaudhary
Indian Council of Agricultural Research – Indian Institute of Farming Systems Research Modipuram, Meerut, Uttar Pradesh, India

N. Balagurusamy, A. K. Chandel (eds.), *Biogas Production*,
https://doi.org/10.1007/978-3-030-58827-4_14

1 Introduction

In the present scenario, most of the energy-related researches are focused on finding the inexpensive, renewable and environment-friendly alternatives of fossil fuel. Solar energy powered technologies such as photovoltaic panels, wind turbines are contributing in electricity production, while biofuel such as biogas, biodiesel, bioethanol and bio-hydrogen are emerging as a sustainable substitute of conventional fuels such as petrol, diesel, natural gas, etc. (Kumar et al. 2018). Production of biofuels depends upon the economic and abundant organic resources, which are available in ample amount as municipal waste, agricultural residues, industrial effluents, food waste, etc. (Kumar et al. 2019).

Biogas is product of complex biochemical degradation of organic residue in the anaerobic environment (Sahota et al. 2018). Biogas mainly consist of methane (CH_4) and carbon dioxide (CO_2) with minor fraction of hydrogen sulphide (H_2S), moisture and hydrogen (H_2) and oxygen (O_2). This composition varies with operating conditions of anaerobic digestion such as temperature, organic loading rate (OLR), pH and type of fed material (Miltner et al. 2017; Kadam and Panwar 2017). Biogas is considered as sustainable renewable energy sources for various applications such as cooking, automobile fuel, power generation, etc.

Biogas in the raw form can directly be utilised for heat production for various thermal application as well as for power production. On the other hand, in pure form (about 90–95% CH_4), upgraded biogas contains the same properties as of natural gas. Only the methane in the raw biogas contributes to combustion and with 60% composition it provides almost 21 MJ of heat from 1.0 cubic metre of biogas. Incombustible constituents of raw biogas such as carbon dioxide, moisture and other traces cause the reduction in energy density. In the process of biogas upgradation, energy density of biogas is enhanced by removing undesirable constituents of biogas such as CO_2, H_2, moisture, and other contaminants with the application of various techniques. Presently, there are number of technologies available for CO_2 separation from the raw biogas. However, it is challenging task to choose the technology as the design and operating conditions vary from manufacturers to manufacturers. Energy demand and CH_4 loss during the upgrading process are considered as the key criteria while selecting a particular upgrading technology.

Upgraded biogas can directly be used in the place of fossil based natural gas, if it matches the standard of natural gas. Therefore, upgraded biogas can be transported through the existing infrastructure of natural gas, utilised as a fuel for vehicular applications and CHP (combined heat and power) systems and as a feedstock for various industrial applications (Sahota et al. 2018).

2 The Need for a Sustainable Alternative of Fossil Fuel

The requirement of energy in transport sectors is mainly fulfilled by petroleum-based fuels like petrol and diesel. These fuels pollute our environment and cause global warming due to the emission of CO_2, NO_x and particulate matter and other exhaust gases. According to the report of IEA (International Energy Agency), 2017, these conventional fuels used in the transportation sector causes almost 24% of global CO_2 emissions (IEA 2017). Moreover, the fuels used in the transport sector contribute to about 50% of the total nitrogen oxides (NO_X) emission and approximately 10% of the total particulate matter ($PM_{2.5}$) emission (IEA 2016). Observing the need for reduction in greenhouse gas (GHG) emission, many policies and incentives have been introduced to cut down the use of fossil fuels. In this regard, Natural gas (NG) in the form of compressed natural gas (CNG) and liquid natural gas (LNG) is effectively replacing the conventional fuels (diesel and petrol), in light and heavy-duty vehicles, respectively. The increasing demand for alternative fuels in the transport sector with the expansion of vehicles is continuously declining the limited reserves of natural gas. Therefore, there is an urgent need for sustainable fuels, which are compatible with natural gas. In this context, biogas in the upgraded form such as CBG (compressed biogas) and LBM (liquid biomethane) is emerging as a renewable substitute of CNG and LNG (Spitoni et al. 2019). Typically natural gas (calorific value ~39 MJ/m^3) contains 85–92% (v/v) methane (CH_4), 0.2–1.5% (v/v) carbon dioxide (CO_2), 1.1–5.9 mg/m^3 hydrogen sulphide (H_2S) and 9% (v/v) heavy hydrocarbons, while biogas produced by anaerobic digestion (calorific value ~22 MJ/m^3) contains 60–70% (v/v.) CH_4, 30–40% (v/v) CO_2 few hundred to 30,000 ppm of H_2S and 0% (v/v) heavy hydrocarbons. In this regard, biogas can be considered as "particular" natural gas of low calorific value with a high fraction of CO_2 and can be converted into natural gas by separation of CO_2 (Ryckebosch et al. 2011; Pellegrini et al. 2018).

3 Upgradation of Biogas as Energy Intense Fuel

Biogas upgradation by removing CO_2 and other traces (H_2S etc.) has been a highlighted research area of recent studies in the field of bioenergy, and many technologies have been developed till date, and some of them are commercially available. The common biogas upgrading technologies are water scrubbing, pressure swing adsorption, chemical absorption, physical absorption, membrane technology, cryogenic separation, biological upgrading methods and in situ upgrading methods. Each biogas upgradation technology has its limitation based upon methane recovery, carbon dioxide separation, energy efficiency and economic feasibility (Sahota et al. 2018). Currently, researches are focused on improving efficiency as well as reducing investment, operational and maintenance cost and making existing technology economical feasible (Sun et al. 2015).

3.1 Pressure Swing Adsorption (PSA)

PSA technology is one of the prevalent technologies for the biogas upgradation. In this technique carbon dioxide is separated from the raw biogas by the absorption on the absorbent material such as activated carbon or zeolites under high pressure. After adsorption the adsorbent material is depressurised to regenerate the material before reloading again. Traces of hydrogen sulphide and water are removed before the adsorption process. In this technique 20–30% of methane loss was reported (Sahota et al. 2018).

3.2 Water Scrubbing

Water scrubbing is one of the favourable techniques for the biogas upgradation. At high pressure carbon dioxide as well as hydrogen sulphide shows higher solubility in water than methane. In this process raw biogas at high pressure is passed through scrubber column packed with high surface area material. In the scrubber column, specifically CO_2 and most of the traces of H_2S are dissolved in water, leaving a concentrate stream of methane as a scrubbed gas. This technology was reported to achieve up to 97% of pure methane with minimal methane loss of 5%.

3.3 Membrane Separation

Dry membranes having permeability of molecular level are also be utilised for biogas upgradation. These membranes are hollow fibre bundled material and made of such material that allow CO_2, ammonia and water and very low amount of methane and nitrogen molecules to permeate through it. Most of the methane part of biogas collected separately and utilised for various purposes (Kadam and Panwar 2017).

3.4 Chemical Scrubbing: Monomethylamine (MEA) System

3.4.1 Chemical Scrubbing Using Monomethylamine

Chemical scrubbing using monomethylamine system is considered as the best biogas upgradation system by achieving up to 99.9% pure methane with negligible loss. This system is being extensively utilised in Germany for upgradation of raw biogas. In this technique impurity of CO_2 is absorbed in chemical as well as it reacts with the

Table 1 Comparison of biogas upgradation technologies

Parameter	Technology: Amine scrubbing	Water scrubbing	PSA/VPSA
Type	Chemical	Physical	Physical
Working principle	Packed tower absorption	Packed tower absorption	Pressure swing adsorption
CO_2 removal	Yes	Yes	Yes
Simultaneous H_2S removal	Not recommended	Yes	No
Common technology for prior removal of H_2S	Biological scrubbing, irreversible chemical reaction with iron oxide, zinc oxide	–	Irreversible chemical reaction with iron oxide or zinc oxide
CH_4 loss	<1%	<5%	20–30%
Plant operation	Complicated	Easy	Easy
Initial cost	High	Low	Low
Recurring cost	High	Low	Low

amine in the chemical. This combined process provides the highly selective separation of concentrated methane with minimal loss (less than 0.1%).

A comparison of the available technologies used for biogas upgradation is shown in Table 1.

Furthermore, Table 2 shows the comparative analysis of two most common biogas purification technologies in India.

3.5 *Standard for Upgraded Biogas in India*

Many developed countries are working on biomethane, either they are developing or they have already developed the benchmark for biomethane which is applicable for most automobiles/transport vehicles. BIS (Bureau of Indian Standards), an Indian agency for formulation of standard for products/standard test methods, has successfully developed the standards for biomethane that can be utilised instead of CNG in automobiles. Various legal authorisations are required before implementation of biomethane for vehicular purpose. BIS published its first standard for biomethane in 2013; later on in 2016 it was revised and published as IS 16087:2016 (for more information please refer to BIS). This standard focuses not only on composition of biomethane but also the sampling technique and its application on piped gas supply network, automobiles, stationary engines and thermal application. Main objective/scope of the standard is to provide guidance for filling of compressed biomethane (CBG). Required composition for biomethane standards for filling it in CNG cylinders at working pressure of 20.0 MPa is shown in Table 3.

Table 2 Comparison of water scrubbing and pressure swing adsorption technology for biogas upgradation

Parameter	Water scrubbing	PSA/VPSA
Working principle		
Carbon dioxide	Absorption of CO_2 in water at ~1.0 MPa pressure in a packed tower and regeneration of water by venting CO_2 at atmospheric pressure	Adsorption of CO_2 in zeolite molecular sieve (ZMS) at ~0.08 MPa pressure and regeneration of ZMS under partial vacuum by venting CO_2 in atmosphere
Moisture	Adsorption of moisture in alumina or ZMS at ~0.4 MPa pressure and regeneration of same at atmospheric pressure by venting moisture in atmosphere	Adsorption of moisture in ZMS at ~0.08 MPa pressure and regeneration of ZMS under partial vacuum by venting moisture in atmosphere
Hydrogen sulphide	Absorption of H_2S in water at ~0.4 MPa pressure in a packed tower and regeneration of water by venting H_2S at atmospheric pressure	Permanent chemical reaction of H_2S in iron oxide or zinc oxide without any regeneration
Working life of reactants		
Carbon dioxide	100% regeneration of water (only evaporative losses) infinite life of water, it never saturates	2–3 years life of ZMS due to attrition of sieves and gradual blocking of nanopores
Moisture	2–3 years life of alumina/ZMS due to attrition of sieves and gradual blocking of nanopores	2–3 years life of ZMS due to attrition of sieves and gradual blocking of nanopores
Hydrogen sulphide	100% regeneration of water (only evaporative losses). Infinite life of water, it never saturates	Regular consumption of iron/zinc oxide due to permanent chemical reaction
Energy consumption	~ 0.25 kWh/Nm^3 of biogas (equivalent)	0.25 kWh/Nm^3 of biogas (equivalent)
Methane loss	<5%	20–30%
Disposable issues	Disposal of very small amount of alumina/ZMS every 2–3 years.	Huge problem of regular disposal of iron/zinc sulphide and disposal of heavy amount of ZMS every 2–3 years
Recurring cost		
Machinery	Regular servicing of moving parts	Regular servicing of moving parts
Chemicals	~ 2% of initial plant cost for the replacement of alumina/ZMS every 2–3 years	~ 50% of initial plant cost for replacement of ZMS every 2–3 years and regular expenditure on purchase of zinc/iron oxide
Design issues		
Purified biogas purity adjustment	Fine control of methane purity is possible by adjusting water flow rate and working pressure for a constant inlet biogas flow rate	No parameters for adjusting purified biogas methane percentage for a constant inlet biogas flow rate

(continued)

Table 2 (continued)

Parameter	Water scrubbing	PSA/VPSA
Operational reliability	Constant performance throughout the plant life	In absence of regular replacement of iron/zinc oxide, H_2S starts adsorbing in ZMS and due to its highly electro negativity and gets permanently adsorbed in ZMS, thus fouling it and severely decreasing its performance and life, resulting in very low methane percentage in purified biogas

Table 3 Benchmarks for biogas composition to be utilised as automobile fuel in India

Biogas component	Content
Methane, %, minimum	90
Moisture, mg/m^3, maximum	5
Total sulphur including hydrogen sulphide, mg/m^3, maximum	20
$CO_2 + N_2 + O_2$, %, maximum (v/v)	10
Carbon dioxide, %, maximum (v/v) (when intended for filling in cylinders)	4.0
Oxygen, %, maximum (v/v)	0.5

Source: Bureau of Indian Standard: Biogas (Biomethane)—Specification; 16087: 2016

4 Liquid Biomethane (LBM) as a Vehicle Fuel

Combustion of liquid methane offers significant environmental benefits compared to diesel fuels: no particle emissions (particulate matter, soot); no SO_x emissions (sulphur oxide); 80–90% less NO_x emissions (nitrogen oxides); > 90% less CO_2 emissions (with liquid biogas/LBG fuel); and 20–25% less CO_2 emissions (with liquid natural gas/LNG fuel). Most of the European countries are very serious about climate change, and they have introduced the policy to replace the 10% of the conventional fuels with the biofuels in the transport sector, by 2020 (Pellegrini et al. 2018). The blending of pure biomethane with natural gas is found to be an economical way to cut down GHG emission, in vehicular applications. In contrast, 20% blending of biomethane with natural gas can produce 39% savings of greenhouse gases (GHG) emission, in comparison of gasoline on the well-to-wheel basis. Biomethane as CBG and LBM can be utilised for fulfilling the fuel requirements of very different vehicles. Furthermore, CBG is a renewable alternative of petrol, in lightweight vehicles, while LBM is a substitute of LNG in heavy-duty vehicles (both of road and of the sea) (EBA 2016).

5 Upgradation of Biogas as Liquid Biomethane (LBM)

Raw biogas can be upgraded to liquid biomethane (LBM) in two ways; the first one is a single step cryogenic process, in which CO_2 removal and CH_4 liquefaction are performed simultaneously. In the second method, raw biogas is upgraded using upgrading technologies (water scrubbing, chemical absorption, pressure swing adsorption) and subsequently liquefied at a cryogenic temperature using small-scale liquefaction plant (Pellegrini et al. 2018; Faramawy et al. 2016). The different constituents of the raw biogas are separated on the basis of their different liquefaction temperature (Yousef et al. 2018). The temperature of raw biogas is decreased sequentially, and liquefied impurities of biogas are removed at different point of temperature (Tan et al. 2017). Generally, the first point is at a temperature of −25 °C, where moisture, H_2S and siloxanes are separated successfully. At the second point, −55 °C, a small fraction of carbon dioxide is liquefied, and gradual fall in temperature till −85 °C facilitates the complete removal of CO_2 in the form of solid CO_2 (Riva et al. 2014).

Tan et al. (2017) compared the cryogenic technologies available for biogas upgrading and suggested that cryogenic technique with carbon capture can be a promising technology for upgrading the biogas. All the cryogenic systems were classified in three categories: the flash liquefaction system, distillation system and liquefaction combined with de-sublimation system. In flash liquefaction system, the biogas is first compressed and then cooled to remove water. The gas stream again cooled in the heat exchanger to obtain liquid CO_2 and vapour of CH_4. In another study, in the liquefaction system, raw biogas with the composition of CH_4 and CO_2 with 65% (v/v) and 35% (v/v) was purified to 90.8% of CO_2 (v/v) (Li et al. 2017). The distillation system was found to give the purified CH_4 and CO_2 with 94.5% (mol.%) and 99.7% (mol.%), respectively, from the biogas consisting 60% (mol.%) CH_4 and 40% (mol.%) CO_2 (Yousef et al. 2016). The liquefaction combined with distillation was reported to give purified CH_4 more than 97% (Hagen et al. 2001). Berg (2017) reported the technology developed by Cryo Pur to upgrade the biogas as BioLNG containing CH_4 approximately 99.4%. This energy-efficient system is based on a combination of cryogenic upgradation and biomethane liquefaction. Spitoni et al. (2016) evaluated a novel cryogenic separation system which recovers liquefied CO_2 and produces liquefied biomethane simultaneously. Yousef et al. (2017) proposed a biogas upgrading method for upgrading biogas by separating CO_2 at low temperatures with low energy penalty. Chang et al. (2009) proposed a simple and novel approach using cryogenic heat exchanger for filtering out the CO_2 as a frost, from landfill gas. They introduced an analytical model for assessing the distribution of CO_2 accretion size of the heat exchanger. The system developed by Pentair Haffmans (2018) can recover 100% of the CH_4 by separating 100% pure CO_2 and sending all impurities including the CH_4 back to the membrane system. Jonsson and dan Westman (2011) reported the effect of variation in different operational parameters such as temperature, pressure, and mass flow rate on the de-sublimation of CO_2 inside a plate heat exchanger. Pellegrini et al. (2018) studied the process of liquefying biogas into biomethane, via cryogenic/low-temperature upgrading technologies used for CO_2 removal from natural gas. Dual pressure low-temperature distillation process, Ryan-Holmes, and the anti-sublimation process were utilised for the removal of CO_2. The Ryan-Holmes process and dual pressure low-temperature process removes CO_2 in liquid form while anti sublimation separate in dry ice form. Nachtmann et al. (2017) proposed a new technique based on low-temperature liquefaction system to separate out CO_2 in dry ice form and CH_4 in liquid form, at normal pressure. They suggested that dry ice can be sold for improving the economic feasibility of the system.

5.1 Cryogenic Technologies for Production of Liquid Biomethane (LBM)

5.1.1 Cryogenic Separation Technology

The equipments used in basic cryogenic technology are mainly compressors, turbines, cooling devices, distillation column, etc. as shown in Fig. 1. Initially the raw biogas is dried to avoid the freezing problem. The dried biogas is compressed to 17–26 bar and cooled at −26 °C using a cooling device (Bauer et al. 2013). Further, it is compressed to 80 bar and cooled to −55 °C to −85 °C. Maximum CO_2 is converted to liquid CO_2 at −55 °C and the remaining CO_2 is solidified at −85 °C. The distillation column separates CH_4 and CO_2, and these products are collected in the pure form. The final product of this process contains CH_4 with the purity more than 97% (Andriani et al. 2014; Kapoor et al. 2019).

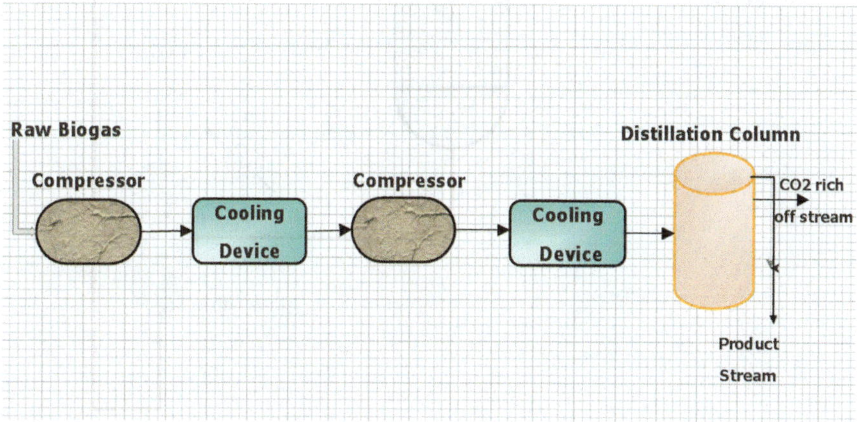

Fig. 1 Schematic diagram of cryogenic separation process

5.1.2 Cryogenic Distillation Technology

Generally, the cryogenic distillation process is less in the application for biogas upgrading, due to its high energy consumption (Langè et al. 2015). Several studies proposed the techniques to reduce the energy consumption, and most of them were based on intensification and integration techniques coupled with hybrid system (Zanganeh et al. 2009; Li et al. 2013; Xu et al. 2014; Ebrahimzadeh et al. 2016). Maqsood et al. (2017) achieved 70% improvement by combining intensification and hybrid cryogenic distillation process, in the process of separation of natural gas. This technique was not used for biogas upgradation and only tested for removal of CO_2 from natural gas (Maqsood et al. 2014a, b, c). Therefore, there is a scope to utilise this process for biogas upgradation. Figure 2 shows the schematic system used for the purification of raw natural gas. In this system raw natural gas is cooled in two single steps prior to sending it to distillation column. The distillation column contains two parts: upper and lower that separate two final products, i.e., CO_2 and

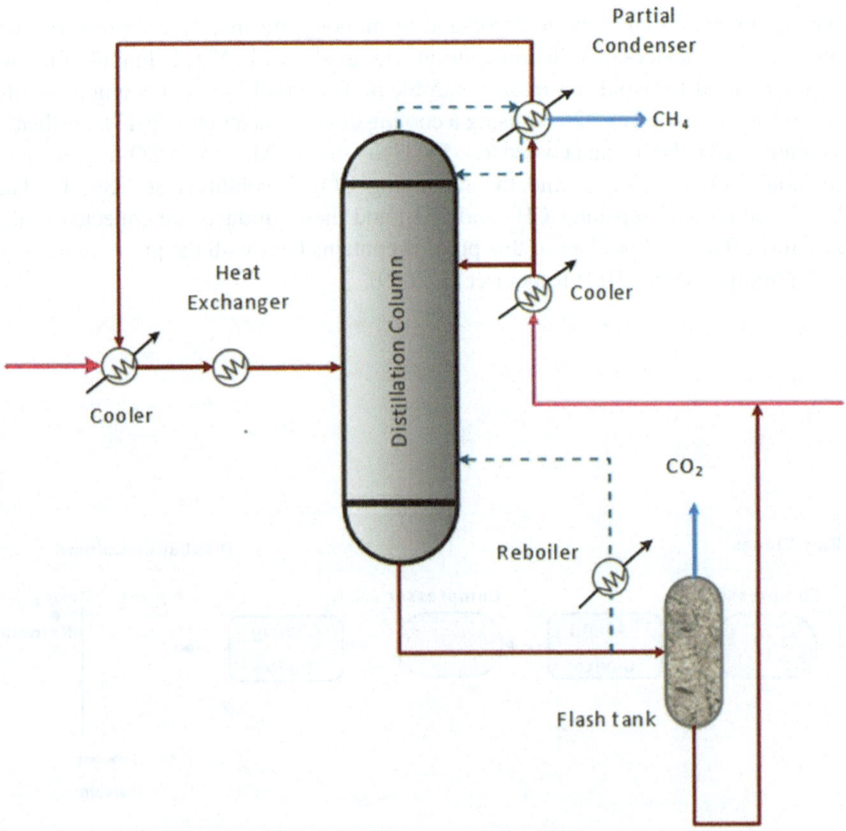

Fig. 2 Schematic diagram for cryogenic distillation process (Baena-Moreno et al. 2019)

CH_4. The product at the top of the column is the high purity vapour of methane, which is taken out from the column by partial condensation process. The bottom of the column contains high purity carbon dioxide, from which major part is extracted and can be sold for enhancing the economics of the system, while the minor part can be reused in the distillation column to maintain the vaporization heat (Baena-Moreno et al. 2019).

5.1.3 Cryogenic Packed Bed Technology

Presently, cryogenic packed bed technology is used to purifying raw natural gas by removing traces of CO_2. However, some researcher tested this technology with numerical simulation for upgrading raw biogas and found positive results. Tuinier and Van Sint Annaland (2012) investigated the base cryogenic packed bed system (Fig. 3), for upgrading biogas, with numerical simulation and also proposed the modified system (Fig. 4) by improving the demerits of the base system. The improved system (with the reverse flow) was found to be more energy efficient with energy consumption of 263.4 kW compared to the base system with energy consumption of 390.7 kW. The requirement of very low temperature (−150 °C) to remove hydrogen sulphide at initial step makes this technology energy intensive. Furthermore, the loss of latent heat during the upgradation process requires thermal insulation facilities to minimize energy losses (Tuinier and Van Sint Annaland 2012; Tuinier et al. 2011). Several studies have been conducted to remove CO_2 from natural gas and with some modification; these techniques can be used to upgrade biogas (Baena-Moreno et al. 2019). In addition, Ali et al. (2014) studied the effect of variation of different working parameters such as temperature, feed flow rate and feed flow composition on counter current switched cryogenic packed bed system. They concluded that this reverse configuration could be suitable for purifying the natural gas with high CO_2 content, and this facility makes this technology favourable for biogas upgrading. In a further study, Ali et al. (2018) investigated the effect of variation in pressure, temperature, and raw gas composition, for optimizing the energy consumption and reducing the hydrocarbon losses, with numerical simulation, in a cryogenic packed bed system. They reported the purity of product up to 94% with hydrocarbon losses of 16%. Finally, they compared their earlier experimental results with recent results from simulation and found better consistency.

5.1.4 Anti-sublimation (AnSU) Process

The anti-sublimation process is utilised to obtain liquefied CO_2 from flue gas. Pan et al. (2013) explored the five stages (as shown in Fig. 5) of this process as follows:

- In the first stage, flue gas is cleaned up and then cooled at −40 °C, followed by removal of trace pollutant and moisture.

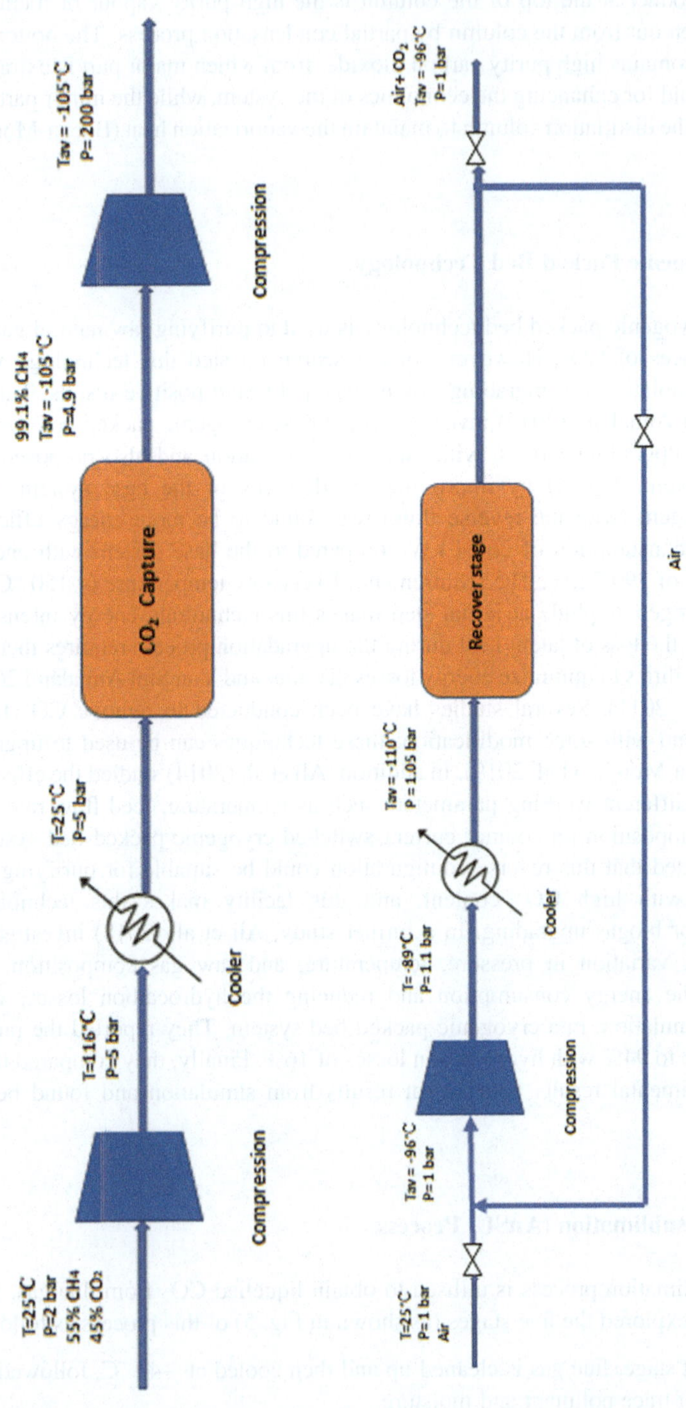

Fig. 3 Basic requirement for case of cryogenic packed bed system (Tuinier and Van Sint Annaland 2012; Tuinier et al 2011)

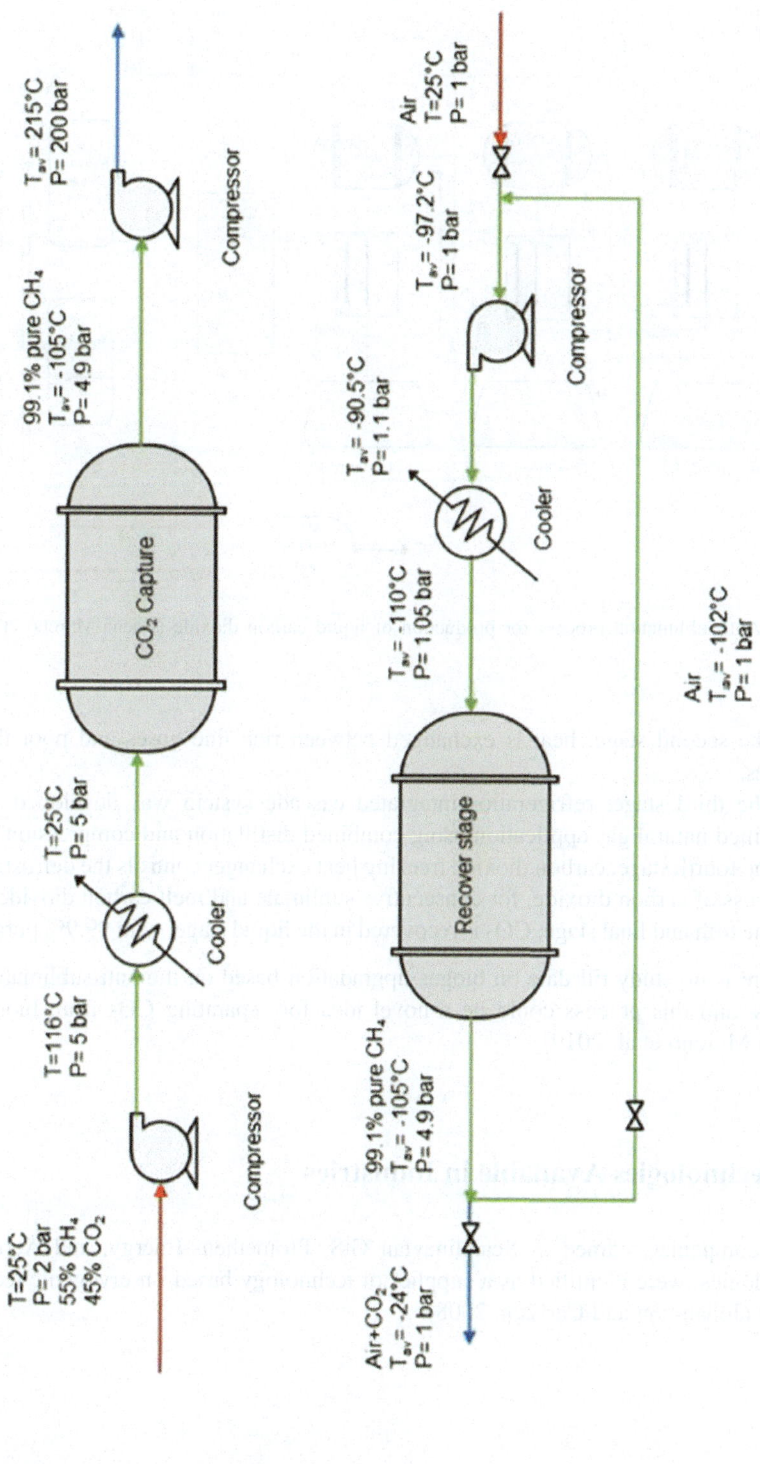

Fig. 4 Modified scenario for cryogenic packed bed system (Tunier and Van Sint Annaland 2012; Tunier et al. 2011)

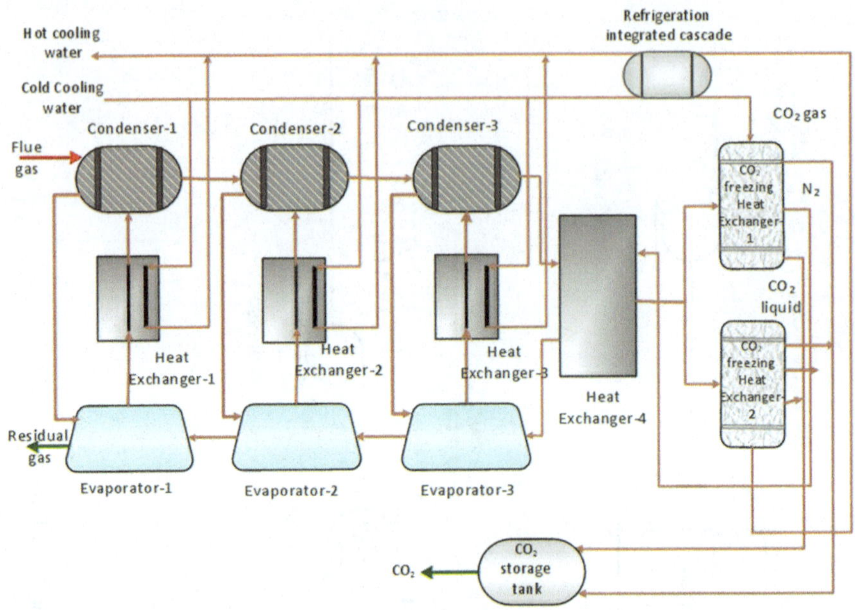

Fig. 5 Anti-sublimation process for production of liquid carbon dioxide (Baena-Moreno et al. 2019)

- In the second stage, heat is exchanged between rich flue gases and poor flue gases.
- In the third stage, refrigeration-integrated cascade system was developed for liquified natural gas application using combined distillation and compression.
- In the fourth stage, carbon dioxide freezing heat exchanger controls the defrosting process of carbon dioxide, for consecutive sublimate and melt carbon dioxide.
- In the fifth and final stage, CO_2 is recovered in the liquid stage, with 99.9% purity.

There is no study till date on biogas upgradation based on the anti-sublimation process, and this process could be a novel idea for separating CO_2 from biogas (Baena-Moreno et al. 2019).

6 Technologies Available in Industries

Three companies, named as Scandinavian GtS, Prometheus-Energy, and Acrion technologies, were identified as a supplier of technology based on cryogenic technology (Johansson and Carlsson 2008).

6.1 Scandinavian GtS

Scandinavian GtS Company is a collaboration of Dutch company Gastreatment Services (GtS) and the Swedish company Scandinavian Biogas. The schematic diagram of the concept used by Scandinavian GtS is shown in Fig. 6. In the first module, the raw biogas gas is cooled to +6 °C to condense most of the contamination and moisture. In module II, the biogas is again cooled to approximately −25 °C to separate the water as ice and condensate siloxanes. Remaining traces of siloxanes and hydrogen sulphide are removed with of SOXSIA®filter. In module III, the gas is chilled to −78 °C to freeze out clean CO_2 in dry form. The liquid CO_2 coming out from the module III can be stored for its further application as refrigerant. Biogas obtained from the module III is clean and moisture free. It can be utilised as compressed biogas (CBG) after compression or it can be further liquified in module IV to obtain liquified biogas (LBG). In module IV the traces of N_2 are separated out and LBG is obtained with methane content more than 99.0% at the temperature of −190 °C (Johansson and Carlsson 2008).

6.2 Prometheus-Energy Process

Prometheus-Energy is an American based company which produces liquid biomethane (LBM) with more than 97.0% pure CH_4. Cost and manufacturing time are being minimized with the integration of different technologies using the basic equipment. An overlook of the modular approach used by this company is shown in Fig. 7. In modules I and II, the gas is compressed and impurities like water and sulphur, and a trace amount of other gases are removed. In module III, instrument air skid is utilised for obtaining high quality dry gas. In module IV, CO_2 is separated from the gas by using a proprietary cryogenic freezing method. Further, CO_2 is removed to the atmosphere. In modules V and VI, the percentage of CH_4 is boosted by dynamic flash evaporation of nitrogen. In module VII, the refrigerant module facilitates cooling to the process through a closed Brayton nitrogen cycle. In 2006, a commercial capacity plant was established to produce LBM from liquefied fuel gas (LFG) that produces 6700 Nm^3/day. The produced LBM was used to run more than 200 buses in California (Johansson and Carlsson 2008).

6.3 Acrion Technologies

Acrion technology is American-based small company that works on the purification of gases containing CO_2 more than 10%. This company uses conventional technology combined with cryogenic technology (Fig. 8). LBM is produced by cleaning raw biogas with two membranes followed by liquefaction, in a distillation column.

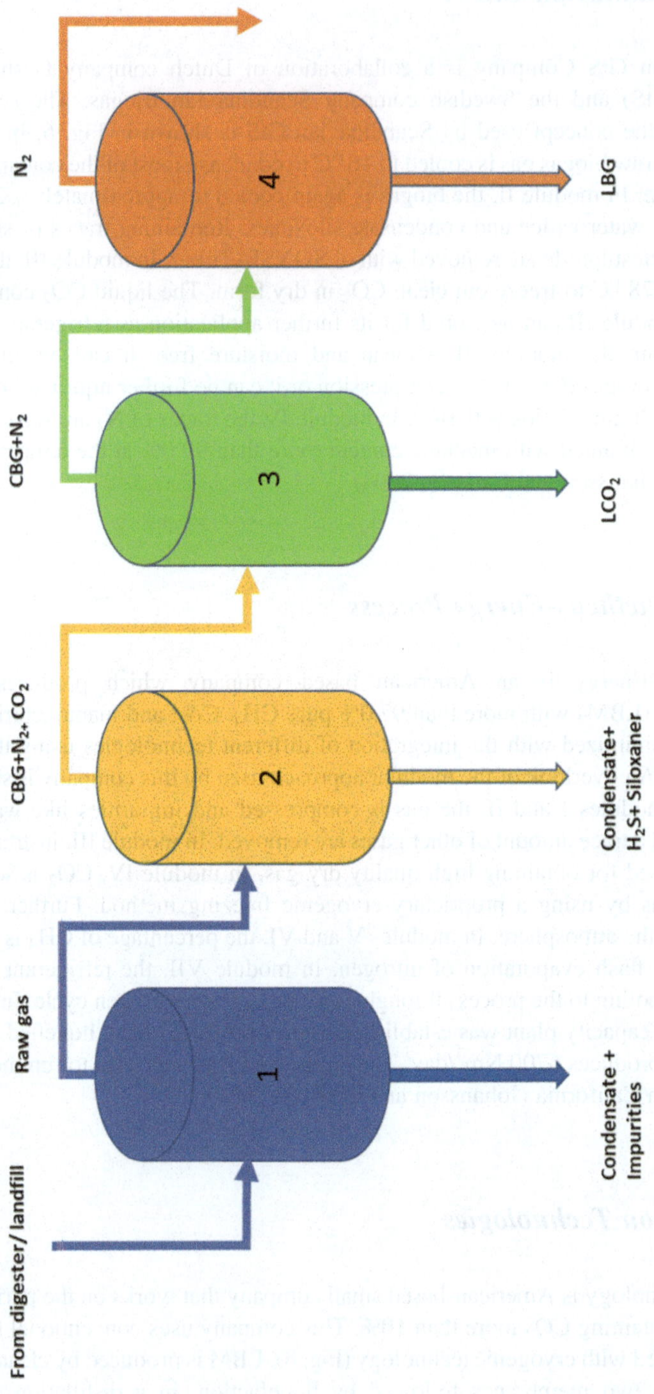

Fig. 6 Scandinavian GtS concept to produce LBG from raw biogas (Johansson and Carlsson 2008)

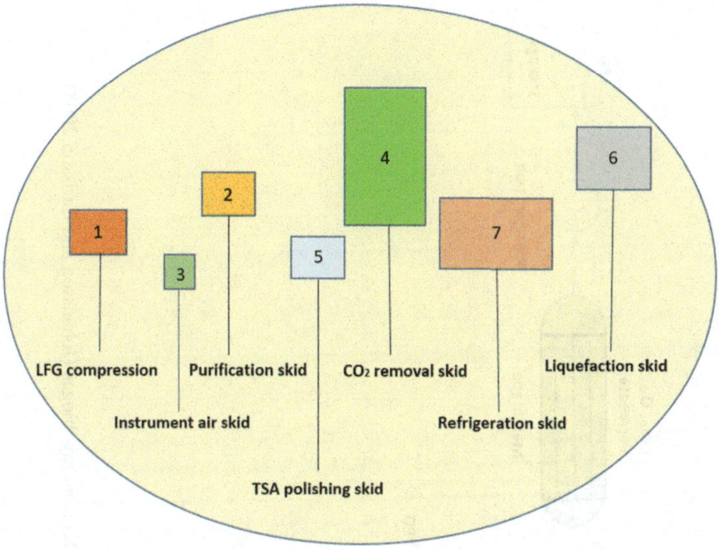

Fig. 7 Schematic diagram of Prometheus-Energy's modular approach (Johansson and Carlsson 2008)

Further, the gas is compressed, de-sulphured and heated before passing through CO_2 wash. A Sulfa Treat which contains iron oxide that removes the hydrogen sulphide. Contamination of CO_2 in the gas stream is condensed in CO_2 Wash column. A major part of condensed CO_2 is collected at top of the column as liquid CO_2. The remaining condensed CO_2 is used to remove the traces such as siloxanes, non-methane organics and halogenated compounds. Gas stream with the composition of CH_4, CO_2, O_2, N_2 and 25% of CO_2 exits from the top of the column and enters into the membrane module. All the O_2 and most of the CO_2 (23–24%) is separated in the membrane module and remaining 1–2% CO_2 with molecular sieve. The remaining traces of N_2 are flashed out in refrigeration plant. In 2005 Acrion examined this technology for producing LBG from landfill gas and obtained the gas with 99.2% of the CH_4 and liquid CO_2, on pilot scale plant. This technology has not been used in the full-scale commercial processes till the date and can be novel method to upgrade biogas (Johansson and Carlsson 2008).

7 Challenges with Technologies to Produce Biomethane

The process by which gas is upgraded and liquefied utilises a complex procedure and number of equipment such as compressor, heat exchangers, coolers, distillation unit, etc. It increases the capital cost which in turn increases the market price of biomethane (Kapoor et al. 2019). Cryogenic technology depends upon the accessibility and form of cryogenic sources. In contrast, packed bed cryogenic technology

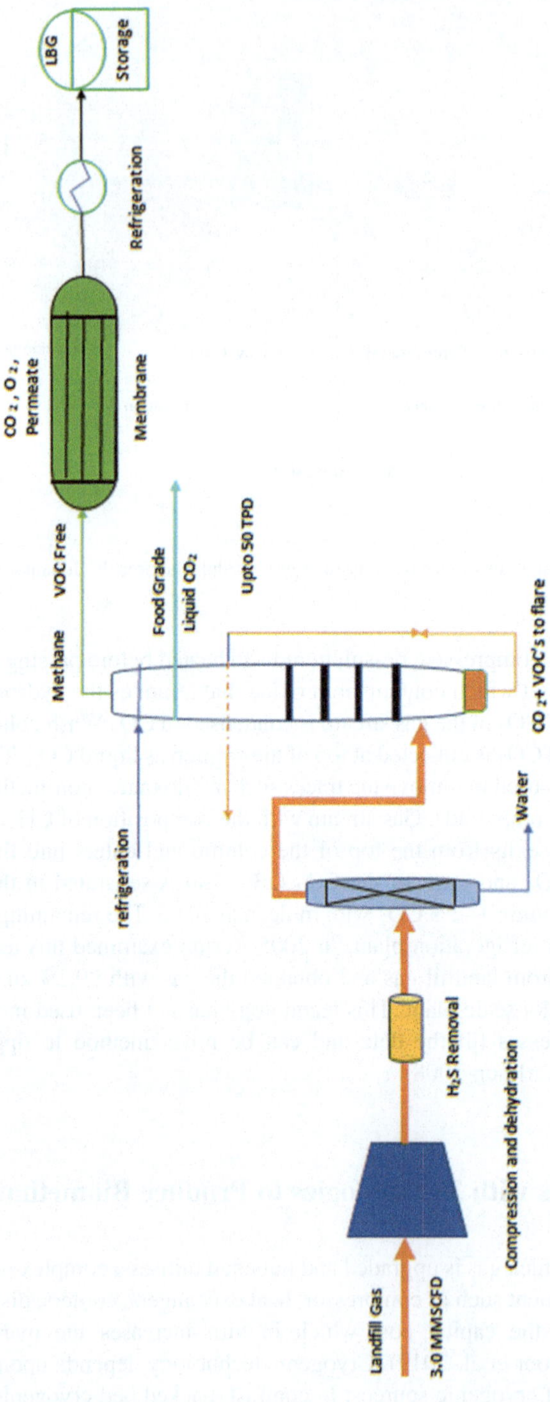

Fig. 8 Process flow diagram of process utilised in Acrion's technology for the production of LBG from raw biogas (Johansson and Carlsson 2008)

depends upon the availability of liquid N_2. Lowering the cost of CO_2 capturing technology is still the challenge both in biogas upgrading systems and other CO_2 capturing technologies. Furthermore, cryogenic technologies involve low-temperature processes which are sensitive to temperature and other environmental parameters. Plants located in cold climate zone are found to be more efficient than in hot climate zone. The minor compounds (H_2S and siloxanes) present in raw biogas are reported as causes of corrosion in plant equipment. This challenge could be solved by using the modified building materials, but it increases the capital cost of the plant (Baena-Moreno et al. 2019).

8 Challenges with Storage Facility and Their Solutions

The variation of temperature and pressure in a cryogenic tank can boil off-gasses. Boil-off is defined as the phenomenon of gas release from the liquefied fuel (LPG/LNG) due to the heat addition during filling and storing operation. This heat can be gained by surroundings along with boil-off (Chen et al. 2004).

Global Warming Potential (GWP) reports that methane emissions will affects the climate 72 times more than CO_2 in a time span of 20 years. As an application of LNG/LBM increases, boil-off gas (BOG) across the LNG/LBM supply chain will be more and if the venting of these gases is not prevented, and then this could arise a question on overall environmental benefits. Determining and quantifying the BOG generation rates across the LNG/LBM supply chain is the first step needed to satisfy the criteria of near-zero emissions from LNG/LBM facilities (Sharafian et al. 2016).

LBM is an inflammable liquid; therefore the transportation vehicle and re-fuelling stations must be equipped with safety facilities such as quick melting plug, an emergency shutdown switch, fireproofing equipment, a nitrogen flushing and filling system, grounding, and a fire hydrant (Lin et al. 2010).

Whether LBM is transported, or it is used as fuel for the vehicle, various issues can arise due to heat being added to LBM, which may be responsible for boil-off gas. Following are the sources that can complicate the process of storing and filling of LBM:

- Heat can leak through the shell of the storage tank.
- Biomethane may return from vehicle tank to bulk storage tank.
- Heat can leak through fuel hose and the dispenser.
- Heat release by the BOG.
- Fuel loss during storage process without fuelling the vehicles.
- Venting rate variation with the number of vehicles that are being filled.
- Due to the high surface area to volume ratio BOG generation rate is high in small tanks.

After fuelling few vehicles, the pressure inside the LBM tank increases the due formation of boil-off gasses. To reduce the pressure, gas must be re-liquefied or vented. In BOG management, it is preferable to use BOG as fuel gas instead of just venting the gas (Chen et al. 2004).

The phenomenon is known as "Rollover" can also be found in the LBM storage tank. When the new liquid is added to the tank, a layer of slightly different temperature is formed within the storage tank. This difference in temperature causes a sudden release of fuel vapours from a storage tank which is known as Rollover. A large number of gasses released during Rollover could develop very high pressure within storage that could lead to a hazardous accident. To prevent the tank from rollover, the storage tank can be equipped with two types of the nozzle: top filling nozzle and the bottom filling nozzle (Kobayashi et al. 2012). Top filling nozzle is used when the density of product LBM is higher than stored LBM; otherwise, the bottom filling nozzle is used. Storing LBM in this manner allows natural mixing and reduces the formation of boil-off gas. Another possible solution for overcoming the Rollover problem is to maintain homogenous temperature and densities by using a recirculation pump within the tank (Bashiri 2002).

9 Conclusions

The chapter has discussed the various cryogenic technologies used for upgradation of biogas as well as natural gas. Liquid natural gas (LNG) and liquid biomethane (LBM) are compatible in terms of energy intensity as the compositions of both the fuels are the same. Liquid natural gas (LNG) is emerging as the alternative fuel of petrol and diesel in many applications, and there is opportunity to introduce liquid biomethane (LBM) as an alternative of liquid natural gas. The liquid biomethane (LBM) can be used for all the applications with existing liquid natural gas facilities. There are similar challenges in the production as well as applications of liquid biomethane, as are in the production and application of liquid natural gas. The processes involved in the production of liquid biomethane (LBM) are energy intensive and costly. Therefore, there is need to invent new technologies as well as modification in existing technologies to optimize the energy and cost for better energy and price compatibility of liquid biomethane.

References

Ali A, Maqsood K, Syahera N, Shariff ABM, Ganguly S (2014) Energy minimization in cryogenic packed beds during purification of natural gas with high co2 content. Chem Eng Technol 37 (10):1675–1685

Ali A, Maqsood K, Shin LP, Sellappah V, Garg S, Shariff ABM, Ganguly S (2018) Synthesis and mixed integer programming based optimization of cryogenic packed bed pipeline network for purification of natural gas. J Clean Prod 171:795–810

Andriani D, Wresta A, Atmaja TD, Saepudin A (2014) A review on optimization production and upgrading biogas through CO2 removal using various techniques. Appl Biochem Biotechnol 172(4):1909–1928

Baena-Moreno FM, Rodríguez-Galán M, Vega F, Vilches LF, Navarrete B, Zhang Z (2019) Biogas upgrading by cryogenic techniques. Environ Chem Lett 17(3):1251–1261
Bashiri A (2002) Modeling and simulation of rollover in LNG storage tanks, vol 2002
Bauer F, Persson T, Hulteberg C, Tamm D (2013) Biogas upgrading - technology overview, comparison and perspectives for the future. Biofuels Bioprod Biorefining 7(5):499–511
Berg B (2017) Efficient liquid biomethane production with cryogenic upgrading
Chang H-M, Chung MJ, Park SB (2009) Cryogenic heat-exchanger design for freeze-out removal of carbon dioxide from landfill gas. J Therm Sci Technol 4:362–371
Chen QS, Wegrzyn J, Prasad V (2004) Analysis of temperature and pressure changes in liquefied natural gas (LNG) cryogenic tanks. Cryogenics (Guildf) 44(10):701–709
EBA (2016) Biomethane in transport
Ebrahimzadeh E, Matagi J, Fazlollahi F, Baxter LL (2016) Alternative extractive distillation system for CO2-ethane azeotrope separation in enhanced oil recovery processes. Appl Therm Eng 96:39–47
Faramawy S, Zaki T, Sakr AAE (2016) Natural gas origin, composition, and processing: a review. J Nat Gas Sci Eng 34:34–54
Kobayashi M, Koyama K, Ebato K, et al (2012) Current status of mixed LNG storage technology development. In: International Gas Union World Gas Conference Paper
Haffmans P (2018) Haffmans biogas upgrading technology. https://foodandbeverage.pentair.com/en/products/haffmans-biogas-upgrading-technology. Accessed 14 May 2019
Hagen M, Polman E, Myken A, Jensen J, Jönsson O, Biomil A, Dahl A (2001) Adding gas from biomass to the gas grid
Johansson N, Carlsson C (2008) Production of liquid biogas, LBG, with cryogenic and conventional upgrading technology – description of systems and evaluations of energy balances. Lund University
Jonsson J, dan Westman S (2011) Cryogenic biogas upgrading using plate heat exchangers
Kadam R, Panwar NL (2017) Recent advancement in biogas enrichment and its applications. Renew Sustain Energy Rev 73:892–903
Kapoor R, Ghosh P, Kumar M, Vijay VK (2019) Evaluation of biogas upgrading technologies and future perspectives: a review. Environ Sci Pollut Res 26(12):11631–11661
Kumar S, Paritosh K, Pareek N, Chawade A, Vivekanand V (2018) De-construction of major Indian cereal crop residues through chemical pretreatment for improved biogas production: an overview. Renew Sustain Energy Rev 90:160–170
Kumar S, Gandhi P, Yadav M, Paritosh K, Pareek N, Vivekanand V (2019) Weak alkaline treatment of wheat and pearl millet straw for enhanced biogas production and its economic analysis. Renew Energy 139:753–764
Langè S, Pellegrini LA, Vergani P, Lo Savio M (2015) Energy and economic analysis of a new low-temperature distillation process for the upgrading of high-CO2 content natural gas streams. Ind Eng Chem Res 54(40):9770–9782
Li H, Hu Y, Ditaranto M, Willson D, Yan J (2013) Optimization of cryogenic CO2purification for oxy-coal combustion. Energy Procedia 37:1341–1347
Li H, Tan Y, Ditaranto M, Yan J, Yu Z (2017) Capturing CO2 from biogas plants. Energy Procedia 114:6030–6035
Lin W, Zhang N, Gu A (2010) LNG (liquefied natural gas): a necessary part in China's future energy infrastructure. Energy 35(11):4383–4391
Maqsood K, Ali A, Shariff ABM, Ganguly S (2014a) Synthesis of conventional and hybrid cryogenic distillation sequence for purification of natural gas. J Appl Sci 14(21):2722–2729
Maqsood K, Mullick A, Ali A, Kargupta K, Ganguly S (2014b) Cryogenic carbon dioxide separation from natural gas: a review based on conventional and novel emerging technologies. Rev Chem Eng 30(5):453–477
Maqsood K, Pal J, Turunawarasu D, Pal AJ, Ganguly S (2014c) Performance enhancement and energy reduction using hybrid cryogenic distillation networks for purification of natural gas with high CO2 content. Korean J Chem Eng 31(7):1120–1135

Maqsood K, Ali A, Shariff ABM, Ganguly S (2017) Process intensification using mixed sequential and integrated hybrid cryogenic distillation network for purification of high CO2 natural gas. Chem Eng Res Des 117:414–438

Miltner M, Makaruk A, Harasek M (2017) Review on available biogas upgrading technologies and innovations towards advanced solutions. J Clean Prod 161(2017):1329–1337

Nachtmann K, Baum S, Fuchsz M, Falk O, Hofmann J (2017) Efficient storage and mobile use of biogas as liquid biomethane. Landtechnik 72(4):179–201

Pellegrini LA, De Guido G, Langé S (2018) Biogas to liquefied biomethane via cryogenic upgrading technologies. Renew Energy 124:75–83

Pan X, Clodic D, Toubassy J (2013) CO2 capture by antisublimation process and its technical economic analysis. Greenh Gases Sci Technol 3:8–20.

Riva M, Campestrini M, Toubassy J, Clodic D, Stringari P (2014) Solid-liquid-vapor equilibrium models for cryogenic biogas upgrading. Ind Eng Chem Res 53(44):17506–17514

Ryckebosch E, Drouillon M, Vervaeren H (2011) Techniques for transformation of biogas to biomethane. Biomass Bioenergy 35(5):1633–1645

Sahota S, Shah G, Ghosh P, Kapoor R, Sengupta S, Singh P, Vijay V, Sahay A, Vijay VK, Thakur IS (2018) Review of trends in biogas upgradation technologies and future perspectives. Bioresour Technol Reports 1:79–88

Sharafian A, Herrera OE, Mérida W (2016) Performance analysis of liquefied natural gas storage tanks in refueling stations. J Nat Gas Sci Eng 36:496–509

Spitoni M, Xiong X, Arteconi A, Bartolini CM, Polonara F, Lin W (2016) Biogas purification and liquefaction by means of a cryogenic upgrading process. Refrig Sci Technol 22:107–115

Spitoni M, Pierantozzi M, Comodi G, Polonara F, Arteconi A (2019) Theoretical evaluation and optimization of a cryogenic technology for carbon dioxide separation and methane liquefaction from biogas. J Nat Gas Sci Eng 62:132–143

Sun Q, Li H, Yan J, Liu L, Yu Z, Yu X (2015) Selection of appropriate biogas upgrading technology-a review of biogas cleaning, upgrading and utilisation. Renew Sust Energ Rev 51:521–532

Tan Y, Nookuea W, Li H, Thorin E, Yan J (2017) Cryogenic technology for biogas upgrading combined with carbon capture-a review of systems and property impacts. Energy Procedia 142:3741–3746

Tuinier MJ, Van Sint Annaland M (2012) Biogas purification using cryogenic packed-bed technology. Ind Eng Chem Res 51(15):5552–5558

Tuinier MJ, Van Sint Annaland M, Kuipers JAM (2011) A novel process for cryogenic CO2 capture using dynamically operated packed beds-An experimental and numerical study. Int J Greenh Gas Control 5(4):694–701

World Energy Outlook (2016) https://webstore.iea.org/world-energy-outlook-2016. Accessed 10 Jan 2020b

World Energy Outlook (2017) https://webstore.iea.org/world-energy-outlook-2017. Accessed 12 Jan 2020a

Xu G, Liang F, Yang Y, Hu Y, Zhang K, Liu W (2014) An improved CO2 separation and purification system based on cryogenic separation and distillation theory. Energies 7 (5):3484–3502

Yousef AMI, Eldrainy YA, El-Maghlany WM, Attia A (2016) Upgrading biogas by a low-temperature CO2 removal technique. Alexandria Eng J 55(2):1143–1150

Yousef AM, Eldrainy YA, El-Maghlany WM, Attia A (2017) Biogas upgrading process via low-temperature CO2liquefaction and separation. J Nat Gas Sci Eng 45:812–824

Yousef AM, El-Maghlany WM, Eldrainy YA, Attia A (2018) New approach for biogas purification using cryogenic separation and distillation process for CO2 capture. Energy 156:328–351

Zanganeh KE, Shafeen A, Salvador C (2009) CO2 capture and development of an advanced pilot-scale cryogenic separation and compression unit. Energy Procedia 1(1):247–252

Nutrient Value of Digestates in Soil Fertility and Crops Productivity

Ayerim Hernández-Almanza, Liliana Londoño-Hernández, Gloria Alicia Martínez-Medina, and Diana Alva-Sánchez

Abstract Organic matter, nitrogen level, ionic strength, and symbiotic relationship between microorganisms as nitrogen bacteria with plants are one of the most significant natural factors controlling soil fertility and can be used for indicators. Nevertheless, the increasing population over the years has demanded large amounts of crops carried out the loss of fertility in soils due to bad conventional cultivation practices using excess fertilizers and pesticides. On the other hand, biogas production is considered one of the most promising alternatives to management of organic residues due this waste can be converted to renewable energy, high added-value material, and organic fertilizers. This residues are called digestates and have some nutrients such as nitrogen, sodium, chloride, and phosphorus among others that can be used like organic fertilizers or soil conditioner. However, the evaluation of real nutritional value as well as the obtaining process and future application is necessary. In this chapter we describe the nutritional value of digestates obtained by biogas production and its impact in soil fertility and crops productivity.

Keywords Crop productivity problems · Digestates · Nutritional value · Biogas production

A. Hernández-Almanza (✉) · D. Alva-Sánchez
School of Biological Science, Universidad Autónoma de Coahuila, Torreón, Coahuila, Mexico
e-mail: ayerim_hernandez@uadec.edu.mx

L. Londoño-Hernández
BIOTICS Group, School of Basic Sciences, Technology and Engineering, Food Chain, Universidad Nacional Abierta y a Distancia UNAD, Boyaca, Colombia

G. A. Martínez-Medina
Food Research Department, Bioprocesses and Bioproducts Group, School of Chemistry, Universidad Autónoma de Coahuila, Saltillo, Coahuila, Mexico

N. Balagurusamy, A. K. Chandel (eds.), *Biogas Production*,
https://doi.org/10.1007/978-3-030-58827-4_15

1 Introduction

The climatic changes, the overexploitation of soils, the decrease of organic matter, and the loss of nutrients are problems that have a strong impact on the fertility of the soils and therefore on crop productivity. These damages have been shown in different countries around the world. Nowadays, governments and various private sector organizations are looking for alternatives to combat this problem. Some of these alternatives include minimum tillage, composting, use of animal manure, and intercalate legume cultivation, among others (Chikopela et al. 2018; Vanlauwe et al. 2020). Coupled whit this, the amount of animal manure produced is greater than limit allowed and represents an important source of ammonia emissions (Valentinuzzi et al. 2020). A possible solution to reduce this environmental damage could be the conversion of animal manure to renewable energy, high added-value material and organic fertilizers through biogas production processes (Guilayn et al. 2019; Waqas et al. 2019). Also, recent reports have been demonstrated that biogas production is an environmentally friendly and low-cost technology which reduce greenhouse gas emission (Valentinuzzi et al. 2020).

The digested residue obtained from biogas production is called digestate and is a nutrient product used as a soil conditioner and biofertilizer (Carlile et al. 2019). There are various reports where the positive effect of digestates on the ground has been tested. Some studies reported in literature will be described in this chapter.

2 Digestates: Obtaining Process and Nutritional Value

Anaerobic digestion is a process widely used for the treatment of organic residues. During anaerobic digestion of manure, conversion results two by-products, liquid and solid digestates that contain significant amounts of minerals such as nitrogen and some potassium, phosphorus, sodium, and chlorine ions, which are essential for plant growth (Carlile et al. 2019; Valentinuzzi et al. 2020).

Slepetiene et al. (2020) evaluated the potential of liquid and solid digestates as fertilizers, and they found that solid digestate is rich in organic carbon; it means that this residue represents a potential for returning carbon to soil. On the other hand, Mórtola et al. (2019) evaluated the composition and pathogen content in a digestate from poultry manure and the effect of its soil application as well as its impact on the growth of *Lactuca sativa*. The authors indicate that the heavy metal and pathogen concentrations in the digestates were below the limit values and the soil presented acceptable characteristics for crop growth. Likewise, Giulio et al. (Cristina et al. 2019) reported the effect of four anaerobic digestates from sewage sludge; the authors indicate that low concentrations of sewage sludge (2.5%) improved germination index of *Lepidium sativum* L. seeds; however, higher dosage showed phytotoxic effects. In recent studies, the use of biofertilizer obtained by this via has proven to be a promising alternative. However, the use of digestates requires an extensive study, mainly about the optimal concentrations, effect on different crops, and improvement of soil fertility among others.

3 Soil Fertility Problems

Soil fertility is described as the ability to produce crops due to the nutrients that exist in the soil like macronutrients and trace elements (Fang et al. 2008; Harmsen 1991). Organic matter, nitrogen level, ionic strength, and symbiotic relationship between microorganisms as nitrogen bacteria with plants are one of the most significant natural factors controlling soil fertility and can be used for indicators (Jacoby et al. 2017; Nájera et al. 2015). Nevertheless, the increasing population over the years has demanded large amounts of crops carried out the loss of fertility in soils due to bad conventional cultivation practices using excess fertilizers and pesticides (Zhou et al. 2019). On other hand, farmers are an important factor in maintaining soil fertility. An example of this is in Cameroon where 91.4% of them maintain fertility using minimum tillage and 18.8% abuse chemical fertilizers (Kome et al. 2018). There are changes in nutrient dynamics with the use of different agricultural activities, when the pH in soil decrease causes a loss of potassium, calcium in soil, C/N affects the growth of all crops, and the type of tillage causes soil pores to clog. The microbial interactions and biogeochemical cycles are affected too (Table 1).

The United Nations Food and Agriculture Organization has estimated 200 million acres of harvested cropland in the United States for field and forage crops. The principal crops that can be lost are barley, rice, potatoes, soya, cotton, and coffee if there is no protection for these crops and soil health is maintained (Deguine et al. 2009; Gomiero 2019). Sirsat et al. (2018) highlights fertility levels to make predictions on fertility map indices in villages and thus help create security policies and proper land management in India. Wang et al. (2019) describe the importance of carrying out an evaluation of the fertility analysis taking into account indicators such

Table 1 Major causes of loss of soil fertility

Principal cause	Effects in soil	Crops	References
Conventional tillage	Declining agricultural productivity The surface layer are affected and physical and biochemical properties	Maize	Kiboi et al. (2019), Wyngaard et al. (2012)
Soil water scarcity	Soil moisture limited Low growth in plants	Rice	Ngetich et al. (2014)
Illegal cultivation	K, Mn, and Si in soil decrease with elevation	Coffee and banana	De Bauw et al. (2016)
Inappropriate fertilization	Degradation, low nutrients, unbalanced ecosystem, decrease the microbial dynamic and high acidity	Rice	Kumar and Yadav (2001)

as Delphi method, multiple linear regression as well as to find a soil with moderate fertility quality in JunXian County according to the TS diffuse neural network model test for rice crops.

4 Crop Productivity Problems

Agricultural activity intensification in favor to attend the worldwide food demand results in highly and increasing soil damage (Banwart 2011), important organic matter, and fertility losses, which directly impact in more elevated investment costs for maintaining the crop productivity (Napoleão et al. 2013).

Soil represents one of the most important factors involved in their successful plant growth, with physicochemical and biological characteristics that could improve or delay their prosperity (Oghode et al. 2015). When soil damage appears, the potential of an agricultural land is diminished, and the main health issues related to loss of health soil are the declining organic matter and the multiple inorganic element deficiency (N, P, K, S, Zn, Fe, and Mn) (Meena et al. 2016). One of the reasons that soil lost the natural properties is the exhaustive cultivation systems that generally cause an important depletion of nutrients from the soil (Khalid et al. 2019) and also some natural phenomena such as extensively rains or winds and also by deforestation procedures (Hurni et al. 2008).

4.1 Growth, Yield, and Quality

Some reports indicate that an infertile or eroded land generate lower productivity rates, typically related to modifications in soil as nutrition depletion and change in water holding capacity, low infiltration rates, microbiota changes, among others (Pimentel et al. 1995).

Even though plants are a group of autotrophic organisms that could take energy and synthesize the prime molecules from a group of chemical species as carbon dioxide, water, and mineral elements in soil for the support of their growth and natural development; deficiencies in some trace elements could impact in the formation of different macromolecules or the intervention of some enzymatic activities generating diverse plant symptoms such as chlorosis, low growth rates, and poor root, tissue, and fruit development (Behboudian et al. 2017; Hopkins and Huner 2008); these characteristics directly impact in the crops qualities and therefore have an economic consequences.

Studies reported by Shunfeng et al. (2018) show how the nutritional status mainly the elemental composition in soil where apple trees are cultivated affects the leaf composition, where Zn deficiencies could be detected in one of the Chinese analyzed regions, recommended a Zn supplementation especially in trees with rosette disease symptoms, and also documented that zinc deficiencies promote an stunted growth and small leaves development in crops (Fu et al. 2015). On the other hand, studies realized by (Wójcik et al. 2019) mentioned that different treatments with supplementation of nutrients during growth stages could enhance flesh firmness. Musacchi and Serra (2018) mention other quality parameters in apple related with agronomical factors as soil nutritional aspects, where not only deficiencies could affect this valuable parameters also the excess; as in the case of nitrogen, considered as one of most valuables components in soils that lead to obtain larger fruit but generate a softer product, with earlier drops and susceptibility to bitter pit; additionally during a long-term and multi-species study from (Cong et al. 2019), demonstrate how the soil composition could affect the herbage yield in grass lands.

4.2 Soil-Borne Diseases

Plant disease attributed to soil-borne pathogens represents a substantial problem at worldwide level for crop development, attached to intensive agriculture practice characterized by poor soils and frequent tillage (Corato 2020), and recorded as one of the most difficult aspects to control, where one strategy to confront this difficulty is enhancing the growth of a group of microorganism which are naturally or not in this ecosystem, viz., bacteria, fungi, and protozoa, and could bring benefits to healthy plant development (Jambhulkar et al. 2015), where specially their heterogenicity exacerbates this benefits (Schloter et al. 2003).

The non-pathogenic soil microbiome allows a set of mechanisms that promote, improve, and support the agricultural systems productivity, with an intervention in the solubilization of some nutritional elements or their nitrogen fixation abilities, biocontrol properties against soil pathogen, and the potential production of different phytohormones (Akhilesh Kumar and Verma 2019).

Nowadays, demands high production rates in agriculture matter generates important weakness in soil balance with respect to physicochemical and microbiome properties; nevertheless some studies are conducted where when different organisms are added at soil level could ameliorate some plant diseases, as (Sennoi et al. 2013), that study the effects against Southern stem root disease caused by pathogen fungi *Sclerotium rolfsii* in Jerusalem artichoke cultivar with a fungus *Trichoderma harzianum* inoculation in soils; also (Hernández-Montiel et al. 2013) study the effect of a combination of *Pseudomonas* sp., and fungal consortia in the antifungal effect against soil pathogen *Fusarium oxysporum* during in vitro *Carica papaya* culture with result in a protection and reduction of the disease caused, while the direct addition of microorganism in soil could decrease plant diseases also the microbiome in soil could be strengthened by other practices as intercropping like (Li et al. 2018) indicate, where in Peanut cultivars *Fusarium* pathogens are suppressed by alternating with the medicinal plant *Atractylodes lancea*, this study attribute this positive effects to changes in soil microorganisms specially in fungal communities.

This remarks how the edaphic factors represent one of the cornerstones during crop culture with the novel point of view in agriculture issues, trying to ensure the food security at worldwide level, for display food derived from crops in enough quantities and with higher quality attributes.

5 Digestate as an Alternative for Soil and Crop Fertility Problems

In order to provide the necessary amount of nutrients (mainly availability of nitrogen, phosphorus, and potassium), to improve the productivity and quality of crops, and to decrease the low fertility of overexploited soils, traditionally mineral fertilizers have been used; however, these can bring some negative environmental effects. Therefore, in recent years the use of organic fertilizers has been proposed to mitigate the environmental impact of mineral fertilizers. Some studies have shown the benefits that the use of digestate can bring to the soil in comparison with mineral fertilizers. While mineral fertilizers present a greater risk of leaching by contaminating water sources, the digestate does not accumulate, is rapidly consumed, is a source of nitrogen, and does not alter the microbial balance of the soils where it is applied (Tsachidou et al. 2019).

In this sense, research evaluating digestate as an alternative to traditional fertilizers and its effects on crop improvement has increased. Alburquerque et al. (2012) evaluated the effect of using digestate during two growing seasons on watermelon and cauliflower crops, compared with the use of conventional mineral fertilizer and organic fertilizer (cattle manure). They found that the addition of digestate provided the soil with nitrogen and phosphorus in the short term, as well as improved biological properties (microbial biomass and enzymatic activities). In terms of yield, it had similar effects to mineral fertilizer, but only in summertime and not in winter, this possibly due to the low temperatures, which cause the microbial activity to be slow. However, due to its efficiency, it is proposed to be used for short-cycle horticultural crops.

The digestate from the anaerobic digestion of organic waste has a high organic matter content and a significant amount of ammonium, characteristics that make its use as an organic fertilizer interesting. The digestate has been used in wheat and corn soils, and it has had a good behavior improving the chemical variables of the soil, the microbial activity, and the development of the plant (García-Sánchez et al. 2015; Verdi et al. 2019).

Experiments on crop soil have shown the efficiency of using digestate as a fertilizer (Slepetiene et al. 2020). To evaluate the efficiency of these and other fertilizers, it is necessary to know the effect of these on some soil properties such as pH, electrical conductivity, the proportion of nutrients, and the type of microorganisms present. Gómez-Brandón et al. (2016) evaluated the effect of digestate application on the chemical and microbiological properties of a cultivable soil compared to untreated soil, soil using fertilizer and soil using vermicompost, finding that the addition of digestate favored the ratio of carbon/nitrogen in the soil, which favors nitrification processes and the growth of beneficial microorganisms. Furthermore, after 60 days of treatment, there was no evidence of the presence of *Escherichia coli* bacteria, and the concentration of *Clostridium perfringens* was low. Sapp et al. (2015) evaluated the use of digestate on the structure and diversity of microorganisms found in spring wheat cropping soil. Forty different bacterial phyla were detected with *Proteobacteria*, *Acidobacteria*, and *Actinobacteria* dominating the communities. These communities were not stable during the study but changed over time as the plant grew, indicating their close relationship. The digestates used significantly influenced the growth of the wheat, and it was also found that the *Planctomycetes*, involved in the nitrogen cycle, decreased in the treatments. An important conclusion of this study is that the digestates favor the microbial growth of the soil and the development of the crops. Similar results have been found in soils where *Beta vulgaris altissima* is cultivated, so it has been indicated that digestates could be substitutes for mineral fertilizers, favoring the concentration of nutrients and the development of microorganisms in the soil (Westphal et al. 2016). Also in hydroponic crops, digestates have been shown to be an alternative source of nutrients, replacing mineral fertilizers (Ronga et al. 2019).

Another important feature of fertilizer-treated soils is the ratio of macro- to micronutrients. Koszel and Lorencowicz (2015) used digestate obtained from biogas in an alfalfa crop, evaluating the content of macro-elements and heavy metals. The results indicate that after addition of the digestate there is little increase in pH (7.56–7.63), which does not affect plant growth. There was also an increase in the content of some macro-elements such as phosphorus, potassium, and magnesium, which are essential for the development of the crop. Later, they evaluated the nutritional content of alfalfa leaves and found that the application of the digestate allowed an increase in the mentioned macro-elements, for which the digestate is proposed as an organic fertilizer. Studies carried out in soils for lettuce cultivation have shown that the application of digestate did not alter the pH, nor other important parameters of the soil; likewise, the relation of microorganisms was maintained. However, it is suggested to stabilize the digestate beforehand to avoid a microbiological imbalance (Mortola et al. 2019).

Despite the beneficial effects that the use of digestate as a fertilizer can show, studies are needed to evaluate what is the maximum concentration allowed to be used given the amount of organic elements, salts, minerals, and pathogenic bacteria present in the digestate and that can cause ecotoxicological problems in the soil. Some studies have revealed that a concentration of between 15% and 20% (w/w dm) of digestate in the soil does not present negative effects on other organisms present in the soil, and on the contrary improves the characteristics and properties of these (Pivato et al. 2016). In relation to this issue, some studies have shown that the use of the digestate increases an environmental risk due to the mobility of metals such as aluminum and chrome (Dragicevic et al. 2018).

It is also important to consider the source of the digestate, because its use and potential as a fertilizer depend on it (Cristina et al. 2019; Czekała et al. 2020; Iocoli et al. 2019). In order to determinate the potential use of digestates, Muscolo et al. (2017) evaluated digests from different types of processes on the characteristics, enzymes, and type and concentration of microorganisms found in the soil. The results indicated that there is a significant influence of the type of raw material used to obtain the digestates and that these directly affect the physicochemical and biochemical characteristics of the soil where they are used. Therefore, in order to use the digestate as a fertilizer, it is initially necessary to define the interaction between the digestate, the soil where it will be used and the type of crop.

6 Final Comments

Over the years, the climatic changes, the overexploitation of soils, and the overpopulation have left havoc in the ecosystems, day by day one looks for to counteract these problematic ones. The use of digestates is an alternative that could reduce problems of low soil fertility and at the same time treat organic waste. However, improvements are needed in the procurement processes that allow the development of large-scale digesters to facilitate industrial application.

References

Alburquerque JA, de la Fuente C, Campoy M, Carrasco L, Nájera I, Baixauli C et al (2012) Agricultural use of digestate for horticultural crop production and improvement of soil properties. Eur J Agron 43:119–128. https://doi.org/10.1016/j.eja.2012.06.001

Banwart S (2011) Save our soils. Nature 474:151–152

Behboudian M, Pickering A, Dayan E (2017) Deficiency diseases, principles. Enc Appl Plant Sci 1:481–492. https://doi.org/10.1056/nejm192506251922602

Carlile WR, Raviv M, Prasad M (2019) Organic soilless media components. In: Raviv M, Lieth JH, Bar-Tal A (eds) Soilless culture (Second). Elsevier, Boston, pp 303–378. doi:https://doi.org/10.1016/B978-0-444-63696-6.00008-6

Chikopela L, Kalinda TH, Kuntashula E, Nicolay GL, Kalala D, Chishala BH (2018) Use of innovation platforms in addressing soil fertility challenges: experiences from rural Zambia. Afr J Rural Dev 3(1):597–615. Retrieved from http://afjrd.org/jos/index.php/afjrd/article/view/358

Cong W, Christensen BT, Eriksen J (2019) Soil nutrient levels define herbage yield but not root biomass in a multispecies grass-legume ley. Agric Ecosyst Environ 276:47–54

Cristina G, Camelin E, Pugliese M, Tommasi T, Fino D (2019) Evaluation of anaerobic digestates from sewage sludge as a potential solution for improvement of soil fertility. Waste Manag 99:122–134. https://doi.org/10.1016/j.wasman.2019.08.018

Czekała W, Lewicki A, Pochwatka P, Czekała A, Wojcieszak D, Jóźwiakowski K, Waliszewska H (2020) Digestate management in polish farms as an element of the nutrient cycle. J Clean Prod 242:118454. https://doi.org/10.1016/j.jclepro.2019.118454

De Bauw P, Van Asten P, Jassogne L, Merckx R (2016) Agriculture, ecosystems and environment soil fertility gradients and production constraints for coffee and banana on volcanic mountain slopes in the East African Rift: a case study of Mt. Elgon. Agric Ecosyst Environ 231:166–175. https://doi.org/10.1016/j.agee.2016.06.036

De Corato U (2020) Disease-suppressive compost enhances natural soil suppressiveness against soil- borne plant pathogens: a critical review. Rhizosphere. doi:https://doi.org/10.1016/j.rhisph.2020.100192

Deguine J-P, Ferron P, Russell D (2009) Crop protection from agrochemistry to (U.S). Taylor & Francis Group, LLC

Dragicevic I, Eich-Greatorex S, Sogn TA, Horn SJ, Krogstad T (2018) Use of high metal-containing biogas digestates in cereal production – mobility of chromium and aluminium. J Environ Manag 217:12–22. https://doi.org/10.1016/j.jenvman.2018.03.090

Fang Y, Wang L, Xin Z, Zhao L, An X, Hu Q (2008) Effect of foliar application of zinc, selenium, and iron fertilizers on nutrients concentration and yield of rice grain in China. J Agric Food Chem 56:2079–2084. https://doi.org/10.1021/jf800150z

Fu C, Li M, Zhang Y, Zhang Y, Yan Y, Wang Y (2015) Morphology, photosynthesis, and internal structure alterations in field apple leaves under hidden and acute zinc deficiency. Sci Hortic 193:47–54. https://doi.org/10.1016/j.scienta.2015.06.016

García-Sánchez M, Siles JA, Cajthaml T, García-Romera I, Tlustoš P, Száková J (2015) Effect of digestate and fly ash applications on soil functional properties and microbial communities. Eur J Soil Biol 71:1–12. https://doi.org/10.1016/j.ejsobi.2015.08.004

Gómez-Brandón M, Juárez MFD, Zangerle M, Insam H (2016) Effects of digestate on soil chemical and microbiological properties: a comparative study with compost and vermicompost. J Hazard Mater 302:267–274. https://doi.org/10.1016/j.jhazmat.2015.09.067

Gomiero T (2019) Soil and crop management to save food and enhance food security. In: Galanakis CM (ed) Saving food, vol 2050. Elsevier Inc., pp 33–88. https://doi.org/10.1016/B978-0-12-815357-4.00002-X

Guilayn F, Jimenez J, Rouez M, Crest M, Patureau D (2019) Digestate mechanical separation: efficiency profiles based on anaerobic digestion feedstock and equipment choice. Bioresour Technol 274:180–189. https://doi.org/10.1016/j.biortech.2018.11.090

Harmsen K (1991) Soil fertility problems in the developed and in the developing world. In: Bolt G, Bood M, Hayes H, McBride M, De Strooper EB (eds) Interactions at the soil colloid - soil solution interface. Springer, Dordrecht, pp 493–506. https://doi.org/10.1007/978-94-017-1909-4_14

Hernández-Montiel LG, Rueda-Puente EO, Cordoba-Matson MV, Holguín-Peña JR, Zulueta-Rodríguez R (2013) Mutualistic interaction of rhizobacteria with arbuscular mycorrhizal fungi and its antagonistic effect on Fusarium oxysporum in Carica papaya seedlings. Crop Prot 47:1–6

Hopkins WG, Huner NP (2008) Plant physiology, 4th edn. Wiley, New York, NY

Hurni H, Herweg K, Portner B, Liniger H (2008) Soil erosion and conservation in global agriculture. In: Braimoh AK, Vlek PLG (eds) Land use and soil resources. Springer, Dordrecht, pp 41–71. https://doi.org/10.1007/978-1-4020-6778-5_4

Iocoli GA, Zabaloy MC, Pasdevicelli G, Gómez MA (2019) Use of biogas digestates obtained by anaerobic digestion and co-digestion as fertilizers: Characterization, soil biological activity and growth dynamic of Lactuca sativa L. Sci Total Environ 647:11–19. https://doi.org/10.1016/j.scitotenv.2018.07.444

Jacoby R, Peukert M, Succurro A, Koprivova A (2017) The role of soil microorganisms in plant mineral nutrition – current knowledge and future directions. Front Plant Sci 8:1–19. https://doi.org/10.3389/fpls.2017.01617

Jambhulkar PP, Sharma M, Lakshman D, Sharma P (2015) Natural mechanisms of soil suppressiveness against diseases caused by fusarium, rhizoctonia, pythium, and phytophthora. In: Meghvansi MK, Varma A (eds) Organic amendments and soil suppressiveness in plant disease management. Springer International Publishing, Cham, pp 95–123. https://doi.org/10.1007/978-3-319-23075-7_5

Khalid S, Khan HA, Arif M, Altawaha AR, Adnan M, Fahad S, Parmar B (2019) Organic matter management in cereals based system: symbiosis for improving crop productivity and soil health, pp 67–92

Kiboi M, Ngetich K, Fliessbach A, Muriuki A, Mugendi D (2019) Soil fertility inputs and tillage in fluence on maize crop performance and soil water content in the Central Highlands of Kenya. Agric Water Manag 217:316–331. https://doi.org/10.1016/j.agwat.2019.03.014

Kome GK, Enang RK, Palmer B, Yerima K (2018) Geoderma Regional Knowledge and management of soil fertility by farmers in western Cameroon. Geoderma Reg 13:43–51. https://doi.org/10.1016/j.geodrs.2018.02.001

Koszel M, Lorencowicz E (2015) Agricultural use of biogas digestate as a replacement fertilizers. Agric Agric Sci Proc 7:119–124. https://doi.org/10.1016/j.aaspro.2015.12.004

Kumar A, Verma JP (2019) The role of microbes to improve crop productivity and soil health. In: Achal V, Mukherjee A (eds) Ecological wisdom inspired restoration engineering. EcoWISE, Singapore, pp 249–265

Kumar A, Yadav DS (2001) Long-term effects of fertilizers on the soil fertility and productivity of a rice ± wheat system. J Agron Crop Sci 54:47–54

Li X, De Boer W, Ding C, Zhang T (2018) Suppression of soil-borne Fusarium pathogens of peanut by intercropping with the medicinal herb Atractylodes lancea. Soil Biol Biochem 116 (2017):120–130

Meena BP, Shirale AO, Dotaniya ML, Jha P, Meena AL, Biswas AK, Patra AK (2016) Conservation agriculture: a new paradigm for improving input use efficiency and crop productivity. In: Bisht JK, Meena VS, Mishra PK, Pattanayak A (eds) Conservation agriculture: an approach to combat climate change in Indian Himalaya. Springer Singapore, Singapore, pp 39–69. https://doi.org/10.1007/978-981-10-2558-7_2

Mortola N, Romaniuk R, Cosentino V, Eiza M, Carfagno P, Rizzo P et al (2019) Potential use of a poultry manure digestate as a biofertiliser: evaluation of soil properties and Lactuca sativa growth. Pedosphere 29(1):60–69. https://doi.org/10.1016/S1002-0160(18)60057-8

Musacchi S, Serra S (2018) Apple fruit quality: overview on pre-harvest factors. Sci Hortic 234 (2017):409–430. https://doi.org/10.1016/j.scienta.2017.12.057

Muscolo A, Settineri G, Papalia T, Attinà E, Basile C, Panuccio MR (2017) Anaerobic co-digestion of recalcitrant agricultural wastes: characterizing of biochemical parameters of digestate and its impacts on soil ecosystem. Sci Total Environ 586:746–752. https://doi.org/10.1016/j.scitotenv.2017.02.051

Nájera F, Tapia Y, Baginsky C, Figueroa V, Cabeza R, Salazar O (2015) Evaluation of soil fertility and fertilisation practices for irrigated maize (Zea mays L.) under mediterranean conditions in central Chile. J Soil Sci Plant Nutr 15:84–97

Napoleão TH, Belmonte BR, Pontual EV, de Albuquerque LP, Sá RA, Paiva LM et al (2013) Deleterious effects of Myracrodruon urundeuva leaf extract and lectin on the maize weevil, Sitophilus zeamais (Coleoptera, Curculionidae). J Stored Prod Res 54:26–33. https://doi.org/10.1016/j.jspr.2013.04.002

Ngetich K, Mucheru-muna M, Mugwe J, Shisanya C, Diels J, Mugendi D (2014) Agricultural and Forest Meteorology Length of growing season, rainfall temporal distribution, onset and cessation dates in the Kenyan highlands. Agric For Meteorol 188:24–32. https://doi.org/10.1016/j.agrformet.2013.12.011

Oghode A, Okorogbona M, Adebola PO (2015) Soil fertility and crop productivity in African sustainable agriculture. https://doi.org/10.1007/978-3-319-09132-7

Pimentel D, Harvey C, Resosudarmo P, Sinclair K, Kurz D, Mcnair M et al (1995) Environmental and economic costs of soil erosion and conservation benefits. Science 267:1117–1123

Pivato A, Vanin S, Raga R, Lavagnolo MC, Barausse A, Rieple A et al (2016) Use of digestate from a decentralized on-farm biogas plant as fertilizer in soils: an ecotoxicological study for future indicators in risk and life cycle assessment. Waste Manag 49:378–389. https://doi.org/10.1016/j.wasman.2015.12.009

Ronga D, Setti L, Salvarani C, De Leo R, Bedin E, Pulvirenti A et al (2019) Effects of solid and liquid digestate for hydroponic baby leaf lettuce (Lactuca sativa L.) cultivation. Sci Hortic 244 (March 2018):172–181. https://doi.org/10.1016/j.scienta.2018.09.037

Sapp M, Harrison M, Hany U, Charlton A, Thwaites R (2015) Comparing the effect of digestate and chemical fertiliser on soil bacteria. Appl Soil Ecol 86:1–9. https://doi.org/10.1016/j.apsoil.2014.10.004

Schloter M, Dilly O, Munch JC (2003) Indicators for evaluating soil quality. Agric Ecosyst Environ 98:255–262. https://doi.org/10.1016/S0167-8809(03)00085-9

Sennoi R, Singkham N, Jogloy S, Boonlue S (2013) Biological control of southern stem rot caused by Sclerotium rolfsii using Trichoderma harzianum and arbuscular mycorrhizal fungi on Jerusalem artichoke (Helianthus tuberosus L.). Crop Prot 54:148–153

Shunfeng G, Zhanling Z, Ling P, Qian C, Yuanmao J (2018) Soil nutrient status and leaf nutrient diagnosis in the main apple producing regions in China. Hort Plant J 4(3):89–93. https://doi.org/10.1016/j.hpj.2018.03.009

Sirsat M, Cernadas E, Barro S (2018) Automatic prediction of village-wise soil fertility for several nutrients in India using a wide range of regression methods. Comput Electron Agric 154:120–133. https://doi.org/10.1016/j.compag.2018.08.003

Slepetiene A, Volungevicius J, Jurgutis L, Liaudanskiene I, Amaleviciute-Volunge K, Slepetys J, Ceseviciene J (2020) The potential of digestate as a biofertilizer in eroded soils of Lithuania. Waste Manag 102:441–451. https://doi.org/10.1016/j.wasman.2019.11.008

Tsachidou B, Scheuren M, Gennen J, Debbaut V, Toussaint B, Hissler C et al (2019) Biogas residues in substitution for chemical fertilizers: a comparative study on a grassland in the Walloon Region. Sci Total Environ 666:212–225. https://doi.org/10.1016/j.scitotenv.2019.02.238

Valentinuzzi F, Cavani L, Porfido C, Terzano R, Pii Y, Cesco S et al (2020) The fertilising potential of manure-based biogas fermentation residues: pelleted vs. liquid digestate. Heliyon *6*(2): e03325. https://doi.org/10.1016/j.heliyon.2020.e03325

Vanlauwe B, Chivenge P, Zingore S (2020) Soil fertility maintenance and nutrient management for agricultural transformation. In: Sikora RA, Terry ER, Chitja VJ (eds) Transforming Agriculture in Southern Africa: constraints, technologies, policies and processes. Routledge, Abingdon

Verdi L, Kuikman PJ, Orlandini S, Mancini M, Napoli M, Dalla Marta A (2019) Does the use of digestate to replace mineral fertilizers have less emissions of N 2 O and NH 3 ? Agric For Meteorol 269–270(January):112–118. https://doi.org/10.1016/j.agrformet.2019.02.004

Wang H, Yao L, Huang B, Hu W, Qu M, Zhao Y (2019) Soil & Tillage Research An integrated approach to exploring soil fertility from the perspective of rice (Oryza sativa L.) yields. Soil Tillage Res 194:104322. https://doi.org/10.1016/j.still.2019.104322

Waqas M, Rehan M, Khan M, Nizami A-S (2019) Conversion of food waste to fermentation products. In: Ferranti P, Berry EM, Anderson JR (eds) Encyclopedia of food security and sustainability. Elsevier, Oxford, pp 501–509. https://doi.org/10.1016/B978-0-08-100596-5.22294-4

Westphal A, Kücke M, Heuer H (2016) Soil amendment with digestate from bio-energy fermenters for mitigating damage to Beta vulgaris subspp. by Heterodera schachtii. Appl Soil Ecol 99:129–136. https://doi.org/10.1016/j.apsoil.2015.11.019

Wójcik P, Filipczak J, Wójcik M (2019) Effects of prebloom sprays of tryptophan and zinc on calcium nutrition, yielding and fruit quality of 'Elstar' apple trees. Sci Hortic 246(October 2018):212–216. https://doi.org/10.1016/j.scienta.2018.10.071

Wyngaard N, Echeverría EH, Rozas SRH, Guillermo DA (2012) Soil & Tillage Research Fertilization and tillage effects on soil properties and maize yield in a Southern Pampas Argiudoll. Soil Tillage Res 119:22–30. https://doi.org/10.1016/j.still.2011.12.002

Zhou B, Sun X, Wang D, Ding Z, Li C, Ma W (2019) ScienceDirect integrated agronomic practice increases maize grain yield and nitrogen use efficiency under various soil fertility conditions. Crop J 7:527–538. https://doi.org/10.1016/j.cj.2018.12.005

Part III
Economics of Biogas Technology

Biogas Commercialization: Commercial Players, Key Business Drivers, Potential Market, and Fostering Investment

Shaileshkumar Sawale, Deepak Patil, Chaitanya Joshi, Basavaraj Rachappanavar, Debadatta Mishra, and Aarohi Kulkarni

Abstract Circular bio-economy with focus on sustainable and renewable energy source has become the need of the hour as mandated by the United Nations. Countries across the globe have united in the quest for deriving value in terms of energy from organic sources earlier considered as waste. A valuable contribution toward sustainability, bio-economy, and maintenance of environmental standards comes from biogas production wherein organic contents are converted under controlled conditions of microbial digestion into biogas that is rich in methane. This chapter delves deep into biogas commercialization over decades considering different models developed, key commercial players who have robust at scale operations, and the market movements that have shaped biogas industries. Current policies across the world and the investment scenarios that are redefining the landscape of the biogas plants have been explained in depth. Government and nongovernment incentives, legislations, policies, and regulatory framework have renewed the focus on biogas considering the overawing advantages of management of waste and environmental credits along with co-product value gain. Details of planning with permissions for setup and operations along with supply chain logistics of pre- and post-biogas processes is covered. Last but not the least is a case study of a large commercial biogas plant that has had huge commercial, social, and environmental impact.

1 Introduction

Sustainable and reliable energy is the main target of United Nations Sustainable Development Goals as the global demand for energy is increasing tremendously. More than 88% of this global energy demand is obtained from the non-renewable

---

S. Sawale · D. Patil · A. Kulkarni (✉)
Praj Matrix – R & D Center (Division of Praj Industries Limited), Pune, Maharashtra, India
e-mail: aarohikulkarni@praj.net

C. Joshi · B. Rachappanavar · D. Mishra
Praj Industries Limited, Pune, Maharashtra, India

N. Balagurusamy, A. K. Chandel (eds.), *Biogas Production*,
https://doi.org/10.1007/978-3-030-58827-4_16

sources such as fossil fuels etc. (Judkins et al. 1991). As per U.S. Energy Information Administration's International Energy Outlook report, the world energy consumption will grow by 28% between 2015 and 2040, to meet the energy demands of the increasing population. Coal is being replaced by natural gas, renewables, and nuclear power for electric power generation and for industrial processes in various countries. Coal share of total world energy consumption will decline significantly from 27% in 2015 to 22% 2040 (IEO 2017). Global awareness about effects of climate change has led to promotion of sustainable and economically feasible programs by various governments to minimize the greenhouse gases (GHGs) emission. As per US Environment Protection Agency (EPA) report, livestock methane emission will contribute about 21% of global non-CO_2 emission in 2030 (EPA 2014). The renewable energy usage is expected to increase up to 20% in energy consumption and 10% in transport by 2020 in European Union. By 2050, renewable energy contribution is projected to increase by 55–75% of total energy consumption (EC 2011). Worldwide, many countries are trying to adapt the anaerobic digestion (AD) systems to prevent the GHGs emission, water pollution, and land pollution due to landfills through the use of sustainable renewable sources. The produced energy from AD system is used for number of application such as compressed heat and power (CHP), cooking, electricity, transportation fuel, etc. (Weiland 2010). In this regards, government bodies are laying down various policies, regulations, and incentives to foster and enhance the biogas project viability at varying scale of operation (Pfay et al. 2017; Scarlat et al. 2015; EC 2009).

The natural production of biogas has been known since the early seventeenth century. Proper construction of biogas plants on the basis of scientific experimentation started in the mid-nineteenth century. The first digester was built in Matunga Leper Asylum in Bombay, India, to produce gas from human waste to light lamps. The idea spread to England and then to China to use produced gas to light street lamps. Van Hemonth in the seventeenth century found that decaying organic material produces flammable gases. The amount of gas produced is proportional to the amount of organic material used, a theory proposed in 1776 by Volta. During 1804–1808, John Dalton and Humphrey Davy proved that the combustible gas is methane (Tietjen 1975). In 1868 Bechamp and in 1890s Omelianski showed that during formation of methane, decomposition of organic material is caused by presence of microorganisms. In 1881, Mouras, a Frenchman, invented a crude version of septic tank named as "automatic scavenger" to treat wastewater. In 1895, Cameron, an Englishman, effectively and successfully treated wastewater using his own constructed "septic tank." Local government of Exeter, UK, in 1897, approved treatment of city's complete wastewater by using "septic tank" due to this success. The biogas produced from the "septic tank" was used for heating and lighting purpose (McCarty 1982). In 1910, Sohngen proposed that fermentation of complex material occurs through oxidation-reduction reactions to form methane, hydrogen, carbon dioxide, and acetic acid (Barker 1956; Omelianski 1906; Sohngen 1906). In the 1970s decade, initially the focus was on farm-based digesters and simplicity of design and operation. However, industrialization of biogas initiated with the focus on waste management of large-scale industrial effluents.

Subsequently, large-scale biogas production plants as standalone industrial units came up. Earlier plants were mostly designed on cow dung, and produced gas was used for localized consumption, and hence the size of plant was small (Bond and Templeton 2011). This limited technology player to small geographical regions and smaller scale of operations. The scale of biogas production all over the world varies from smaller and simpler construction for household use to commercial plants using tremendous waste obtained from various sources (Vogeli et al. 2014). In present scenario, commercial biogas production has a great future potential as a sustainable solution in combination with other renewable energy efficient technologies and reduction of present use of energy obtained from non-renewable sources such as fossil fuels (Scarlat et al. 2015; Lyytimaki 2018). During the early nineteenth century, the digesters were commonly used to decompose and reduce organic content of wastewater. The gas generated during the process was used to heat the digester and obtain sludge as a product used for agricultural sector (Kumar et al. 2015; Zheng et al. 2017). Shortly after World War II, biogas industry increased substantially in Germany, the UK, and France for agriculture and energy sector. In 1920, Germany constructed the first sewage treatment plant for production and supply of biogas. Similarly, in 1950, it also built the first large-scale agricultural biogas plant (Bond and Templeton 2011). In 1921, Guorui Luo of China constructed an 8 m^3 biogas plant using household waste (He 2010). During 1950s the biogas development was hampered due to reduction in fossil fuel prices. However, once again due to high oil prices during the 1970s, the biogas industry growth was accelerated in Asia, Latin America, and Africa (Ni and Nyns 1996). During the nineteenth century, many countries were considering biogas as an alternate renewable energy source using animal manure as the main raw material source followed by industrial wastewater, agricultural waste, etc. However, the biogas development was mostly linked to the stabilization of putrescible solids from domestic wastewater during the nineteenth century (Hartman et al. 1979). Development of heated fully mixed digesters was widely used during the nineteenth century, and many of these digesters are being used even today for the treatment of livestock manure and municipal waste for biogas production. Industrial wastewater treatment using anaerobic digester systems was aggravated due to increased stringent pollution control policies and regulations to reduce environmental pollution. Advanced AD having high efficiency was need of time compared to conventional mixed digesters for treatment of wastewater, livestock manure, agricultural waste, etc. (Chawla 1986).

In India, S.V. Desai and N.V. Desai of Indian Agriculture Research Institute, New Delhi, have worked initially for development of anaerobic digester. Their scientific experiments on AD for biogas production in 1939 lead to development of first Indian biogas plant in 1951, the Gramlaxmi plant of Khadi and Village Industries Commission (KVIC 1993). KVIC was the first to introduce biogas plants among the farmers in rural India (Meynell 1976). In 1962, KVIC design became standardized. Two other models which became popular were the Janata biogas plant introduced in 1978, and its successor, the Deenbandhu digester, developed by Action for Food Production (AFPRO) in 1984. National Biogas and Manure Management Programme (NBMMP) was implemented in India in 1981–1982 for biogas plant promotion based on cattle dung and other organic waste as a feedstock source (Singh and Sooch 2004). Under the NBMMP, about 49.6 lakh Household Size Biogas Plants have been installed since the inception of the biogas programme in India (IBEF 2018). In 1999, three million family-sized biogas plants were constructed in India. In 2007, Indian central and state government provided subsidy ranging from 30% to 100% for about four million household-sized biogas plant construction (Tomar 1995). In India, over three million domestic digesters and 3000 community and institutional plants were constructed by the end of 2002 (Aggarwal 2003), and since 2005, more than 100,000 biodigesters have been disseminated annually (Myles 2008).

In China, India, and Nepal, household and institutional biodigester have gained widespread acceptance. Since 2001, China has disseminated over two million household digesters annually; in addition, the Chinese government supported over 200 large and medium livestock farms for large and advanced biogas units (Jingming 2006). From 2001 to 2007, over 18 million households adopted AD technology leading to the production of over seven billion m^3 of biogas. Moreover, 87 million tons of animal waste was treated by 3556 biogas plants, and more than 300 Clean Development Mechanism (CDM) projects involving biogas power generation, with a total capacity of 1 GW and an annual emission reduction of over 20 metric ton of CO_2, were also prevented (Schwegler 2007). Biogas can play an important role in current scenario as a renewable energy source in the fast-growing and game-changing thrust to minimize the use of fossil fuel. The natural and complex mechanism of biochemical processing of waste by anaerobic digestion under proper conditions will generate biogas. Biogas is complex in nature due to mixture of gases such as methane, 50–60%; CO_2, 35–40%; nitrogen, 2–4%; oxygen, 0–0.5%; H_2S, 0.1–4%; ammonia, 1–2%; and moisture, 1–3% (Weiland 2010). Additionally, the type of feedstock and processing technology used affects the nature of biogas produced (Gigot et al. 2012). The produced biogas can be used as combustible producing heat and electricity in a combined heat and power plant (CHP). It can also be used in transport vehicles as fuel and can be incorporated in gas grids replacing natural gas. Biogas also helps for the reduction of greenhouse gas emission as compared to other power sources (Chynoweth et al. 2001; Hiloidhari et al. 2014).

The success of any biogas plant majorly depends on the cost and quality of feedstock that affects the methane generation potential, digestibility, microbial load, and contamination load due to chemical, biological, or physical technical operations of conversions (Theurel et al. 2019). The feedstocks include animal manure, agricultural residues, vegetable residues, farm waste, food waste, municipal solid waste, etc. The size of biogas plant depends on the amount of biogas generated per ton of feedstock and the continuous supply of feedstock. Various agricultural biogas plants co-digest industrial waste or municipal waste to achieve higher biogas yield. Securing continuous feedstock supply is an important aspect to run a successful commercial biogas plant. Thus, it is important to implement a long-term contract with feedstock supplier/suppliers including payment conditions for the guaranteed quality and quantity of feedstock (Ammenberg et al. 2017). The biogas plant location is an important consideration during setup considering future aspects. The site must be analyzed for quality of land and soil before plant erection. It should be located at a proper distance from residential area, with ease of accessibility for the electricity grid, proper road transport to reduce the time and cost for feedstock supply, and sale of biogas and digestate (Scarlat et al. 2018; Brahma et al. 2016).

As per India brand equity foundation reports, the Government of India has set an ambitious target of achieving 175 GW of renewable energy capacity by 2022 as a part of Paris Agreement commitments. India added record 11,788 MW of renewable energy capacity during 2017–2018. It is expected that India will overachieve its Paris Agreement goals. Renewable sources are expected to help meet 40% of India's power needs by 2030. The renewable energy space in India has become very attractive from investors' perspective and has received FDI inflow of US$ 6.39 billion between 2000 and 2018. More than US$ 42 billion has been invested in India's renewable energy sector since 2014. India has also ranked second in the Renewable Energy Attractiveness Index 2017 as there is ample push from the government and the economics of the market is improving. Installed renewable power generation capacity has increased steadily over the years, posting a CAGR of 9.29% during 2008–2018 (IBEF 2018).

2 Commercial Players in Biogas Sector

As per the European Biogas Agency (EBA), the total number of biogas plants in Europe increased from 6227 to 17,662 installations from 2009 to 2016. In 2016, nearly 12,496 plants were running on agriculture substrates followed by biogas plants running on sewage sludge (2838 plants), landfill waste (1604 plants), and various other types of waste (688 plants) (EBA 2018b). The German biogas energy economy accounts nearly one quarter of the total global installed capacity due to the uninterrupted policy support. As per Global Data Report, biogas power generation is forecast to increase from 18,244 Gigawatt hours (GWh) in 2012 to 28,265 GWh in 2025, representing a Compound Annual Growth Rate (CAGR) of 3.4%. By comparison, the USA is the second most productive biogas power producer and is expected to increase generation from 9072 GWh to 20,936 GWh in 2025, at a CAGR of 6.6%. Global Data forecasts moderate growth for the global biogas power market between 2012 and 2025, expecting it to rise from 50,516 GWh to 130, 321 GWh at a CAGR of 7.6% (Global Data Report 2013). In India, the biogas production is estimated upto 20.757 Lakh M^3 2014–2015 which is *equivalent to 6.6 crore domestic LPG cylinders (5% of total LPG consumption). In India*, more than 49.6 lakh household-sized biogas plants already installed under National Biogas and Manure Management Program (NBMMP) and under had set a target to install 6.5 lakh biogas plants across the nation (MNRE 2018). Global and Indian players providing biogas technology are mentioned in Table 1.

Table 1 List of various industries in biogas sector including India

Company	Nature or business	Feedstock
GasCon, Denmark	EPC and consultancy	Landfill gas and CHP production
BIOGEST Energie, Austria	Plant manufacturer, engineering, own plant operations	Biogas plant with dairy waste, poultry, landfill, biomass
CH4 Biogas LLC, Greenwich, CT, USA	EPC	Animal slurry from local farmers and industrial waste from local food processing companies
IES BIOGAS, Italy	EPC	Agricultural and agro-industrial sectors, sugar industry waste
ARCHEA New Energy GmbH, Germany	EPC	Organic waste materials of agricultural, commercial or industrial production
Biogas Hochreiter, Germany	EPC, O and M	Agricultural and agro-industrial sectors, cattle waste
BIOGAS Equity 2, Inc., USA	Develops, finances, builds and operates biogas plants	Manure food waste and biosolids
Xergi A/S, Denmark	EPC, Gas cleaning Technology	Agriculture and agro-industrial sector waste, farm waste
DVO Inc., Chilton, WI, USA	EPC	Agriculture and agro-industrial sector waste, farm waste
Indian companies in biogas sector		
Kirloskar Integrated Technologies Ltd (KTIL), Pune, Maharashtra	EPC	Agricultural waste
Sampurn Agri Ventures Pvt. Ltd, Chandigarh, Punjab	Technology development	Agricultural Waste
AgroGas-Primove Pvt. Ltd, Pune, Maharashtra	Development of biogas plant	Agricultural waste, food waste
Spectrum Renewable Energy Pvt. Ltd Gurgaon, Haryana	Biogas plant	Agricultural waste
Lars Enviro, Nagpur, Maharashtra	Biogas plant	Wastewater
Praj Industries Limited, Pune, Maharashtra	EPC, Technology provider, Plant supplier	Agricultural waste, sugar industry waste
Mailhem Ikos Environment Pvt. Ltd. Pune, Maharashtra	Biogas plant	Food waste, leather waste, wastewater

3 Planning and Building a Biogas Plant

Establishment of any biogas plant is aimed at environment protection and renewable energy production from waste including financial and non-financial incentives. Government authorities, local farmers, waste producers, energy producers, equipment manufacturers, and suppliers are involved to set up a biogas plant. Thus, to set up a biogas plant, the investor or financing bodies must have a clarity on biogas project before erection including feasibility study, technological aspects, detailed planning of biogas plant, technology providers, EPC players, etc. (USDA 2014).

3.1 Project Idea and Feasibility Study

The biogas project developer/investor must take into consideration the techno-economic feasibility of biogas project, planning and permission procedure, capacity of project, continuous and uniform supply of feedstock, location of biogas plant, funding from government schemes, subsidies and soft bank loan schemes, operation and maintenance, investment costs, refurbishment, safety, location of biogas plant, and selling of end products such as biogas, biomethane, digestate as fertilizer, etc. Along with project developer, the feedstock supply chain management companies, biogas project technology provider, engineering, procurement and construction (EPC) players, oil and gas companies, government agencies, and financial institutions are involved as stakeholders in biogas projects. Detailed planning for the construction of biogas plant, its operation and maintenance, etc. must be taken into consideration for running a successful biogas plant (EAEF 2006).

Once the project feasibility study is completed, detailed planning and finalization steps need to be looked into. These includes identification of suitable technology provider providing lower conversion cost of feedstock used, patented and proprietary technology and permission for royalty fees, agreement with guarantees of yield and productivity per ton consumption of feedstock including engineering guarantees based on equipment life under defined process conditions, operational guarantees after dew testing and commissioning, safety guarantees, analytical guarantees including specification sheet of feedstocks, and in-process samples and product (Paterson et al. 2015). The design of biogas plants varies in configuration depending on feedstock nature (Teng et al. 2014; Claudius 2013). Process temperature and pH influence biogas production from any feedstock. In a well-established biogas plant, neutral to slightly alkaline pH and mesophilic or thermophilic temperature and a carbon nitrogen ratio of 20:1 to 30:1 are essential for continuous and constant production of biogas (Wang et al. 2014). Solid concentration of feed material also plays a key role to ensure biogas production as well as mixing and handling. In addition, hydraulic retention time (HRT) is the most significant factor to determine the digester volume which in turn determines the plant cost. The longer the HRT, the higher is the cost of construction which results in delayed return of investment (Eliasson et al. 2017). A few general schematic flowcharts are outlined for generic biogas plants that are at commercial scale of operations.

3.1.1 Kitchen Waste to Biogas

Organic matter of kitchen waste has high calorific value and nutritive value due to which during anaerobic digestion, high conversion efficiency to biogas can be obtained. Due to inadequate management, kitchen waste is discarded on landfill causing various health hazards such as cholera, typhoid, malaria, etc. Additionally, it causes pollution of land and water due to leaching, unpleasant odor, and methane released from such landfills adds to greenhouse gas emission contributing to global warming (Abeliotis et al. 2015). Figure 1 shows a general flow diagram from production of biogas from kitchen waste.

3.1.2 Animal Waste to Biogas

Large amount of animal waste generally collected is usually left to decompose in open area. Animal waste such as cow manure is high potential to harness the untapped renewable energy potential to convert organic waste content to biogas. This waste produces methane, nitrous oxide, ammonia, hydrogen sulfide, and volatile organic components that cause serious health hazards and contribute to global warming (Koneswaran and Nirenberg 2008). AD is thus a unique solution for value addition to animal waste as it provides a constant source of renewable energy and reduces health hazards and water, land, and environmental pollution (Lv et al. 2018; Raj et al. 2014). Figure 2 shows a general flow diagram for the production of biogas from animal waste.

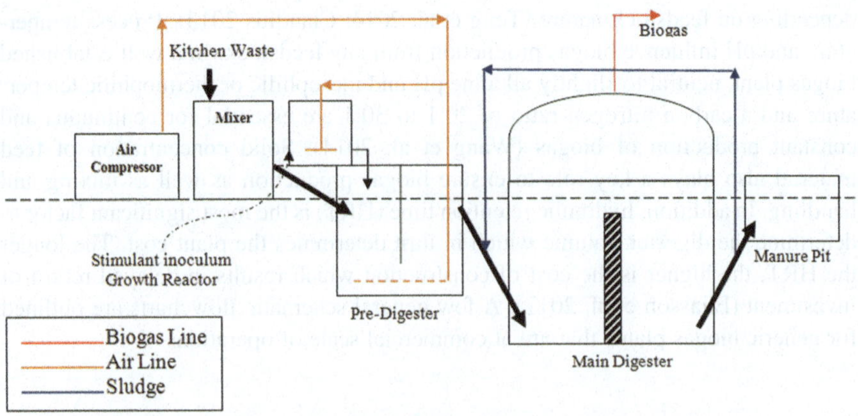

Fig. 1 Process flow diagram of biogas generation from kitchen waste

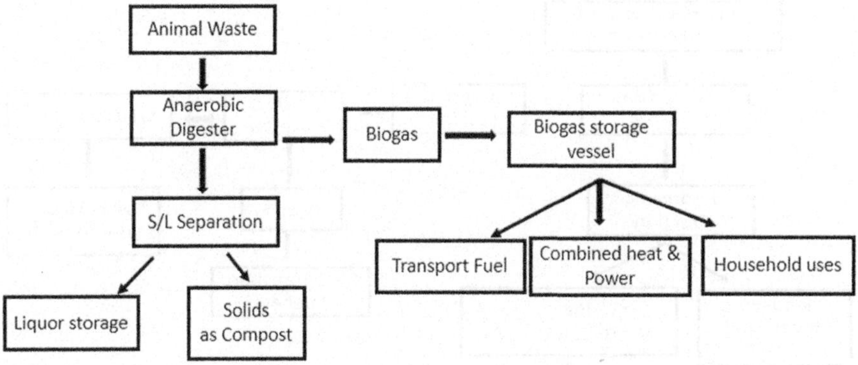

Fig. 2 Process flow diagram of biogas generation from animal waste

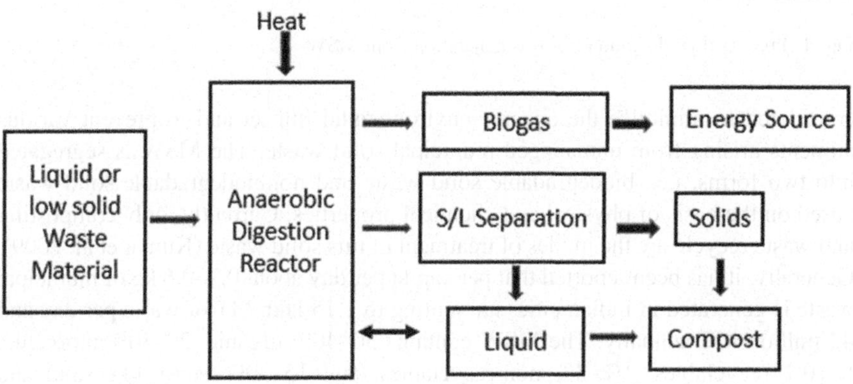

Fig. 3 Process flow diagram of biogas generation from liquid waste

3.1.3 Liquid Waste to Biogas

The spent wash obtained from distillery is a high pollutant of land and water resources due to its high odor and color. Nearly 6–15 M^3 of waste (spent wash) is produced per M^3 of alcohol. Land disposal of distillery spent wash leads to groundwater contamination (Beltran et al. 2001). The spent wash can be used for biogas production (Bhoite and Vaidya 2018). Figure 3 shows a general flow diagram for the production of biogas from liquid waste.

3.1.4 Municipal Waste to Biogas

Municipal solid waste management (MSWM) all over world is looked as a major challenge. MSMW is an utmost important parameter toward sustainable development of any country. Collection, segregation, processing, storage, and disposal to be

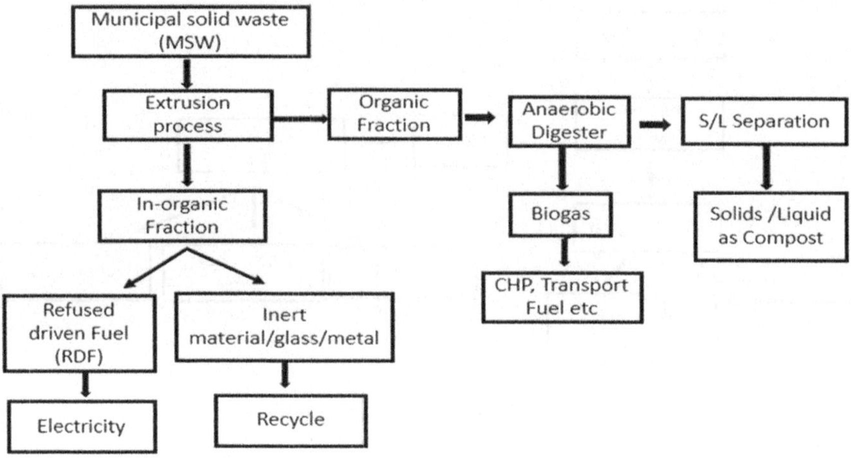

Fig. 4 Process flow diagram of biogas generation from MSW

considered to minimize the adverse environmental impact and to prevent various ailments arising from unmanaged municipal solid waste. The MSW is segregated into two forms, i.e., biodegradable solid waste and nonbiodegradable solid waste based on the basis of physical and chemical properties. Currently, only composting and waste recycle are the modes of treatment of this solid waste (Kumar et al. 2009). Generally, it has been reported that per capita per day about 0.2–0.6 kg of municipal waste is generated in Indian cities amounting to 1.15 lakh MT of waste per day and 42 million MT annually. The MSW contains 30–40% organic, 20–30% inorganic, 5–10% recyclables, 1%–3% non-recyclables, and 15–30% inerts like sand and stones (Kumar et al. 2009; Sharholy et al. 2008; NEERI 1996). MSW contains about 50–80% moisture and calorific value of 600–800 Kcal/Kg (Kumar and Gaikwad 2004). Biodiesel, BioCNG, fuel ethanol, and solid and liquid fertilizer can be produced from biodegradable solids. 1600 MT of MSW has a potential to generate biodiesel of up to 18–20 MT/day, fuel ethanol up to 7–9 M^3/day, BioCNG of 10–12 MT/day, and in addition 1100–1200 M^3 liquid manure per day (Sony et al. 2016). Figure 4 shows a general flow diagram for the production of biogas from MSW.

3.1.5 Biomass Waste to Biogas

Agricultural residue is abundant and is widely distributed in nature. Mostly it is being used as an energy source in kitchen for cooking purpose which may cause respiratory health hazards as well as environmental hazards due to smoke obtained after burning (DeKoning et al. 1985; Sigsgaard et al. 2015). To avoid such health hazards due to agricultural residue, it can be used as a feedstock source for biogas production using anaerobic digestion technology. In most of the cases, the biomass

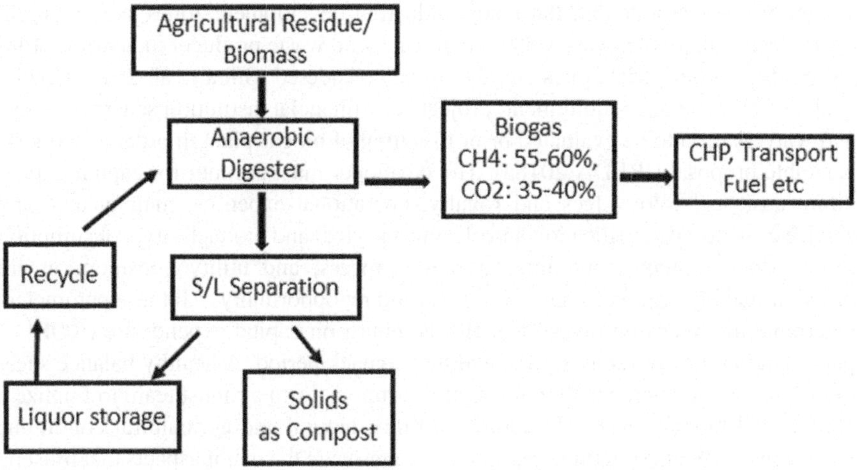

Fig. 5 Process flow diagram of biogas generation from biomass

is pretreated using biological, chemical, or physiochemical method to loosen the fibrous material (Agbor et al. 2011). The pretreated feedstock is then transferred to AD process. The process parameters such as pH, temperature, solid loading ratio, etc. are critical and thus maintained to achieve high volume of biogas containing high proportion of methane (Meegoda et al. 2018). A generic scheme for the biogas from agriculture residue is shown in Fig. 5. The solid residue and liquid obtained after anaerobic digestion from all waste material can be used as a source of fertilizer to improve the soil condition.

3.2 *Planning and Permission Procedure*

Careful planning is of utmost important to build a successful commercial biogas plant. Unsatisfactory output of a biogas plant is mostly observed in unplanned biogas units. Unfortunately, such mistakes are only seen after several years of operation and are costly. Unplanned plants are a high liability and are difficult to recover. Thus, at the project initiation of any biogas plant customer's expectations, appropriate T-EPC and funding availability must be taken into consideration to lay down proper functioning sustainable plant. A planner must explain all the steps including cost for erection of a biogas plant mentioning all extension services, risks, mitigation, and troubleshooting initiatives to be followed (IEA 2011; Szymanski et al. 2013). The investor must take into consideration quantitative and qualitative energy requirement. For agri-residues the biogas-generating potential is to be calculated on the basis of biomass production and energy demand during planning phase. Nature of a feedstock and its availability, biogas plant location criteria, land area available, water availability, industrial zone details along with utilities such as industrial water, assembly lines, production, flat lands, industrial development zones such as food, pharmaceutical, textile, wastewater, chemical, solid waste producer industries, effluent producing industrial zones are also to be considered (Silba et al. 2014; Coelho et al. 2017). Financial requirement, project cost, financial institution schemes, loans, and waivers need to be evaluated prior to setting of biogas plant in order to work on bankable proposal (IREDA 2018a). The developer must account for capital expenditure cost, technology fees and royalty, operational expenses, maintenance and service cost, cost of certification and clearances, yield and productivity, consumption of feedstock, schemes and interest rates, process, and utility consumption are decision-making criteria for encashing on funding opportunity. All these parameters determine the return on investment (ROI), return on capital expenditure (ROE) of plant, internal rate of return (IRR), and the payback period. A healthy balance sheet of developer and approachable financial institutions form a robust team to finalize a commercial biogas project. In addition to this, natural and agricultural conditions including social and economic aspects of a country are the major aspects that must be considered before building a biogas plants (FAO 2014; Bonazzi and Lotti 2015; Carlini et al. 2017; Mohammed et al. 2017).

Government permission, policies, and subsidies are subject to geographical constraint. Most of the countries sanction these project under green projects. The rules and regulation including the procedure, criteria, and documentation for the erection of any biogas plant vary from country to country. In general aspects, the investor must adhere to national legislation issues like feedstock used, its handling and recycle, limit values for emission, noise and odor, effect on groundwater, land protection, work and building safety, etc. It is important to involve government authorities from the beginning to obtain complete and clear information to be produced at different stages of plant erection to receive required permits. Involvement of an experienced company in getting all such permits is useful depending on the regional conditions. Approval certificates required for setting up biogas plant in India are listed below:

(a) Consent to Establish/Operate—State Pollution Control Board
(b) The Petroleum and Explosives Safety Organization (PESO) approval
(c) Factory Inspector's Certificate
(d) Labor/Commissioner permission for employing contracted persons/agencies
(e) Weigh-bridge stamping from weights and measured
(f) State electricity approval for co-generation of power, if required
(g) Natural Organic Certification Agro (NOCA) certification for solid and liquid manure disposal in a scientific matter

3.3 Design and Construction of a Biogas Plant

Globally, the designs of biogas plant are developed considering the specific climatic and socioeconomic conditions. It is important to be familiar with the basic design considerations during construction for any household and commercial plant. The performance of a biogas plant depends on local soil and climatic conditions including raw material for digestion and used material for building at reliable cost. Insulation and heating devices are important part of a digester in case of lower temperature. The inlet and outlet construction of digester must take into consideration the amount and type of substrate to be digested. Standardization of standard procedures between biogas plant planner, raw material provider, biogas purchaser, etc. are required to avoid mistakes and misunderstanding and save time and money. Additionally, a number of skilled labor are required for a successful running of a biogas plant. Biogas plants are short-cycle project for construction involving mostly civil work as digester is constructed in civil (MNRE 2015a). Continuous stirred tank reactor (CSTR)-based designs can be cast in metal for small-scale plants and liquid effluent-based plants. Different unit designs of digesters like CSTR, plug flow reactor (PFR), up-flow anaerobic sludge blanket (UASB), reactor and vortex designs are evolved over the years of biogas history. CSTR is easy to operate and most commonly used for wastewater treatment with high chemical oxygen demand (COD) and high solid content (Chan et al. 2009; Wang et al. 2005). CSTR stabilizes the sludge by conversion of biodegradable fractions into biogas (Massoud et al. 2007). To increase process rates, CSTR are operated at high temperatures. Mixing is performed mechanically or by flow recycle or from produced biogas. Perfect mixing is difficult in big CSTR volumes. Hence, mixing efficiency is an important factor to be considered during modeling of CSTR in regard to solid transport in reactor and evaluation of solid retention time. Materials with high COD loading rates (30 $kg/M^3/day$) can be treated using CSTR with an adequate treatment at lower HRT (Wang et al. 2005). In case of CSTR, a removal efficiency of 85–95% of COD of inlet material, and a produced methane content of 80–85% has been reported (Chan et al. 2009; Wang et al. 2005).

The PFR produces biogas at a variable pressure with a constant working volume. These digesters consist of a narrow and long tank with an average length to width ratio of 5:1. Both inlet and outlet are kept above the ground and are in opposite directions, and the complete biodigester is built underground in inclined position. Due to inclination, the feed passes into the inlet and digestate passes toward outlet. The inclined position separates acidogenesis and methanogenesis phases longitudinally producing a two-phase system (Rajendran et al. 2012). The HRT under thermophilic conditions ranges between 15 and 20 days. The solid concentration of the feed is between 11 and 24% (Abbasi et al. 2012). PFR can be operated in mesophilic conditions even though they maintain optimal thermophilic conditions (Strezov and Evans 2015).

Globally, UASB are mostly used for wastewater treatment (Abbasi et al. 2012). UASB biodigesters operate in up-flow mode. The material is from the bottom; it passes through dense sludge bed having high microbial activity and gas-liquid-solid separation device. The separator device separates liquid effluent from solid sludge that remains in the digester while biogas is collected (Strezov and Evans 2015). UASB process depends on the natural immobilization of anaerobic bacteria (Chan et al. 2009). In countries, where landfilling is commonly utilized for waste management, use of batch reactors is good option to treat such biowaste due to its simplicity and portability. In batch reactor operations, the biomass is loaded once and discharged on the process completion (Abu-Reesh 2014). The biogas yield in case of batch reactor is 50–100% higher compared to landfills due to proper biowaste recirculation. Additionally, it is possible to recover the digestate material in a batch reactor (Mogal 2013). Extra safety is required to prevent the explosion at the time of unloading batch reactor on the completion of the reaction. Every design has a capacity limitation leading to single reactor or multiple reactor designs in series. Maintenance of anaerobic nature of digesters, maintenance of critical gas pressure, dead zones in solid or liquid phase, and gas zone are key criteria in designing type of agitator and digesters. A number of agitators and its design are governed by type of feedstock, type of digester, and capacity of digester (Lemmer et al. 2013). Temperature maintenance of a digester is a crucial criterion for deciding type of digester, location of digesters and its material of construction (MOC). Slightly alkaline nature of digester or process defines MOC of digesters, while acidic pH determines MOC of pre-methanogenic phase. Special coating of surfaces in gas cleaning systems are critical to avoid erosion and corrosion (Naegele et al. 2013). Globally as per requirement batch mode, continuous mode and semi-continuous mode of biogas plants are available. Depending on the space, existing structure, cost and substrate plug flow digester, continuous stirred rank reactor, balloon plant, earth pit plant fixed dome, and floating drum biogas plants are available (Amaratunga 1986). In India floating digesters such as Khadi Village Industries Commission (KVIC) model and Pragati and Ganesh model and fixed dome digesters such as Janta and Deenbandhu models are used for biogas production (Bhol et al. 2011; Sooch and Gautam 2013).

3.3.1 KVIC Type Biogas Plant

The biogas plant contains a digester for anaerobic fermentation and a floating drum for gas collection as shown in Fig. 6. The depth and diameter of digester are 3.5–6.5 m and 1.2–1.6 m, respectively. The partition wall divides digester vertically after complete slurry filling. Dung mixed with water (4:5 ratio) and filled into digester thorough inlet pipe serves as the starter culture passes through outlet pipe. The outlet is mostly connected to a compost pit. Gas holder is cylindrical in shape with concave top. The gas holder sinks into slurry due to its weight and rests on a ring constructed for this purpose.

3.3.2 Janta-Type Biogas Plant

Gobar Gas Research Station has introduced Janta-type biogas plant to reduce the investment cost compared to KVIC model. Janta plant does not contain any steel during construction, and it does not have any moving part due to which maintenance cost is low. To avoid any structural damage and gas leakage in the long run, the plant must be constructed using good quality bricks and cement material. Such type of biogas model has higher capacity and longer life span with respect to KVIC model.

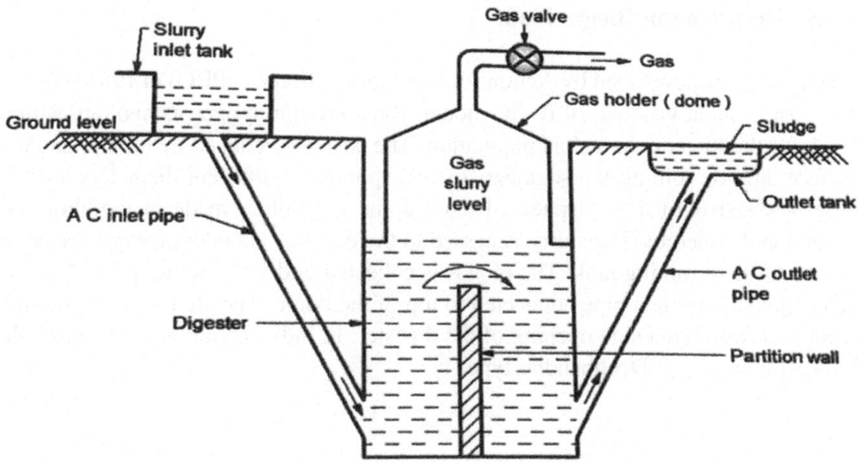

Fig. 6 KVIC biogas model

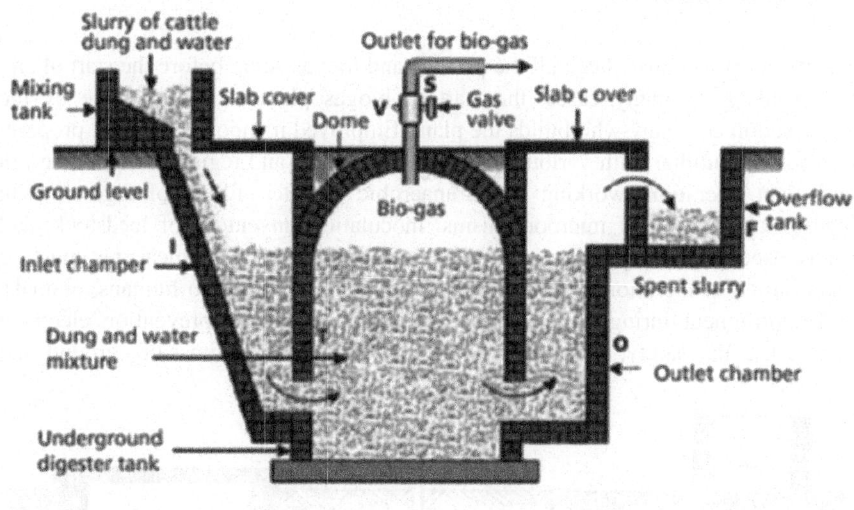

Fig. 7 Janta biogas model

Substrates such as municipal waste and plant residue along with cattle dung can be used. Two rectangular openings act as inlet and outlet, while the dome-shaped roof remains below ground level. The gas outlet pipe is fitted on top of the dome-shaped roof. The biogas is collected in restricted space of fixed dome leading to high pressure of gas (Fig. 7).

3.3.3 Deenbandhu Biogas Plant

The model was developed by Action for Food Production (AFPRO) in 1984 (Fig. 8). The cost of plant was half of KVIC model which brought biogas technology within reach for the lower economical population. The cost was reduced by minimizing the surface area by joining the segments of two spheres of different diameters at their base. It consists of a hemispherical fixed dome gas holder made of prefabricated cement and concrete. The slurry is passed to the digester via inlet pipe connected to digester from a mixing tank. The biogas is collected under the dome space. The gas can be taken out via a pipe connected to top of the dome. The sludge, a co-product, comes out through a side opening of the digester. In India, about 90% of household biogas plants are of Deenbandhu type.

3.4 Operation, Maintenance, and Safety Parameters of Biogas Plant

The plant owner must check all the permits and biogas setup before the start of any biogas plant. In many cases, the start of biogas plant is always done by the construction company who builds the plant. Employed manpower must be properly trained and familiar with various aspects such as rules and regulation, plant design and plant operation, working of an anaerobic digester, filling of digester with feedstock, handling of microorganisms, inoculation, inventory of feedstock, and safety aspects of biogas plant. Proper precautions and safety measures must be taken into consideration to avoid potential risks and hazards to humans, animals, and environment during biogas plant building. Fire, explosion prevention, electrical safety, thermal safety, noise emission level, air pollution emission, ground and

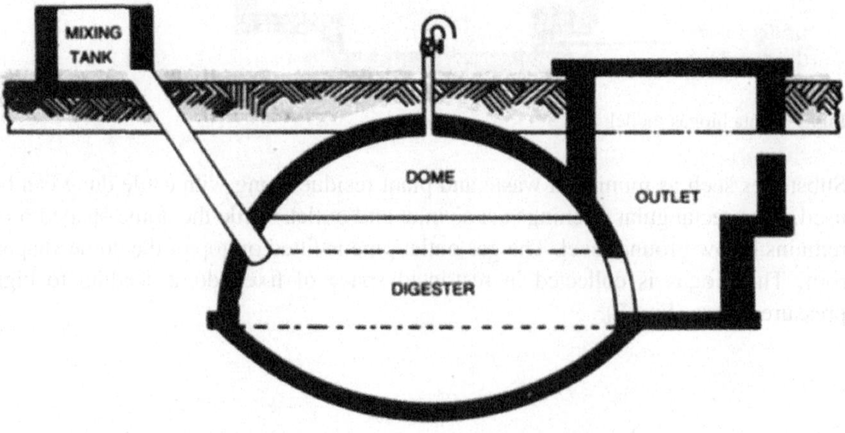

Fig. 8 Deenbandhu biogas plant model

surface water penetration prevention, reduction in released pollutants, maintenance of hygiene, avoidance of poisoning due to inhaling, etc. must be considered before the start of a biogas plant (Salvi et al. 2011; Bragatto 2013). In European countries, European Directive 1992/92/EC classifies explosion safety in terms of zones, based on frequency and duration of occurrence of an explosive atmosphere (EC Directive 1999). These norms being highest in safety levels, following or achieving these is considered as the best industrial norms.

4 Key Business Drivers

Different driving forces are important for biogas technology at local and commercial levels. Project financing is one of the major business drivers in order to ensure project viability due to long payback periods. Options of funding agencies, funding at subsidized rate or lower interest rates for longer period, boost investors for biogas plant initiation (Menind and Olt 2009). Biogas plant erection initiative can be taken by private investor and/or local people interested to create a sustainable environmental system. Memorandum of agreement or third party agreement for the uptake of produced biogas and manure with right price can make the project more attractive and financially viable. Project financing bodies will require a higher degree of project risk reduction. Such projects must focus for the reduction of risk associated with feedstock availability, its providers, and long-term pricing (Lauer et al. 2018; Szymanski et al. 2013). Initiatives at government level for regulation and proper policy framework to encourage investors to build a biogas plant as a renewable source of energy is of utmost important. The proper lay down of legislation policies has seen a tremendous growth in implementation of biogas plants in countries such as Canada, UK, and USA including the European Union (Scarlat et al. 2018). The European Union has been pioneering in the setting of biogas plants on multitude of feedstocks with Italy and Germany playing a key role in engineering and technological advances. Global Methane Initiative (GMI) report for "global perspective of anaerobic digestion (AD)" provides a broad overview of policies, regulations, and government incentives that are used in around the world to promote AD technology (GMI 2014). AD has not realized its full potential due to high cost of investment for design and construction, limited incentives for investors, non-availability of proper structural setup for connection to electrical grid, price uncertainty for biogas uptake, and time frame to realize on return on investment (Mugodo et al. 2017; Yousuf et al. 2016). Government of European countries is leading the promotion of various policies for AD. Similarly, many other countries are now adapting European norms and policies for AD to reduce emission of greenhouse gases, to provide sustainable growth to agriculture and as an alternative source of renewable energy. The biogas projects are more feasible and attractive to investors when government provides financial inventive to offset the capital and operation cost. Incentives such as to generate extra revenue by selling excess electricity to electrical grid, charging fees for waste intake, selling of biogas at premium prices, etc. Policies can be structured to maximize saving on heat and electricity bills by allowing the use of formed biogas to replace plant's existing electricity, gas, and heat consumption either partially or completely. Policies can be modified to acquire more benefits due to environmental protections. Regulations and policies are focused on comprehensive agriculture, livestock production, air emission, water discharge, manure storage, nutrient management, and renewable energy regulations (EBA 2018a; Konigsberger et al. 2018; MoPNG 2018; Sapp 2017; GMI 2014). The best management practices in case of agriculture sector are to be implemented for feedstock collection, handling and storage, digestate storage, and strategies to reduce odor and dust (Vac and Popita 2015; Alayi et al. 2016). In Europe, a Common Agriculture Policy (CAP) was formed in 1962 and recently again reformed in 2013. The CAP focuses on farmers to meet the need of 500 million of European population, to ensure a good standard of living, and to provide a stable and safe food at affordable prices. It focuses on viable food production, sustainable management of natural resources, and balanced development in rural areas (EU 2018).

Policies must be in place to prevent air pollution due to formed biogas such as discharge of gases via flares, boilers, or from cogeneration equipment. Indonesian Sustainable Palm Oil (ISPO) policy focuses on reduction of GHG emission, improving sustainability, and increasing global market competition for Indonesian palm oil. ISPO mandates palm oil plant to register for emission reduction plan, including methane capture for treatment of palm oil mill effluent (Hidayat et al. 2018).

Animal wastes from livestock are leading contributor to water pollution. Manure management policies related to its collection, storage and processing to prevent water contamination are being adapted by many countries (Teenstra et al. 2014). The liquid digestate obtained from AD can be used as fertilizer. However, the water discharged via AD plant if not properly managed may lead to contaminate ground water and surface water quality; hence policies and regulation that limit effluent discharge need to be considered to maintain water quality. In USA, Federal Water Pollution Control Act or Clean Water Act (CWA) policy focuses to protect and restore water by regulating allowable discharge (Copeland 2016).

Many EU nations have renewable energy target specific for biogas, and biomethane European Renewable Energy Council countries have policy target to generate 20% more energy from renewable sources by 2020 (EREC 2011), of which 28 EU member states have already archived 2020 renewable energy targets (WEF 2019). Similarly, US target about 16% by 2020 and New Zealand has set a goal of 90% renewable energy by 2025 (Finalayson 2011). Mexico, India, Japan, and Brazil also set to increase total renewable capacity by nearly four times by 2030 (Ross and Damassa 2015).

4.1 Incentives

The biogas projects all over the world present a financial challenge even after considering its major benefits due to high initial investment cost. These financial problems must be resolved at the initial stages of biogas plant implementation. The cost is associated with securing investment capital cost for design and construction of biogas system. Operation and maintenance (O&M) cost includes equipment cost, labor and training cost, etc. Unfamiliarity of biogas technology for financial institutions results in the perception of project as high risk, capital cost intensive, and low reward compared to other conventional energies such as wind and solar technology. To increase economic feasibility and project attractiveness, many countries such as the USA, UK, EU region, China, India, Thailand, etc. provide incentives and subsidies to offset the cost and generate revenue from biogas plants. Incentives are given to assure the biogas premium value to involve market forces to use the environmental benefits due to biogas produced from renewable sources, while subsidies enable developers and financial institutions to offset capital expenditure risks. Globally, different types of incentives provided to biogas produces are mentioned below.

4.1.1 Feed-in Tariff (FIT)

Feed-in tariffs (FIT) is offered in developed and developing countries such as the USA, UK, Thailand, China, Germany, France, Argentina, and Canada for the development of renewable energy projects (IEA 2014a). It is a long-term contract between biogas producer and government. FITs boost investors' confidence by creating more certainty in the value of electricity generated and thus ensures the project ROI required to build the project. A premium price can be charged in cases where the electricity is supplied to grid from a biogas plant which assures developers to generate additional revenue. However, biogas project investors need to have a long-term contract for FITs. Feed-in premium (FIP) is an additional premium payment on top of natural gas market price is paid. Fixed revenue is guaranteed and the premium varied as a function of natural gas market price (Szyszczak 2014).

4.1.2 Carbon Credits

Carbon credits are measured as carbon dioxide equivalents units (CO_2e). These are major incentives given to biogas plants for the environmental benefits in terms of reduction in emission of greenhouse gases leading to reduced use of non-renewable energy such as fossil fuels. The biogas investors can sell these carbon credits in the market to offset the investment cost of developed biogas project. Emission trading scheme of New Zealand, EU emission trading system, Mexican carbon platform (MEXICO2) allows biogas owners to earn carbon credits and sell them in the market and thus generate revenue (Mardiatmoko 2018; Leining and Kerr 2016; Ripley 2013; Conte and Kotchen 2010; EU-ETS n.d.).

4.1.3 Tax Exemptions

Tax exemption on instruments and material required for biogas plant construction is a boon in many places. Tax exemption on biogas produced and electricity generated can help investors to generate extra revenue and offset the invested cost. These incentives can be re-invested for higher energy efficiency and reduction in carbon emission and to generate interest of other investors to build biogas plants. Germany, Sweden, Switzerland, UK, USA, China, France, etc. are the largest producer of vehicle fuel. These countries have promoted and supported biogas production and utilization through combination of tax-exemptions, investment subsidies, and incentives for biogas injection into the natural gas grid (IRENA 2017a). In Finland, the biomethane generation is exempted from excise taxes (GMI 2014).

4.1.4 Credits for Renewable Energy (REC)

Credits for renewable energy are earned for the environmental protections and energy produced from renewable sources. These credits generate revenue for biogas investors and offset the non-renewable cost. RECs thus help and maximize chances of more investors entering for biogas plant installations (GMI 2014). In the UK, Renewable Obligation Certificates (ROC) or Green Certificates were introduced in 2002 to earn credits for the energy generated from renewable sources (UK 2015).

4.1.5 Credits for Renewable Transportation Fuel

The USA, UK, Belgium, and Mexico are the few countries that provide credits to biogas investors for renewable transportation fuel to encourage the minimum usage of fossil fuel to generate energy. In such cases, the government prepares a mandate to use the energy of biogas plants as a source of transportation fuel. The biogas produced can be upgraded to use in transportation vehicles and help transportation fuel suppliers to earn these credits. In Belgium, Law of Blending Obligation; in the UK, Renewable Transport Fuel Obligation; and in the USA, Renewable Identification Number (RIN) system were introduced as a part of energy act policy to use energy produced from renewable sources as a fuel for transportation vehicles (Poloncarz et al. 2019; Thompson et al. 2010).

4.1.6 Payments for Producing Renewable Heat

Renewable heat incentives (RHI) policy can be given to the owners who install and supply biogas as energy into natural gas distribution pipeline. Such incentives are provided in the UK and Italy. The UK government encourages uptake of renewable heat technologies among householders, communities, and businesses through financial incentives. The UK expects the RHI to contribute toward the 2020 ambition of 12% of heating coming from renewable sources (Simpson 2010). In the UK, property owners can receive cash payment upon installation of heating equipment that use renewable energy sources. Thus, the investment cost of that equipment can be earned over the designated period of time (UK 2015; Kim et al. 2019).

4.1.7 Credits for Nutrient Load Reduction

Nutrients load reduction credits are similar to carbon credit. The nutrient credits can be earned by replenishment of nutrients such as nitrogen, phosphorous, etc. from environment. Use of digestate manure obtained from biogas plants can earn nutrient credits. These credits can help farmers and biogas investors to generate additional revenue. Canada, Italy, and the USA offer such incentives. Countries in EU and the USA are trying to prepare a mandatory requirement for the diversion of organic materials from landfills to attract various biogas developers by providing various incentives as mentioned before. Such act will help to save the landfill land for other purpose. Increased organic feedstock availability will improve the financial viability of biogas plants (Movafaghi et al. 2013; Dennison et al. 2012). EU countries have implemented Landfill Directive in 1991 to reduce the landfill waste up to 35% which will help to reduce GHGs (EC Directive 1999).

Many countries are promoting public-private partnership (PPP) including government authorities, investors, agricultural industry, and researchers for the development of biogas projects. These groups bring together the knowledge, expertise, and resources for the development. Such PPPs have a positive mindset by communities for the biogas plant technology acceptances and contribute for protection of environment (Heldeweg et al. 2015; Huang et al. 2018; GMI 2014).

4.2 Funding Agency

As per the National Biofuel Policy of India, NABARD, IREDA, SIDBI, financing agencies, and commercial banks can finance for various activities of entire biofuel value chain at different stages. As per policy, 100% FDI can be approved for biofuel technologies and related projects, provided that produced biofuel will be used only for domestic purpose (Narayanan and Hamsalakshmi 2014; MNRE 2015b; IREDA 2018b; IBEF 2018).

4.2.1 Personal Loan from Banks

In this case the banks give loan to the investor. The bank checks the financial background of the investor in order to decide on the reliability and risk of the engagement. Typical loan periods are about 10–15 years. Also in many countries, such projects, which are categorized under renewable sector, have a preference with low interest rates. In Germany, for example, the KfW IPEX-Bank GmbH (Kreditanstalt fur Wiederaufbau) with NIBC Bank and Raiffeisenlandesbank Niederosterreich-Wien provided a loan of EUR 80 million to world's largest biogas plant (Press 2009).

4.2.2 Project Financing by Banks

In this case banks fund a specific biogas project in consideration. The banks look at repayment from cash flow generation from the project. Banks usually take the assets of the project as security and normally fund up to 70% of the cost of the project. Balance amount has to be funded by the investor. Bank in this case does not have access to investor's personal assets. A separate legal company has to be created for this project. This type of funding requires higher due diligence. Typically, all commercial biogas projects are funded through this method. All major commercial banks provide such project financing (IREDA 2017).

4.2.3 Investment Funds

There are equity investment funds that invest in the project and take significant equity stake. Normally such funds are created to fund multiple projects. Such funds also can be created by pooling money from many investors. One example of such arrangements can be farmers' cooperative. Advantage of such funding is that project does not require any loan. But it requires the investors to take all the risk. Hence long-due diligence is required (Rutz and Ferber 2011).

4.2.4 Developer Model

In this model a developer conceives the project. He then ties up with raw material such as feedstock suppliers and the uptake partner [customer for the plant output, i.e., electricity, gas for national grid or transportation fuel (compressed biogas)]. Developer may take up funding from such customers in lieu of committing to provide output for long term (Rutz and Ferber 2011).

4.3 Legislation and Policy Framework

European countries including Sweden, Germany, and Spain have realized the potential of biogas as a source of renewable energy. These countries have prepared a policy framework for waste management, agricultural policies, and renewable policies. Germany policies are seeming to be more efficient for the implementation of biogas projects. Flexible and multifaceted policy support measures are important and necessary for the uninterrupted development of biogas sector. The success of a biogas plant depends on many factors such as consideration of local conditions in individual projects. The local, national, and international policies must be taken into consideration for the biogas project development. Public opposition for the biogas projects or technology has to be solved at local level. In addition, it is a difficult task to recommend single technology as the best. Investment in current technology and research and development should be done (Capodaglio et al. 2016; Scarlat et al. 2018).

In China, the biogas project development was started in the late 1970s. Agricultural Ministry of China issued the "Energy and environmental engineering construction plan of medium and large Livestock farm's" at national planning level in 2000. The Ministry of Environmental Protection of China (MEPC) announced "Regulation on Contamination Control and Management of Livestock Farms" and Discharge Standard of Pollutants for Livestock and Poultry Raising Industry" (GB18596-2001) in 2001. MAC in 2005 issued "Biogas Construction Plan for the whole China in 2006-2010" in which construction of 5000 biogas projects were planned. Additionally, "Well-off Environmental Protection Action Plan" was declared by MPEC in 2006. Laws and regulations are critical for the development of biogas projects. After issue of "Renewable Resources Law of the People's Republic of China" in 2005, other related regulations were issued such as "Tentative Management Measures for Allocation of Prices and Expenses for Generating Electricity by Renewable Energy," "Renewable Energy Relevant Management Regulations," and "Renewable Energy Development Interim Measures for the Administration of the Special Fund." The "Agricultural Ecological Environmental Management Regulations" were implemented in 20 province/cities in China (Wang et al. 2012).

4.3.1 Government Policy Framework in India

Government focus has changed over the period of time toward energy production from renewable sources compared to energy production from fossil fuels. The Ministry of Petroleum and Natural Gas of India has drafted the National Biofuel Policy in 2018. India is among the fastest-growing economy in the world, and its energy consumption is slated to increase rapidly. The country currently imports nearly 77% of its crude oil requirements and about 50% of natural gas requirement, leading the Government of India to set a target of reducing this import by at least 10% by 2022. Further, it has set a target of increasing the contribution of gas in India's energy mix from existing 6.5% (global average is 23.5%) to 15% by 2022 (MoPNG 2018).

The National Policy on Biofuels, 2018, emphasized on active promotion of advanced biofuels, including compressed biogas (CBG). The government also launched the GOBAR-DHAN (Galvanising Organic Bio-Agro Resources) scheme to convert cattle dung and solid waste produced in agricultural farms to CBG and compost. The scheme proposed to cover 700 projects across the country in

2018–2019. Use of biogas is expected to reduce natural gas and crude oil imports and also act as buffer against oil and gas price fluctuations. Biogas uptake is likely to help replace CNG in automotive, industrial, and commercial use. The Indian Government is now providing guarantees and assurance on de-risking CNG prices over the longer period of time, developing the expression of interest in investors. The Government of India has come up with the program on energy from urban, industrial, and agriculture waste/residues for 3 years starting from 2017 till 2020. The approval is accorded the biogas and BioCNG as the components under waste to energy program with Central Financial Assistance with the target of 57.0MWeq. This is in line to the schemes where the Ministry of Road Transport and Highways, Government of India, had permitted usage of bio-compressed natural gas (BioCNG) for motor vehicles as an alternate composition of the compressed natural gas (MORTH 2015).

The objectives of the government program are as follows

(a) To promote setting up of projects for the recovery of energy in the form of biogas/BioCNG/power urban, industrial, and agriculture waste/residues.
(b) To create a conducive condition and environment, with fiscal and financial regime, to develop, demonstrate, and disseminate utilization of waste and residues for recovery of energy.
(c) To achieve Sustainable Alternative Towards Affordable Transportation (SATAT).
(d) To provide Central Financial Assistance provides financial support in the form of capital subsidy and Grant-in-Aid for biogas produced from industrial waste, wastewater, municipal solid waste, agricultural waste by biomethanation process.
(e) To assist promotional activities including research and development, resources assessment, technology up gradation and performance evaluation, etc.

4.4 Feedstock Supply

Uninterrupted sourcing of high-quality feedstock at right quantity, right price, and right place is a major challenge for the biogas projects involved with biomass. The risk can be minimized by a long-term contract with multiple biomass suppliers. Cow dung as a traditional feedstock for the biogas generation has been utilized for domestic cooking in rural area. However, considering the increased interest in biogas technology and to extend its application for power generation, the search for alternative feedstock becomes mandatory. Various competitive feedstock projects deal with waste feedstock such as municipal solid waste algae, food waste, cow manure, and agricultural solid waste, because they can be sourced at lower cost making biogas projects profitable (Hills and Roberts 1981; IEA 2010; Dueblein and Steinhauser 2008). In such cases the risk of increased price due to transportation and handling and inconsistency of contracts must be considered while financing such projects. A variety of organic waste sources obtained from society can be used a feedstock source for the biogas production. Use of such organic waste for biogas production can be mentioned as a resource instead of waste (Mustafa et al. 2016). It is time consuming to change the perception of society, and huge amount of organic waste is still being disposed of by different means such as composting and land filling (Muvhiiwa et al. 2017; Mittal et al. 2018).

The agricultural feedstocks, such as press mud, rice straw, corn cob, sugarcane trash, cotton stalk, etc. as well as municipal solid waste, spent wash must have high levels of convertible organic matter such as simple and complex carbohydrates, fats, and proteins. A high level of high convertible organic matter can be readily converted to biogas directly (Zubr 1986; Braun 2007; Achinas et al. 2017). Globally, all feedstocks are characterized using proximate and ultimate analysis. Proximate analytical parameters such as moisture, ash, volatile matter, and fixed carbon, while ultimate analysis parameters such as mineral matter, carbon, hydrogen, sulfur, nitrogen, and oxygen must be analyzed as a selection criterion for any feedstock before its use into a biogas plant. Feedstock must also be analyzed for phenolic components, butyric acid, ammonical nitrogen, and sulfur as its high level can inhibit the formation of biogas during anaerobic digestion process (Aramrueang et al. 2017; Akkoli et al. 2018).

Hence, feedstocks must be analyzed for above compositional parameters to maintain uniformity in gas yields throughout the year round during the process. Thus, it is important to explore different sustainable energy sources (Ravindranath and Hall 1995). Cost of spent wash or municipal solid waste is lower compared to agriculture residue such as press mud or biomass. The municipal solid waste is not properly segregated due to which variation in compositional parameters is more. Similarly, in case of agriculture residue such as rice straw, cotton stalk, corn cob, etc., the values of compositional parameters keep on changing due to variety, seasonal, and geographical variation (Mohammed et al. 2018). In case of press mud as feedstock, high degradation occurs over the long period of storage due to which press mud becomes inefficient for constant yield and productivity of biogas. India has surplus agricultural and forest area which comprises about 500 million metric tons of biomass availability per year. Biomass accounts up to 35% as primary energy utilization for developing countries, contributing to 14% of global primary energy utilization (Balat and Ayar 2005). Since this process is at no or little production costs, therefore they are ignored and not utilized efficiently like major amounts of leafy wastes are burnt and cause air pollution (Jain et al. 2014; Junpen et al. 2018). The concept of reduce, reuse, and recycle is not new for European Union countries. The Council of European Communities in 1975 declared that member states shall take appropriate steps to encourage prevention, recycling and processing of waste, and the extraction of raw materials and possibly of energy there from and any other process for the reuse of waste (EU Commission 2008). Biogas projects are financed with the awareness of multiple feedstock suppliers with varying feedstock reducing the supply and competitor risk. Seasonal changes affect the composition of feedstock and thus affecting its price. The feedstock must meet the regional regulatory requirements. The presence of hazardous chemicals or substances may cause rejection of feedstock creating a trouble to the project. Third party assessment for feedstock resource including competition for feedstock and pricing information can be useful (SBP Report 2015; Ammenberg et al. 2017).

4.5 Regulatory Framework

The biogas technology has a tremendous potential as a source of renewable energy involving waste management and providing an alternative for fossil fuel. The produced biogas can be used as an alternative of natural gas for heat production, combined heat and power (CHP) production, and transport fuel applications. The biogas technology is at advanced stage most in the European region. The biogas production, means of production, and utilization varied significantly from country to country due to the implemented policies for biogas development and its end use application. In Europe, Germany is the highest biogas producer and have a share of more than half of the total biogas produced in Europe (Scarlat et al. 2018). A high and stable electricity price is guaranteed to the producer has led to a boost of small-scale agricultural biogas production compared to Sweden or Spain. On other hand, Sweden promotes use of biogas as transport fuel. Different sectors of society need to be involved in such projects. Type of technology under consideration or plant installation size depends on local conditions. Different types of initiation are required for the successful implementation and running of the project with its full potential. Regulations and proper policy framework and uninterrupted government support are mandatory for any local or commercial biogas plant. Understanding of the implemented technology, political strategies, national strategies, political incentives, and other driving forces for the continue development of biogas sector are seeming to be most effective for driving a commercial success of any biogas plant. Renewable energy production is always promoted as an energy sector at national and international level. The renewable energy development helps to decrease the depending on import of petroleum items, provides an opportunity for job creation at local level, prevents emission of greenhouse gases. However, these ambitions are rarely achieved due to various known and unknown reasons. Hence, such reasons need to be evaluated and solved before the actual implementation of developmental stage only. The major focus areas are renewable energy policies, waste management policies, and agriculture policies for the success of a biogas plant (Engdahl 2010; Capodaglio et al. 2016).

In 2005, the Biomass Action Plan of Europe shown an increase from 800 TWh (2900 PJ) to 2200 TWh from 2003 to 2010, respectively, as a great potential for biomass utilization for renewable energy production. Recently, AEBIOM, the European Biomass Association, shows a biogas potential of around 460 TWh (1700 PJ) by 2020 in the EU-27 for which agriculture products (energy crops and manure) and waste (biodegradable waste and sewage sludge) are recommended as substrates. Landfill gas recovery is also included in this potential. The produced biogas from above substrates are equivalent to third of natural gas production in Europe (Engdahl 2010).

To promote the development of biogas sector, European Commission states in Renewable Energy Road Map of 2006 that "The EU has compelling reasons for setting up an enabling framework to promote renewables. They are largely indigenous, they do not rely on uncertain projections on the future availability of fuels, and their predominantly decentralised nature makes our societies less vulnerable." Additionally, in this report, the large investments in coal and nuclear power made in the past are compared to the large investments in renewable energy needed today. Report also mentions the fundamental changes in policies are necessary for desired transition to a sustainable society (Ragwitz et al. 2005).

4.6 Risk

Oil prices dropped to US$28 per barrel from US$112 in January 2016, a decrease of more than 65% leading to maximum oil usage compared to other bioenergy sources such as natural gas. According to the World Bioenergy Association (WBA) oil price survey in 2015, equipment suppliers in renewable energy sector were affected leading to lower investment, lower profit margins, and less financial resources for future development due to lesser oil prices. However, on the other hand, lower prices of oil increased economic condition in agriculture with increased farm productivity. The transportation costs were decreased due to lower project costs for bioenergy producers (WBA 2015; World Energy Council Report 2016; Owusu and Asumado-Sarkodie 2016). Strategy of strong blending mandates were also affected. Inflation was on rise in developing countries, which were highly dependent on fossil imports. Thus, globally policy makers need to reduce fossil fuel subsidies and increase investments in renewable energy. In addition, mandatory blends of biofuel need to increase to help the renewable energy sector. In the USA, the current government has pulled out of PARIS Agreement; additionally no major push has been seen in terms of promoting use of BioCNG, policy frame, etc. (Zhang et al. 2017). In the UK, recent policy fluctuations, tariff reduction, and funding issues have created negative impacts in the biogas market. In Holland, the local fossil gas exploration has been closed down, due to which Holland has a great potential in biogas production from biomass and cattle waste. However, the regulatory procedure and related approvals are too lengthy due to which may affect the mindset of biogas plant investors (Thran et al. 2015; IRENA 2018; Mittal et al. 2018).

5 Potential Market

Germany has reached to its maximum potential in this sector, thus limiting further growth. Denmark, Italy, and Flanders are having good potential with supportive policy frameworks, thus proving to be lucrative markets. Ukraine has opportunities with potential to get converted in recent future. In Saudi Arabia, potential of setting up plant from poultry and food waste exists and currently under exploration. The Southeast Asia (SEA) has a good potential to generate biogas from palm oil waste. India's total biogas production is 2.07 billion m^3 per year which is quite low compared to its estimated potential of 29–48 billion m^3 per year. Schemes such as NBMMP, off-grid biogas power generation program, and waste to energy program have been started by the Indian government for biogas development in India (MNRE 2015a; Shukla 2007).

5.1 Biogas Products and Co-products

To successfully run a biogas plant by economical means, it is important to generate revenue from main product (biogas/bio-methane) as well as from the co-products (such as CO_2, H_2S, liquid and solid digestate, etc. obtained during process). Till the decade of 1990, the available technologies were used only for the purpose of industrial and agricultural waste disposal, i.e., landfill and storage (Robert 1993). Over the period of time, leaching of heavy metals and organics due to stored waste material lead to land, air, and water pollution. Increased socio-economical concerns due to pollution forced the concept of recycle, renew, and reuse of any kind of waste to obtain value-added products. Due to which various researchers and industries were in the search of novel sustainable technology for valorization. Methane (55–70%) is the main product of biogas anaerobic digestion including CO_2 (25–45%), H_2S (0.002–2%), and other gases such as nitrogen (<2%), water vapor (2–7%), hydrogen, and ammonia (<1%) (Hashimoto and Varriel 1978). The biogas composition differs depending on the feedstock used during anaerobic digestion. Biogas as such can be used to provide energy in the form combined heat and power (CHP). The biogas purification is important prior to use as fuel as it contains H_2S, which may lead to formation of sulfuric acids resulting into failure and damage of engines. The purified biogas, i.e., biomethane can be sold to various industries in the form of compressed natural gas (CNG) or as compressed biogas (CBG), liquefied natural gas (LNG) to be used for transport and cooking applications.

The biogas byproducts, i.e., CO_2 can be used for construction materials (such as concrete and aggregates), chemical intermediates (such as methanol, syngas and formic acid), fuels (such as methane and liquid fuels), algal biomass (to create biofuels or food additives), polymers (such as polycarbonates, polyurethane and poly-hydroxy acids), novel materials (such as carbon fiber), as well as in agriculture sector as microbial fertilizer and bio-pesticide (Batzias et al. 2005; Bereketidou and Goula 2012; Moeller et al. 2004). According to Global CO_2 initiative (GCI), carbon-based products industry (CBPI) can significantly contribute to reduce carbon emissions upto 10% of annual CO_2 emissions. The CBPI products represents an annual revenue opportunity of $800billion to $1.1 trillion. In 2016, GCI at World Economic Forum in Davos announced to drive substantial economically based change by developing and harnessing market demand for products that capture and reuse

CO_2 to reduce the impact of greenhouse gases on climatic change as per PARIS agreement (GCI 2016).

The solid residue obtained after anaerobic digestion in the form of solid can be used as a high-value fertilizer for agricultural and horticulture as an alternative to common chemical fertilizers not allowed in ecological practice to increase the yield. The residue contains various micro- and macronutrients including nitrogen, phosphorous, and potassium. Organic matter of digestate can help to increase humus content in soil, which is beneficial for organic fertilizers of arid and semi-arid lands with low carbon content (Sogn et al. 2018; Herrmann et al. 2017). The economic viability of any biogas plant crucially depends on the revenue obtained from fertilizers and other components generated during process. H_2S obtained from biogas plant after scrubbing can be used to produce sulfuric acid and elemental sulfur. Inorganic sulfides obtained from H_2S can be used in pesticide, dyes, and pharmaceutical industry. It is used in metallurgy and to produce heavy water for nuclear power plants. Hydrogen sulfide is used to manufacture dyes, rubber chemicals, pesticides, polymers, plastic additives, leather, and pharmaceuticals. It is also used for the purification of nickel, manganese, hydrochloric acid, and sulfuric acid and as a source of hydrogen. Elemental sulfur as fertilizer is used to reduce soil pH (Beauchamp et al. 1984; Dimitris et al. 2018).

Increasing environmental regulations to reduce greenhouses gases help to drive waste derived biogas market globally to focus on renewable energies. According to Future Market Insights, the global biogas market revenue was about $24.5 billion in 2015. The global biogas revenue is estimated to reach $48762.2 million growing at a 6.5 CAGR. The revenues are projected to double between 2019 and 2026 with Asia-Pacific and Latin America as a key region fueling demand. Over $22 billion global revenues will be accounted by biogas sourced from agriculture waste by 2026, which is about $3 billion more than biogas produced from municipal waste (Future Market Insights 2017). According to BCC Research Report, the global biogas production technologies market is expected to grow at a CAGR of 10.6% upto 2022, worth of $10.1 billion. Wastewater/sludge and industrial application will lead the market with a CAGR of 11.4%, and landfill gas, agriculture, municipal waste, and food will capture a market of $5.8 billion by 2022 (BCC Research Report 2018).

5.2 *Barrier to Biogas Commercialization*

The major economic barriers to biogas commercialization are high investment including high capital and installation cost, unavailability of long-term financing options, high interest rate, high level of bureaucracy, delay in receiving funds from financial sources, procurement, price, and income are the market barriers for biogas commercialization. Lower profitability after selling for electricity or heat utilization, additional cost when connecting to grid, logistic difficulties for raw material collection and dispersal. Easy accessibility to alternative fuels, i.e., firewood, cow dung, solid biomass, liquid petroleum gas (LPG), kerosene, etc., and other factors such as fuel supply assurance, easy procurement are additional deterrents (Bansal et al. 2013; Rao et al. 2010).

Lack of social awareness about the biogas acceptance from waste and lower participation of women due to gender discrimination (Nelsan and Kuriakose 2017; Mittal et al. 2018). Inadequate supply of feedstock, lack of labor, inadequate maintenance leading to lower biogas generation affecting the economy of biogas plant in developing countries (Zuzhang 2013). Lack of awareness of the best available technologies, government policies or bureaucracy, difficulty in obtaining permission, unstable policies, cost of feedstock marketing of produced biogas, manure and other bi-products are challenges to be overcome. Additionally, initial cost of investment, lack of infrastructure and storage, enough filling station, costly upgradation, and competition with natural gas are supply chain challenges (Kemausuor et al. 2018; Mittal et al. 2018).

6 Fostering Investment

Germany, Italy, the UK, France, and Switzerland are providing legal framework, encouraging developed technologies for the biogas production (Kreeft 2018; Benato and Macor 2019; Theurel et al. 2019). In Germany, the biogas plants are increased to 9000 from 4136 in year 2010. Majority of the plants are being run by farmer cooperatives using maize or turnip as feedstock and produces about 6.6 Million Metric Ton (MMT) of biogas. In India, about 62MMT of CBG can be produced from various sources with a bio-manure generation capacity of 370 MMT (EOI 2018). The bio-manure includes biomass/waste generated from agricultural residue, MSW, press mud, distillery spent wash, cattle dung, kitchen waste, animal waste, and sewage treatment plant waste. In India, the task force of NITI Aayog has decided to promote technologies which are providing higher BioCNG yield per unit of waste processed (MoPNG 2018). Countries like Germany, China, and India are leading the production of biogas. Bolivia implemented National Programme for biogas in 2013 and aiming installation of 6500 domestic biogas digesters plants as well as a viable market for supply and maintenance of these plants in the long run. In case of Turkey, municipal waste is a viable and sustainable alternative for fossil fuels. Sixty-four biogas plants are running on municipal waste with a total capacity of 322 Mwe. In Kenya, Gorge Farm AD power plant is the largest plant delivering 2.4 MW output to farm and local grid. The power can be used to light up 8000 households. The plant produces 35Kilo ton of natural fertilizer as by-product using 50 kilo tons or organic crop waste annually. The fertilizer is used as an alternative of synthetic fertilizer to improve soil conditioning and crop yield (World Energy Council Report 2016). Thus with the support of multiple initiatives, policies, frameworks, etc., investment in this sector is seeing traction and pull.

6.1 *Strengthening Market for Biogas*

Development of a holistic market for biogas systems and products is a chicken and egg story. Unless there is an established ecosystem, investments will not come. Similarly, without many investments in the sector, developing ecosystems is not fruitful. In view of this, generally governments facilitate to help in strengthening the market. The critical aspects of the biogas market that requires development are mentioned below.

6.1.1 Biogas Use: Policy Directions

Depending on the requirements which are case specific, countries generally promote the use of biogas through various policies. For example, in Germany and other European countries, biogas was used to generate power. Government gave special feed-in tariffs. In the USA, policies were made to overcome barriers to integrate biogas into electricity and natural gas market through interconnection and pipeline injection standards, fee structures, incentives, etc. (Smolinksi and Cox 2016; Scarlat et al. 2015). In India initially small biogas plants were promoted to be used to regenerate gas for cooking and lighting purpose. In India, under National Biofuel Policy 2018, the biogas is considered under advance biofuels category. Currently India is promoting compressed biogas as an alternate for transport fuel use under SATAT scheme (MoPNG 2018). Such policies create an atmosphere of stability which promotes investments. A snapshot of the Indian policies regarding biogas is shown in Table 2 (MoPNG 2018; IREDA 2018a; Mittal et al. 2018; MNRE 2015a, 2018).

Table 2 Indian biogas policy implementation

Year	Policy
1981	The first program of biogas development was launched. Capital subsidy for installing small biogas plants
1995	National program to recover energy from municipal solids, industrial and agricultural waste launched
2006	National Biogas and Manure Management programme (NBMMP) by Ministry of New and Renewable Energy (MNRE) MNRE offered financial incentives for setting up biogas power or cogeneration plants or production of BioCNG using biomethnation technology
2016	New Tariff Policy mandated electricity distribution companies to procure 100% power from waste to energy plants
2018	New Biofuel Policy categorized biogas as advance biofuel SATAT scheme announced by Ministry of Petroleum and Natural Gas to use compressed biogas for transportation sector

6.1.2 Promotion of Biogas Co-products

The major co-product of any biogas plants is solid slurry, and liquid wash needs to be classified as fertilizers or soil amendments as they have suitable nutrients and fiber. Creation of channel outlets to buy such by products from the plants and sell them as value added products is crucial. The Government of India is encouraging use of such organic or biofertilizer through various schemes such as National Mission for Sustainable Agriculture (NMSA), Paramparagat Krishi Vikas Yojana (PKVY), Rashtriya Krishi Vikas Yojana (RKVY), Mission for Integrated Development of Horticulture (MIDH), National Mission on Oilseeds and Oil Palm (NMOOP), National Biogas and Manure Management Programme (NBMMP), Network Project on organic farming by Indian Council of Agricultural Research (ICAR), and National Programme on Organic Production (NPOP) of Agricultural and Processed Food Products Export Development Authority (APEDA) (MoPNG 2018; IREDA 2018a; MNRE 2015a). In order to promote renewable fuels, it is imperative to give credits to biogas systems and products. In the USA according to Energy Policy Act of 2005, such projects are eligible for Renewable Identification Number (RIN) and Low Carbon Fuel Standard (LCFS) credits. Such projects are also covered under CDM (Carbon Offset) credit mechanism (USDA 2015).

6.1.3 Evacuation

Biogas can be used to generate power and natural gas or as an alternate to transportation fuel. In all such scenarios, it is very critical to establish an effective evacuation channel. Biogas cannot be economically stored in large volumes. So to manage the demand and supply an effective evacuation channel is necessary. The evacuation channel could be created by connecting with electricity grid, connecting with natural gas pipelines or nearby gas pump stations. European countries have created a national gas pipeline network (Aryal and Kvist 2018; Daniel-Gromke et al. 2018; IEA 2014b). India is trying to create an ecosystem to use biogas in transportation sector, which has a huge potential. Under SATAT scheme, the Indian Government has mandated the oil marketing companies (OMCs) to buy compressed biogas (CBG) from all such producers. The scheme also provides for long-term commitment of gas uptake and price. The OMCs will put up the infrastructure to build nearby pump stations, if required. Thus effective evacuation strategies are being implemented to boost the sector (MoPNG 2018).

6.1.4 Subsidies and Funding Support

Ample subsidy and funding support for waste to energy project exists. To channelize investments in this sector initially subsidies and funding support (low interest capital) are needed. Subsidy in the form of income tax breaks, low tax on the output product, CAPEX subsidy, or soft loans are given provided globally.

6.2 Establish a Biogas Opportunities Roadmap

To implement programs to support and expand biogas industry, countries generally create an opportunity roadmap and form working groups. Such working groups then take up the execution task. In 2014, the USA released the Climate Action Plan—Strategy to Reduce Methane Emissions. U.S. Department of Agriculture (USDA), the U.S. Environmental Protection Agency (EPA) and the U.S. Department of Energy (DOE) prepared a biogas roadmap for biogas industry. The roadmap outlined the actions for Federal government to enhance the biogas use such as promoting biogas utilization through existing agency programs, fostering investments in biogas systems, strengthening markets for biogas systems and system products, and improving communication and coordination among investors and government bodies (USDA 2014). The main task of the working group was to collaborate with industry to identify and prioritize policies. Existing and potential interagency cooperative structures and initiatives were also included, specifically:

- EPA's and USDA's AgSTAR program
- DOE's and USDA's biomass research and development initiative
- EPA, USDA, and DOE "Biodigesters and Biogas" work group
- EPA, USDA, and USGS integrated nutrient management strategy

In India key departments driving the biogas program are the Ministry of New and Renewable Energy (MNRE), Ministry of Petroleum and Natural Gas (MoPNG), Oil Marketing Companies (OMCs), Agricultural and Processed Food Products Export Development Authority (APEDA) (MNRE 2015a; MoPNG 2018).

6.3 Marketing Mode of Biogas Commercialization

Biogas produced at these CBG plants would be transported in cylinders to the fuel station network of OMCs for marketing as a green transport fuel alternative. In India, 1500 CNG stations network currently serves about 32 lakh gas-based vehicles. CBG networks can be integrated with city gas distribution (CGD) networks. Later, the retailing from OMC fuel stations could be injected into CGD pipelines for efficient distribution and access. The potential for CBG production from various sources in India is estimated at about 62 million tons per annum. Additionally, the entrepreneurs would be allowed to separately market other co-products from these plants to enhance returns on investment (EOI 2018). The Government of India is planning to double city gas networks to 400 districts, quadruple piped cooking gas connection to households to 2 crores and also set up 10,000 CNG dispensing station to create a gas based economy. The share of biogas in energy basket will be increased from 6.2% to 15% by 2030, will create lakhs of jobs, and will help meet India's COP-21 commitment of cutting emission intensity by 33–35% (Press Report 2018a). The Petroleum and Natural Gas Regulatory Board (PNGRB) recently reformed to a large extent the CGD authorization regulations. CGD network is turning out to be the big investment opportunity in India with an investment of as much as 1.1 trillion expected over the next decade (Press Report 2018b).

6.4 Commercial and Local Biogas Case Study

Biogas can improve lives of people and communities. The chapter mentions few success stories related to biogas production showing how biogas helped to improve health conditions and preventing environmental damage with better financial gains (Press Report 2016).

1. Khamtara village of Madhya Pradesh, India, during 1995–2005 under government schemes built about 150 biogas units ranging from 2 to 4 M^3 in size depending on the animal number and cow dung obtained from these animals per house. Even today, these units are operational even today. The produced biogas has replaced about 20 tons (100 bullock carts) of firewood costing about 1000/– Rs. per cart being used yearly resulting in better utilization of saved money. People involved in firewood business have now shifted to other occupations. Replacement of firewood with produced biogas for kitchen work reduced about 80% of air pollution/smoke resulting in lesser number of patients related to asthma, eye, nose, cancer, etc. and thus improving overall health status of people. In addition, the fights on petty issue have dropped leading to villager's healthy co-operation with each other.
2. In 1987, Gujarat Energy Development Agency and Dudh Sagar Dairy built the largest biogas system in Methan Village, Gujarat, based on cow dung. The plant provided biogas to 236–500 household via underground pipes at a marginal cost per family per month. Each day about 2.5 tons of cow dung transferred to eight giant biogas units produce about 630 M^3 of biogas. The produced biogas saves about 500 tons of firewood each year and provides clean fuel with health benefits. Firewood replacement prevented carbon dioxide emission of 860 tons per year reducing impact on climate change.
3. In Sirsi Village in Karnataka state of India, 100% of biogas plants built on dung are functioning satisfactorily and 85% of households met all their cooking energy needs with biogas, improving quality life of women. Competitive participation of various investors for household plant construction, commissioning, procuring subsidy, guaranteed performance and free servicing, and presence of multiple agencies in dissemination network contributes toward high rate of biogas plant success (Bhat et al. 2001).

6.5 Commercial Biogas Plant of China as Case Study

In China the biogas projects are based on rural household waste, agricultural waste, and industrial organic waste based. Till 2014, about 35,533,000 rural household biogas plants were in operation accounting for 30% of suitable household purpose. Total annual biogas production by 2014 in China reached to 15.5 billion M^3. Agricultural waste biogas plant produced about 2 billion M^3 of biogas and industrial waste plant generated 0.25 billion M^3 of biogas (Minister of Agriculture 2015).

Deqingyuan Agricultural Scientific Company Ltd. (DQY), Beijing, is the first large-scale biogas project of China which runs only on chicken manure as a feedstock. It is the first agricultural company to be approved by Clean Development Mechanism (CDM) of China introduced by Kyoto protocol. The DQY ecological garden has provided a job opportunity to more than 400 workers. DQY is the largest egg farm in Asia. It recycles about 80,000 Ton of chicken manure and 1 Lac Ton of sewage produced per year for biogas generation. DQY employed an advanced technology to convert obtained waste into biogas to prevent environmental pollution creating significant economical, social, and environmental benefits. DQY promotes renewable energy and also contributes to sustainable development. DQY closely connects livestock farming, plantations, surrounding households, and markets, forming a demonstrative model of a circular economy. The plant focuses on maximization of resource use, conversion of maximum livestock manure to energy and simultaneously reducing manure pollution and providing a prototypical model of contract agriculture. In terms of technology, DQY biogas project uses CSTR, high manure concentration, and medium temperature. The plant is divided into five main components such as pre-treatment of raw materials, biological desulphurization, a power grid, biogas for household use, and biogas digestate for fertilization of agricultural crops. CSTR allows mixing and heating of raw material and microorganisms uniformly and increases the rate of methane formation. The technology consumes lower energy providing high processing capacity, easy operation and management, low production cost, and maximum biogas production. The approximate annual DQY plant operating cost is RMB six million yuan. Currently, DQY produces 14 million KWh of power annually and earns a profit of RMB 8.4 million yuan each year from power generation. In addition, DQY also generates profit of RMB two million yuan by selling of 6600 Ton annually. DQY improved discharged water quality to reduce discharge penalties and thus saved another RMB 2 Lac yuan each year. The total economic benefits of project are about RMB 10.6 million yuan per year. Considering the operating cost, the net economic profit of DQY is RMB 4.6 million yuan per year, and anticipated static payback period is about 14.13 years. As per the World Bank, Dutch government and DQY company analysis, the project helps for annual reduction in GHG emission which are equivalent about 84,000 Ton of CO_2. In this regards, DQY gains a subsidy of RMB eight million annually contributing to the economic profit upto RMB 12.6 million yuan annually and significantly shortening investment payback period to 5.16 years of the biogas project. DQY shows a circular economy which combines project with ecological agriculture. The DQY biogas plants represent a successful commercial example of rural renewable energy strategy of China (Chen et al. 2017).

6.6 Socioeconomical and Environmental Impact

Development and installation of biogas plants will majorly improve the environment conditions, increase farm productivity and improve agricultural land quality due to use of fertilizer, create jobs, and improve the health conditions of women who use to cook daily food using agri-residue. The income can be generated through the sale of fertilizer is a major benefit. Reduction in landfill space can save the land cost. BioCNG obtained from biogas used in transport vehicles will lead to reduction in carbon footprint, improving climatic conditions and reduction in greenhouse gases (Buadit et al. 2013; Paolini et al. 2018). Global CO_2 emissions can be reduced by 2050 and completely phased out by 2060 by the energy sector with a net positive economic outlook (IRENA 2017b). Due to installation of biogas plants, the local people get employment leading to improved rural and regional economics creating additional income with farming. The sector will also provide jobs to people in transport of feedstock and products. Improved health condition of biogas users indirectly increases income by preventing expenses during ill-conditions (Scarlat et al. 2018; Lauer et al. 2018). With holistic policies, the transition may greatly boost overall employment in the renewable sector. Globally more than 7.7 million people are employed in the renewable energy sector. An 18% increase compared to last year's 6.5 million jobs. China, Brazil, USA, India, Japan, and Germany accounts for most renewable energy jobs. Globally, 62% of jobs were found in Asia. China is the largest employer with 44% of world's jobs followed by Brazil (0.9 million), USA (0.7 million), India (0.4 million), and Germany (0.3 million) in renewable energy sector (IRENA 2017c).

Biogas in India is of strategic importance with the ongoing initiatives of the Government such as Make in India and Swachh Bharat Abhiyan and offers great opportunity to integrate with the ambitious targets of doubling of farmer's income, import reduction, employment generation, and waste to wealth creation. Simultaneously, the existing biodiversity of the country can be put to optimum use by utilizing dry lands for generating wealth for the local populous and in turn contribute to the sustainable development. India is expected to roll out more than 5000 CBG plants across nation in a phased manner, with 250 plants by the year 2020, 1000 plants by 2022, and 5000 plants by 2025. These plants are expected to produce 15 million tons of CBG per annum, which is about 40 percent of current CNG consumption of 44 million tons per annum in the country. The move would attract investment of around Rs. 1.7 lakh crore and is expected to generate direct employment for 75,000 people and produce 50 million tons of bio-manure for crops, aimed to provide a sustainable fuel alternative that would benefit vehicle users, farmers, and entrepreneurs (MoPNG 2018).

Praj Industries Limited Praj Industries, ranked eighth in the list of TOP 50 hottest companies in Advanced Bio-economy for year 2019 by Biofuel Digest, is India's most successful company in the field of bio-based technologies and engineering with headquarters in Pune, Maharashtra, India. Praj has spread its presence across the globe with more than 750 references in more than 75 countries. Today Praj is a globally leading company with a basket of sustainable solutions for bioenergy, high purity water, critical process equipment, breweries, and industrial wastewater treatment. Praj has carved a unique position in the world of ethanol technology by virtue of its expertise which cuts across a variety of sugar to starch based feedstock, collectively called as first-generation feedstock. Praj is one of the handful of companies in the world to successfully develop and demonstrate second-generation ethanol technology using agri-residue. Praj has developed technologies for several clean, renewable fuels and chemicals, viz., BioCNG, Bio-butanol, etc. which have the potential to redefine the global energy matrix. Right from conceptualization, technology, design, and plant engineering up to project installation and commissioning—Praj offers complete solution backed by expertise and experience. The backbone of Praj's technology development is Praj Matrix—R&D Center (division of Praj Industries) designed along the principles of sustainability and innovation with the goal of providing environmentally friendly solutions for a future-perfect world. Praj Matrix applies multidisciplinary experience and expertise, deploying world class laboratory, pilot, and scale up facilities to accelerate the development of bio-based technologies. These facilities ensure development of robust technology packages that exceed customer expectations of performance, cost, and quality. Located in Pune (India), Praj Matrix has received status of Private Sector BioTech Park by the Government of Maharashtra. The authors of this chapter are engaged in the Compressed Biogas Business of Praj in diverse capacities spanning the entire value chain from R&D to commercial.

References

Abbasi T, Tauseef SM, Abbasi SA (2012) Low-rate and high-rate anaerobic reactors/digesters/ fermenters. Biogas energy. Springer, New York

Abeliotis K, Lasaridi K, Costarelli V, Chroni C (2015) The implications of food waste generation on climate change: the case of Greece. Sustain Prod Consump 3:8–14

Abu-Reesh IM (2014) Kinetics of anaerobic digestion of labaneh whey in a batch reactor. Afr J Biotechnol 13:1745–1755

Achinas S, Achinas V, Euverink GJW (2017) A technological overview of biogas production from biowaste. Engineering 3(3):299–307

Agbor VB, Cicek N, Sparling R, Berlin A, Levin DB (2011) Biomass pretreatment: fundamentals toward application. Biotechnol Adv 29:675–685

Aggarwal D (2003) Biogas plants based on nightsoil. The Energy and Resources Institute (TERI), New Delhi

Akkoli KM, Gangavati PB, Ingalagi MR, Chitgopkar RK (2018) Assessment and characterization of agricultural residues. Mater Today Proc 5(9):17548–17552

Alayi R, Shamel A, Kasaeian A, Harasii H, Topchlar M (2016) The role of biogas to sustainable development. J Chem Pharm Res 8(4):112–118

Amaratunga M (1986) Structural behaviour and stress conditions of fixed dome types of biogas units. In: El-Halwagi MM (ed) Biogas technology, transfer and diffusion, London, pp 295–301

Ammenberg J, Bohn I, Feiz R (2017) Systematic assessment of feedstock for an expanded biogas production-A multi-criteria approach. Biogas Research Center Report

Aramrueang N, Zicari SM, Zhang R (2017) Characterization and compositional analysis of agricultural crops and residues for ethanol production in California. Biomass Bioenergy 105:288–297

Aryal N, Kvist T (2018) Alternative of biogas injection into the Danish gas grid system – A study from demand perspective. Chem Eng 2:43

Balat M, Ayar G (2005) Biomass energy in the world, use of biomass and potential trends. Energy Sources 27(10):931–840

Bansal M, Saini RP, Khatod DK (2013) Development of cooking sector in rural areas in India – a review. Renew Sust Energy Rev 17(C):44–53

Barker H (1956) Bacterial fermentations. Wiley, New York

Batzias FA, Sidiras DK, Spyrou EK (2005) Evaluating livestock manures for biogas production: a GIS based method. Renew Energy 30:1161–1176

BCC Research Report (2018) Waste derived biogas: global market for anaerobic digestion equipment. https://www.bccresearch.com/market-research/energy-and-resources/waste-derived-biogas-global-markets-for-anaerobic-digestion-equipment-report.html

Beauchamp RO, Bus JS, Popp JA, Boreiko CJ, Andjelkovich DA (1984) A critical review of the literature on hydrogen sulfide toxicity. Crit Rev Toxicol 13(1):25–97

Beltran FJ, Alvarez PM, Rodriguez EM, Garcia-Araya JF, Rivas J (2001) Treatment of high strength distillery wastewater (cherry stillage) by integrating aerobic biological oxidation and ozonation. Biotechnol Prog 17:462–467

Benato A, Macor A (2019) Italian biogas plants: tends, subsidies, cost, biogas composition and engine emissions. Energies 12:979

Bereketidou OA, Goula MA (2012) Biogas reforming for syngas production over nickel supported on cera-alumina catalyst. Catal Today 195(1):93–100

Bhat PR, Chanakya HN, Ravindranath NH (2001) Biogas plant dissemination: success story of Sirsi, India. Energy Sustain Dev 5(1):39–46

Bhoite GM, Vaidya PD (2018) Improved biogas generation from biomethanated distillery wastewater by pretreatment with catalytic wet air oxidation. Ind Eng Chem Res 57(7):2698–2704

Bhol J, Sahoo BB, Mishra CK (2011) Biogas digesters in India: a review. National Conference on Renewable and New Energy Systems

Bonazzi G, Lotti M (2015) Evaluation of biogas plants by the application of an internal rate of return and debt service coverage approach. Am J Environ Sci 11(1):35–45

Bond T, Templeton MR (2011) History and future of domestic biogas plants in the developing world. Energy Sustain Dev 15(4):347–354

Bragatto P (2013) Safe operation of biogas plants in Italy. Chem Eng Trans 32:199–204

Brahma A, Saikia K, Hiloidhari M, Baruah DC (2016) GIS based planning of a biomethnation power plant in Assam, India. Renew Sust Energ Rev 62:596–608

Braun R (2007) Anaerobic digestion: a multi-faceted process for energy, environmental management and rural development. In: Ranalli P (ed) Improvement of crop plants for industrial end uses. Springer, Dordrecht, pp 335–145

Buadit T, Aroonsrimorakotb BK, Thavipoke P (2013) Biogas production and greenhouse gases reduction from wastewater at Mahidol University, Salaya campus, Thailand. APCBEE Procedia 5:169–174

Capodaglio AG, Callegari A, Lopez MV (2016) European framework for the diffusion of biogas uses: emerging technologies, acceptance, incentive strategies and institutional-regulatory support. Sustainability 8(4):298

Carlini M, Mosconi EM, Castellucci S, Villarini M, Colantoni A (2017) An economical evaluation of anaerobic digestion plants fed with organic agro-industrial waste. Energies 10:1165
Chan YJ, Chong MF, Law CL, Hassell DG (2009) A review on anaerobic–aerobic treatment of industrial and municipal wastewater. Chem Eng J 155:1–18
Chawla OP (1986) Advances in biogas technology. Publications and Information Division, Indian Council of Agricultural Research, New Delhi
Chen L, Cong R, Shu B, Mi Z (2017) A sustainable biogas model in China: the case study of Beijing Dequingyuan biogas project. Renew Sust Energ Rev 78:773–779
Chynoweth DP, Owens JM, Legrand R (2001) Renewable methane from anaerobic digestion of biomass. Renew Energy 22(1–3):1–8
Claudius DCG (2013) Biogas as an energy option: an overview, The Biogas Handbook, Woodhead Publishing Series in Energy, pp 1–16
Coelho LMG, Lange LC, Coelho HMG (2017) Multi-criteria decision making to support waste management: a critical review of current practices and methods. Waste Manag Res 35(1):3–28
Conte MN, Kotchen MJ (2010) Explaining the price of voluntary carbon offsets. Clim Chang Econ 1(2):93–111
Copeland C (2016) Clean water act: a summary of the law
Daniel-Gromke J, Rensberg N, Denusenko V, Stinner W, Schmalfu T, Scheftelowitz M, Nelles M, Liebetrau J (2018) Current developments in production and utilization of biogas and biomethane in Germany. Chemie Ingenieur Technik 92(1–2):17–35
DeKoning HW, Smith KR, Last JM (1985) Biomass fuel combustion and health. Bull World Health Organ 63(1):11–26
Dennison W, Helfrich M, Michelsen E, Pritzlaff R, Tutman F (2012) Senior scientists and policymakers for the bay nutrient trading subcommittee. Chesapeake Bay Action Plan. https://www.epa.gov/sites/production/files/2015-07/documents/nutrient_trading_final_report.pdf
Dimitris LB, Styliani NC, Miltiadis M, Georgios IS, Petros PS (2018) Effect of elemental sulfur as fertilizer ingredient on the mobilization of iron from the iron pools of a calcareous soil cultivated with durum wheat and the crop's iron and sulfur nutrition. Agriculture 8:20
Dueblein D, Steinhauser A (2008) Biogas from waste and renewable resources. KGaA; Wiley-VCH Verlag GmbH and Co
EAEF (2006) Guidelines for the preparation of feasibility studies for the generation of renewable energy from organic waste and biomass for Vietnam and other ASEAN countries. http://homepages.hs-bremen.de/~office-ikrw/renew/Download/Guideline.pdf
EBA (2018a) Overview on key EU polices for the biogas sector. http://european-biogas.eu/2019/02/06/overview-on-key-eu-policies-for-the-biogas-sector/
EBA (2018b) Statistical report of the European Biogas Association 2018. Brussels, Belgium, December 2018
EC (2009) DIRECTIVE 2009/28/EC of the European Parliament and of the Council on the Promotion of the Use of Energy from Renewable Sources and Amending and Subsequently Repealing Directives 2001/77/EC and 2003/30/EC, 2009
EC (2011) Energy roadmap 2050. Communication from the Commission to the European Parliament, the Council, the European Economic and Social Committee and the Committee of the Regions, COM (2011) 885 final. 2011
EC Directive (1999) EC Directive (1999/31/EC) Landfill Directive. http://ec.europa.eu/environment/waste/compost/EC Directive 1999/92/EC
Eliasson KA, Nadeau E, Lev L, Schnurer A (2017) Production efficiency of Swedish farm-scale biogas plants. Biomass Bioenergy 97:27–37
Engdahl K (2010) Biogas policies, incentives and barriers – a survey of the strategies of three European countries. Master Thesis. Lund University. http://lup.lub.lu.se/luur/download?func=downloadFile&recordOId=4468127&fileOId=4469065
EOI (2018) EOI for production and supply of compressed bio gas (CBG). https://www.bharatpetroleum.com/pdf/EOICBGFINAL-46978-4fd490.pdf

EPA (2014) Global mitigation of non-CO2 greenhouse gases: 2010–2030 executive summary. April 2014. EPA report 430S14001. http://www.epa.gov/climatechange/EPAactivities/economics/nonco2mitigation.html

EREC (2011) Mapping renewable energy pathways towards 2020. http://www.eufores.org/fileadmin/eufores/Projects/REPAP_2020/EREC-roadmap-V4.pdf

EU (2018) The common agricultural policy at a glance. https://ec.europa.eu/info/food-farming-fisheries/key-policies/common-agricultural-policy/cap-glance

EU Commission (2008). The European Commission (2008a) COM (2008) 19 final, Directive of the European parliament and of the council: on the promotion of the use of energy from renewable sources, Brussels

EU-ETS (n.d.). https://ec.europa.eu/clima/policies/ets_en

FAO (2014) Bioenergy and food security rapid appraisal. http://www.fao.org/3/a-bp844e.pdf

Finalayson R (2011) New Zealand targets 90% renewable energy by 2025. https://www.industrialinfo.com/news/abstract.jsp?newsitemID=186373

Future Market Insights (2017) Biogas market: Asia Pacific anticipated to witness significant growth till 2020: Global industry analysis and opportunity assessment 2016-2026. https://www.prnewswire.com/news-releases/global-biogas-market-to-increase-at-65-cagr-through-2026-610821185.html

GCI Report (2016) A roadmap for the global implementation of carbon utilization technologies transforming CO_2 from a liability to an asset at significant market scale

Gigot C, Ongena M, Fauconnier ML, Muhovski Y, Wathelet JP, du Jardin P, Thonart P (2012) Optimization and scaling up of a biotechnological synthesis of natural green leaf volatiles using Beta vulgaris hydroperoxide lyase. Process Biochem 47(12):2547–2551

Global Data Report (2013). https://www.process-worldwide.com/germany-remains-the-worlds-leading-biogas-energy-producer-a-391244/

GMI (2014) Finland biogas update of the Global Methane Initiative (GMI)

Hartman RB, Smith DG, Bennett LKD (1979) Sludge stabilization through aerobic digestion. Water Pollut Control Fed 51(10):2353–2365

Hashimoto G, Varriel H (1978) Factors affecting methane yield and production rate. Am Soc Agric Eng 49085:68

He PJ (2010) Anaerobic digestion: an intriguing long history in China. Waste Manag 30 (4):549–550

Heldeweg MA, Sanders M, Harmsen M (2015) Public-private or private-private energy partnerships? Toward good energy governance in regional and local green gas projects. Energy Sustain Soc 5:9

Herrmann A, Kage H, Taube F, Sieling K (2017) Effect of biogas digestate, animal manure and mineral fertilizer application on nitrogen flows in biogas feedstock production. Eur J Agron 91:63–73

Hidayat NK, Offermans A, Glasbergen P (2018) Sustainable palm oil as public responsibility? On the governance capacity of Indonesian standard for sustainable palm oil (ISPO). Agric Hum Values 35(1):223–242

Hills DJ, Roberts DW (1981) Anaerobic digestion of dairy manure and field crop residues. Agric Wastes 3(3):179–189

Hiloidhari M, Das D, Baruah DC (2014) Bioenergy potential from crop residue biomass in India. Renew Sust Energy Rev 32:504–512

Huang J, Liu S, Plaza A, Zhou W (2018) Creating and enabling environment for public-private partnerships in waste to energy projects. No. 58 ADB Sustainable Development Working Paper Series

IBEF (2018). https://www.ibef.org/download/Renewable-Energy-Report-July-2018.pdf

IEA (2010) Energy technologies perspectives: scenarios and strategies to 2050. International Energy Agency, Paris

IEA (2011) Renewable energy technology deployment. Risk quantification and risk management in renewable energy projects. Altran GmbH and Co

IEA (2014a) IEA bioenergy task 37 – energy from biogas. http://task37.ieabioenergy.com/files/member-upload/Countryreportsummary2013.pdf
IEA (2014b) A perspective on the potential role of biogas in smart energy grids. http://task37.ieabioenergy.com/files/daten-redaktion/download/Technical%20Brochures/Smart_Grids_Final_web.pdf
IEO (2017). https://www.eia.gov/outlooks/archive/ieo17/exec_summ.php
IREDA (2017). https://economictimes.indiatimes.com/news/economy/finance/ireda-finances-rs-10000-crore-green-projects-during-2016-7/articleshow/58099563.cms?from=mdr
IREDA (2018a). https://ireda.in/forms/contentpage.aspx?lid=1943
IREDA (2018b) Financing norms and schemes. http://www.ireda.in/writereaddata/Financing_norms14-05-2018.pdf
IRENA (2017a) Biogas for road vehicles
IRENA (2017b) Perspectives for the energy transition
IRENA (2017c) Renewable Energy and Jobs
IRENA (2018) Renewable energy policies in a time of transition. https://www.irena.org/media/Files/IRENA/Agency/Publication/2018/Apr/IRENA_IEA_REN21_Policies_2018.pdf
Jain N, Bhatia A, Pathak H (2014) Emission of air pollutants from crop residue burning in India. Aerosol Air Qual Res 14:422–430
Jingming L (2006) Rural biogas development in China. Division of energy, ecology and environment, Center for science and technology development, Ministry of agriculture, Beijing
Judkins RR, Fulkerson W, Sanghvi MK (1991) Conference: the dilemma of fossil fuel use and global climate change. American Chemical Society National Meeting, Atlanta, GA. AC05-84OR21400
Junpen A, Pansuk J, Kamnoet O, Cheewaphogphan P, Garivait S (2018) Emission of air pollutants from rice residue open burning in Thailand. Atmos 9:449
Kemausuor F, Adaramola MS, Morken J (2018) A review of commercial biogas systems and lessons for Africa. Energies 11:2984
Kim H, Lim S, Yoo S (2019) Residential consumer's willingness to pay price premium for renewable heat in South Korea. Sustainability 11:1234
Koneswaran G, Nirenberg D (2008) Global farm animal production and global warming: impacting and mitigating climate change. Environ Health Perspect 16(5):578–582
Konigsberger S, Deremince B, Mertens Y, Stambasky J, Wellinger A (2018) Promotion of sustainable biogas production in the EU. http://european-biogas.eu/wp-content/uploads/2018/09/BA-PROMO-BROCHURE-FINAL-AS.pdf
Kreeft GJ (2018). Legislative and regulatory framework for power to gas in Germany, Italy and Switzerland. STORE&GO Project, Deliverable 7.3
Kumar S, Gaikwad SA (2004) Municipal solid waste management in Indian urban centres: an approach for betterment. In: Gupta KR (ed) Urban development debates in the new millennium. Atlantic Publishers and Distributors, New Delhi, pp 100–111
Kumar S, Bhattacharyya JK, Vaidya AN, Chakrabarti T, Devotta S, Akolkar AB (2009) Assessment of the status of municipal solid waste management in metro cities, state capitals, class I cities, and class II towns in India: an insight. Waste Manag 29:883–895
Kumar S, Malav LC, Malav MK, Khan SA (2015) Biogas slurry: source of nutrients for eco-friendly agriculture. Int J Ext Res 2:42–46
KVIC (1993) Khadi and VI Commission and its non-conventional energy programmes. KVIC, Bombay. https://www.ibef.org/download/Renewable-Energy-Report-July-2018.pdf
Lauer M, Hansen J, Lamers P, Thran D (2018) Making money from waste: the economic viability of producing biomethane in the Idaho dairy industry. Appl Energy 222:621–636
Leining C, Kerr S (2016) Lessons learned from the New Zealand emission trading scheme. Motu Economic and Public Policy Research. http://motu-www.motu.org.nz/wpapers/16_06.pdf
Lemmer A, Naegele HJ, Sondermann J (2013) Howe efficient are agitators in biogas digesters? Determination of the efficiency of submersible motor mixers and incline agitators by measuring nutrient distribution in full-scale agricultural biogas digesters. Energies 6(12):6255–3273

Lv Z, Feng L, Shao L, Kou W, Liu P, Gao P, Dong X, Yu M, Wang J, Zhang D (2018) The effect of digested manure on biogas productivity and microstructure evolution of corn stalks in anaerobic cofermentation. BioMed Res Int. Article ID 5217369, 1–10

Lyytimaki J (2018) Renewable energy in the news: environmental, economic, policy and technology discussion of biogas. Sustain Prod Consump 15:65–73

Mardiatmoko G (2018) From forest biomass to carbon trading. In: Lopez EJ, Zepka LQ (eds) Renewable resources and biorefineries. IntechOpen Limited, London

Massoud K, George T, Robert CB (2007) Biomass conversion processes for energy recovery. Energy conversion. CRC Press

McCarty PL (1982) One hundred years of anaerobic treatment. In: Hughes DE, Stafford DA, Wheatley BI (eds) Anaerobic digestion, 1981: proceedings of the second international symposium on anaerobic digestion. Elsevier Biomedical, Amsterdam, pp 3–22

Meegoda JN, Li B, Patel K, Wang LB (2018) A review of the processes, parameters, and optimization of anaerobic digestion. Int J Environ Res Public Health 15:2224

Menind A, Olt J (2009) Biogas plant investment analysis, cost benefits and main factors. 8th International scientific conference engineering for rural development 28–29.05.2009, Jelgava, Latvia

Meynell PJ (1976) Methane: planning a digester. Schocken Books, New York

Minister of Agriculture (2015) China agricultural statistics. China Agriculture Press, Beijing

Mittal S, Ahlgren EO, Shukla PR (2018) Barriers to biogas dissemination in India: a review. Energy Policy 112:361–370

MNRE (2015a) Designs approved under National Biogas and Manure Management Programme (NBMMP). Energetica India

MNRE (2015b) Annual report, 2015–16. In: Energy, M.O.N.A.R. New Delhi

MNRE (2018). http://pib.nic.in/newsite/PrintRelease.aspx?relid=177871

Moeller HB, Sommer SG, Ahring BK (2004) Methane productivity of manure, straw and solid fractions of manure. Biomass Bioenergy 26:485–495

Mogal PR (2013) Energy generation from food wastage. Int J Emerg Technol Adv Eng 3:37–42

Mohammed M, Egyir S, Donkor AK, Amoah P, Nyarko S, Boateng KK, Ziwu C (2017) Feasibility study for biogas integration into waste treatment plants in Ghana. Egypt J Pet 26(3):695–703

Mohammed NI, Kabbashi N, Alade A (2018) Significance of agricultural residues in sustainable biofuel development. In: Aladjadjiyan A (ed) Agricultural waste and residues. IntechOpen Limited, London

MoPNG (2018). http://petroleum.nic.in/sites/default/files/biofuelpolicy2018_1.pdf

MORTH (2015). http://egazette.nic.in/WriteReadData/2015/164547.pdf

Movafaghi OS, Stephenson K, Taylor D (2013) Farmer response to nutrient credit trading opportunities in the coastal plain of Virginia. Agricultural and Applied Economics Association's 2013 AAEA Annual Meeting, Washington, DC. https://ageconsearch.umn.edu/record/150414/files/Movafaghi_Olivia_AAEA.pdf

Mugodo K, Magama PP, Dhavu K (2017) Biogas production potential from agricultural and agro-processing waste in South Africa. Waste Biomass Valori 8(7):2383–2392

Mustafa MY, Calay RK, Roman E (2016) Biogas from organic waste – a case study. 8th International Cold Climate HVAC 2015 Conference. Procedia Eng 146:310–317

Muvhiiwa R, Hildebrandt D, Chimwani N, Ngqbevana L, Matambo T (2017) The impact and challenges of sustainable biogas implementation: moving towards a bio-based economy. Energy Sustain Soc 7:20

Myles R (2008) Practical pictorial field guide on Grameen Bandhu biogas plant. Second and revised edition. INSEDA, New Delhi. http://www.inseda.org/Manuals/Grameen%20BandhuBiogas%20plant%20-%20A%20Pictorial%20Field%20Guide.pdf

Naegele HJ, Linder J, Merkle W, Lemmer A, Jungbluth T, Bogenrieder C (2013) Effect of temperature, pH and O_2 on the removal of hydrogen sulphide from biogas by external biological desulfurization in a full scale fixed-bed trickling bioreactor (FBTB). Int J Agric Biol Eng 6 (1):69–81

Narayanan R, Hamsalakshmi R (2014) FDI opportunities in Indian renewable energy sector. Paripex-Indian J Res 3(4):38–39
NEERI (1996) Strategy paper on solid waste management in India. National Environmental Engineering Research Institute, Nagpur, pp 1–7
Nelsan S, Kuriakose AT (2017) Gender and renewable energy: entry points for women's livelihoods and employment. Climate Investment Funds. https://www.climateinvestmentfunds.org/sites/cif_enc/files/gender_and_re_digital.pdf
Ni JQ, Nyns EJ (1996) New concept for the evaluation of rural biogas management in developing countries. Energy Convers Manag 37(10):1525–1534
Omelianski W (1906) Uber Methanbildung in der Natur bei biologischen Prozessen. Zentralblatt fuer Bakteriol. Parasitenkd. II
Owusu PA, Asumado-Sarkodie S (2016) A review of renewable energy sources, sustainability issues and climate change mitigation. Cogent Eng 3:1167990
Paolini V, Petracchini F, Segreto M, Tomassetti L, Naja N, Cecinato A (2018) Environmental impact of biogas: a short review of current knowledge. J Environ Sci Health A 53(10):899–906
Paterson M, Kayser K, Bonhomme S, Majewsk E, Amrozy M, Berruto R, Parola F, Bijnagte JW, Gysen M (2015) Implementation guide for small-scale biogas plants, BioEnergy Farm II Publication, KTBL, Germany. http://www.bioenergyfarm.eu/wp-content/uploads/2015/05/D3.2_EN-Implementation-guideline.pdf
Pfay SF, Hagens JE, Dankbaar B (2017) Biogas between renewable energy and bio-economy policies-opportunities and constraints resulting from a dual role. Energy Sustain Soc 7:17
Poloncarz K, Humenik S, Dean J, King K, Brugato T (2019) An analysis of the renewable fuel standard's RIN market. Covington & Burling LLP. https://www.api.org/~/media/Files/Policy/Fuels-and-Renewables/2019/RIN-market-paper.pdf
Press (2009). https://www.kfw-ipex-bank.de/Presse/News/Pressemitteilungsdetails_10305-2.html
Press Report (2016) Six success stories in India that will show you how biogas can save money, the environment and more. https://www.thebetterindia.com/79053/biogas-digester-usage-india-environment/
Press Report (2018a). //economictimes.indiatimes.com/articleshow/66752011.cms?from=mdr&utm_source=contentofinterest&utm_medium=text&utm_campaign=cppst
Press Report (2018b). http://www.sustainabilityoutlook.in/news/city-gas-distribution-see-big-investments-763567
Ragwitz M, Schleich J, Huber C, Resch G, Faber T, Voogt M, Coenraads R, Cleijne H, Bodo P (2005) FORRES 2020: analysis of the renewable energy sources' evolution upto 2020. Karlsruhe
Raj A, Jhariya MK, Toppo P (2014) Cow dung for eco-friendly and sustainable productive farming. Environ Sci 3(10):201–202
Rajendran K, Aslanzadeh S, Taherzadeh M (2012) Household biogas digesters – a review. Energies 5:2911–2942
Rao PV, Baral SS, Dey R, Mutnuri S (2010) Biogas generation potential by anaerobic digestion for sustainable energy development in India. Renew Sust Energy Rev 14(7):2086–2094
Ravindranath NH, Hall DO (1995) Biomass energy and environment: a developing country perspective from India. Oxford University Press, Oxford
Ripley (2013) Mexico launches voluntary carbon credit exchange. https://sdg.iisd.org/news/mexico-launches-voluntary-carbon-credit-exchange/
Robert KH (1993) Solid waste management in the 1990s: trends and facilities in the USA and Europe. Ekistics 60:26–29
Ross K, Damassa T (2015) Assessing the post-2020 clean energy landscape. Technical note. World Resources Institute, Washington, DC. https://www.wri.org/sites/default/files/WRI-OCN_Assessing-Post-2020-Clean-Energy-Landscape.pdf
Rutz D, Ferber E (2011) Options for financing biogas plants. EFF Project "BiogasIN"
Salvi O, Delsinne S, Evanno S (2011) Biogas – a European perspective on safety and regulation. J Risk Anal Crisis Response 1(1):1–19

Sapp M (2017) State of biogas in the world:2017. http://www.globalbioenergy.org/fileadmin/user_upload/gbep/docs/2017_events/9_GBEP_WGCB_30_November_2017/GBEP_CESC_biogas_report.pdf

SBP Report (2015) SBP framework standard 1: feedstock compliance standard. https://sbp-cert.org/docs/2015-03/sbp-standard-1-feedstock-compliance-standard-v1-0.pdf

Scarlat N, Dallemand JF, Monforti-Ferrario F, Banja M, Motola V (2015) Renewable energy policy framework and bioenergy contribution in the European Union an overview from national renewable energy action plans and progress reports. Renew Sust Energy Rev 51:969–985

Scarlat N, Fahl F, Dallemand J-F, Monforti F, Motola V (2018) A spatial analysis of biogas potential from manure in Europe. Renew Sust Energ Rev 94:915–930

Schwegler M (2007) Promoting investments for renewable technology and biogas through carbon financing in China. Biogas Markets Asia, Singapore

Sharholy M, Ahmad K, Mahmood G, Trivedi RC (2008) Municipal solid waste management in Indian cities – a review. Waste Manag 28:459–467

Shukla PR (2007) Biomass energy strategies for aligning development and climate goals in India. Netherlands Environmental Assessment Agency.

Sigsgaard T, Forsberg B, Maesano IA, Blomberg A, Bolling A, Boman C, Bonlokke J, Brauer M, Bruce N, Heroux M, Hirvonen M, Kelly F, Kunzli N, Lundback B, Moshammer H, Noonan C, Pagels J, Sallsten G, Sculier J, Brunekreef B (2015) Health impacts of anthropogenic biomass burning in the developed world. Eur Respir J 46:1577–1588

Silba S, Almedia LA, Dias LC (2014) Biogas plants site selection integrating multicriteria decision aid methods and GIS techniques: a case study in a Portuguese region. Biomass Bioenergy 71:58–68

Simpson A (2010) Renewable heat incentive, pp 1–11. https://friendsoftheearth.uk/sites/default/files/downloads/renewable_heat_incentive.pdf

Singh KJ, Sooch SS (2004) Comparative study of economics of different models of family size biogas plants for state of Punjab, India. Energy Convers Manage 45:329–1341

Smolinksi S, Cox S (2016) Policies to enable bioenergy deployment: key consideration and good practices. NREL/TP-7A40-66322. https://www.nrel.gov/docs/fy16osti/66322.pdf

Sogn TA, Dragicevic I, Linjordet R, Krogstad T, Eijsink VGH, Eich-Greatorex S (2018) Recycling of biogas digestates in plant production: NPK fertilizer value and risk of leaching. Int J Recycl Org Waste Agric 7(1):49–58

Sohngen N (1906) Uber bakterien, welche methan als kohlenstoffnahrung und energiequelle gebrauchen. Zentrabl Bakteriol Parasitenk Infekt

Sony A, Patil D, Aragade K (2016) Municipal solid waste management. Procedia Environ Sci 35:119–126

Sooch SS, Gautam A (2013) PAU Kutcha – Pucca model of biogas plants in Punjab – a case study. Agric Eng Today 37(3):6–11

Strezov V, Evans TJ (2015) Biomass processing technologies. CRC Press

Szymanski SJ, Devine MA, Lee JCY (2013) Managing risks, reaping rewards through biogas applications. Caterpillar Inc, Peoria, IL

Szyszczak (2014) Commission communication on guidelines on state aid for environmental protection and energy 2014–2020

Teenstra E, Vellinga T, Aektasaeng N, Amatayakul W, Ndambi A, Pelster D, Germer L, Jenet A, Opio C, Andeweg K (2014) Global assessment of manure management policies and practices. Livestock Research Report. https://www.wur.nl/upload_mm/a/2/f/8a7d1a1e-2535-432b-bab5-fd10ff49a2b1_Global-Assessment-Manure-Management.pdf

Teng Z, Hua J, Wang C, Lu X. (2014) Design and optimization principles of biogas reactors in large scale applications, Chapter 4. In: Reactor and process design in sustainable energy technology, pp 99–134

Theurel S, Herrmann C, Heiermann M, Grundmann P, Landwehr N, Kreidenweis U, Prochnow A (2019) The future agricultural biogas plant in Germany: a vision. Energies 12:396

Thompson W, Meyer S, Westhoff P (2010) The new markets for renewable identification numbers. Appl Econ Perspect Policy 32(4):588–603

Thran D, Dotzauer M, Lenz V, Liebetrau J, Ortwein A (2015) Flexible bioenergy supply for balancing fluctuating renewables in the heat and power sector-a review of technologies and concepts. Energy Sustain Soc 5:35

Tietjen C (1975) From biodung to biogas – a historical review of the European experience. In: Jewell WJ (ed) Energy, agriculture, and waste management: proceedings of the 1975 Cornell Agricultural Waste Management Conference. Ann Arbor Science, Ann Arbor, pp 207–260

Tomar SS (1995) Status of biogas plants in India – an overview. Energy Sustain Dev 1(5):53–56

UK (2015) Policy paper 2010 to 2015 government policy: low carbon technologies. https://www.gov.uk/government/publications/2010-to-2015-government-policy-low-carbon-technologies/2010-to-2015-government-policy-low-carbon-technologies#appendix-5-the-renewables-obligation-ro

USDA (2014) Biogas opportunities roadmap. https://www.usda.gov/oce/reports/energy/Biogas_Opportunities_Roadmap_8-1-14.pdf

USDA (2015) Biogas opportunities roadmap. https://www.energy.gov/sites/prod/files/2015/12/f27/biogas_opportunites_roadmap_progress_report_0.pdf

Vac S, Popita G (2015) Biomass economical, social, and environmental aspects in biogas plants implementation. J Environ Protect Ecol 16(3):1212–1220

Vogeli Y, Riu C, Gallardo A, Diener S, Zurbrugg C (2014) Anaerobic digestion of biowaste in developing countries: practical information and case studies. Swiss Federal Institute of Aquatic Science and Technology (Eawag), Dubendorf. https://www.eawag.ch/fileadmin/Domain1/Abteilungen/sandec/publikationen/SWM/Anaerobic_Digestion/biowaste.pdf

Wang LK, Hung YT, Lo HH, Yapijakis C (2005) Waste treatment in the food processing industry. CRC Press

Wang A, Li W, Yu H (2012) Advances in biogas technology. In: Bai FW, Liu CG, Huang H, Tsao GT (eds) Advances in biochemical engineering/biotechnology: biotechnology in China III-biofuels and bioenergy, vol 128, pp 119–141

Wang X, Lu X, Li F, Yang G (2014) Effect of temperature and carbon-nitrogen (C/N) ratio on the performance of anaerobic co-digestion of dairy manure, chicken manure and rice straw: focusing on ammonia inhibition. PLoSOne 9(5):e97265

WBA (2015). https://worldbioenergy.org/uploads/WBA_Annual%20Report_2015_public_low.pdf

WEF (2019). https://www.weforum.org/agenda/2019/02/these-11-eu-states-already-meet-their-2020-renewable-energy-targets/

Weiland P (2010) Biogas production: current state and perspectives. Appl Microbiol Biotechnol 85:849–860

World Energy Council Report (2016). https://www.worldenergy.org/wp-content/uploads/2016/10/World-Energy-Resources-Full-report-2016.10.03.pdf

Yousuf A, Khan MR, Pirozzi D, Wahid ZA (2016) Financial sustainability of biogas technology: barriers, opportunities, and solutions. Energy Sources Part B Econ Plann Policy 11(9):841–848

Zhang YX, Chao QC, Zhen QH, Huang L (2017) The withdrawal of the U.S. from the Paris agreement and its impact on global climate change governance. Adv Clim Chang Res 8 (4):213–219

Zheng X, Fan J, Xu L, Zhou J (2017) Effects of combined application of biogas slurry and chemical fertilizer on soil aggregation and C/N distribution in an Ultisol. PLoS One 12(1):e0170491

Zubr J (1986) Methanogenic fermentation of fresh and ensiled plant materials. Biomass 11 (3):159–171

Zuzhang X (2013) Domestic biogas in a changing China: Can biogas still meet the energy needs of China's rural households? International Institute for Environment and Development, London

Biogas in Circular Bio-Economy: Sustainable Practice for Rural Farm Waste Management and Techno-economic Analyses

José Alberto Silva-González, Anuj Kumar Chandel, Silvio Silvério da Silva, and Nagamani Balagurusamy

Abstract Geopolitical concerns over increased energy consumption, excessive use of fossil fuels, and the urgency to meet the target of 2 °C as per COP 21 are pushing the international communities toward sustainable energy alternatives. In particular, developing economies and rural communities can take advantage of the renewable energy technologies to improve and support social and economic welfare, apart from diminishing their contribution in the emission of greenhouse gases. In this regard, anaerobic digestion (AD) of organic wastes aids in carbon recovery in the form of methane, a renewable energy source. Apart from traditional feedstocks available in rural and urban areas, such as animal manures and sewage sludge, rich lignocellulosic and agro-industrial wastes offer additional means of increasing renewable energy production and can substitute fossil energy requirements in rural areas. In addition, the use of digestate from AD as an organic fertilizer can potentially reduce pollution due to use of fertilizers in agriculture. An integrated management of wastes and biogas production is particularly well-suited for circular economy and could contribute to rural economic development, especially in developing countries. This review focuses on biogas production in relation to in-paradigm and new-paradigm as well as the benefits of biogas technology, such as the ability to improve access to clean energy in rural communities and the uplifting of rural economic development through circular economy approach.

J. A. Silva-González
Facultad de Ciencias Biológicas, Laboratorio de Bioremediación, Universidad Autónoma de Coahuila, Torreón, Coahuila, Mexico

A. K. Chandel · S. S. da Silva
Department of Biotechnology, Engineering School of Lorena (EEL), University of Sao Paulo (USP), Lorena, São Paulo, Brazil

N. Balagurusamy (✉)
Laboratorio de Biorremediación, Facultad de Ciencias Biologicas, Universidad Autonoma de Coahuila, Torreón, Coahuila, Mexico
e-mail: bnagamani@uadec.edu.mx

N. Balagurusamy, A. K. Chandel (eds.), *Biogas Production*,
https://doi.org/10.1007/978-3-030-58827-4_17

Keywords Agro-wastes · Biogas · Circular bio-economy · Renewable energy · Sustainable practice

1 Introduction

The energy sector has become one of the most important indicators of successful economic growth, due to the close relationship between energy consumption and gross domestic product (Gozgor et al. 2018). But decreasing fossil fuel reserves and emission of greenhouse gases are highlighting the need for implementing renewable energy technologies for a sustainable future. Further, most part of the world have access to one or more sources of renewable energies such as solar, wind, hydro, or biomass, and the suitable renewable energy technology for a specific area depends on the balance between their availability and economic cost involved. Levelized cost of energy (LCOE) studies is a useful tool to determine the economic advantage of a renewable energy source over another source, while life cycle assessment (LCA) provides a comparison of the environmental impact between different processes of renewable energy production and aids in taking suitable decisions in the implementation of the technology. Among the different sources of bioenergy, biogas technology is forecasted to play as a keystone in circular economy by providing the means for sustainable management of agro-industrial residues. Moreover, wastes are considered as resources and carbon is recovered in the form of methane, an energy, and generation of nutrient-rich organic fertilizer as a by-product during this process enables the return of wastes back into the economic productive chain (Begum et al. 2018; Lybæk and Kjær 2019). In brief, due to the ubiquitous availability of biomass suitable for anaerobic digestion (AD) in most rural areas, biogas technology could be an integral solution to tackle two different priority social issues: energy poverty and climate change. The impact of anaerobic digestion and consequent biogas production on climate change is straightforward as they aid in minimizing GHG emissions from decaying agro-wastes. In addition, the use of biogas also reduces GHG emissions in comparison with the use of fossil fuel (Styles et al. 2016; Thomas et al. 2017).

It is predicted that up to 1 billion people will directly suffer from energy poverty in 2030 and another 2.6 billion due to lack of access to clean fuel to cover basic necessities, as heating and cooking (Birol et al. 2015). Dale and Ong (2012) estimated a minimum of 4 kW per capita power consumption would permit people to attain a high Human Development Index (HDI) by overcoming energy poverty. Gozgor et al. (2018) reported that to achieve high HDI, the actual world power requirement should be as high as 28 TW, which is approximately twice that of current energy consumption of 15 TW. Although on the global scale fossil fuels are the main energy source, their dwindling price patterns and their negative impact on environment highlight the importance to deploy a sustainable renewable energy source (Lynd et al. 2017; Soares et al. 2016). Of the various renewable energy sources, biogas from anaerobic digestion technology is considered a carbon-neutral technology (Reinelt et al. 2017). In addition, biogas has several advantages over conventional and renewable sources of energy. It provides an integral solution for waste management of human and animal wastes by eliminating pathogens and unpleasant odors (Zareei 2018). Moreover, biogas can easily be burned in households and industrial units without the need for additional complex infrastructure (Cuéllar and Webber 2008; Mittal et al. 2018) and can significantly improve indoor air quality when it is used as replacement of fossil fuels, and wood and peat as biogas significantly reduce emission of harmful particles (Lukehurst and Bywater 2015). Moreover, upgraded biogas also called biomethane (purified biogas containing 95% methane) can be injected and distributed via natural gas grids for household use (Angelidaki et al. 2018), and as supplement to balance power demand fluctuations in the general electrical grid or to supply power in small rural energy plants. Further, the capabilities of biogas for long-term storage in the form of highly compressed gas greatly increased its flexibility as fuel and its potential use in a wide range of applications, including as transport fuel (Strzalka et al. 2017). In contrast, efficient storage systems are still a big concern for other renewable energy sources like solar or wind energy (Park et al. 2014). Furthermore, the digestate, which is the remaining material after anaerobic biodigestion is rich in nitrogen and other nutrients and is proved to be an efficient organic fertilizer. Thus the application of digestate minimizes the use of inorganic fertilizers and consequently reduces soil, water, and air pollution. These factors collectively make biogas technology attractive and suitable for farms in rural areas to implement circular economy scheme, where bioenergy production can have multiple benefits besides energy security, such as generation of employment, food security, and sustainable socioeconomic development (Acosta-Michlik et al. 2011; Surendra et al. 2014; Chen and Liu 2017).

In this review, we endeavor to appraise crucial parameters involved in maximizing the biogas production and the role of biogas technology in circular bio-economy. Moreover, comparison of different technologies of biogas production, commercialization of biogas, and an overview on life cycle assessment are presented.

2 Importance of Energy Access and Techno-Economic Comparison of Renewable Energy Generation

Access to clean and affordable energy is a key factor to achieve social and economic growth in developing countries, especially when it comes to rural household development (Hamid and Blanchard 2018). As mentioned previously, burning of fossil fuels (natural gas, oil, coal) in rural households and elsewhere has not only resulted in air, soil, and water pollution but also has significantly contributed to climate change (van der Ploeg and Rezai 2017). Generation of low-cost energy with minimal environmental impact is a difficult task (Divya and Gopinath 2015). Therefore, it is common for rural areas in developing countries to suffer from energy poverty, which is a major threat for economic progress of those communities (Li et al. 2016a, b; Hamid and Blanchard 2018), and it is imperative to meet burgeoning energy demand and reduce greenhouse gas emissions (Wang et al. 2017).

A common misconception is that the increasing share of renewables in the mix grid can result in expensive energy due to the relative higher renewable energy production cost. More recently, however, Atems and Hotaling (2018) addressed the positive effect on economic growth of renewable energy when they are paired with policies supporting the transition to cleaner energy. Furthermore, according to the data from the International Renewable Energy Agency (www.irena.org) and the International Energy Agency (IEA, www.iea.org), an unprecedented growth in the deployment of renewable energy generation technologies is being witnessed nowadays.

The share of renewable energy production has steadily grown worldwide over the last decades increasing from 0.6% of the total supply share, equivalent to 36 TWh in 1973, up to 7.1% (1722 TWh) in 2015 (IEA 2017). It is observed that more than half of the total shares of the new energy production projects since 2012 are that of renewable energies (Birol et al. 2015).

The fall in the prices of renewable energies as well as a steady growth of their share sets the scenario to achieve an affordable and sustainable solution to the energy needs of rural communities. Biogas production in OECD member countries is gradually outgrowing other forms of renewable energy sources, increasing from 3.7 TWh in 1990 to 81.3 TWh in 2016, with an average annual growth rate of 12.7% since 1990 (IEA 2016). When it comes to power generation, start-up phases of biogas production are relatively short and less cumbersome as compared with bioethanol, biodiesel, solar energy, etc. Small gas turbines of 100 kW or less show full cycles of start-up and shutdown of 5–6 min duration, costing in average 3.75 dollars for each full cycle (Berkeley 2001). These small gas turbines are suitable to be used in small biogas plants for self-consumption or to supply power to rural household complex. Harnessing the potential of cheap and surplus available biomass (agro-industrial, municipal and forestry wastes) into biogas will push the renewable energy drive. However, most of the times, the availability of other renewable energy sources apart from biogas, such as biofuels (bioethanol, biodiesel and bio-butanol), photovoltaic solar power or wind generation makes it hard to choose the most suitable technology.

Levelized cost of electricity (LCOE) is one of the most useful tools to measure the cost of electricity generation from different sources (IRENA 2012a, b, c). The average LCOE for generation of renewable energy from different sources is presented in Table 1. Though LCOE can be a useful tool to select the type of renewable energy based on economics, it would be better to use it in combination with other indicators. For instance, with reference to photovoltaics, Sims et al. (2003) stated that seasonal to daily fluctuations in weather conditions, along with their geographical locations, are limiting factors in LCOE calculation. Some of the other variations that could affect are the type and use of solar concentrating techniques or the fact that areas near the equator receive approximately twice as much annual solar radiation as in 60° latitudes. Furthermore, wind power generation is not suitable for most areas and is subject to similar restrictions as solar power generation due to the stochastic nature of weather conditions and in consequence, results in fluctuating power output, which must be treated with special care during its integration into the electricity system grid (Siyal et al. 2015).

Table 1 Cost comparison of different renewable energy systems

Renewable energy	Investment costs USD/kW	LCOE range USD/kWh	References
Bio-digesters	2574–6104	0.06–0.15	IRENA (2012a)
Landfill gas	1917–2436	0.09–0.12	IRENA (2012a)
Co-firing	140–850	0.04–0.13	IRENA (2012a)
PV system	3800–5800	0.25–0.65	IRENA (2012b)
PV system with battery storage	5000–6000	0.36–0.71	IRENA (2012b)
Onshore wind turbine (Europe)	1850–2100	0.08–0.14	IRENA (2015)
Offshore wind turbine (Europe)	4000–4500	0.14–0.19	IRENA (2015)

In contrast to solar and wind energy, which are heavily affected by geography and stochastic weather phenomenon, biomass power generation is affected mainly by the cost of feedstocks and in a lesser extent by seasonal weather fluctuations. In the case of rural farms and other agro-industrial applications, biogas feedstock cost is often negligible apart from recollection-associated cost. However, power generation cost from biogas is still higher than most fossil fuels and is in the range from 0.06 to 0.15 USD per kWh, and this price can only be achieved only in cogeneration units (IRENA 2017). However, electricity generation from fossil fuel typically fluctuates around 0.025 USD per kWh (IRENA 2017). This difference in price range can be attributed to project scale and technology available. On average, LCOE values tend to be higher in North America and Europe than other regions, reflecting the use of more sophisticated technology with strict emission control as well as higher feedstock costs of nearly 0.085 USD per kWh. In brief, if capital costs stay relatively low and low-cost feedstocks are available, bioenergy can be offered at competitive prices, with LCOE as low as around USD 0.04 per kWh (IRENA 2015). This suggests the need for innovation in reactor design and optimization of process parameters for an efficient management of biodigesters in terms of energy yield.

It is noteworthy that the most competitive bioenergy plants do share some common characteristics, such as low cost of agro-residues and their high availability near the sites suitable for the deployment of combined heat and power (CHP) systems, which permit LCOE to reach values as low as USD 0.03 per kWh (IRENA 2015). Operation and maintenance cost also known as "fixed cost" in bioenergy power plants typically falls in the range of 2–6% of the total installed costs per year (Lee 2017; Silva et al. 2017). This cost usually includes labor, maintenance, and routine component/equipment replacement and is not higher than maintenance cost of fossil fuel plants.

Another key factor to be considered is the subsidies given for renewable energy production as a short-term strategy to achieve economic viability in European countries such as Sweden, Denmark and Germany (Ahlberg-Eliasson et al. 2017). However, there are no uniform patterns of subsidies among the countries. For example, Sweden offered a subsidy of 30% covering installation costs, but since 2015, it has tightened its policy due to environmental concerns related to GHG emissions from manure. Currently, the subsidy is based on the total biogas produced in relation to the quantity of manure used in terms of wet basis. Unfortunately, the policy does not consider co-digestion or digestate quality that would otherwise improve efficiency and nutrient recovery from the effluent (Ahlberg-Eliasson et al. 2017). It would also be effective to include methane content to decide on the incentives or subsidies to promote biogas technology. Yasar et al. (2017) reported that subsidy for installation of household and small farm-scale biodigesters in terms of their capacity helped biogas replace non-environmental friendly, traditional fuels as well as cover energy demands of rural community in Pakistan. In the case of Mexico, Hernandez-De Lira et al. (2015) reported that anaerobic digestion of animal manure could generate approximately 410.41GWh of electricity and simultaneously reduce methane emission by 2240.64Gg CO_2 Eq. Gutierrez et al. (2016) calculated that a subsidy of 0.45 USD per m^3 of biomethane is necessary to achieve economic viability of biogas as a transport fuel in Mexico. Patrizio et al. (2015) suggested that a subsidy on wholesale prices of biomethane would favor its use as a transport fuel. Shea et al. (2020) observed that biofuels' obligation certificate is an effective incentive for industries in Ireland to use their wastes for biogas production. They also reported that compared to an incentive of 38 € per MWh, 106 € per MWh increased biomethane production and consequently increased energy production from 1.4 GWh to 508 GWh per year. Although subsidies favor the application of AD technology for renewable energy production, the type and level of subsidies should be governed by specific local requirements and conditions, and no universal subsidy pattern can be recommended.

3 Bioconversion of Agro-Industrial Wastes for Biogas Production

Anaerobic digestion (AD) is widely used for the treatment of organic wastes, viz., animal manure, municipal solid wastes (MSW), and other agro-residues. However, lack of technical and practical knowledge can lead to poor operational practices, especially in households and small-scale biogas plants. Most of biogas plants often fail after a short period due to a number of factors, for instance, inadequate C/N and dilution ratio of the feedstock or lack of access to skilled workers for operation and maintenance (Mittal et al. 2018; Ullah and Martin 2016). Training or providing operation manual to households' biogas owners is important for its efficient functioning. On the other hand, microbiome and their role in AD process are a black box as well as a bottleneck to improve the performance of large-scale anaerobic digesters. At least 11 different trophic groups of microorganisms belonging to bacteria and archaea domains participate in AD process. The process in general is divided into four steps: hydrolysis, fermentation or acidogenesis, acetogenesis, and methanogenesis (Alvarado et al. 2014; Johanna et al. 2009). In the first step of hydrolysis, complex organic molecules (cellulose, proteins, and fats) are broken down into their simpler forms (sugars, amino acids, and fatty acids) by hydrolytic bacteria using a mix of intracellular and extracellular hydrolases like cellulase, hemicellulase, pectinase, ligninase, etc., depending on the physicochemical composition of the feedstock. In the next step, fermentative bacteria act upon simpler forms and produce short-chain volatile fatty acids (e.g., formate, acetate, propionate, butyrate, caproic, and valeric), alcohols (e.g., ethanol, propanol and butanol), H_2, and CO_2. This process is followed by acetogenesis where acetogenic microorganisms oxidize short-chain fatty acids $\geq C_3$ to acetate, at the same time producing hydrogen as an obligatory product (Nagamani and Ramasamy 1999). In the final stage, methane is formed under strict anaerobic conditions by two different pathways: CO_2 reduction and acetoclastic reaction (Alvarado et al. 2014; Liu and Whitman 2008). Nevertheless, microbial groups and participating species do not remain constant, and their dynamics depend on feedstocks and operating conditions (Wang et al. 2018). Currently, an important share of AD research is aimed to increase methane yields as well as to enhance process stability by improving reactor designs and understanding of the complex interactions among diverse microorganisms of bacterial and archaeal domains.

3.1 Operational Factors Affecting AD Process

A wide variety of agro-wastes are used for biogas production; however, their chemical composition influences the total biogas yield and its composition. In general, high carbohydrate and protein contents in organic wastes increase the rate of degradation, but fat content provides higher biogas yields (Ware and Power 2016; Weiland 2010). On the other hand, accumulation of fatty acids during the process also inhibits methanogenesis (Hamawand 2015). In summary, more often biogas production from animal manure and agro-wastes cannot be utilized at its full potential due to the imbalance of carbon to nitrogen (C/N) ratio or carbon to sulfur ratio (C/S) (Neshat et al. 2017; Tan et al. 2019). In general, high carbon and high nitrogen contents affect the AD process due to accumulation of toxic intermediates, result in inhibition of methanogenesis, and finally cause failure. It been reported that C/N ratio in the range of 25–30 favored maximum methane potential and reduced inhibitory effects of ammonia at 35° and 55 °C (Wang et al. 2014). In addition, other factors as organic load rate (OLR) and hydraulic retention time (HRT) greatly influence AD performance (Ahlberg-Eliasson et al. 2017; Naik et al. 2014). Conventional AD reactors usually operate around 15 to 30 days of HRT and at an OLR < 2.5 g VS L^{-1} d^{-1} (Gou et al. 2014; Stuckey 2012). However, longer HRT is applied for higher OLR of 7 g VS L^{-1} d^{-1} to provide enough time for the degradation of recalcitrant compounds and to prevent the washout of the slow-growing methanogens (Gou et al. 2014; Ratanatamskul et al. 2014; Santos et al. 2017). Lim et al. (2008) reported that high OLR of 13 g L^{-1} d^{-1} led to low methane yields and moreover resulted in instability of AD process. However, there is no consensus for the optimal HRT in AD reactors, mainly due to the characteristics of different feedstocks as well as different requirements of diverse microbial groups involved in AD. Eslami et al. (2018) also observed that a sudden increase in OLRs negatively affects methane production through the accumulation of fatty acids, which inhibit methanogenic activity. Nevertheless, HRT of AD under thermophilic temperatures (55–60 °C) is less than those at mesophilic temperatures (Zupančič and Grilc 2012). One of the variations employed to overcome the limitations of OLR is the use of two-phase anaerobic reactors to treat high OLR effluent. A variant of anaerobic digestion process with high OLR can be carried out in two-phase anaerobic reactors, where during the first phase, hydrolysis-acidogenesis activity is favored, while the second phase allows low-growing methanogens to carry out the methanogenic stage of anaerobic digestion (Demirer and Chen 2005). Demirer and Chen (2005) compared the performance of one-phase and two-phase anaerobic reactors and concluded that the two-phase reactor recorded 50% and 67% higher biogas yield than one-phase reactor at an OLR of 5 and 6 g VS/L day and also showed a higher tolerance to OLR up to 12.6 g VS/L day. Zupančič and Grilc (2012) suggested thermophilic conditions for the first phase and mesophilic condition for the second phase to reduce HRT and observed that this facilitated an increase of OLR by 20%.

Finally, anaerobic digestion can also be negatively affected by other factors, for instance, presence of toxic compounds in the feedstock or accumulation of secondary metabolites produced along the process. A better understanding of these inhibition mechanisms during AD can provide insights to overcome or minimize negative effects and further develop possible solutions (Chen et al. 2014). For instance, co-digestion of different substrates favor the obtention of optimum C:N ratios, and increase biomethanation potential of feedstocks (Chen et al. 2008).

3.2 *Use of Lignocellulosic Substrates for Biogas Production*

Although AD systems traditionally have used animal manure as primary feedstock, mainly because of its high availability and low related cost, it can also be employed for lignocellulosic feedstocks derived from different agro-industrial activities. Lignocellulosic residues can be divided into two types according to their origin: agricultural and agro-industrial wastes (Kabir et al. 2015). Agro-industrial residues are not considered as primary feedstocks for AD due to their recalcitrant nature and as well as wider C:N ratio (Alkanok et al. 2014; Toop et al. 2017). However, abundant availability of lignocellulosic biomass offers the possibility to improve the energy security, employment opportunities, and living standards in rural areas

Table 2 Biogas yield in the co-digestion of different of feedstocks

Manure type	Co-substrate	Ratio	Yield	References
Cattle manure	–	1	0.166–0.25 L/g VS	Fuchsz and Kohlheb (2015)
Cattle manure	Palm-pressed fiber (C/N 57.03)	1:3	0.342 L/g VS	Bah et al. (2014)
Cattle manure	Corn stover (C/N 59)	1:3	0.194 L/g VS	Li et al. (2016a, b)
Cattle manure	Food wastes, sewage sludge	70:20:10	0.603 L/g VS	Marañón et al. (2012)
Dairy manure	Chlorella	8:2	0.238 L/g VS	Li et al. (2017)
Dairy manure	Kitchen wastes (C/N 31.18)	1:1	0.18 L/g VS	Zhai et al. (2015)
Dairy manure	Corn stover, tomato residues	54:33:13	0.415 L/g VS	Li et al. (2016a, b)
Cow manure	Fish ensilage	15:85	0.729 L/g VS	Vivekanand et al. (2018)
Chicken manure	Chlorella 1067	8:2	0.238 L/g VS	Li et al. (2017)
Chicken manure	Apple pulp waste	2: 1	0.34 L/g VS	Li et al. (2018a, b)
Pig manure	Sugar beet by-product	N/A	0.362 L/g VS	Aboudi et al. (2015)
Sheep dung	Waste paper	2:3	0.198 L/g VS	Li et al. (2018a, b)
Sludge	Microalgae mix	9:1	0.560 L/g VS	Thorin et al. (2018)
Sludge mud	Vinasse	1:3	0.365 L/g VS	López-González et al. (2017)

(Peidong et al. 2009). One of the ways to overcome lignocellulosic biomass associated problems is by the implementation of co-digestion systems, which involves mixing of different feedstock to adjust C:N ratios to improve their biodegradability and consequently increase biogas production (Sandberg et al. 2018). However, Hagos et al. (2017) observed that co-digestion process increased the complexity of the system and that further research on feedstock characterization, their biomethane potential, advanced simulation models, and innovative strategies such as the use of environmental friendly nanoparticles to improve the stability and performance of the process could yield positive results. Velásquez Piñas et al. (2018) compared the size, electricity, and heat consumption of mono- and co-digestion systems and concluded that the performance of the latter was better. Methane yields from various anaerobic co-digestion processes are summarized in Table 2.

One of the other variations of AD systems is employing thermophilic conditions (temperature in the range of 55–70 °C) to reduce retention time, to increase rate of biodegradation, and in some cases for elimination of pathogens present in feedstocks, such as human wastes. Cieślik et al. (2016) concluded that thermophilic fermentation (55 °C) of maize straw silage resulted in shorter retention times (17%

shorter) than under mesophilic conditions for the same feedstock and further recorded a slight increase in biogas production. Terboven et al. (2017) observed a more stable methane content in shorter retention time under thermophilic conditions even after shock loads were applied. They further observed that methane content of biogas increased to values above 40% within 3 to 4 h for thermophilic conditions, while it was 5 to 6 h for mesophilic conditions, when fed with sugar beet silage. Several studies of thermophilic AD have focused on technical feasibility of biogas production from paper mill sludge's, which are more recalcitrant. Bayr et al. (2013) reported that hydrothermal treatment in combination with enzymatic treatment or with ultrasonic treatment increased methane yield from paper mill secondary sludge by 31%. Earlier, Bayr and Rintala (2012) observed that anaerobic digestion of primary sludge required a hydraulic retention time (HRT) of 16–32 days at an organic loading rate of 1–1.4 kg VS/m^3/day, but co-digestion of primary and secondary sludge required more HRT, of 25–31 days for an organic loading rate of 1 kg VS/m^3/day. However, when it comes to full-scale biogas plant do Carmo Precci Lopes et al. (2018) reported that the energy produced by AD of kraft pulp mill secondary sludge did not meet the energy demands to maintain the digester under thermophilic conditions.

3.2.1 Dry Anaerobic Digestion (DAD)

Another variation of AD process is based on the content of total solids (TS), viz., wet AD (WAD) where typically TS is <20% and dry AD (DAD) in which the content of TS usually exceeds 20%. Recently, DAD has been studied due to its lower water and energy requirement for heating and mixing. Further, there are other advantages such as higher volumetric methane productivity, limited wastewater generation, and an easy to handle digestate due to its low water content (Arelli et al. 2018). Chiumenti et al. (2018) also observed that DAD of cow manure and agricultural products in a full-scale plant demonstrated comparable methane yields to that of wet AD. Similar results were achieved by Qian et al. (2015), and they reported a specific methane yield of 270 m^3/t VS from municipal solid waste-fed full-scale DAD system, with an average of total and volatile solid contents in the range of 43.5–54.0%. They concluded that despite the good yield, the process performance can be improved through inoculation and improving the recirculation and digestate mixing. Patinvoh et al. (2017) calculated that the cost of installation of DAD projects is in the range of $7000 (small, 3 t/year) to $9,800,000 USD (large-scale biodigesters, 51,000 t/year) fed with agricultural wastes and requires a maximum payback period of 5 years by using a microbial inoculum acclimatized for a long period.

4 Key Drivers Influencing Biogas Commercialization

Though AD is a cost-effective technology to provide a sustainable solution for organic waste management as well as to decrease GHG emissions, the marketability needs to be improved for wider applications. Apart from producing methane as a marketable product (Lauer et al. 2018), AD process also generates other marketable by-products such as digestate which can be used as organic fertilizer for sustainable agriculture (Nayal et al. 2016). The coupling of AD and waste management of industrial wastes and residues is increasingly appreciated because of both environmental and economic benefits such as reduction in waste management costs and generation of renewable energy for selling or self-consumption. These features have significantly attracted the interests of many companies (Table 3) and are consistent with the major biogas-producing countries, according to the International Energy Agency (Table 4).

Another major factor that influences methane yield is the type of anaerobic reactors employed for biogas production. Although continuous-fed anaerobic reactor

Table 3 Biogas industries and installation capacity of energy generation for different feedstocks

Company	CCapacity (kW)	Feedstock	Website
Phaidon Energy, Munich, Germany	500–2500	Mixed organic wastes	http://www.phaidon-energy.de/
Archea New Energy, Hess, Germany	40–1000	Mixed organic wastes	http://www.archea-biogas.com/
Binowa, Freyburg, Germany	100–500	Cattle manure	http://www.binowa.de/
CH4 Biogas, Connecticut, USA	1000–3000	Mixed organic wastes	http://www.ch4biogas.com/
Xergi, Denmark	300–3000		https://www.xergi.com/
Biogest Energie, Klosterneuburg, Austria	100–2000	Cattle manure, agro-industrial residues	http://www.biogest.at/home/en
IES Biogas, Pordenone, Italy	100–1000	Cattle manure, agro industrial residues	http://www.iesbiogas.it/en
Puxin, USA	250–2800	Mixed organic wastes	http://www.planet-biogas-usa.com/
Greenlane Biogas, British Columbia, Canada	2000–5000	Food wastes	http://greenlanebiogas.com/contact.html
MT-Energie, Germany	500–837	Agro-industrial wastes	https://mte-service.com/
Schmack Biogas, Regensburg, Germany	500–2800	Mixed organic wastes	http://www.schmack-biogas.com/en.html
streisal GmbH, Wagen, Germany	50–600	Agro-industrial wastes	https://www.streisal.de/en/
AB Energy, Orzinuovi BS, Italy	50–1500	Agro-industrial wastes	https://www.gruppoab.it/en/
BTS Biogas, Brunico, Italy	25–1500 kW	Agro-industrial wastes	www.bts-biogas.com

Table 4 Gross electricity (GE) and gross heat (GH) generation from biogas in different countries

	2000		2005		2010		2015	
Country	GE (Gwh)	GH (TJ)	GE (Gwh)	GH (TJ)	GE (Gwh)	GH (TJ)	GE (Gwh)	GH (TJ)
Austria	61	0	308	223	647	307	624	145
Australia	449	0	593	0	1016	0	1491	0
Belgium	98	41	223	153	566	263	955	388
Canada	608	0	542	529	792	2582	972	2140
Chile	0	0	0	0	0	0	0	0
Czech Republic	135	384	160	103	635	256	2611	623
Denmark	209	903	281	1158	357	1148	485	2099
Estonia	0	61	14	43	10	64	50	112
Finland	31	163	63	870	106	366	358	763
France	308	0	480	8	1005	349	358	763
Germany	1683	0	3862	823	17,430	1508	33,073	9285
Greece	0	0	122	0	190	0	230	0
Hungary	0	0	25	4	118	121	293	131
Iceland	0	0	4	0	0	0	0	0
Ireland	95	0	122	0	204	0	202	0
Italy	567	0	1198	1084	2055	1029	8212	8604
Japan	0	0	0	0	0	0	0	0
Korea	11	0	130	292	532	515	588	368
Latvia	0	0	36	43	57	50	391	892
Luxembourg	4	0	28	0	55	33	62	80
Mexico	17	0	28	0	118	0	161	0
Netherlands	286	44	294	68	1028	282	1036	48
New Zealand	286	44	294	68	1028	282	1030	48
Norway	0	12	0	8	13	78	7	118
Poland	31	37	111	43	398	106	906	436
Portugal	2	0	35	0	100	0	294	0
Spain	318	0	623	0	848	0	982	0
Sweden	32	1042	54	866	36	731	11	274
Switzerland	149	50	146	19	209	2	304	0
Turkey	21	0	29	0	296	213	1208	1815
UK	2555	0	4766	0	5839	0	7189	0
USA	5230	2191	6449	125	9806	1883	13,674	1388
Brazil	0	0	0	0	138	0	682	0
China	0	0	0	0	0	0	0	0
India	0	0	0	0	765	0	1014	0
OECD total	13,121	4931	20,940	6597	44,864	12,163	79,153	31,918
Non-OECD total	0	0	45	18	1358	133	3330	1030
OECD Europe	6427	2678	12,827	5642	31,929	7000	60,922	26,632
OECD America	5955	2191	7019	654	10,716	4465	14,819	3528
OECD Asia and Oceania	569	0	930	292	1814	515	2395	368

Note: Data from IEA (2018). Retrieved from https://www.iea.org/countries/

is the most commonly used bioreactor, other types such as continuous stirred-tank reactor, anaerobic piston flow reactor, anaerobic contact reactor, fluidized bed and upflow anaerobic sludge blanket reactors have been employed based on the specific character of the feedstocks or in specific cases to improve the process efficiency and to increase methane yield.

Biogas is composed mainly of methane (CH_4), carbon dioxide (CO_2), and trace amounts of hydrogen sulfide (H_2S), ammonia (NH_3), hydrogen (H_2), carbon monoxide (CO), and water vapor. Carbon dioxide and water vapor are the two components that mainly affect the calorific value of biogas. Another major problem is the presence of hydrogen sulfide in biogas, which can be removed by a complex process of adsorption, biofiltration or precipitation using iron shreds. The removal of hydrogen sulfide is of major importance as it is corrosive and can easily damage inner engine parts, considerably reducing the life span of the motor (Weiland 2010).

On the other hand, technologies for conversion of renewable energy generated into electricity, heat, and steam are making substantial progress in the last two decades (Whiting and Azapagic 2014). The global amount of electricity produced from biogas is projected to grow up to 63.9 TWh mainly due to the recent improvements in power generation from biogas (Scarlat et al. 2018). In Europe, Germany is the leader of bioenergy production from biogas, exceeding the USA in biogas power generation (Table 4).

Sweden is an example for the successful deployment of commercial biogas plants. For example, Stockholm municipal wastewater treatment plant in collaboration with Swedavia, state-owned company produces biogas and promote their use in public transport system. A significant amount of the biogas produced (up to 40%) is upgraded to biomethane and is used in the transport sector (Fallde and Eklund 2015). Despite low prices and taxes levied on natural gas in Sweden, a combination of biomethane (up to 40%) and natural gas (50–60%) is used as transport fuel by providing a full exemption from energy and carbon dioxide tax (Larsson et al. 2016). Thus, high penetration of biogas use in the transportation sector can be achieved by strong political support and the participation of renewable energy agencies, which could result in the shift from fossil-dependent vehicle fleet to environmental friendly cars by 2030 (Ammenberg et al. 2018; Larsson et al. 2016). This also implies that a strong research and development is essential for development and application of biogas technology for commercial use. A list of patents on the technological advancement of biogas production is presented in Table 5.

5 The Implicit Role of Biogas in Circular Economy

Circular economy is a sustainable alternative to traditional linear economy, where resources are made, used, and disposed (Lieder and Rashid 2016). However, in circular economy, resources are maintained in the production chain as long as possible by recovery, recycling, or transformation into different products or materials, which can later be used in subsequential production steps (Winkler 2011).

Table 5 Patents on biogas production at industrial scale

Patent number	Title	Issue date	Patent owner
DE102016014103A1	A process for the recycling of industrial and agricultural biomass and biogenic residues	30/05/2018	Rainer Gottschalk, Waldemar Reule
WO2018091603A1	System for producing biogas from solid biomass and corresponding biogas method	24/05/2018	YANNCO
US9969949B1	Method and system for providing upgraded biogas	15/05/2018	Iogen Corp
1.0202E+14	Purification technology used for industrial biogas	22/08/2017	HEFEI ZEJUN Electrical Equipment Co., Ltd
DE102015114510A1	Apparatus and reducing the formation of foam in a biogas fermenter	02/03/2017	Agraferm Technologies AG
CN106316525A	Biogas slurry enhanced liquid organic fertilizer preparation method	11/01/2017	Forest Academy of the Guangxi Zhuang Autonomous Region
DE202015104848U1	Apparatus for producing biogas	14/12/2016	Pro Agri GmbH
EP3162898A1	Process for the production of biogas from a solid digestate	27/10/2015	TIRSI S R L
DE102014003618A1	A process for the production of biogas and integrated process water treatment	17/09/2015	Klaus Doelle, Dieter, Lorenz Freinecker
US8470567B2	Apparatus and process for production of biogas	25/06/2013	Gemini Corp
WO2013060338A1	Method for anaerobic fermentation and biogas production	02/05/2013	Xergi Technology A/S
EP2529022A1	Biogas production process with enzymatic pretreatment	05/12/2012	Novozymes AS
US7883884B2	Concept for slurry separation and biogas production	08/02/2011	Green Farm Energy AS
EP1757562B1	Device and process for treating residues of biogas production, manure, and sludge	07/06/2005	Winfried Hitze International GmbH
DE10034279A1	Reactor for biomass methanization comprises a sealable and heatable container which is provided with means for biogas withdrawal and liquid drainage	21/02/2002	BEKON GmbH

Biogas production through AD of agro-industrial residues such as cattle manure and agro-industrial residues is well-suited for circular economy, as it combines renewable energy production from the wastes generated by the farm to meet the energy needs of the same farm. Additionally, the process generates nutrient-rich digestate as an organic fertilizer for sustainable agricultural production from the farm that generates waste. Nevertheless, the biogas technology is still not an economically viable process on its own in most of the countries due to several factors including low prices of fossil fuels and the lack of financial investments from the private sector or policies favoring the inclusion of biogas.

Nonetheless, the integration of biogas production into circular economy could add additional benefits and turn AD as an economically viable business option of sustainable bioenergy production from organic wastes. Rural farming communities are well-suited for the application of the circular economy (Blades et al. 2017). Lignocellulosic materials are generated from wide range of agricultural processes, and their high availability and relatively low price make them as a potential feedstock for the development of bio-circular economy. Nevertheless, the use of lignocellulosic residues at rural locations is inadequate because of their complex structure and recalcitrant nature of some of their components, which make them unsuitable for conventional AD (Anwar et al. 2014). Implementation of pretreatment strategies and co- digestion processes can overcome this limitation (Sanna 2020). Bio-based economies face enormous challenges that demand interdisciplinary cooperation to offer unique solutions to minimize environmental carbon footprint and fight climate change through revalorization of wastes in the value chain (Begum et al. 2018). Biogas-derived energy from these wastes can be used by local community as hearing source in household applications and as fuel for farm vehicles as well as to provide electricity to local farms, households, and industries (Winkler 2011). The energy produced in the CHP unit could cover completely or at least partially the energy requirements of the farm itself or generate income by way of selling the energy produced through the power grid (Blades et al. 2017). Additionally, the AD digestate can be used as a fertilizer in crop production, and the cycle can be continued (Tampio et al. 2017). Furthermore, the use of the digestate as organic fertilizer helps in reducing the consumption of inorganic fertilizers and minimizes soil and water contamination, eutrophication of water bodies, and GHG emissions (Tampio et al. 2017).

Thus, circular economy of biogas production opens the possibility of exploiting waste to create extra income as well as an opportunity to cut the current expenses involved with farming and its waste management. The combination of circular economy and biogas production as a common practice throughout the farms would represent an important step toward the strengthening of green economy. However, steps to upgrade biogas to biomethane and coupling of liquid or solid digestate for cultivation of traditional crops for cattle feed or energy crops for further bioenergy production are required to implement this sustainable technology into circular bio-economy by way of strengthening the inclusive financial growth in rural and remote areas (Jin et al. 2015). A scheme of biogas production in circular bio-economy is presented in Fig. 1.

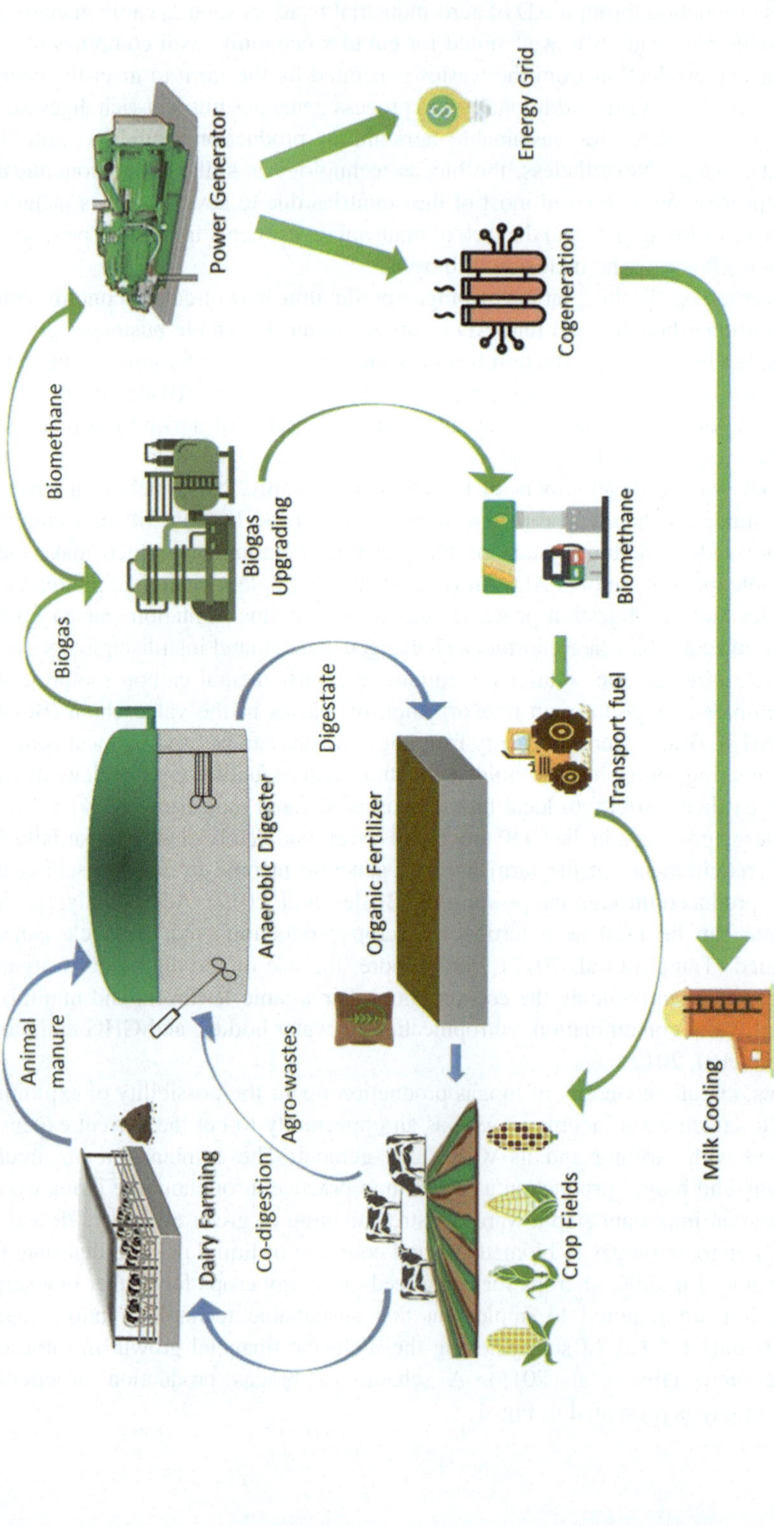

Fig. 1 Scheme of integrated biogas production in circular bio-economy concept

6 Life Cycle Assessment of Biogas Production: Key Factors

Implementation of biogas technology is a key strategy to meet the sustainability needs of rural communities and to fight GHG emissions and the consequent global warming effect from agro-industrial wastes (Bacenetti et al. 2016). However, it is extremely important to accurately assess the environmental impact of different configurations of AD in order to pick the eco-friendliest solution to minimize the negative effects on the environment (Hamelin et al. 2014). Life cycle assessment (LCA) is a suitable tool to quantify the environmental impact associated with each stage of biogas production process. The technical framework for the LCA methodology, as defined in ISO 14040, consists of four phases, namely, (1) goal and scope definition, (2) inventory analysis, (3) characterization, and (4) interpretation (Pryshlakivsky and Searcy 2013). One of the major advantages of LCA is process integration, which allows the analyst to increase the level of details according to the results of the previous scenarios (Fig. 2).

The general goal of LCA of biogas energy system is to determine the environmental impact during their production, use, GHG emissions and as well as to assess energy and material requirements for unit of energy produced. Different LCA studies related to biogas production more often include one or more of the following specific goals: (1) comparison of different alternatives of energy production using different feedstocks and production schemes with respect to cost/efficiency and power generation and (2) assessment of environmental impact of biogas production using a common unit for comparison of the different scenarios. Recently, Lansche and Müller (2017) used LCA to study the environmental impact of two processes: methane production using AD of animal dung and burning dung as a household cooking fuel, which is a common practice in rural areas of Africa and Asia. The results indicated that for every mega joule (MJ) of heat produced through AD, emissions of 0.51 kg of CO_2 eq., 0.06 g PO_4- eq., and 0.28 g SO_2 eq. can be prevented. Further, they estimated that an emission of about 13,542 t of CO_2 eq. could be reduced annually in Ethiopia from more than 8000 biogas plants installed during 2009 to 2013. Furthermore, they observed that the digestate could be used as an organic fertilizer to increase crop productivity.

More recently, Pérez-Camacho et al. (2017) carried out an LCA to compare energy production from biogas production from cattle slurry and food wastes with energy production from fossil fuel. This study showed that 1 MWh energy generation from biogas produced from cattle slurry could result in reducing 296 kg CO_2 eq. in terms of GWP, 0.24 kg SO_2-eq. in AP category, and 17.0 kg 1,4 dichlorobenzene eq. in terms of human toxicity potential (HTP) in comparison with fossil fuel production of 1 MWh in Northern Ireland. Furthermore, in the same study, the use of food wastes as feedstock for biogas production was also assessed, which resulted in savings of 458 kg CO_2 eq. in GWP, 2.22 SO_2-eq in AP, and 102.0 kg DCB eq. in HTP category. Despite the higher environmental benefit from using food wastes as feedstock, better availability and more predictable supply of cattle slurry make it as a more suitable feedstock than food wastes for AD in rural communities.

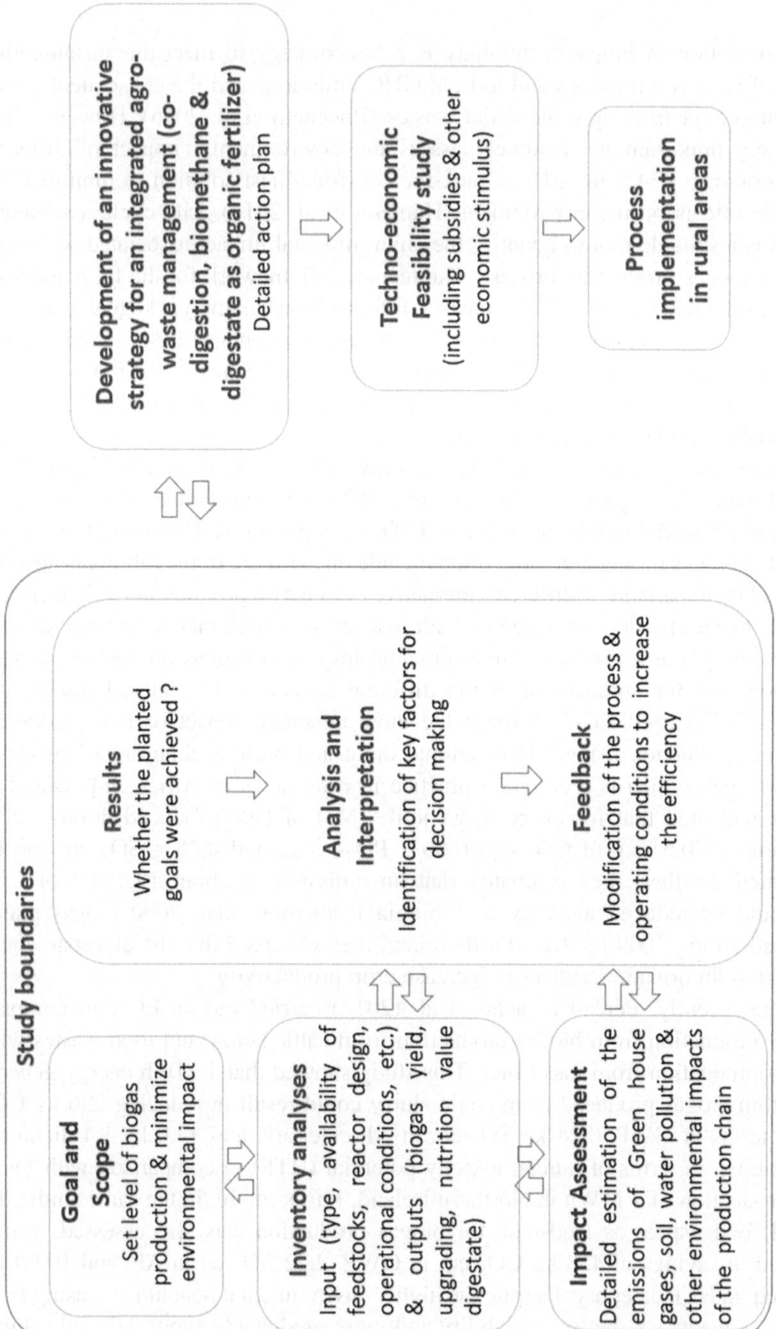

Fig. 2 Framework for techno-economic evaluation of biogas to achieve circular bio-economy

Ramírez-arpide et al. (2018) evaluated the feasibility of using nopal (*Opuntia ficus-indica*) in co-digestion with dairy cow manure by measuring the environmental effect associated with the process using the LCA methodology. The outcome of this study showed a reduction of 1.72 g CO_2-eq. for every MJ of electricity produced. Nevertheless, an increase of 0.0253 g PO_4-eq. and an increase in 0.041 g SO_2-eq in terms of acidification potential were observed.

Soam et al. (2017) evaluated the potential use of rice straw to produce biogas from an environmental perspective using LCA to compare with the open field straw burning practice in India. The results showed reductions of 1023 kg CO_2 eq., 0.4 kg PO_4-eq., and 3.4 kg SO_2-eq. in terms of GWP, AP, and EP respectively, for every ton of dry straw as a functional unit. They concluded that biogas production can result in sustainable management of straw, providing clean energy and environmental benefits. In order to assess environmental friendliness, Ertem et al. (2017) evaluated the substitution of energy crops by marine macroalgae as a feedstock in an industrial-scale biogas plant. They reported that for every MJ of energy produced from co-digestion of macroalgae (60%) with chicken litter (40%), reduced the global warming potential and acidification by 52% and 83% respectively, in comparison to that of the same quantity of energy produced from energy crops. They concluded that sustainable bioenergy production can be achieved by the co-digestion of chicken manure and macroalgae, but not from macroalgae alone as substrate.

Unfortunately, the lack of homology in the functional units and scenarios proposed by different authors limits the use of LCA to iterations of the given scenario and derives definitive conclusions (Ingrao et al. 2019).

7 Conclusions

Anaerobic digestion offers the scope of integrated and sustainable agro-waste management along with renewable energy production and reduced greenhouse gas emissions. Constant improvements in the design of biodigesters and operating conditions are essential to employ a wide range of non-traditional feedstocks. Economic competitiveness of biogas can be achieved by developing cost-effective technologies for upgradation as biomethane, use as heat energy in rural areas, as well as biofuel in the transportation sector, and use of digestate for organic agriculture. An integrated biogas production could be a key strategy in developing circular bio-economy for a sustainable rural development in terms of energy, environment, and economy.

Acknowledgments JASG thank the Mexican National Council of Science and Technology (CONACYT) for awarding a scholarship for his graduate studies, Master's program on Biochemical Engineering (PNPC reference No. 002439). NB, AKC & SSS thank the grant from SEP-Programa de Fortalecimiento de la Investigación para el Desarrollo de la Educación y la Sociedad (PROFIDES) 2017, Government of Mexico. AKC is grateful to USP-CAPES for its visiting researcher and professor program. SSS thank CNPq, CAPES, and FAPESP for their financial support.

References

Aboudi K, Álvarez-Gallego CJ, Romero-García LI (2015) Semi-continuous anaerobic co-digestion of sugar beet byproduct and pig manure: effect of the organic loading rate (OLR) on process performance. Bioresour Technol 194:283–290. https://doi.org/10.1016/j.biortech.2015.07.031

Acosta-Michlik L, Lucht W, Bondeau A, Beringer T (2011) Integrated assessment of sustainability trade-offs and pathways for global bioenergy production: Framing a novel hybrid approach. Renew Sust Energ Rev 15:2791–2809. https://doi.org/10.1016/j.rser.2011.02.011

Ahlberg-Eliasson K, Nadeau E, Levén L, Schnürer A (2017) Production efficiency of Swedish farm-scale biogas plants. Biomass Bioenergy 97:27–37. https://doi.org/10.1016/j.biombioe.2016.12.002

Alkanok G, Demirel B, Onay TT (2014) Determination of biogas generation potential as a renewable energy source from supermarket wastes. Waste Manag 34:134–140. https://doi.org/10.1016/j.wasman.2013.09.015

Alvarado A, Montañez-Hernández LE, Palacio-Molina SL, Oropeza-Navarro R, Luévanos-Escareño MP, Balagurusamy N (2014) Microbial trophic interactions and mcrA gene expression in monitoring of anaerobic digesters. Front Microbiol 5:1–14. https://doi.org/10.3389/fmicb.2014.00597

Ammenberg J, Anderberg S, Lönnqvist T, Grönkvist S, Sandberg T (2018) Biogas in the transport sector—actor and policy analysis focusing on the demand side in the Stockholm region. Resour Conserv Recycl 129:70–80. https://doi.org/10.1016/j.resconrec.2017.10.010

Angelidaki I, Treu L, Tsapekos P, Luo G, Campanaro S, Wenzel H, Kougias PG (2018) Biogas upgrading and utilization: current status and perspectives. Biotechnol Adv 36:452–466. https://doi.org/10.1016/j.biotechadv.2018.01.011

Anwar Z, Gulfraz M, Irshad M (2014) ScienceDirect agro-industrial lignocellulosic biomass a key to unlock the future bio-energy: a brief review. J Radiat Res Appl Sci 7:163–173. https://doi.org/10.1016/j.jrras.2014.02.003

Arelli V, Begum S, Anupoju GR, Kuruti K, Shailaja S (2018) Dry anaerobic co-digestion of food waste and cattle manure: impact of total solids, substrate ratio and thermal pre treatment on methane yield and quality of biomanure. Bioresour Technol 253:273–280. https://doi.org/10.1016/j.biortech.2018.01.050

Atems B, Hotaling C (2018) The effect of renewable and nonrenewable electricity generation on economic growth. Energy Policy 112:111–118. https://doi.org/10.1016/j.enpol.2017.10.015

Bacenetti J, Sala C, Fusi A, Fiala M (2016) Agricultural anaerobic digestion plants: What LCA studies pointed out and what can be done to make them more environmentally sustainable. Appl Energy 179:669–686. https://doi.org/10.1016/j.apenergy.2016.07.029

Bah H, Zhang W, Wu S, Qi D, Kizito S, Dong R (2014) Evaluation of batch anaerobic co-digestion of palm pressed fiber and cattle manure under mesophilic conditions. Waste Manag 34:1984–1991. https://doi.org/10.1016/j.wasman.2014.07.015

Bayr S, Rintala J (2012) Thermophilic anaerobic digestion of pulp and paper mill primary sludge and co-digestion of primary and secondary sludge. Water Res 46:4713–4720. https://doi.org/10.1016/j.watres.2012.06.033

Bayr S, Kaparaju P, Rintala J (2013) Screening pretreatment methods to enhance thermophilic anaerobic digestion of pulp and paper mill wastewater treatment secondary sludge. Chem Eng J 223:479–486. https://doi.org/10.1016/j.cej.2013.02.119

Begum S, Anupoju GR, Sundergopal S, Bhargava SK, Jegatheesan V, Eshtiaghi N (2018) Significance of implementing decentralized biogas solutions in India: a viable pathway for biobased economy. Detritus 1:75–82

Berkeley L (2001) Behavior of capstone and honeywell microturbine generators during load changes

Birol F, Cozzi L, Gould T, Bromhead A, Priddle R (2015) World outlook energy 2015. doi:https://doi.org/10.1787/weo-2014-en

Blades L, Morgan K, Douglas R, Glover S, De Rosa M, Cromie T, Smyth B (2017) Circular biogas-based economy in a rural agricultural setting. Energy Procedia 123:89–96. https://doi.org/10.1016/j.egypro.2017.07.255

Chen Q, Liu T (2017) Biogas system in rural China: upgrading from decentralized to centralized? Renew Sust Energ Rev 78:933–944. https://doi.org/10.1016/j.rser.2017.04.113

Chen Y, Cheng JJ, Creamer KS (2008) Inhibition of anaerobic digestion process: a review. Bioresour Technol 99:4044–4064. https://doi.org/10.1016/j.biortech.2007.01.057

Chen JL, Ortiz R, Steele TWJ, Stuckey DC (2014) Toxicants inhibiting anaerobic digestion: a review. Biotechnol Adv 32:1523–1534. https://doi.org/10.1016/j.biotechadv.2014.10.005

Chiumenti A, da Borso F, Limina S (2018) Dry anaerobic digestion of cow manure and agricultural products in a full-scale plant: efficiency and comparison with wet fermentation. Waste Manag 71:704–710. https://doi.org/10.1016/j.wasman.2017.03.046

Cieślik M, Dach J, Lewicki A, Smurzyńska A, Janczak D, Pawlicka-Kaczorowska J, Boniecki P, Cyplik P, Czekała W, Jóźwiakowski K (2016) Methane fermentation of the maize straw silage under meso- and thermophilic conditions. Energy 115:1495–1502. https://doi.org/10.1016/j.energy.2016.06.070

Cuéllar AD, Webber ME (2008) Cow power: the energy and emissions benefits of converting manure to biogas. Environ Res Lett 3. https://doi.org/10.1088/1748-9326/3/3/034002

Dale BE, Ong RG (2012) Energy, wealth, and human development: Why and how biomass pretreatment research must improve. Biotechnol Prog 28:893–898. https://doi.org/10.1002/btpr.1575

Demirer GN, Chen S (2005) Two-phase anaerobic digestion of unscreened dairy manure. Process Biochem 40:3542–3549. https://doi.org/10.1016/j.procbio.2005.03.062

Divya DLR, Gopinath PMC (2015) A review on current aspects and diverse prospects for enhancing biogas production in sustainable means. Renew Sust Energy Rev 42:690–699

do Carmo Precci Lopes A, Mudadu Silva C, Pereira Rosa A, de Ávila Rodrigues F (2018) Biogas production from thermophilic anaerobic digestion of kraft pulp mill sludge. Renew Energy 124:40–49. https://doi.org/10.1016/j.renene.2017.08.044

Ertem FC, Neubauer P, Junne S (2017) Environmental life cycle assessment of biogas production from marine macroalgal feedstock for the substitution of energy crops. J Clean Prod 140:977–985. https://doi.org/10.1016/j.jclepro.2016.08.041

Eslami H, Hashemi H, Fallahzadeh RA, Khosravi R, Fard RF, Ebrahimi AA (2018) Effect of organic loading rates on biogas production and anaerobic biodegradation of composting leachate in the anaerobic series bioreactors. Ecol Eng 110:165–171. https://doi.org/10.1016/j.ecoleng.2017.11.007

Fallde M, Eklund M (2015) Towards a sustainable socio-technical system of biogas for transport: € ping in Sweden the case of the city of Link o. J Clean Prod 98:17–28. https://doi.org/10.1016/j.jclepro.2014.05.089

Fuchsz M, Kohlheb N (2015) Comparison of the environmental effects of manure- and crop-based agricultural biogas plants using life cycle analysis. J Clean Prod 86:60–66. https://doi.org/10.1016/j.jclepro.2014.08.058

Gou C, Yang Z, Huang J, Wang H, Xu H, Wang L (2014) Chemosphere effects of temperature and organic loading rate on the performance and microbial community of anaerobic co-digestion of waste activated sludge and food waste. Chemosphere 105:146–151. https://doi.org/10.1016/j.chemosphere.2014.01.018

Gozgor G, Lau CKM, Lu Z (2018) Energy consumption and economic growth: new evidence from the OECD countries. Energy 153:27–34. https://doi.org/10.1016/j.energy.2018.03.158

Gutierrez EC, Xia A, Murphy JD (2016) Can slurry biogas systems be cost effective without subsidy in Mexico? Renew Energy 95:22–30. https://doi.org/10.1016/j.renene.2016.03.096

Hagos K, Zong J, Li D, Liu C, Lu X (2017) Anaerobic co-digestion process for biogas production: progress, challenges and perspectives. Renew Sust Energ Rev 76:1485–1496. https://doi.org/10.1016/j.rser.2016.11.184

Hamawand I (2015) Anaerobic digestion process and bio-energy in meat industry: a review and a potential. Renew Sust Energ Rev 44:37–51. https://doi.org/10.1016/j.rser.2014.12.009
Hamelin L, Naroznova I, Wenzel H (2014) Environmental consequences of different carbon alternatives for increased manure-based biogas. Appl Energy 114:774–782. https://doi.org/10.1016/j.apenergy.2013.09.033
Hamid RG, Blanchard RE (2018) An assessment of biogas as a domestic energy source in rural Kenya: developing a sustainable business model. Renew Energy 121:368–376. https://doi.org/10.1016/j.renene.2018.01.032
Hernandez-De Lira IO, Huber DH, Espinosa-Solares T, Balagurusamy N (2015) Methane emission and bioenergy potential from livestock manures in Mexico. J Renew Sust Energy 7. https://doi.org/10.1063/1.4934564
Ingrao C, Bacenetti J, Adamczyk J, Ferrante V, Messineo A, Huisingh D (2019) Investigating energy and environmental issues of agro-biogas derived energy systems: a comprehensive review of life cycle assessments. Renew Energy 136:296–307. https://doi.org/10.1016/j.renene.2019.01.023
IEA (2016) World Energy Outlook 2016, IEA, Paris. https://doi.org/10.1787/weo-2016-en
IEA (2017) World Energy Outlook 2017, OECD Publishing, Paris/IEA, Paris. https://doi.org/10.1787/weo-2017-en
IEA (2018) World Energy Outlook 2018, IEA, Paris. https://doi.org/10.1787/weo-2018-en
IRENA (2012a) Renewable energy technologies: cost analysis series. Biomass Power Gener 1:60. https://doi.org/10.1016/B978-0-08-098330-1.00011-9
IRENA (2012b) Renewable energy technologies cost analysis series: solar photovoltaics. Compr Renew Energy 4:595–636. https://doi.org/10.1016/B978-0-08-087872-0.00319-X
IRENA (2012c) Renewable energy technologies: cost analysis series: wind power. Power Gener Technol 1:223–242. https://doi.org/10.1016/B978-0-08-098330-1.00011-9
IRENA (2015) Renewable power generation costs in 2014: an overview. IRENA 92. doi:https://doi.org/10.1007/SpringerReference_7300
Jin Y, Chen T, Chen X, Yu Z (2015) Life-cycle assessment of energy consumption and environmental impact of an integrated food waste-based biogas plant. Appl Energy 151:227–236. https://doi.org/10.1016/j.apenergy.2015.04.058
Johanna C, Lozano S, Mendoza MV, De Arango MC, Fernando E, Monroy C (2009) Microbiological characterization and specific methanogenic activity of anaerobe sludges used in urban solid waste treatment. Waste Manag 29:704–711. https://doi.org/10.1016/j.wasman.2008.06.021
Kabir MM, Forg G, Ilona S (2015) Lignocellulose-based bioproducts. doi:https://doi.org/10.1007/978-3-319-14033-9
Lansche J, Müller J (2017) Life cycle assessment (LCA) of biogas versus dung combustion household cooking systems in developing countries – a case study in Ethiopia. J Clean Prod 165:828–835. https://doi.org/10.1016/j.jclepro.2017.07.116
Larsson M, Grönkvist S, Alvfors P (2016) Upgraded biogas for transport in Sweden - effects of policy instruments on production, infrastructure deployment and vehicle sales. J Clean Prod 112:3774–3784. https://doi.org/10.1016/j.jclepro.2015.08.056
Lauer M, Hansen JK, Lamers P, Thrän D (2018) Making money from waste: the economic viability of producing biogas and biomethane in the Idaho dairy industry. Appl Energy 222:621–636. https://doi.org/10.1016/j.apenergy.2018.04.026
Lee DH (2017) Evaluation the financial feasibility of biogas upgrading to biomethane, heat, CHP and AwR. Int J Hydrog Energy 42:27718–27731. https://doi.org/10.1016/j.ijhydene.2017.07.030
Li F, Cheng S, Yu H, Yang D (2016a) Waste from livestock and poultry breeding and its potential assessment of biogas energy in rural China. J Clean Prod 126:451–460. https://doi.org/10.1016/j.jclepro.2016.02.104

Li Y, Li Y, Zhang D, Li G, Lu J, Li S (2016b) Solid state anaerobic co-digestion of tomato residues with dairy manure and corn stover for biogas production. Bioresour Technol 217:50–55. https://doi.org/10.1016/j.biortech.2016.01.111

Li R, Duan N, Zhang Y, Liu Z, Li B, Zhang D, Lu H, Dong T (2017) Co-digestion of chicken manure and microalgae Chlorella 1067 grown in the recycled digestate: nutrients reuse and biogas enhancement. Waste Manag 70:247–254. https://doi.org/10.1016/j.wasman.2017.09.016

Li K, Liu R, Cui S, Yu Q, Ma R (2018a) Anaerobic co-digestion of animal manures with corn stover or apple pulp for enhanced biogas production. Renew Energy 118:335–342. https://doi.org/10.1016/j.renene.2017.11.023

Li W, Siddhu MAH, Amin FR, He Y, Zhang R, Liu G, Chen C (2018b) Methane production through anaerobic co-digestion of sheep dung and waste paper. Energy Convers Manag 156:279–287. https://doi.org/10.1016/j.enconman.2017.08.002

Lieder M, Rashid A (2016) Towards circular economy implementation: a comprehensive review in context of manufacturing industry. J Clean Prod 115:36–51. https://doi.org/10.1016/j.jclepro.2015.12.042

Lim SJ, Kim BJ, Jeong CM, Choi J d r, Ahn YH, Chang HN (2008) Anaerobic organic acid production of food waste in once-a-day feeding and drawing-off bioreactor. Bioresour Technol 99:7866–7874. https://doi.org/10.1016/j.biortech.2007.06.028

López-González LM, Pereda Reyes I, Romero Romero O (2017) Anaerobic co-digestion of sugarcane press mud with vinasse on methane yield. Waste Manag 68:139–145. https://doi.org/10.1016/j.wasman.2017.07.016

Liu Y, Whitman WB (2008) Metabolic, phylogenetic, and ecological diversity of the methanogenic archaea. Ann N Y Acad Sci 1125:171–189. https://doi.org/10.1196/annals.1419.019

Lukehurst C, Bywater A (2015) Exploring the viability of small scale anaerobic digesters in livestock farming. Waste Manag

Lybæk R, Kjær T (2019) Pre-assessment of the circular economic benefits and challenges of biogas production in Denmark when utilizing sand bedding in dairy cow stables. J Clean Prod 219:268–277. https://doi.org/10.1016/j.jclepro.2019.01.241

Lynd LR, Liang X, Biddy MJ, Allee A, Cai H, Foust T, Himmel ME, Laser MS, Wang M, Wyman CE (2017) Cellulosic ethanol: status and innovation. Curr Opin Biotechnol 45:202–211. https://doi.org/10.1016/j.copbio.2017.03.008

Marañón E, Castrillón L, Quiroga G, Fernández-Nava Y, Gómez L, García MM (2012) Co-digestion of cattle manure with food waste and sludge to increase biogas production. Waste Manag 32:1821–1825. https://doi.org/10.1016/j.wasman.2012.05.033

Mittal S, Ahlgren EO, Shukla PR (2018) Barriers to biogas dissemination in India: a review. Energy Policy 112:361–370. https://doi.org/10.1016/j.enpol.2017.10.027

Nagamani B, Ramasamy K (1999) Biogas production technology: An Indian perspective. Curr Sci 77:44–55

Naik L, Gebreegziabher Z, Tumwesige V, Balana BB, Mwirigi J, Austin G (2014) Factors determining the stability and productivity of small scale anaerobic digesters. Biomass Bioenergy 70:51–57. https://doi.org/10.1016/j.biombioe.2014.01.055

Nayal FS, Mammadov A, Ciliz N (2016) Environmental assessment of energy generation from agricultural and farm waste through anaerobic digestion. J Environ Manag 184:389–399. https://doi.org/10.1016/j.jenvman.2016.09.058

Neshat SA, Mohammadi M, Najafpour GD, Lahijani P (2017) Anaerobic co-digestion of animal manures and lignocellulosic residues as a potent approach for sustainable biogas production. Renew Sust Energ Rev 79:308–322. https://doi.org/10.1016/j.rser.2017.05.137

Park SR, Pandey AK, Tyagi VV, Tyagi SK (2014) Energy and exergy analysis of typical renewable energy systems. Renew Sust Energ Rev 30:105–123. https://doi.org/10.1016/j.rser.2013.09.011

Patinvoh RJ, Osadolor OA, Sárvári Horváth I, Taherzadeh MJ (2017) Cost effective dry anaerobic digestion in textile bioreactors: experimental and economic evaluation. Bioresour Technol 245:549–559. https://doi.org/10.1016/j.biortech.2017.08.081

Patrizio P, Leduc S, Chinese D, Dotzauer E, Kraxner F (2015) Biomethane as transport fuel - a comparison with other biogas utilization pathways in northern Italy. Appl Energy 157:25–34. https://doi.org/10.1016/j.apenergy.2015.07.074

Peidong Z, Yanli Y, Yongsheng T, Xutong Y, Yongkai Z, Yonghong Z, Lisheng W (2009) Bioenergy industries development in China: Dilemma and solution. Renew Sust Energ Rev 13:2571–2579. https://doi.org/10.1016/j.rser.2009.06.016

Pérez-Camacho MN, Curry R, Cromie T (2017) Life cycle environmental impacts of substituting food wastes for traditional anaerobic digestion feedstocks. Waste Manag 73:140–155. https://doi.org/10.1016/j.wasman.2017.12.023

Pryshlakivsky J, Searcy C (2013) Fifteen years of ISO 14040: a review. J Clean Prod 57:115–123. https://doi.org/10.1016/j.jclepro.2013.05.038

Qian MY, Li RH, Li J, Wedwitschka H, Nelles M, Stinner W, Zhou HJ (2015) Industrial scale garage-type dry fermentation of municipal solid waste to biogas. Bioresour Technol 217:82–89. https://doi.org/10.1016/j.biortech.2016.02.076

Ramírez-arpide R, Demirer N, Gallegos-v C, Espinosa-Solares T (2018) Life cycle assessment of biogas production through anaerobic co-digestion of nopal cladodes and dairy cow manure 172. doi:https://doi.org/10.1016/j.jclepro.2017.11.180

Ratanatamskul C, Onnum G, Yamamoto K (2014) A prototype single-stage anaerobic digester for co-digestion of food waste and sewage sludge from high-rise building for on-site biogas production. Int Biodeterior Biodegrad 95:176–180. https://doi.org/10.1016/j.ibiod.2014.06.010

Reinelt T, Delre A, Westerkamp T, Holmgren MA, Liebetrau J, Scheutz C (2017) Comparative use of different emission measurement approaches to determine methane emissions from a biogas plant. Waste Manag 68:173–185. https://doi.org/10.1016/j.wasman.2017.05.053

Sandberg T, Crist J, Espinosa C, Thorin E, Gr S (2018) Large-scale biogas generation in Bolivia e A stepwise reconfiguration. J Clean Prod 180:494–504. https://doi.org/10.1016/j.jclepro.2018.01.174

Sanna A (2020) Advanced biofuels from thermochemical processing of sustainable biomass in Europe. doi:https://doi.org/10.1007/s12155-013-9378-4

Santos FS, Ricci BC, França Neta LS, Amaral MCS (2017) Sugarcane vinasse treatment by two-stage anaerobic membrane bioreactor: effect of hydraulic retention time on changes in efficiency, biogas production and membrane fouling. Bioresour Technol 245:342–350. https://doi.org/10.1016/j.biortech.2017.08.126

Scarlat N, Dallemand JF, Fahl F (2018) Biogas: developments and perspectives in Europe. Renew Energy 129:457–472. https://doi.org/10.1016/j.renene.2018.03.006

Shea RO, Wall D, Kilgallon I, Murphy JD (2020) Assessment of the impact of incentives and of scale on the build order and location of biomethane facilities and the feedstock they utilise. Appl Energy 182:394–408. https://doi.org/10.1016/j.apenergy.2016.08.063

Silva S, Alçada-Almeida L, Dias LC (2017) Multiobjective programming for sizing and locating biogas plants: a model and an application in a region of Portugal. Comput Oper Res 83:189–198. https://doi.org/10.1016/j.cor.2017.02.016

Sims REH, Rogner HH, Gregory K (2003) Carbon emission and mitigation cost comparisons between fossil fuel, nuclear and renewable energy resources for electricity generation. Energy Policy 31:1315–1326. https://doi.org/10.1016/S0301-4215(02)00192-1

Siyal SH, Mörtberg U, Mentis D, Welsch M, Babelon I, Howells M (2015) Wind energy assessment considering geographic and environmental restrictions in Sweden: a GIS-based approach. Energy 83:447–461. https://doi.org/10.1016/j.energy.2015.02.044

Soam S, Borjesson P, Sharma PK, Gupta RP, Tuli DK, Kumar R (2017) Bioresource technology life cycle assessment of rice straw utilization practices in India. Bioresour Technol 228:89–98. https://doi.org/10.1016/j.biortech.2016.12.082

Soares LCSR, Chandel AK, Pagnocca FC, Gaikwad SC, Rai M, da Silva SS (2016) Screening of yeasts for selection of potential strains and their utilization for in situ microbial detoxification (ISMD) of sugarcane bagasse hemicellulosic hydrolysate. Indian J Microbiol 56:172–181. https://doi.org/10.1007/s12088-016-0573-9

Strzalka R, Schneider D, Eicker U (2017) Current status of bioenergy technologies in Germany. Renew Sust Energ Rev 72:801–820. https://doi.org/10.1016/j.rser.2017.01.091

Stuckey DC (2012) Recent developments in anaerobic membrane reactors. Bioresour Technol 122:137–148. https://doi.org/10.1016/j.biortech.2012.05.138

Styles D, Dominguez EM, Chadwick D (2016) Environmental balance of the of the UK biogas sector: an evaluation by consequential life cycle assessment. Sci Total Environ 560–561:241–253. https://doi.org/10.1016/j.scitotenv.2016.03.236

Surendra KC, Takara D, Hashimoto AG, Khanal SK (2014) Biogas as a sustainable energy source for developing countries: opportunities and challenges. Renew Sust Energ Rev 31:846–859. https://doi.org/10.1016/j.rser.2013.12.015

Tampio E, Lehtonen E, Kinnunen V, Mönkäre T, Ervasti S, Kettunen R, Rasi S, Rintala J (2017) A demand-based nutrient utilization approach to urban biogas plant investment based on regional crop fertilization. J Clean Prod 164:19–29. https://doi.org/10.1016/j.jclepro.2017.06.172

Tan L, Cheng Q, Sun Z, Tang Y, Kida K, Al TANET, Ioeng JBIB (2019) Effects of ammonium and/or sul fi de on methane production from acetate or propionate using biochemical methane potential tests. J Biosci Bioeng 127:345–352. https://doi.org/10.1016/j.jbiosc.2018.08.011

Terboven C, Ramm P, Herrmann C (2017) Demand-driven biogas production from sugar beet silage in a novel fixed bed disc reactor under mesophilic and thermophilic conditions. Bioresour Technol 241:582–592. https://doi.org/10.1016/j.biortech.2017.05.150

Thomas P, Soren N, Rumjit NP, George James J, Saravanakumar MP (2017) Biomass resources and potential of anaerobic digestion in Indian scenario. Renew Sust Energ Rev 77:718–730. https://doi.org/10.1016/j.rser.2017.04.053

Thorin E, Olsson J, Schwede S, Nehrenheim E (2018) Co-digestion of sewage sludge and microalgae – biogas production investigations. Appl Energy 227:64–72. https://doi.org/10.1016/j.apenergy.2017.08.085

Toop TA, Ward S, Oldfield T, Hull M, Kirby ME, Theodorou MK (2017) AgroCycle - developing a circular economy in agriculture. Energy Procedia 123:76–80. https://doi.org/10.1016/j.egypro.2017.07.269

Ullah E, Martin AR (2016) Review of biogas digester technology in rural Bangladesh. Renew Sust Energ Rev 62:247–259. https://doi.org/10.1016/j.rser.2016.04.044

van der Ploeg F, Rezai A (2017) Cumulative emissions, unburnable fossil fuel, and the optimal carbon tax. Technol Forecast Soc Change 116:216–222. https://doi.org/10.1016/j.techfore.2016.10.016

Velásquez Piñas JA, Venturini OJ, Silva Lora EE, Calle Roalcaba OD (2018) Technical assessment of mono-digestion and co-digestion systems for the production of biogas from anaerobic digestion in Brazil. Renew Energy 117:447–458. https://doi.org/10.1016/j.renene.2017.10.085

Vivekanand V, Mulat DG, Eijsink VGH, Horn SJ (2018) Synergistic effects of anaerobic co-digestion of whey, manure and fish ensilage. Bioresour Technol 249:35–41. https://doi.org/10.1016/j.biortech.2017.09.169

Wang X, Lu X, Li F, Yang G (2014) Effects of temperature and carbon-nitrogen (C/N) ratio on the performance of anaerobic co-digestion of dairy manure, chicken manure and rice straw: focusing on ammonia inhibition. PLoS One 9(5):e97265. https://doi.org/10.1371/journal.pone.0097265

Wang J, Feng L, Tang X, Bentley Y, Höök M (2017) The implications of fossil fuel supply constraints on climate change projections: a supply-side analysis. Futures 86:58–72. https://doi.org/10.1016/j.futures.2016.04.007

Wang P, Yu Z, Zhao J, Zhang H (2018) Do microbial communities in an anaerobic bioreactor change with continuous feeding sludge into a full-scale anaerobic digestion system? Bioresour Technol 249:89–98. https://doi.org/10.1016/j.biortech.2017.09.191

Ware A, Power N (2016) Biogas from cattle slaughterhouse waste: energy recovery towards an energy self-sufficient industry in Ireland. Renew Energy 97:541–549. https://doi.org/10.1016/j.renene.2016.05.068

Weiland P (2010) Biogas production: current state and perspectives. Appl Microbiol Biotechnol 85:849–860. https://doi.org/10.1007/s00253-009-2246-7

Whiting A, Azapagic A (2014) Life cycle environmental impacts of generating electricity and heat from biogas produced by anaerobic digestion. Energy 70:181–193. https://doi.org/10.1016/j.energy.2014.03.103

Winkler H (2011) CIRP journal of manufacturing science and technology closed-loop production systems – a sustainable supply chain approach. CIRP J Manuf Sci Technol 4:243–246. https://doi.org/10.1016/j.cirpj.2011.05.001

Yasar A, Nazir S, Rasheed R, Tabinda AB, Nazar M (2017) Economic review of different designs of biogas plants at household level in Pakistan. Renew Sust Energ Rev 74:221–229. https://doi.org/10.1016/j.rser.2017.01.128

Zareei S (2018) Evaluation of biogas potential from livestock manures and rural wastes using GIS in Iran. Renew Energy 118:351–356. https://doi.org/10.1016/j.renene.2017.11.026

Zhai N, Zhang T, Yin D, Yang G, Wang X, Ren G, Feng Y (2015) Effect of initial pH on anaerobic co-digestion of kitchen waste and cow manure. Waste Manag 38:126131. https://doi.org/10.1016/j.wasman.2014.12.027

Zupančič GD, Grilc V (2012) Anaerobic treatment and biogas production from organic waste. In: Management of organic waste, p 198. doi:https://doi.org/10.5772/32756

Biogas Technology in Africa: An Assessment of Feedstock, Barriers, Socio-Economic Impact and the Way Forward

Ashira Roopnarain, Busiswa Ndaba, Maryam Bello-Akinosho, Emomotimi Bamuza-Pemu, Mashudu Mukhuba, Rosina Nkuna, and Rasheed Adeleke

Abstract In spite of the abundance and diversity of renewable energy resources in Africa, a dire energy crisis currently prevails. The adoption of biogas technology in Africa is promising due to its multitude of benefits including energy generation in the form of biogas, waste management and soil ameliorant production. Furthermore, biogas technology is particularly attractive in Africa due to the economic, environmental and social benefits that are associated with the technology as well as the favourable environmental conditions on the African continent that is conducive to the anaerobic digestion process. Anaerobic digestion of the array of organic waste streams that are available in Africa, which range from agricultural, municipal and industrial waste, could aid in the provision of a renewable energy source (biogas), thereby helping to mitigate energy poverty. The socio-economic benefits of adoption of biogas technology are vast. Technology adoption could result in improved health and agriculture, economic prosperity and social benefits. However, even though the technology is widely exploited in selected developing countries, adoption remains in its infancy on the African continent. This has been attributed to numerous factors

A. Roopnarain (✉) · B. Ndaba · M. Bello-Akinosho · E. Bamuza-Pemu
Microbiology and Environmental Biotechnology Research Group, Agricultural Research Council, Institute for Soil, Climate and Water, Pretoria, South Africa
e-mail: RoopnarainA@arc.agric.za

M. Mukhuba
Microbiology and Environmental Biotechnology Research Group, Agricultural Research Council, Institute for Soil, Climate and Water, Pretoria, South Africa

Department of Environmental Science, College of Agriculture and Environmental Science, Pretoria, South Africa

R. Nkuna
Department of Environmental Science, College of Agriculture and Environmental Science, Pretoria, South Africa

R. Adeleke
Unit for Environmental Sciences and Management, North-West University, Potchefstroom, South Africa

N. Balagurusamy, A. K. Chandel (eds.), *Biogas Production*,
https://doi.org/10.1007/978-3-030-58827-4_18

that range from elevated initial capital costs, absence of support structures, lack of awareness as well as negative perceptions of the technology cast by past failures. Such hurdles need to be overcome to improve technology adoption and reap the multitude of benefits associated with the technology. This chapter provides a brief overview of the current status of biogas technology adoption in Africa with reference to both household and commercial anaerobic digesters. In addition, the feedstock that may be utilised and its availability for the anaerobic digestion process are highlighted. The societal perception of biogas technology and socio-economic impact associated with technology adoption are also outlined. Lastly, barriers impeding biogas technology adoption in Africa and strategies to promote its adoption are discussed.

Keywords Biogas technology · Africa · Feedstock · Socio-economic impact · Barriers

1 Introduction

Development, economic growth, urbanisation, industrialisation and quality of life are directly affected by energy availability and supply. Energy is therefore indispensable in modern society and is a good indicator of socio-economic development (Surendra et al. 2014). Globally, the importance of sustainable energy supply as a key factor for development is acknowledged (Heegde and Sonder 2007). Despite this acknowledgement, the energy sector in Africa is underdeveloped which is a major threat to the realisation of the region's economic hopes (IEA 2014). African countries are faced with energy supply challenges with majority of communities in different countries depending mainly on biomass fuel such as wood, crop residues and cow dung for energy (Akinbami et al. 2001). The energy demand in sub-Saharan Africa grew by about 45% between 2000 and 2012, a growth that represents only 4% of the world's total (IEA 2014). With rapidly growing populations and economies in Africa, a lack of reliable supply of electricity is stunting development. This is because many Africans spend a huge proportion of their income on expensive, environmentally unhealthy and unsustainable alternatives in the form of household generator sets (IEA 2014; Philip 2014). For example, in Nigeria, the domestic sector relies on privately owned and maintained generators, and businesses are powered by diesel or fuel generator sets resulting in an estimated annual expenditure of $13.35M (Nnaji et al. 2013). The African continent seen from space at night is dark and unlit (The Economist 2007) as there is a relatively poor supply of electricity in most African countries, with the exception of South Africa. As a result of insufficient electricity supply, an estimated 2.7 billion people in Africa use traditional solid biomass to meet their primary energy needs such as cooking and heating (REN21 2017; Kamp and Bermúdez Forn 2016). An estimated 70–90% of households in Africa depend on biomass fuel mainly for cooking (Clemens et al. 2018; Tumwesige et al. 2011; Heegde and Sonder 2007). However, this form of energy is associated with numerous health and environmental drawbacks. Due to overdependence on traditional cooking fuels such as wood, agricultural residue and dried dung, there is a decline in their availability, whilst commercial fuels such as liquid petroleum gas (LPG) and paraffin in most African countries are too expensive and with inconsistent/unreliable supply (Heegde and Sonder 2007). Furthermore, collection of traditional fuels is primarily done by women and children in households, thus taking away from the time available for productive activities like school attendance for children and business activities for women. Other negative impacts of burning biomass fuels include exposure to indoor environmental air pollution which results in complications, especially respiratory diseases and eye ailments (Clemens et al. 2018; Heegde and Sonder 2007; Akinbami et al. 2001). Therefore, there is an urgent need in Africa for more sustainable alternative energy sources to meet the growing demand for sustainable energy (Ouedraogo 2017).

Several global and local programmes sponsored by different governments, non-governmental organisations (NGOs) and collaborative projects have been put in place to provide adequate electricity to the estimated 700 million people in Africa without electricity. These include the US-sponsored $7B Power Africa programme (USAID 2019), $765M World Bank-sponsored Nigeria electrification project (World Bank 2019) and Energy Africa campaign sponsored by UK's Department for International Development (UK Gov 2015). Others include Light UP and Power Africa—a new deal for energy in Africa sponsored by the African Development Bank (ADB 2019) and Electrification Financing Initiative sponsored by the European Commission of the European Union (EU) (EDFI 2019). The Sustainable Energy Fund for Africa (SEFA) is a $95 M facility sponsored by governments of the USA, UK, Denmark and Italy (SEFA 2019). The biggest and most ambitious programme is former US president Obama's Power Africa initiative. This programme incorporates the American and African governments, international and African businesses, World Bank, European Union (EU) and the African Development Bank and aims 'to increase access to electricity in sub-Saharan Africa by adding more than 30,000 megawatts (MW) of cleaner, more efficient electricity generation capacity and 60 million new home and business connections'. The programme has helped facilitate 120 private-sector power transactions which are generating (or soon to generate) a total of 10,095 MW (USAID 2019). Such initiatives provide the platform for job creation in communities through a flourishing private sector, impact positively in the lives of students allowing for longer study hours after dark and allow more efficient healthcare delivery.

Alternative energy sources include nuclear power; renewable energy such as solar, wind, biomass, geothermal, hydropower and tidal (wave) energy; and converting waste to energy (Jain and Jain 2017). Energy challenges faced by several African countries can be alleviated by exploring sustainable alternative energy sources, especially renewable energy (Sawyerr et al. 2019; Jain and Jain 2017). Of the available alternative sources of energy, biogas production during anaerobic digestion stands out as a feasible process for Africa because of the relatively uncomplicated process and portable systems that make installations easy even in rural areas.

1.1 Biogas Production

Biogas is produced from a variety of organic waste types (Achinas et al. 2017) by harnessing degradative pathways under anaerobic conditions controlled by a consortium of microorganisms (Roopnarain and Adeleke 2017; Bond and Templeton 2011). The breakdown of biodegradable organic waste produces a mixture of gases, mainly methane (60–70%), carbon dioxide (30–40%) and a few other gases in minute quantities (IRENA 2017; Roopnarain and Adeleke 2017; Surendra et al. 2014). Four key metabolic stages are involved in biogas production including hydrolysis, acidogenesis, acetogenesis and methanogenesis. During hydrolysis, complex organic compounds are broken down into simpler compounds like fatty acids, sugars, glycerol and amino acids by hydrolytic bacteria. Simple monomers produced during hydrolysis are fermented, converting them into volatile fatty acids (VFAs) and carbon dioxide (CO_2) by fermentative bacteria in the acidogenesis stage of the process. Hydrogen-producing acetogenic bacteria oxidise alcohols and VFAs in the system to acetate, CO_2 and water (H_2O) during acetogenesis. A second process produces acetate from hydrogen (H_2) and CO_2 in the stage by hydrogen-utilising bacteria. Finally, methanogenic bacteria convert acetate, CO_2 and H_2 into biogas (mixture of methane and CO_2 during methanogenesis) (Roopnarain and Adeleke 2017; Surendra et al. 2014).

1.2 General Overview of Biogas Technology Adoption in Africa

Biogas in Africa is a resource that is barely tapped but its potential is enormous. In many African countries, biogas technology is in its infancy, although good progress has been made in East Africa, especially Kenya (Clemens et al. 2018). Unlike in Africa, other developing countries in Asia have thoroughly explored and exploited the potential of domestic biogas, especially China and India. Its implementation in many Asian countries has been supported with establishment of national programmes for domestic biogas in Nepal, in Vietnam and, more recently, in Cambodia and Bangladesh (Heegde and Sonder 2007). India registered 3.67 million domestic biogas installations by 2010, but Africa is yet to experience such a propagation of domestic biogas installation and acceptance (Clemens et al. 2018). Despite several biogas dissemination initiatives, the total number of registered, constructed biogas installations for the whole of Africa is still in the order of thousands, and a large number of these are not functional, mainly due to lack of use or maintenance (Heegde and Sonder 2007). Although the use of this technology is still limited, there is progress in the dissemination of domestic biogas plants in Africa, mainly through the implementation of different programmes from government and non-governmental organisations (NGOs), which promote the technology (Lwiza et al. 2017). A summary of biodigesters installed in selected African

Table 1 Number of domestic biodigesters installed in selected African countries

Country	Number of biogas installations as of 2012 (Surendra et al. 2014)	Number of biogas installations as of 2014
Kenya	6749	14,110
Ethiopia	5011	10,680
Tanzania	4980	11,100
Uganda	3083	5700
Rwanda	2619	1700
Burkina Faso	2013	N/a
Senegal	334	N/a
Cameroon	159	N/a
Benin	42	N/a

N/a Data not available

countries is presented in Table 1. Other countries with few or several domestic biogas plants are listed by Roopnarain and Adeleke (2017), Mshandete and Parawira (2009) and Akinbami et al. (2001).

A study by the Netherlands Development Organisation (SNV) and the International Institute of Tropical Agriculture (IITA) in 2007 reported the technical feasibility of the generation and use of biogas for cooking for 18.5 million families in 24 African countries. Their assessment was based on feedstock availability using ownership of livestock as a baseline criteria, population, availability of water and climatic conditions (Heegde and Sonder 2007). With irregular energy supply problems facing the African continent, renewable energy is emerging as a viable alternative, especially with the launch of the Biogas Africa Initiative in 2005 in Nairobi (van Veenhuizen and Dubbeling 2017).

However, the reality is that biogas technology has not received wide acceptance, and its potential is underexploited in Africa. This could be associated with the high cost of initial investment and a lack of political will. These teething problems were discussed in the 'Biogas for Better Life—An African Initiative' conference in 2007, and as a follow-up, Rwanda launched the first national biogas initiative, installing 2400 digesters from 2007 to 2011. These were increased to over 9000 by 2018 (Clemens et al. 2018; Kamp and Bermúdez Forn 2016; Bedi et al. 2015; Mwirigi et al. 2014; Landi et al. 2013). Examples of other programmes across Africa include the Africa Biogas Partnership Program (ABPP), National Biogas Programme of Ethiopia (NBPE), Southern African Biogas Industry Association (SABIA) and South African National Energy Development Institute (SANEDI) (Clemens et al. 2018; Kamp and Bermúdez Forn 2016).

1.2.1 East Africa

Biogas in East Africa is ably championed by the Africa Biogas Partnership Program (ABPP), a partnership between SNV (Netherlands) and Hivos (Clemens et al. 2018) providing an integrated approach to dissemination of biogas in the region. The ABPP supports national programmes in four member countries in East Africa including Kenya, Tanzania, Ethiopia, Uganda and Burkina Faso in West Africa (ABPP 2019). The aim of the programme is to construct 100,000 biogas digesters in the five participating countries, thereby providing an estimated half a million people with access to a sustainable energy source (ABPP 2019). Implementation of the biogas programme in each country is supervised by a National Implementing Agency (NIA). Challenges encountered during implementation include poor development and participation by the private sector, credit facility unavailability in some regions, poor commitment to the programme by stakeholders at the regional level and the rising cost of materials needed for the construction of the bioreactors. Despite these challenges, the ABPP has made significant progress and has installed 18,560 digesters in Kenya, 18,534 in Ethiopia, 7628 in Uganda and 6441 in Tanzania (ABPP 2019). Biogas implementation in Kenya has been highly successful, and the model can be adopted in other African countries. The George Farm in Kenya became the first biogas producer in Africa to sell excess electricity produced to the national grid (Reuters Foundation 2017).

1.2.2 West Africa

Biogas in West Africa is yet to achieve wide recognition as there are no integrated programmes such as the ABPP in East Africa, with the exception of Burkina Faso which is the only non-East African country in that programme where 10,310 digesters have been installed by ABPP. The Economic Community of West African States (ECOWAS) Centre for Renewable Energy and Energy Efficiency (ECREEE) suggests that West Africa could base up to 54% of its power supply on renewable energy by 2030 (ECREEE 2016). This assessment includes the use of hydropower. With ready availability of organic waste feedstock, there is high potential for biogas production in the region (van Veenhuizen and Dubbeling 2017).

However, the wide use of biogas technology in sub-Saharan Africa is crippled by key obstacles, including (1) the usual high initial investment (construction) costs, (2) challenges with mobilisation of organic waste feedstocks, (3) lack of stability of national biogas programmes and (4) technological, institutional as well as sociocultural barriers. Growth of the sector requires an integrated approach with innovation and capacity building to give the necessary support to governments, the private sector and civil society (van Veenhuizen and Dubbeling 2017).

In Nigeria, the technology remains at a research level in institutions. Urban centres in Nigeria are plagued by heaps of open waste dumped indiscriminately. Utilisation of this waste for biogas production would not only contribute to alleviate the poor energy supply problem but would serve as a good waste management regime, thereby providing the citizens with a healthier environment (Ngumah et al. 2013). Lack of awareness has been identified as a major factor limiting its implementation in Nigeria. A survey by Uzoma et al. (2010) revealed a high degree of ignorance about biogas technology with only two non-functional biogas installations identified in the eastern part of Nigeria where the study was conducted. In Ghana, Biogas Ghana designs and constructs biodigesters and has installed at least 400 of them (Biogas Ghana 2019). Outside of Burkina Faso and Ghana, a few other countries in West Africa have some biogas installations including Senegal, the Republic of Benin and Cameroon (Table 1).

1.2.3 Southern Africa

South Africa (SA) is one of the few countries in Africa with a reliable electricity supply and consumes about 45% of Africa's energy supply. Energy provision is monopolised by the government-owned Eskom (Energy Supply Commission), generating electricity from coal combustion (Roopnarain and Adeleke 2017). With the rising cost of electricity and the negative environmental impact of coal combustion, renewable energy including biogas is becoming a viable option, although the need in SA is still not critical due to the ready supply of electricity by Eskom. Biogas was first experimented in SA in 1957 by John Fry, but the technology has experienced limited penetration (Munganga 2013). This has been attributed to a number of factors including low cost of electricity, high cost of installation of biogas facilities, lack of financial incentives, cheap cost of landfill for waste disposal and public perception (Roopnarain and Adeleke 2017; Munganga 2013). There are no integrated biogas programmes and no established biogas industry in SA, and an estimated 400 biodigesters are installed in the country (Roopnarain and Adeleke 2017). No data is available for biogas installations in Namibia although there are a few on commercial farms (Roopnarain and Adeleke 2017). Anaerobic digestion of *Phragmites australis* (a common reed of environmental concern) has been suggested to augment energy supply in Namibia to reduce energy imported from SA (von Oertzen 2009). In Botswana, biogas technology was adopted in the early 1980s but has not grown since due to failures in earlier attempts and limited water supply in the country (Roopnarain and Adeleke 2017).

2 Application of Biogas Technology in Africa

Biogas technology is considered an appropriate source of renewable energy in Africa due to the simplicity of the process, enabling its production at both a small scale, with feasibility for household production, and large scale, for commercialisation purposes (Luostarinen et al. 2011; Heegde and Sonder 2007). A study conducted in over 24 African countries revealed the potential technical feasibility of biogas used for cooking purposes to be over 18 million households. The findings were based on two basic requirements which are availability of cattle dung and water (Clemens et al. 2018; Heegde and Sonder 2007).

2.1 Small-Scale Biodigesters

Digesters in rural or peri-urban areas range from small to medium scale with biogas production of less than 100 cubic metres (Akinbami et al. 2001). They are referred to as domestic biogas plants and are mainly used for household purposes. Different designs of the biodigesters are available, but two common digester designs are widely used in Africa: the floating drum digester (Indian digester technology) and the fixed concrete dome (Chinese digester technology) (Akinbami et al. 2001) (Fig. 1). The fixed dome is more popular because it was found to be more sustainable than the floating drum, which resulted in more process failures compared to the fixed dome (Mwirigi et al. 2014). Although the design and size may differ, the principle is similar. Domestic biogas plants consist of an inlet, used for feeding, and an outlet, from which the by-products are collected (Kamp and Bermúdez Forn 2016). The digester itself is a sealed structure in which microbiological processes take place to convert organic matter into biogas.

Apart from cooking purposes, biogas can be used in biogas lamps, to power internal combustion engines, radiant heaters and refrigerators (Tumwesige et al. 2011). However, the energy crisis and need to deviate from the uncontrolled use of traditional biomass fuel in African countries steered the use of biogas mainly for cooking. This improves the lives of poor people in many parts of Africa, especially from health problems caused by the use of traditional biomass fuel (Clemens et al. 2018; Tumwesige et al. 2011). Thus, surveys for the feasibility and estimation of the biogas industry in Africa have been conducted, but, to date, only limited use of the technology has been reported in comparison to other developing regions. For instance, Mohammed et al. (2017) conducted a biogas production feasibility study for cooking in Ghana. In their findings, it was suggested that adoption of biogas technology for cooking purposes is the most plausible option. This was also a trend for a study conducted in South Africa, whereby several substrates were recommended to be used in combination for a smooth running biodigester for household cooking (Msibi and Kornelius 2017).

2.2 Large-Scale Biodigesters

Large-scale commercial biogas plants (Fig. 2) are used to provide the required energy for processing/manufacturing or for financial benefits through the conversion of biogas to electricity and heat (Kemausuor et al. 2018; Mittal et al. 2018). The technology for producing biogas is similar to that on a small-scale, but the large-scale technology is more advanced with more automated systems to prevent process

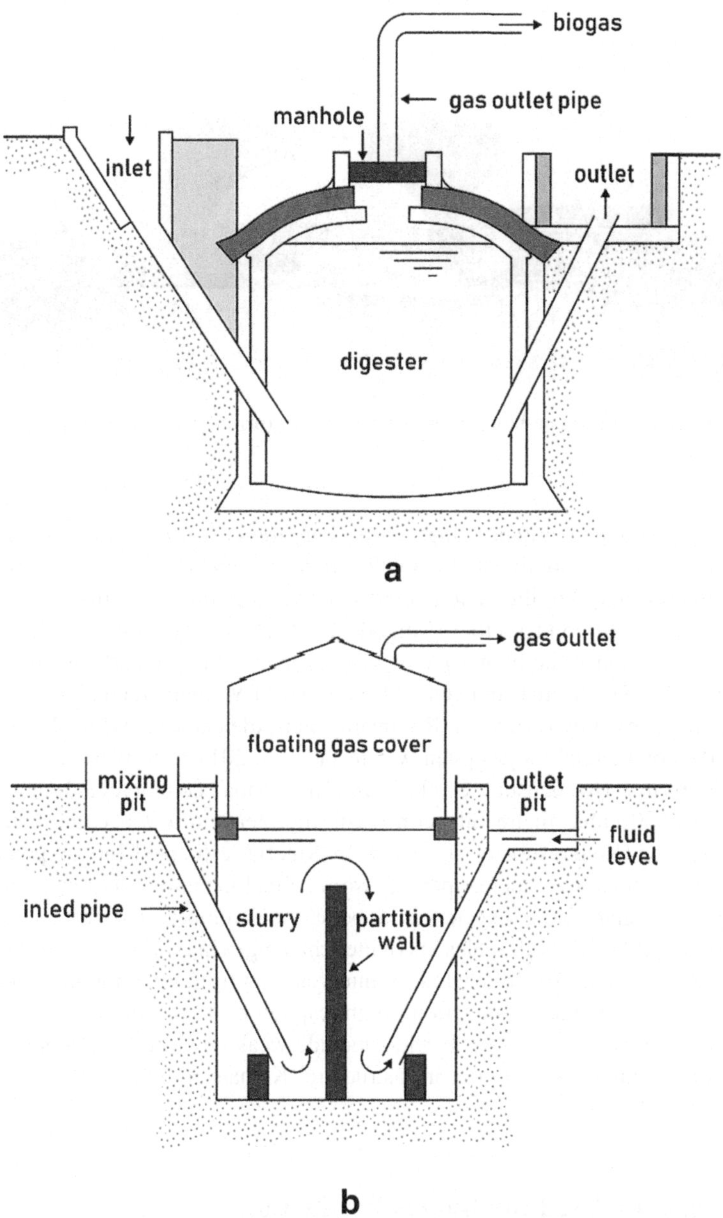

Fig. 1 Schematic representation of small-scale fixed dome (**a**) and floating drum biodigesters (**b**) (adapted from Nzila et al. 2012)

failure (Kigozi et al. 2014). Although adoption of large-scale biogas technology in African countries is still low, the digesters that are available are generally operated successfully (Rupf et al. 2016; Kigozi et al. 2014).

Fig. 2 Bio2Watt large-scale biodigester in South Africa (permission to use image provided by Thomas, 2019)

Although still in the infancy stage, there is progress in the operation of large-scale biogas plants in the continent. In South Africa, commercial biogas production is estimated at over 100 million cubic metres, based on estimates of different available feedstock from different sectors (Mugodo et al. 2017). According to Kemausuor et al. (2018), commercial biogas plants with a capacity between 30 kW and 19 MW are operated in SA (Kemausuor et al. 2018). A 4.4 MW commercial plant is operated in Bronkhorstspruit by Bio2Watt (Roopnarain and Adeleke 2017) (Fig. 2). In Kenya, a 2.2 MW commercial biogas plant was installed in 2017 in Naivasha, operated by Biojoule Kenya (Kemausuor et al. 2018). In Ghana, commercial biogas plants include a 2000 cubic metre oil palm waste digester and a 900 cubic metre fruit waste digester (Kemausuor et al. 2018). In Nigeria, estimated biogas potential is around six million cubic metres per day from animal manure (Aliyu et al. 2015).

Progress in small-scale digesters is far better than commercial digesters (Clemens et al. 2018). Although both scales encounter challenges in cost, the cost of installing small-scale digesters was alleviated by interventions from government and NGOs. For commercial biogas production, high capital cost is still a critical barrier (Kemausuor et al. 2018). Other barriers include weak environmental policies, poor institutional framework and poor infrastructure (Kemausuor et al. 2018).

3 Feedstock Used for Biogas Production

In any anaerobic digestion (AD) process, the most important parameter to be considered is the type of feedstock to be used. Biogas feedstock is defined as any organic waste that can undergo AD with the action of microorganisms and produce biogas. Millions of tons of different types of organic wastes are generated annually, including animal manure, wastewater sludge, food waste, agricultural residues and

Table 2 Substrates that may be used for biogas production in selected African countries

Country	Available biogas substrates	Estimated waste generated	Estimated biogas that can be produced
South Africa	Agricultural crop waste, manure (poultry, cattle and dairy farms/feedlots), abattoir waste, municipal solid waste and sewage (Msibi 2015)	Not recorded	The agricultural and agro-processing sector of SA has the potential to generate about 85.61×10^6 m^3 biogas/year (Mugodo et al. 2017)
Nigeria	Industrial wastes, household waste, animal dung, crop residues (Akinbami et al. 2001)	Nigeria produces about 227,500 tons of fresh animal waste daily	Nigeria has the potential to produce about 6.8 million m^3 of biogas every day from animal waste only (Mshandete and Parawira 2009)
Zimbabwe	Municipal solid waste (MSW), industrial and agricultural waste sources (Mshandete and Parawira 2009)	The total MSW produced is 552,975 tons annually	31,025,730 m^3 of biogas may be produced from MSW (Jingura et al. 2013)
Republic of Congo	Raw cassava peels	Unutilised cassava peel was estimated at 175,000 tons annually (Cuzin et al. 2001)	Not recorded
Senegal	Animal waste, crop residues, faecal sludge	In tons of dry matter per year: 2.311 million animal waste; 2.426 billion crop residues; 0.216 million household biowastes; 8.713 million faecal sludge	Not recorded

municipal solid wastes (Ali Shah et al. 2014). Anaerobic digestion of different feedstocks results in the production of various yields of methane which is dependent on feedstock composition.

The composition of waste differs spatially and seasonally (Leung and Wang 2016). Waste composition and characterisation is one of the most important steps in AD technology, and the choice of feedstock to be used for the process depends on its availability. Knowing the composition of the substrate in the digester is essential as this helps in determining the level of substrate pretreatment (Mukhuba et al. 2018), calculating the amount of biogas that can be produced as well as the amount of energy that the feedstock contains (Curry and Pillay 2012). The feedstock should be biodegradable and contain important nutrients (Kothari et al. 2014) such as carbohydrates, fats, cellulose, hemicellulose and proteins that can support the growth of the microorganisms involved in the AD process (Weiland 2009). Furthermore, the feedstock must not be entirely decomposed naturally as this leads to low methane production (NNFCC 2016). The various biogas feedstocks used in different African countries are shown in Table 2.

3.1 Agricultural Waste

The main source of organic waste in the agricultural sector is animal manure (Sibiya et al. 2017). Improper disposal of such waste can result in unpleasant odours, contamination of the environment with pathogens and associated health problems (Abubakar and Ismail 2012). The use of animal manure such as cow dung as a substrate for the AD process has been reported to be successful. According to Mukhuba (2017), approximately 52% of the total biogas generated at day 30 in a semi-continuous stirred tank reactor using cow dung was composed of methane. The digestate from cow dung was also shown to have the ability to enhance soil fertility (Mukhuba et al. 2018).

3.2 Plant Material and Crop Residues

Plants and crop residues are an important source of biomass used for biogas production (Sibiya et al. 2017). The biodegradability of plant material is challenging because it is lignocellulosic in nature, e.g. grass consists of lignin (10–13%), hemicellulose (15–50%) and cellulose (24–40%) (Nizami et al. 2009). Several pretreatment methods have been used to come up with a solution to improve AD of plant material (Nkuna et al. 2019). A study conducted by Amon et al. (2007) used maize for biogas production, whereas Hutňan et al. (2010) produced biogas using maize grain and maize silage and reported 0.72 m^3 biogas/kg of volatile suspended solids (VSS). In Zimbabwe, *Jatropha curcas* has been used as a nonfood, energy crop for biogas production (Jingura et al. 2013).

3.3 Municipal Solid Waste (MSW) and Food Waste

Millions of tons of MSW are generated in South Africa annually. The waste is normally collected as a mixed stream and disposed in landfill sites (Sibiya et al. 2017). This act is considered as a waste of energy and nutrients since most of the organic fraction of MSW has the potential to be used as substrates for AD (Sibiya et al. 2017). Food waste accounts for a large portion of the MSW in both developed and developing countries (Zhang et al. 2014). Leftovers from restaurants and cafeterias, food lost in the manufacturing process, uneaten food and any food that is no longer fit for human consumption are considered as food waste (Zhang et al. 2007). Most of these losses occur during handling, distribution, packaging and storage (van Schie 2013). Unfortunately, the quick deterioration of food waste causes environmental pollution and bad smells (Hartmann and Ahring 2006).

3.4 Challenges Associated with Anaerobic Digestion of Various Feedstocks and Co-digestion

Feedstocks that are primarily composed of cellulose, hemicellulose and lignin often lengthen the hydrolysis stage, making it the rate-limiting step in the AD process (Khanal 2011). In contrast, substrates such as mixed food waste which contain carbohydrates undergo rapid hydrolysis which creates greater chances of AD inhibition (Mukhuba 2017). The digestion of single substrates is usually challenging owing to an imbalance of nutrients, poor buffering capacity and a low pH (Mukhuba 2017; Demirel and Scherer 2008). To overcome the challenges associated with mono-digestion, several authors reported optimal biogas production with co-digestion (Sebola et al. 2015; Wang et al. 2014; Gashaw and Teshita 2014; Khalid et al. 2011; Iyagba et al. 2009). Co-digestion refers to the breaking down of more than one substrate or waste type in the same digester and is considered profitable in terms of CH_4 production (Gashaw and Teshita 2014; Khalid et al. 2011; Iyagba et al. 2009). Molinuevo-Salces et al. (2010) reported improved CH_4 yield when vegetable processing waste was co-digested with animal waste, which was attributed to a balanced C/N ratio. Furthermore, studies conducted by Sebola et al. (2015) and Wang et al. (2014) found that co-digestion of animal manure and any substrate with a high C/N ratio can increase the yield of the biogas (Gashaw and Teshita 2014). Co-digestion is also advantageous as it eliminates or reduces challenges associated with bacterial inhibition during the AD process (Sebola et al. 2015).

4 Socio-Economic Impact of the Adoption of Biogas Technology

Socio-economic development is hinged on advancements in the energy sector which drives the production of all goods and services. Although Africa is deemed a net energy exporter (Keho 2016), especially non-renewable energy, providing 8.9% and 10.2% of the world's gas and oil exports, respectively (BP Statistical Review 2018), many African populace remain with limited access to energy. Energy independence is championed as the seventh goal among the seventeen UN Sustainable Development Goals (SDGs), which advocates for access to affordable, reliable, sustainable and modern energy for all. Adoption of biogas technology in Africa will therefore ensure achievement of that SDG, the provision of green energy as well as a reduction in electricity consumption from the national grid and also serves as replacement or supplementation of fossil fuels. Biogas energy can be used domestically for various functions such as cooking, lighting and heating, whilst it can also be used industrially. Over-reliance on fossil fuel-based energy as well as negative environmental impacts due to fossil fuel usage has necessitated the use of alternative forms of energy such as renewables. Biogas energy, adopted on either a small or large-scale, will bring several socio-economic benefits to Africa. Unlike first-generation liquid biofuels, biogas produced from organic waste materials does not compete with food crops for land, water and fertilisers (Mshandete and Parawira 2009). Biogas, an important biomass energy, was introduced for domestic use in some African countries about four decades ago (Mulinda et al. 2013). Its use improves quality of life and health of individuals previously using traditional biomass energy such as wood, whilst its energy is indeed disruptive to the presently dominating non-renewable energy sources. Beyond fuel production, biogas technology improves sanitation, reduces greenhouse gas (GHG) emissions from biomass burning, reduces the need for wood and charcoal and provides a high-quality organic soil ameliorant (Wang et al. 2019; Yu et al. 2010). Furthermore, biogas technology assists in improving the management of solid and liquid wastes, from an environmental point of view. Biogas technology indeed harnesses the natural process of decomposition in converting organic waste into a methane-rich gas which is useful as an energy source in cooking, heat generation and lighting. Biogas can be stored and used on demand, thus serving as an energy source able to meet baseload demand (Kemausuor et al. 2018). Many African countries have vast biomass resources that could serve as feedstock for biogas plants; therefore, adopting biogas technology opens up many opportunities for diverse socio-economic benefits.

4.1 Health and Environmental Benefits

Adopting biogas technology will replace the burning of different traditional biomass in the process of domestic cooking. The combustion of traditional biomass results in the production of inefficient energy, and the heat release rate is difficult to control (Msibi and Kornelius 2017). Furthermore, burning biomass results in high carbon monoxide (CO), toxic hydrocarbons and particulate matter emissions (Apte and Salvi 2016). In most cases, biomass burning for domestic cooking is done indoors, thereby exposing inhabitants to prolonged inhalation of those gases which then results in several deleterious health issues such as indoor air pollution (IAP). Use of traditional biomass energy has also been associated with the recurrence of a number of diseases such as child pneumonia, pulmonary diseases and lung cancer (Dherani et al. 2008). Other health issues that have been implicated with IAP include asthma, cataracts, tuberculosis, high blood pressure, low birth weight and stillbirth (Apte and Salvi 2016). Globally, 4.3 million people die annually due to IAP (Apte and Salvi 2016). Even when the burning of biomass is done outdoors, GHGs are released to the atmosphere. Extensive use of firewood biomass energy severely impacts on local forests causing deforestation which, worldwide, accounts for 17–25% of anthropogenic GHG emissions (Strassburg et al. 2009). Deforestation is also a major contributing factor in soil erosion and land degradation.

Associated benefits of using biogas energy in many rural settings include provision of toilets in every home since many funding bodies for biogas subsidy programmes, for example, in India, specify the provision of a toilet in each home where biogas technology was funded. This eliminates issues related with improper disposal of human waste. Africa can learn from this to increase the adoption of biogas technology. India launched the National Project on Biogas Development (NPBD) which focussed on the promotion of biogas and provision of financial support for the implementation as well as provision of incentives to link toilets to biogas digesters. This helped to reduce health issues associated with improper sewage disposals (Surendra et al. 2014; Kurian 2004). Undoubtedly, two major issues in many African settings are energy generation and waste disposal. Both of these are addressed by biogas technology.

4.2 Social/Gender Perspective

Women and children are more at risk of the deleterious health effects of biomass burning as they spend more time indoors and are involved in domestic cooking. They are majorly involved in wood and dung collection as a source of biomass energy which takes a toll on the time the women and youth would have otherwise spent on other economic or educational purposes. Generally, many hours are wasted in the course of searching for wood, much manpower is used and health effects such as back pain could result. Therefore, adopting biogas technology will improve the quality of life of women and children who are most vulnerable in African society.

4.3 Climate Change and Protection

Many African countries such as the Democratic Republic of Congo, Mozambique, Zambia, Angola, Cameroon and Sudan rely on hydroelectric energy supply which is dependent on water availability. Reliance on hydropower is threatened by climate change-induced weather variability which affects rainfall patterns and causes drought situations as well as reduced river flow (Cole et al. 2014). Furthermore, reliance on hydro-energy threatens the other uses that water is meant to serve. Adoption of biogas energy supplements hydro-energy and therefore meets the sixth SDG which advocates for availability and sustainable management of water. Furthermore, it ensures reduction in GHG emissions and elimination of unpleasant odours emanating from organic compounds. Water resources are also protected from invasive aquatic weeds such as water hyacinth which can be used as a substrate for biogas production (Nkuna et al. 2019).

4.4 Economic Perspective

On a commercial scale, prices of fossil fuels are increasing, including the taxes on them. This volatility in fossil fuel prices continuously affects economic development of any country which is even worse for non-oil-producing countries (Isa and Ganda 2018). Furthermore, on a small scale, rising fossil fuel prices affect the individual's economy. Adoption of biogas technology, which has been proven to be cost-effective, is, therefore, a more viable and sustainable economic substitute to using fossil fuel-based and traditional biomass energy sources.

4.5 Household Energy Independence and Agricultural Improvement

Optimal biogas production from small-scale digesters is capable of producing energy on which individual households can depend, especially for cooking purposes (Walekhwa et al. 2014; Smith et al. 2013). Adopting biogas technology, therefore, guarantees energy access to remote households as well as independence to households from reliance on the national grid.

Beyond fuel provision, biogas technology also provides a source of soil ameliorant resulting from biogas slurry which contains optimal carbon and nitrogen as well as organic nutrients that could be available immediately, and released either rapidly or slowly (Smith et al. 2014). Biogas slurry has been used successfully for improving agricultural production (Wang et al. 2019; Walekhwa et al. 2014).

5 Major Hurdles in Adoption of Biogas Technology

Biogas technology adoption depends on the level at which the technology is chosen to be used by a person or an organisation. Although biogas has potential for energy generation and plays an important role in the quest towards sustainable development, there is still a huge gap in commercialisation of the technology in Africa (Amigun et al. 2008). This is mainly due to factors such as political, financial, social and technical restrictions (Roopnarain and Adeleke 2017) which contribute positively or negatively towards technology adoption. The acceptance by society coupled with the political interests of the stakeholders involved is usually considered a key factor for the implementation of any project. In addition, environmental and energy impacts determine acceptability of any renewable energy project (Pagnussatt et al. 2018).

The relevance of any energy technology may be evaluated financially and according to the satisfaction of the society (Mengistu et al. 2015). A technology that is viewed to be more advantageous than the existing one has the potential to be adopted faster and by more people. A technology that is easy to understand and use also has a greater probability of being adopted quickly as compared to one that is more complex. All the above considerations, including cost, contribute significantly to the rejection or adoption of a technology (Mwirigi et al. 2014).

5.1 Financial Implications

Development of a biogas sector in Africa requires effort from various stakeholders to establish different working mechanisms such as subsidy schemes, policies and reliable credit facilities to ease adoption and adaptability for both small- and large-scale AD adoption. In assessing financial implications, the areas of biogas application should be considered: individual household units, community/institutional plants or large-scale commercial operations. It should also be noted that financial support to small enterprises and individuals who are keen to set up effective biogas technology seems to be neglected in most parts of Africa (Clemens et al. 2018). Certainly, there will be an immediate negative effect on the progress of biogas dissemination if there are no appropriate measures to tackle such challenges.

5.1.1 Large-Scale Biogas Technology

About 65% of the cost of initiating a biogas project goes into construction. This was reported in Nigeria, whereby most of the amount went on purchasing cement and steel (Amigun and von Blottnitz 2010). However, the cost can vary depending on the size and materials used during construction. The design, economics and construction of biogas plants are explained in the literature (e.g. Widodo et al. 2009; Wilkie et al. 2004). The challenge is that all these expenses are encountered prior to installation, and some installation components are imported to Africa, which makes the technology even more expensive (Kemausuor et al. 2018). This becomes a problem in many African countries where poverty is endemic. Although biogas feedstock is considered to be zero to low cost, costs may be encountered during transportation of the feedstock. Another important factor is labour costs, but these will depend on the size of the plant as well as the feedstock required to run the plant.

5.1.2 Small-Scale Biogas Technology

Government subsidy availability is important for technology adoption, especially by rural individuals or community households. Although the cost incurred during small-scale biogas installation is lower than for large-scale digesters, cost of digester construction is considered. The initial investment cost is probably the major bottleneck to the adoption of biogas technology in Africa where a significant proportion of the population live under the poverty line (Clemens et al. 2018). This would result in the requirement of a larger anaerobic digestion system to support the needs of the family.

Such a system will have associated cost implications due to costs associated with additional contruction material as well as greater installation costs (Sooryamoorthy and Makhoba 2016).

5.2 Technical Barriers

Adequate supply of water and substrate are two crucial factors for the effective functioning of biogas plants. Underfeeding of inputs or feeding in the wrong ratios either results in suboptimal performance of a biogas plant or the formation of scum, making the plant completely dysfunctional (Rupf et al. 2015). Such failures can discourage potential users. Furthermore, feedstock amounts can fluctuate, resulting in the under-collection of waste which eventually leads to improper functioning of the plant. Moreover, the collection of feedstock in disaggregated places could also increase cost or labour intensity (Mittal et al. 2018).

5.3 Labour Intensity

Biogas technology is labour intensive because continuous feeding of the digesters is required. This has led to the technology not being widely adopted, despite its advantages (Yadvika et al. 2004). For example, in cases where collection of the feedstock is required, the owners of the digester may need to request for assistance from community members for collecting cow dung. Therefore, the labour required for acquisition of feedstock, collecting of water for mixing with the feedstock as well as regular maintenance have increased the stress on the owners of biogas digesters. These are seen as drawbacks to African biogas technology (Yadvika et al. 2004).

5.4 Uncertainties in the Availability of Biogas Feedstock

A variety of natural resources, found in most regions, can be used as substrates for biogas generation. After construction of a digester, it must be fed regularly with sufficient amounts of organic waste (Clemens et al. 2018). Failure to supply adequate amounts of substrate may lead to biogas technology failure. For example, in Uganda, biogas technology failure was due to unavailability of dung and water to start up the digesters or, if started, for the continuous feeding of the digesters (Lwiza et al. 2017). The quantity of substrate and the commitment of the user to continuously feed the digester are key to ensuring long-term operation. For example, large biodigesters installed on farms based on cattle dung input substrate require substantial amounts of cow dung. This is a big concern for owners when cattle number is reduced due to mortality, resulting in less dung being collected as biogas feedstock.

5.5 Biogas Technology Pilot Phase Failure

Installation of biogas plants with the aim of providing safe and renewable energy has been accompanied by challenges, for instance, low maintenance of large plants (Kemausuor et al. 2018; Wamwea 2017; Jonušauskait 2010). The first demonstration and pilot phase in Africa took place about four decades ago, but it has been reported that most of the digesters constructed during that period showed various malfunctions that discouraged biogas dissemination and adoption (Kemausuor et al. 2018; Amigun and Von Blottnitz 2010). Technical capacity for the construction and maintenance of commercial biogas plants is generally lacking in many developing countries. However, there are currently successful systems reported in African countries such as South Africa, Kenya and Ghana (Kemausuor et al. 2018).

5.6 Lack of Awareness of Biogas Technology

Lack of awareness of biogas technology has also hindered its progress despite the benefits in comparison to fossil fuels. In addition, there are no systems in place to inform African communities about the benefits of biogas over fossil fuels which have detrimental effects on the environment (Muvhiiwa et al. 2017; Abdulkarimm et al. 2013). Despite the existence of a wide range of business opportunities in rural areas in Africa, experience has shown that local entrepreneurs have yet to take advantage of such opportunities due to lack of information to develop a robust business plan for this technology (Muvhiiwa et al. 2017).

5.7 Social Perceptions of Biogas Technology

Many people consider handling animal or food waste for biogas production as unhygienic, leading to a reluctance to use these materials out of concern for sanitation (Brown 2006; Adeyemo 2002). Farmers are also reluctant to use the digestate as fertiliser due to possible contamination with pathogens that might be present in the utilised waste (Brown 2006). Some consider the technology to be expensive because the installation of biodigesters requires a source of funding as well as labour (Mittal et al. 2018). Biodigesters are normally installed outdoors which makes them prone to vandalising, depending on the material used to build them. In addition, the availability of other energy sources has made people hesitant in adopting biogas technology. For instance, in South Africa, most people source their energy from the Eskom grid at subsidised rates for poor communities (NEDLAC Trade and Industry Chamber 2010). Acceptance and adoption of new technologies is explained in the technology acceptance model (TAM) (Davis et al.

1989) which states that people can only adopt a new technology if they are fully convinced about its usefulness.

5.8 *Political Issues Involving Biogas Technology*

Another bottleneck which prevents the adoption of biogas technology is the lack of government incentives that are intended to support communities in adopting this technology in countries such as Cameroon and Namibia (Roopnarain and Adeleke 2017). Lack of co-ordination among institutions involved in renewable energy development and commercialisation is also another challenge in many African countries. These institutions include government ministries of energy/science and technology, research institutes, NGOs and financial institutions (Yousuf et al. 2016). For example, in 1983, Ghana established the National Energy Board (NEB) with the mandate to implement renewable energy in that country. However, the NEB terminated its operation in 1991. In addition, very few African countries have clear strategies and targets for renewable energy development in place (Amigun et al. 2008). This has created a setback for many potential biogas agencies.

Therefore, solutions to challenges in adoption of biogas in Africa are left for individual countries to develop and implement innovative policy frameworks, including subsidy policies that allow wider biogas technology adoption for small- and industrial-scale plants. This will ensure viability and affordability of the technology for everyone.

6 The Way Forward: Opportunities and Recommendations

The strong correlation between access to energy and education, urbanisation, employment, income generation, health, empowerment and overall improvement of quality of life is well-known (Nepal and Amatya 2006). Access to renewable energy, particularly in remote areas, can be achieved by the adoption of biogas technology. Africa has been identified as one of the regions with the highest global biogas potential (Kemausuor et al. 2018; Yousuf et al. 2016).

6.1 *Opportunities for Biogas Technology in Africa*

The total biogas production potential during anaerobic digestion of feedstock that is available to households, communities and industry in sub-Saharan Africa has been reported to be equivalent to 270 TWh of heat energy (Rupf et al. 2016). Biogas generation is particularly attractive in Africa and other developing nations due to the multitude of economic, environmental and social benefits that are associated with the technology (Rupf et al. 2015).

A number of favourable conditions exist on the African continent that promote the adoption of biogas technology. One of the major contributing factors is the comparably warm climate with average monthly temperatures mostly above 18 °C throughout the year. Such temperatures are well-suited for the anaerobic digestion process (Rupf et al. 2015). Another factor that motivates adoption of biogas technology in Africa is the widespread practice of animal rearing since animal excreta may be utilised as a feedstock for the AD process. This is of particular importance in countries where zero grazing is practised or where livestock are herded overnight in camps (semi-zero grazing) as this ensures ease of dung collection (Mwirigi et al. 2014). Such practices are common in various African countries including Tanzania, Uganda, Kenya, Malawi and South Sudan (Rupf et al. 2015; Walekhwa et al. 2009). The increasing costs and negative environmental impacts associated with fossil fuels have also motivated the adoption of biogas technology in Africa (Rupf et al. 2015). Fossil fuel sources are concentrated in a few countries such as Nigeria and Angola; hence, importation of fossil fuels or energy derived from them is necessary in most African countries, thus contributing to the elevated cost. Unlike with fossil fuels, the waste material necessary to drive the AD process is widely available in Africa, thereby ensuring energy provision to all geographic locations without the need for importation (Mwirigi et al. 2014). Biogas-derived energy is also more viable and environmentally friendly than the use of fuelwood in isolated communities that do not have access to grid electricity (Rupf et al. 2015).

The AD process results in the production of two valuable products, i.e. biogas and a nutrient-rich soil ameliorant. Whilst much attention has been placed on the generated biogas, in some African countries, such as Ethiopia, the soil ameliorant is of greater value than biogas energy, particularly due to the high cost of chemical fertilisers (Mwirigi et al. 2014). Chemical fertiliser production is restricted to certain areas in southern and North Africa implying that, much like fossil fuels, importation is necessary in other African countries, with associated cost implications. This may be a contributing factor to the limited fertiliser use in Africa which is equivalent to approximately 10% of global average usage (Roopnarain and Adeleke 2017). The soil ameliorant by-product of the AD process would be available in any region where biogas technology is adopted, enabling its widespread use. Overall, the adoption of biogas technology could enhance energy and food security by the provision of a stable, decentralised energy supply and soil ameliorant from indigenous waste resources (Rabezandrina 1990).

Apart from the generation of a renewable, sustainable energy source and nutrient-rich soil ameliorant, biogas technology has associated environmental benefits due to its contribution to improved waste management and indirect reduction in deforestation (Surendra et al. 2014; Rabezandrina 1990). Furthermore, adoption of the technology in Africa has numerous economic benefits through job creation, skills development and technological advancements from local construction of anaerobic digesters and digester components (Roopnarain and Adeleke 2017; Rupf et al. 2015). With the multitude of benefits associated with the adoption of biogas technology in Africa, it is imperative that hurdles to the adoption of the technology be identified and strategies to improve implementation be initiated and established.

6.2 Strategies to Improve Adoption of Biogas Technology in Africa

Various strategies to improve adoption of biogas technology in Africa have been identified. Such strategies are targeted at key areas that range from regulatory framework and policies to technological advances, funding and overcoming societal challenges. Table 3 presents an overview of strategies, proposed actions, target stakeholders and intended outcomes (Kemausuor et al. 2018; Roopnarain and Adeleke 2017; Rupf et al. 2015; Mwirigi et al. 2014; Rabezandrina 1990). Implementation of such strategies could aid in maximising biogas potential in Africa.

7 Conclusion

As indicated by Yousuf et al. (2016), 'the most powerful future biogas market is in Africa'. Adoption of biogas technology has the potential to address numerous challenges in Africa including energy security, food security through the use of soil ameliorant from the digester effluent as well as waste management and environmental preservation. The prevalent climatic conditions and substrate availability in Africa are major factors that promote adoption of biogas technology. However, successful implementation of the technology requires intervention from numerous stakeholders such as government, financial institutions, NGOs, research and training institutions as well as biogas entrepreneurs. Such interventions should be targeted at key areas that include the development of regulatory framework and policies that are conducive to biogas technology adoption, technological advancements of anaerobic digesters and the provision of funding. Societal challenges may be overcome by adequately informing society of the technology and its associated advantages. Furthermore, it is of utmost importance that a bridge be constructed between research and implementation, and this can be achieved by the provision of grants for the development of pilot plants and adequate communication between scientists and the public sector. The concerted efforts from all stakeholders will inevitably ease the adoption of the technology. Furthermore, it is important to mention that any action aimed at technology dissemination needs to be carefully planned in an effort to prevent repetition of past mistakes which may cast a negative outlook on the technology.

Table 3 Collation of strategies to improve adoption of biogas technology in Africa (adapted from Kemausuor et al. 2018; Roopnarain and Adeleke 2017; Rupf et al. 2015; Mwirigi et al. 2014; Rabezandrina 1990)

Key area	Proposed strategy	Target stakeholder (s)	Proposed actions	Intended outcome
Regulatory framework and policies	Development of local policy frameworks that are conducive to the growth and establishment of biogas technology	• National and regional governments	• Inclusion of biogas technology in energy policies • Inclusion of biogas technology in waste management policies • Development of national standards and guideline for use of digestate as soil ameliorant • Development of national standards and guidelines for bioreactor construction • Development of national standards and safety guidelines for biogas harvesting, storage, sale and use • Policy development to encourage zero grazing • Policy development to restrict tree felling for fuelwood • Policy development to restrict indiscriminate waste disposal • Implementation of policy mechanisms such as feed-in tariffs and net-metering • Where policies exist, strict implementation of policies	Promotion of uptake of biogas technology by creating an enabling environment with a governmental support structure
Social impact—awareness and environmental sustainability	Create awareness on the multitude of environmental benefits provided by the technology	• National, regional and local government • Schools, research councils and universities • NGOs • Biogas companies • Biogas entrepreneurs	• Information-sharing events and workshops • Establishment of demonstration units in various locations such as schools, smallholder farms and tourist sites • Interviews and demonstrations aired on local radio/TV channels emphasising the numerous benefits of the technology such as waste management, renewable energy generation and organic soil ameliorant production • Inclusion of biogas technology in school/university curricula • Establishment of biogas technology interest groups • Hands-on training on digester construction, feeding and maintenance	Promoting adoption of technology due to awareness of associated benefits and acquisition of necessary skills to operate and maintain digesters

(continued)

Table 3 (continued)

Key area	Proposed strategy	Target stakeholder (s)	Proposed actions	Intended outcome
Financial and economic	Improve the affordability of biogas technology	• Financial institutions/banks • Biogas companies • Government • NGOs	• Provision of loans at reasonable repayment rates or special 'interest free' renewable energy loans • Provision of government subsidies • Provision of incentives such as 'gate-fees' for waste collection and treatment • Increase awareness and aid in application for international funding schemes that promote biogas technology in Africa • Provision of plant insurances • Improve land tenure security—title deeds may be utilised as collateral for bank loans • Provision of incentives for adoption of biogas technology	Increased uptake of biogas technology by improving accessibility due to financial aid
Research and development	Optimise anaerobic digestion process to maximise biogas yield and prevent system collapse	• Schools, research councils and universities • NGOs • Biogas companies • Biogas entrepreneurs	• Ongoing interdisciplinary research on methods of process optimisation • Development and commercialisation of products that may be utilised to augment biogas yield • Development and commercialisation of products that may be utilised to stabilise the anaerobic digestion process and prevent process failure • Development of optimisation strategies that are conducive to African environmental conditions to prevent process failure, promote African technology and create local job opportunities	Promotion of uptake of technology by preventing negative outlook caused by dysfunctional systems and provision of solutions when problems occur to prevent digester abandonment

Technical and infrastructural	Development of user-friendly anaerobic digester systems that are specifically designed for local conditions and feedstock	• Government • NGOs • Biogas companies • Biogas entrepreneurs • Research councils and universities	• Design of affordable digesters that may be constructed from locally available material • Standardisation of small-scale digesters and biogas appliances • Development of prefabricated digesters • Design of digesters should be closely linked to local environmental conditions, water and feedstock availability • Digester components should be locally produced and easily available • Maintenance personnel should be well-trained and available • Provision of warranties and after-service support on digesters • Establishment of knowledge hubs to provide assistance in the case of digestion failures	Increase uptake of digester systems due to presence of proven, well-established AD systems on the market and availability of after-service support

References

Abdulkarimm I, Lambu I, Ahmed B et al (2013) People's awareness and attitude on biogas as an alternative domestic energy in urban Kano. Acad Res Int 4(6):625

ABPP (African Biogas Partnership Program) (2019). https://www.africabiogas.org/. Accessed 14 May 2019

Abubakar B, Ismail N (2012) Anaerobic digestion of cow dung for biogas production. ARPN J Eng Appl Sci 7(2):169–172

Achinas S, Achinas V, Euverink G (2017) A technological overview of biogas production from biowaste. Engineering 3:299–307. https://doi.org/10.1016/j.eng.2017.03.002

Adeyemo O (2002) Unhygienic operation of a city abattoir in South Western Nigeria: environmental implication. Afr J Environ Assess Manag 4(1):23–28

African Development Bank Group (2019) Light up and power Africa – a new deal on energy for Africa. https://www.afdb.org/en/the-high-5/light-up-and-power-africa-%E2%80%93-a-new-deal-on-energy-for-africa/. Last Accessed 7th June 2019

Akinbami J, Ilori M, Oyebisi T et al (2001) Biogas energy use in Nigeria: current status, future prospects and policy implications. Renew Sust Energ Rev 5:97–112. https://doi.org/10.1016/s1364-0321(00)00005-8

Ali Shah F, Mahmood Q, Maroof Shah M et al (2014) Microbial ecology of anaerobic digesters: the key players of anaerobiosis. Sci World J 2014:1–21. https://doi.org/10.1155/2014/183752

Aliyu A, Dada J, Adam I (2015) Current status and future prospects of renewable energy in Nigeria. Renew Sust Energ Rev 48:336–346. https://doi.org/10.1016/j.rser.2015.03.098

Amigun B, von Blottnitz H (2010) Capacity-cost and location-cost analyses for biogas plants in Africa. Resour Conserv Recycl 55:63–73. https://doi.org/10.1016/j.resconrec.2010.07.004

Amigun B, Sigamoney R, von Blottnitz H (2008) Commercialisation of biofuel industry in Africa: a review. Renew Sust Energ Rev 12:690–711. https://doi.org/10.1016/j.rser.2006.10.019

Amon T, Amon B, Kryvoruchko V et al (2007) Biogas production from maize and dairy cattle manure – influence of biomass composition on the methane yield. Agric Ecosyst Environ 118:173–182. https://doi.org/10.1016/j.agee.2006.05.007

Apte K, Salvi S (2016) Household air pollution and its effects on health. F1000Res 5:2593. https://doi.org/10.12688/f1000research.7552.1

Bedi A, Pellegrini L, Tasciotti L (2015) The effects of Rwanda's biogas programme on energy expenditure and fuel use. World Dev 67:461–474. https://doi.org/10.1016/j.worlddev.2014.11.008

Biogas Ghana (2019). https://www.biogasghana.org/. Accessed 14 May 2019

Bond T, Templeton M (2011) History and future of domestic biogas plants in the developing world. Energy Sustain Dev 15:347–354. https://doi.org/10.1016/j.esd.2011.09.003

BP Statistical Review (2018). https://www.bp.com/content/dam/bp/business-sites/en/global/corporate/pdfs/energy-economics/statistical-review/bp-stats-review-2018-africa-insights.pdf. Accessed on 7 May 2019

Brown V (2006) Biogas: a bright idea for Africa. Environ Health Perspect 114(5):A300–A303. https://doi.org/10.1289/ehp.114-a300

Clemens H, Bailis R, Nyambane A et al (2018) Africa biogas partnership program: a review of clean cooking implementation through market development in East Africa. Energy Sustain Dev 46:23–31. https://doi.org/10.1016/j.esd.2018.05.012

Cole M, Elliott R, Strobl E (2014) Climate change, hydro-dependency, and the African dam boom. World Dev 60:84–98. https://doi.org/10.1016/j.worlddev.2014.03.016

Curry N, Pillay P (2012) Biogas prediction and design of a food waste to energy system for the urban environment. Renew Energy 41:200–209. https://doi.org/10.1016/j.renene.2011.10.019

Cuzin N, Labat M, Garcia J et al (2001) *Methanobacterium congolense* sp. nov., from a methanogenic fermentation of cassava peel. Int J Syst Evol Microbiol 51:489–493. https://doi.org/10.1099/00207713-51-2-489

Davis F, Bagozzi R, Warshaw P (1989) User acceptance of computer technology: a comparison of two theoretical models. Manag Sci 35:982–1003. https://doi.org/10.1287/mnsc.35.8.982

Demirel B, Scherer P (2008) The roles of acetotrophic and hydrogenotrophic methanogens during anaerobic conversion of biomass to methane: a review. Rev Environ Sci Biotechnol 7:173–190. https://doi.org/10.1007/s11157-008-9131-1

Dherani M, Pope D, Mascarenhas M et al (2008) Indoor air pollution from unprocessed solid fuel use and pneumonia risk in children aged under five years: a systematic review and meta-analysis. Bull World Health Organ 86:390–398

ECREEE (2016) Biogas technology in Burundi. Regional progress report on renewable energy, energy efficiency and energy access in Ecowas Region. Energypedia. https://energypedia.info/wiki/Biogas_Technology_in_Burundi. Accessed 27 February 2019

EDFI (2019) Impact investment in sustainable energy. https://www.electrifi.eu/. Accessed 14th May 2019

Gashaw A, Teshita A (2014) Co-digestion of Ethiopian food waste with cow dung for biogas production. Int J Res 1(7):475–500

Hartmann H, Ahring B (2006) Strategies for the anaerobic digestion of the organic fraction of municipal solid waste: an overview. Water Sci Technol 53:7–22. https://doi.org/10.2166/wst.2006.231

Heegde F, Sonder K (2007) Domestic biogas in Africa; a first assessment of the potential and need. Biogas for a better life: an African initiative. http://www.snv.org/public/cms/sites/default/files/explore/download/20070520_biogas_potential_and_need_in_africa.pdf

Hutňan M, Špalková V, Bodík I et al (2010) Biogas production from maize grains and maize silage. Pol J Environ Stud 19:2

IEA (2014) World energy outlook 2014. https://www.oecd-ilibrary.org/docserver/weo-2014-en.pdf

IRENA (2017) Biogas for domestic cooking: technology brief. http://irena.org/-/media/Files/IRENA/Agency/Publication/2017/Dec/IRENA_Biogas_for_domestic_cooking_2017.pdf

Isa Y, Ganda E (2018) Bio-oil as a potential source of petroleum range fuels. Renew Sust Energ Rev 81:69–75. https://doi.org/10.1016/j.rser.2017.07.036

Iyagba E, Mangibo I, Mohammad Y (2009) The study of cow dung as co-substrate with rice husk in biogas production. Sci Res Essays 4(9):861–866

Jain S, Jain P (2017) The rise of renewable energy implementation in South Africa. Energy Procedia 143:721–726. https://doi.org/10.1016/j.egypro.2017.12.752

Jingura R, Musademba D, Kamusoko R (2013) A review of the state of biomass energy technologies in Zimbabwe. Renew Sust Energ Rev 26:652–659. https://doi.org/10.1016/j.rser.2013.05.036

Jonušauskait G (2010) Perceived benefits and challenges of biogas adoption among households in Kenya. IVM Institute for Environmental Studies, Amsterdam, pp 1–59

Kamp L, Bermúdez Forn E (2016) Ethiopia's emerging domestic biogas sector: current status, bottlenecks and drivers. Renew Sust Energ Rev 60:475–488. https://doi.org/10.1016/j.rser.2016.01.068

Keho Y (2016) What drives energy consumption in developing countries? The experience of selected African countries. Energy Policy 91:233–246. https://doi.org/10.1016/j.enpol.2016.01.010

Kemausuor F, Adaramola M, Morken J (2018) A review of commercial biogas systems and lessons for Africa. Energies 11:2984. https://doi.org/10.3390/en11112984

Khalid A, Arshad M, Anjum M et al (2011) The anaerobic digestion of solid organic waste. Waste Manag 31:1737–1744. https://doi.org/10.1016/j.wasman.2011.03.021

Khanal S (2011) Anaerobic biotechnology for bioenergy production. Wiley

Kigozi R, Muzenda E, Aboyade A (2014) Biogas technology: current trends, opportunities and challenges. 6th International conference on green technology, renewable energy and environmental engineering, pp 311–317

Kothari R, Pandey A, Kumar S et al (2014) Different aspects of dry anaerobic digestion for bio-energy: an overview. Renew Sust Energ Rev 39:174–195. https://doi.org/10.1016/j.rser.2014.07.011

Kurian P (2004) Socio-economic and environmental impact of bio gas programme. Kerala Research Programme on Local Level Development, Centre for Development Studies, Thiruvananthapuram. www.cds.ac.in/krpcds/report/pkkurian.pdf

Landi M, Sovacool B, Eidsness J (2013) Cooking with gas: policy lessons from Rwanda's National Domestic Biogas Program (NDBP). Energy Sustain Dev 17:347–356. https://doi.org/10.1016/j.esd.2013.03.007

Leung D, Wang J (2016) An overview on biogas generation from anaerobic digestion of food waste. Int J Green Energy 13:119–131. https://doi.org/10.1080/15435075.2014.909355

Luostarinen S, Normak A, Edström M (2011) Overview of biogas technology. Baltic manure WP6 energy potentials 49. doi:https://doi.org/10.1016/j.addbeh.2004.02.033

Lwiza F, Mugisha J, Walekhwa P et al (2017) Dis-adoption of household biogas technologies in Central Uganda. Energy Sustain Dev 37:124–132. https://doi.org/10.1016/j.esd.2017.01.006

Mengistu M, Simane B, Eshete G et al (2015) A review on biogas technology and its contributions to sustainable rural livelihood in Ethiopia. Renew Sust Energ Rev 48:306–316. https://doi.org/10.1016/j.rser.2015.04.026

Mittal S, Ahlgren E, Shukla P (2018) Barriers to biogas dissemination in India: a review. Energy Policy 112:361–370. https://doi.org/10.1016/j.enpol.2017.10.027

Mohammed M, Egyir IS, Donkor AK, Amoah P, Nyarko S, Boateng KK, Ziwu C (2017) Feasibility study for biogas integration into waste treatment plants in Ghana. Egypt J Pet 26:695–703. https://doi.org/10.1016/j.ejpe.2016.10.004

Molinuevo-Salces B, García-González M, González-Fernández C et al (2010) Anaerobic co-digestion of livestock wastes with vegetable processing wastes: a statistical analysis. Bioresour Technol 101:9479–9485. https://doi.org/10.1016/j.biortech.2010.07.093

Mshandete M, Parawira W (2009) Biogas technology research in selected sub-Saharan African countries – a review. Afr J Biotechnol 8(2):116–125

Msibi S (2015) Potential for domestic biogas as household energy supply in South Africa. MSc (Applied Science): environmental technology dissertation. University of Pretoria

Msibi S, Kornelius G (2017) Potential for domestic biogas as household energy supply in South Africa. J Energy South Afr 28:1. https://doi.org/10.17159/2413-3051/2017/v28i2a1754

Mugodo K, Magama P, Dhavu K (2017) Biogas production potential from agricultural and agro-processing waste in South Africa. Waste Biomass Valor 8:2383–2392. https://doi.org/10.1007/s12649-017-9923-z

Mukhuba M (2017) Ecological guild of microbes that drive production of biogas from multiple feedstock. MSc dissertation, University of South Africa

Mukhuba M, Roopnarain A, Adeleke R et al (2018) Comparative assessment of bio-fertiliser quality of cow dung and anaerobic digestion effluent. Cogent Food Agric 4(1):1435019. https://doi.org/10.1080/23311932.2018.1435019

Mulinda C, Hu Q, Pan K (2013) Dissemination and problems of African biogas technology. Energy Power Eng 05:506–512. https://doi.org/10.4236/epe.2013.58055

Munganga G (2013) Overview of biogas market in SA. Greencape waste economy sector. http://refhub.elsevier.com/S0959-6526(18)34021-6/sref43

Muvhiiwa R, Hildebrandt D, Chimwani N et al (2017) The impact and challenges of sustainable biogas implementation: moving towards a bio-based economy. Energy Sustain Soc 7(1):20. https://doi.org/10.1186/s13705-017-0122-3

Mwirigi J, Balana B, Mugisha J et al (2014) Socio-economic hurdles to widespread adoption of small-scale biogas digesters in Sub-Saharan Africa: a review. Biomass Bioenergy 70:17–25. https://doi.org/10.1016/j.biombioe.2014.02.018

NEDLAC Trade and Industry Chamber (2010) A study into approaches to minimise the impact of electricity price increases on the poor. http://www.dti.gov.za/industrial_development/docs/fridge/NEDLAC_final_report.pdf

Nepal G, Amatya V (2006) Understanding rural energy programme and poverty reduction linkage. Energy Sector Assistance Programme, Alternative Energy Promotion Centre, Lalitpur

Ngumah C, Ogbulie J, Orji J, Amadi E (2013) Biogas potential of organic waste in Nigeria. J Urban Environ Eng 7:110–116. https://doi.org/10.4090/juee.2013.v7n1.110116

Nizami A, Korres N, Murphy J (2009) Review of the integrated process for the production of grass biomethane. Environ Sci Technol 43:8496–8508. https://doi.org/10.1021/es901533j

Nkuna R, Roopnarain A, Adeleke R (2019) Effects of organic loading rates on microbial communities and biogas production from water hyacinth: a case of mono- and co-digestion. J Chem Technol Biotechnol 94:1294–1304. https://doi.org/10.1002/jctb.5886

Nnaji C, Chukwu J, Nnaji M (2013) Electricity supply, fossil fuel consumption, CO_2 emissions and economic growth: implications and policy options for sustainable development in Nigeria. Int J Energy Econ Policy 3(3):262–271

NNFCC (National Non-Food Crops Center) (2016) The official information portal on anaerobic digestion. http://www.biogas-info.co.uk/. Accessed 4 Apr 2019

Nzila C, Dewulf J, Spanjers H et al (2012) Multi criteria sustainability assessment of biogas production in Kenya. Appl Energy 93:496–506. https://doi.org/10.1016/j.apenergy.2011.12.020

Ouedraogo N (2017) Africa energy future: alternative scenarios and their implications for sustainable development strategies. Energy Policy 106:457–471. https://doi.org/10.1016/j.enpol.2017.03.021

Pagnussatt D, Petrini M, Santos A et al (2018) What do local stakeholders think about the impacts of small hydroelectric plants? Using Q methodology to understand different perspectives. Energy Policy 112:372–380. https://doi.org/10.1016/j.enpol.2017.10.029

Philip S (2014) Electrify Africa: bringing light to the Dark Continent. https://www.brandsouthafrica.com/investments-immigration/economynews/electrify-africa-bringing-light-to-the-dark-continent. Accessed 11 Apr 2019

Rabezandrina R (1990) Biogas: Evolution of actions and prospects for the rural environment in Africa. Ambio:424–426

REN21 (Renewable Energy Policy Network for the 21st Century) (2017) Renewables 2017 Global Status Report, REN21, Paris. http://www.ren21.net/wp-content/uploads/2017/06/17-8399_GSR_2017_Full_Report_0621_Opt.pdf

Reuters Foundation (2017) Africa's first grid-connected biogas plant powers up. https://reader.elsevier.com/reader/sd/pii/. Accessed 20 May 2019

Roopnarain A, Adeleke R (2017) Current status, hurdles and future prospects of biogas digestion technology in Africa. Renew Sust Energ Rev 67:1162–1179. https://doi.org/10.1016/j.rser.2016.09.087

Rupf G, Bahri P, de Boer K et al (2015) Barriers and opportunities of biogas dissemination in Sub-Saharan Africa and lessons learned from Rwanda, Tanzania, China, India, and Nepal. Renew Sust Energ Rev 52:468–476. https://doi.org/10.1016/j.rser.2015.07.107

Rupf G, Bahri P, de Boer K et al (2016) Broadening the potential of biogas in Sub-Saharan Africa: an assessment of feasible technologies and feedstocks. Renew Sust Energ Rev 61:556–571. https://doi.org/10.1016/j.rser.2016.04.023

Sawyerr N, Trois C, Workneh T et al (2019) An overview of biogas production: fundamentals, applications and future research. Int J Energy Econ Policy 9(2):105–116

Sebola R, Tesfagiorgis H, Muzenda E (2015) Production of biogas through anaerobic digestion of various waste: review. In: Proceedings of international conference on chemical, integrated waste management and environmental engineering (ICCIWEE), Johannesburg, pp 196–201

SEFA (2019) Sustainable energy fund for Africa. https://www.se4all-africa.org/seforall-in-africa/financing-opportunities/sustainable-energy-fund-for-africa/. Accessed 7th June 2019

Sibiya N, Muzenda E, Mbohwa C (2017) Evaluation of potential substrates for biogas production via anaerobic digestion: a review. Proc World Congr Eng Comput Sci 2:583–588

Smith J, Apsley A, Avery L et al (2013) The potential of small scale biogas digesters to improve livelihoods and long term sustainability of ecosystem services in Sub-Saharan Africa. Department for International Development, United Kingdom

Smith J, Abegaz A, Matthews R et al (2014) What is the potential for biogas digesters to improve soil fertility and crop production in Sub-Saharan Africa? Biomass Bioenergy 70:58–72. https://doi.org/10.1016/j.biombioe.2014.02.030

Sooryamoorthy R, Makhoba M (2016) The family in modern South Africa: insights from recent research. J Comp Fam Stud 47:309–321. https://doi.org/10.3138/jcfs.47.3.309

Strassburg B, Turner R, Fisher B et al (2009) Reducing emissions from deforestation - the "combined incentives" mechanism and empirical simulations. Glob Environ Chang 19:265–278. https://doi.org/10.1016/j.gloenvcha.2008.11.004

Surendra K, Takara D, Hashimoto A et al (2014) Biogas as a sustainable energy source for developing countries: opportunities and challenges. Renew Sust Energ Rev 31:846–859. https://doi.org/10.1016/j.rser.2013.12.015

The Economist (2007) Electricity in Africa – the dark continent. https://www.economist.com/middle-east-and-africa/2007/08/16/the-dark-continent. Accessed 10 Apr 2019

Tumwesige V, Avery L, Austin G (2011) Small-scale biogas digester for sustainable energy production in Sub-Saharan Africa. In Proceedings of the 1st world sustainability forum. Molecular Diversity Preservation International 1. doi: https://doi.org/10.3390/wsf-00587

UK Gov (2015). https://www.gov.uk/government/news/energy-africa-campaign. Accessed 7th June 2019

USAID (2019) Power Africa energy transactions and projects. https://www.usaid.gov/powerafrica/newsroom/power-africa-energy-transactions-and-projects. Accessed 14 May 2019

Uzoma C, Ibeto C, Okpara C (2010) Level of awareness and utilization of biogas technology in South-Eastern Zone of Nigeria. Int J Appl Agric Res 5(5):569–574

van Schie K (2013) SA wastes 10 million tons of food a year. https://www.iol.co.za/lifestyle/food-drink/sa-wastes-10-million-tons-of-food-a-year-1528945

van Veenhuizen R, Dubbeling M (2017) Integrated development of biogas in West Africa. Western Africa biowastes for energy and fertilizer. https://agritrop.cirad.fr/585817/7/ID585817_EN.pdf

von Oertzen D (2009) Green energy in Namibia. https://www.voconsulting.net/pdf/energy/Green%20Energy%20in%20Namibia%20-%20VO%20CONSULTING.pdf. Accessed 5 Apr 2015

Walekhwa P, Mugisha J, Drake L (2009) Biogas energy from family-sized digesters in Uganda: critical factors and policy implications. Energy Policy 37:2754–2762. https://doi.org/10.1016/j.enpol.2009.03.018

Walekhwa P, Lars D, Mugisha J (2014) Economic viability of biogas energy production from family-sized digesters in Uganda. Biomass Bioenergy 70:26–39. https://doi.org/10.1016/j.biombioe.2014.03.008

Wamwea S (2017) Success and failure of biogas technology systems in rural Kenya: an analysis of the factors influencing uptake and the success rate in Kiambu and Embu counties. Master's thesis, Norwegian University of Life Sciences, Ås

Wang C, Zuo J, Chen X et al (2014) Microbial community structures in an integrated two-phase anaerobic bioreactor fed by fruit vegetable wastes and wheat straw. J Environ Sci 26:2484–2492. https://doi.org/10.1016/j.jes.2014.06.035

Wang J, Gu D, Yu X et al (2019) Application effects of biogas slurry partly substituting for chemical fertilizer on autumn tomato production in winter-solar greenhouse. J Appl Ecol 30(1):243–250

Weiland P (2009) Biogas production: current state and perspectives. Appl Microbiol Biotechnol 85:849–860. https://doi.org/10.1007/s00253-009-2246-7

Widodo T, Asari A, Ana N et al (2009) Design and development of biogas reactor for farmer group scale. Indones J Agric 2(2):121–128

Wilkie A, Smith P, Bordeaux F (2004) An economical bioreactor for evaluating biogas potential of particulate biomass. Bioresour Technol 92:103–109. https://doi.org/10.1016/j.biortech.2003.08.007

World Bank (2019) Nigeria electrification project. http://projects.worldbank.org/P161885?lang=en. Last Accessed 7 June 2019

Yadvika S, Sreekrishnan T et al (2004) Enhancement of biogas production from solid substrates using different techniques – a review. Bioresour Technol 95:1–10. https://doi.org/10.1016/j.biortech.2004.02.010

Yousuf A, Khan M, Pirozzi D et al (2016) Financial sustainability of biogas technology: Barriers, opportunities, and solutions. Energ Source Part B Econ Plann Policy 11:841–848. https://doi.org/10.1080/15567249.2016.1148084

Yu F, Luo X, Song C et al (2010) Concentrated biogas slurry enhanced soil fertility and tomato quality. Acta Agr Scand Sect B Plant Soil Sci 60:262–268. https://doi.org/10.1080/09064710902893385

Zhang R, El-Mashad H, Hartman K et al (2007) Characterization of food waste as feedstock for anaerobic digestion. Bioresour Technol 98:929–935. https://doi.org/10.1016/j.biortech.2006.02.039

Zhang C, Su H, Baeyens J et al (2014) Reviewing the anaerobic digestion of food waste for biogas production. Renew Sust Energ Rev 38:383–392

Governmental Policies to Promote Biogas Production, Boosting Role of Biogas in Economic Growth of Developing Nations

Mónica María Cruz-Santos, Mariel Rubí Macías-Franco, Karla Sofia Herrera-Medina, Erick Diaz-Ruiz, Anuj Kumar Chandel, Júlio César dos Santos, Silvio Silvério da Silva, Paulian Thankanadathi Ramesh, and María Alejandra Sánchez-Muñoz

Abstract Anaerobic digestion (AD) is a renewable process more studied around the world. This process allows the use of agro-industrial waste, such as manure, sewage water, sorghum, corn, and cane waste, among others, to be converted into heat and electrical energy. AD has seen as an alternative form for being a sustainable bioenergy practice, aiming health, pollution, energy, and sanitation concerns. In developed countries, the biogas system has focused on large-scale biogas plants, based on farms, electricity, and heat in commercial and industrial form. On the other hand, in developing countries, the biogas is used for cooking and lighting, which is produced in small or domestic-scale digesters. However, in these last countries, the government has not bet on the implementation of biogas systems, thus generating a series of challenges for the implementation of these systems, being the main challenges due to technical deficiency, lack of adequate infrastructure, low economic budget, and low support from the government. However, high levels of global pollution and the development of new techniques have been attributed to consider implementing these systems around the world. Looking forward to this, positive prospects for pollution reduction, waste utilization, energy recovery, and economic gains for developed and developing countries will be discuss throughout this chapter based on the current governmental policies that enhances the economic growth to seek biogas production.

M. M. Cruz-Santos · M. R. Macías-Franco · K. S. Herrera-Medina · E. Diaz-Ruiz
Facultad de Ciencias Biológicas, Laboratorio de Biorremediación, Universidad Autónoma de Coahuila, Torreón, Mexico

A. K. Chandel · J. C. dos Santos · S. S. da Silva
Department of Biotechnology, Engineering School of Lorena (EEL), University of Sao Paulo (USP), Lorena, São Paulo, Brazil

P. T. Ramesh
Agricultural College and Research Institute, Killikulam, Tamil Nadu Agricultural University, Coimbatore, India

M. A. Sánchez-Muñoz (✉)
Laboratorio de Biorremediación, Facultad de Ciencias Biologicas, Universidad Autonoma de Coahuila, Torreón, Coahuila, Mexico

N. Balagurusamy, A. K. Chandel (eds.), *Biogas Production*,
https://doi.org/10.1007/978-3-030-58827-4_19

1 Introduction

The concern to mitigate the environmental impact of fossil fuels with the interest in the reduction of greenhouse gases (GHG) emissions and several strategies have been developed in various areas. One of them is the generation of clean energy from a range of technologies. Anaerobic digestion (AD) is an alternative that involves the production of biogas, which has taken great interest for decades, being a highly competitive and studied process. After the energy crisis in 1970, anaerobic digestion underwent rapid development (Vasco-Correa et al. 2018). The generation of energy from biogas increased by 85% in the last two decades.

Countries in Europe and Asia, as well as the United States, have adopted the AD as an environmental mitigation method governed by policies and regulations, which allow the implementation of biogas production for energy generation. These countries are leaders in biogas production policies, and other countries from developing countries have joined this strategy, adopting policies of developed countries in the field, as well as creating their own policies.

A particular case is the European Parliament who promoted policies for energy production, thus estimating biogas production increased to 20,000 m^3 by 2030 (Lara Grando et al. 2017). Other investment programs promoting regulations in energy production, in the case of China and Germany, estimated a generation for 2020 of 6 GW_{ton} and a projection of 25 GW_{tot} for 2050, the case of India, its policies promoted the transformation of biomass into energy, with the creation of approximately 5000 biogas plants by the Ministry of New and Renewable Energy (MNRE). Other countries in Latin America and Africa are still developing strategies to mitigate climate change through AD. The Mexican government estimates that in 2050 greenhouse gas emissions will decrease approximately 50% with the implementation of biogas systems, and 4.8% of the energy will be produced from renewable sources.

However, AD systems still present great challenges for both developed and policy developing countries; therefore, in recent years, various strategies have been proposed for greater control of the AD process in biodigesters. In this chapter, some of the regulations and policies of example countries that carry out the strategies for the production of biogas and the reduction of greenhouse gases, as well as some limitations of the process and perspectives, are disclosed.

2 Brief History of Anaerobic Digestion Technology and Biogas

Anaerobic digestion was developed as a wastewater treatment process in configuration with a septic tank. This design was known as "Mouras' Automatic Scavenger" and later was improved by Donald Cameron in 1895, in Exeter, England. Because of its success, the government of Exeter approved the treatment of the entire city's wastewater by these septic tanks. Recognizing the value of the methane gas as value-added product was collected and used for heating and lighting (Rittmann and McCarty 2001).

The application of anaerobic digestion for biogas production gained popularity in 1900, but decreased after the 1950s, because of the low fossil fuels prices. Then after the energy crisis in the 1970, the biogas digestion experienced a rapid development (Vasco-Correa et al. 2018), for being a sustainable bioenergy practice, aiming health, pollution, energy, and sanitation concerns (Yousuf et al. 2016). In developing countries, the biogas is used for cooking and lighting, which is produced in small or domestic-scale digesters. Meanwhile in developed countries, digesters are larger, and farm based, where the biogas produced is used for electricity generation and heat biogas plants (Scarlat et al. 2018).

2.1 Asia

In 1988 China had 4.7 million household biogas plants; meanwhile by the late 1990s, India had over three million family-sized biogas plants. In 2007, China reached about 26.5 million biogas plants and India nearly four million family-sized biogas plants (Bond and Templeton 2011). Of late, an estimated of 100,000 modern biogas plants and 43 million residential-scale digesters were viable in China by 2014, generating about 15 billion m^3 of biogas (Scarlat et al. 2018). In recent years, Asia generated 4268 GWh, which represents only 4% of the energy produced in the world from biogas (IEA 2019).

2.2 European Union

France in the mid-1990 had over 1000 anaerobic digestion plants (Abbasi et al. 2012). Germany one of the European leaders in biogas production had 850 on-farm anaerobic digester plants with 500 m^3 of capacity in 2000 (Weiland 2010). The electrical capacity in Europe by 2013 was 7852 MW with 282 biomethane plants. Europe in that year had the greatest biogas production with 282 plants, estimating 9.4 TWh of biomethane. Most plants in the European Union are located in Germany (154 plants), Sweden (54 plants), and the Netherlands (23 plants) (REN21 2015). Germany is one of the leaders in energy, involving the biogas generation, had 10,971 large-scale biogas plants (WBA 2019).

2.3 *Africa*

In 2016, Africa accounted for 68,000 household biogas digesters installed by African Biogas Partnership Program, with an energy biogas capacity of 1494 MW. Kenya and Ethiopia had growing on biogas plants between 2014 and 2016, being in the more important countries in installation of new plants (Dubois et al. 2019). By 2017, it improved their production in 80% in 10 years (IEA 2019).

2.4 *Central and Latin America*

In 2007, Mexico, Cuba, and Colombia had 1050, 79, and 60 household plants, respectively. In 2009 Central America had a biogas energy capacity of 4 MW, producing 1GWh, but by the end of 2017, it had a capacity of 36 MW and a generation of 112 GWh from biogas (IEA 2019) (Table 1). Brazil had and average production of biogas of 582.7 m^3/day with a total 121,453 kW in 2019 (Freitas et al. 2019). The USA had over 1528 anaerobic digester plants and Canada more than 100 biogas systems in the same year (Viancelli et al. 2019).

Table 1 Global installed capacity of energy from bioenergy/biogas (MW)

Area	Technology	2011	2012	2013	2014	2015	2016	2017
Africa	Bioenergy	1104	1196	1290	1444	1485	1495	1501
	Biogas	9	10	11	11	23	29	34
Asia	Bioenergy	15,290	16,325	18,641	21,303	24,217	29,743	33,780
	Biogas	408	509	650	846	955	1097	1287
C America + Carib	Bioenergy	1482	1580	1657	1801	2098	2440	2495
	Biogas	11	11	15	14	23	32	36
Europe	Bioenergy	29,302	30,693	31,706	33,211	34,643	35,739	36,662
	Biogas	8474	9661	10,145	10,738	11,218	11,667	12,073
Middle East	Bioenergy	65	67	72	80	91	98	98
	Biogas	32	34	39	47	58	65	65
N America	Bioenergy	12,511	13,279	14,312	15,030	16,109	16,247	17,194
	Biogas	1946	2257	2425	2547	2524	2574	2724
Oceania	Bioenergy	1004	1009	1001	1012	1013	1009	1020
	Biogas	272	277	269	279	280	276	275
S America	Bioenergy	10,936	12,228	14,605	15,108	15,767	16,590	17,031
	Biogas	184	221	222	243	273	537	349

Source: IRENA (2019), Renewable Energy Statistics 2019, The International Renewable Energy Agency, Abu Dhabi

2.5 Global Generation

In 2000 the global energy production from biogas was about 13,184 GWh (Fig. 1) and with 2455 MW energy capacity (Table 1). Germany the leader in the European Union produces 38.5% of the biogas produced in the world and 6% of the bioenergy produced from only biogas in the world. The top five countries in production in 2017 were Germany, the USA, the UK, Italy, and China (IEA 2019). In general, the world generation of power from biogas improved 85% from 2000 to 2017. The USA, for example, produced 13,723 GWh in 2017, by 2005 it produced 6449 GWh of electricity. Another example is Italy that produced by 2000 only 567 GWh but by 2017 improved their production to 8299 GWh. The total production in 2017 was 87,932 GWh of electricity generation from biogas, showing the importance of this technology for energy production.

3 Mitigation Strategies for Climate Change

Climate change mitigation along with the reduction GHG emissions has evolved as a major challenge that worldwide society is facing nowadays (Ramanathan and Feng 2008). This is mainly due by the excessive amounts of anthropogenic emissions, where GHG-intensive from the energy and industry sectors are the largest contributors (Cadez and Czerny 2016). Hence, rapid actions to climate change mitigation and the urgent reduction of GHG are being done.

As mentioned before, biomass has emerged as an important source of carbon, mainly for systems that aim to meet strict climate targets. This energy source has been used for heat and electricity production, as transport fuel and a feedstock for

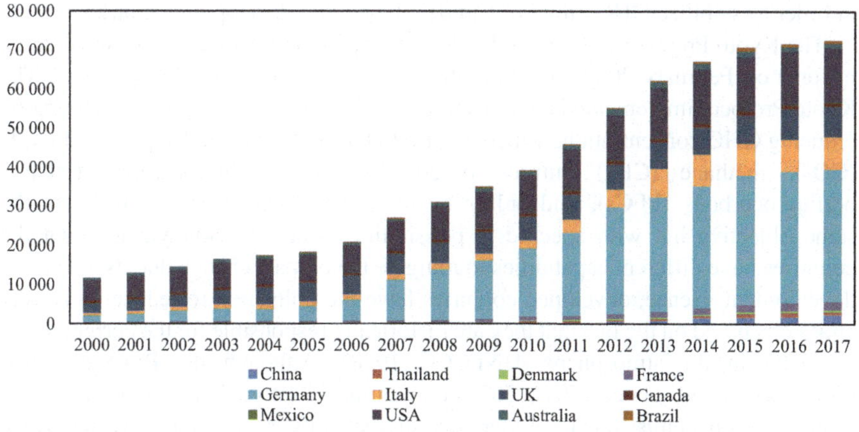

Fig. 1 Electricity generation from 2000 to 2017 (GWh). Source: IRENA (2019), Renewable Energy Statistics 2019, The International Renewable Energy Agency, Abu Dhabi

chemical compounds. From a policy perspective, the use of bioenergy has been widely implemented in big cities to reduce greenhouse gas GHG emissions and to improve energy security. However, in 2018 global energy-related CO_2 emissions rise 1.7% to a historic high of 33.1 Gt CO_2. Fossil fuels emissions accounted for 18.6 Gt CO_2 of emission growth, while coal use as power alone emissions exceeded to 10.1 Gt CO_2 for the first time, mostly in Asia. While China, India, and the accounted for 85% of the net increase in emissions (IEA 2019).

The most viable strategies to reduce GHG emissions depend on the nature of the gases itself in a particular firm. For example, firms with combustions emissions, which result from burning fossil fuels, implement rules to reduce fossil fuel consumption. Hence, strategies via GHG reduction depend on a specific characteristic of industrial processes.

Generally, four main options are taken for GHG-intensive firms to reduce climate change, which involved GHG restriction. The first one involves replacing conventional carbon-based materials (e.g., plastic) with eco-friendly (e.g., biodegradable), or recycled materials (e.g., recycled aluminum) (Boiral 2006; Jeswani et al. 2008). The second choice involves the replacement of carbon-based products by noncarbon-based products (e.g., steel products are replaced by wood products). The third alternative involves the usage of low carbon technologies (e.g., generating energy from renewable resources) (Jeswani et al. 2008; Kolk and Pinkse 2005). The fourth option is related to changes and improvements in industrial processed (e.g., utilization of inert electrodes for aluminum manufacture) (Cadez and Czerny 2016).

Globally, the United Nations Framework Convention on Climate Change (UNFCCC) was opened since 1992 where 154 nations signed the treaty in order to accomplish the main objective to "stabilize GHG concentrations in the atmosphere at a level that would prevent dangerous anthropogenic interference with earth's climate system" (UNFCCC 1992). Parties to the UNFCCC have met at conferences to discuss treaty goals which leads to Kyoto Protocol and Paris Agreement, among others, which have been signed in order to discuss, set, and adequate some objectives in order to stabilize GHG emissions in developed and developing countries.

The Kyoto Protocol was adopted in Kyoto, Japan, on December1997 and entered in force on February 2005, and currently, 192 parties are within the protocol. The Kyoto Protocol implemented the objective to reduce the onset of global warming by reducing GHG concentrations which applies to the six GHG listed as carbon dioxide (CO_2), methane (CH_4), nitrous oxide (N_2O), hydrofluorocarbons (HCFs), perfluorocarbons (PFCs), and sulfur hexafluoride (SF_6). This protocol has the same objective but with specific responsibilities since it acknowledges that the countries have different capabilities to mitigate the climate change due its economic development. Hence, developed countries have the obligation to reduce its current emissions based on the fact that they are historically responsible for the current levels of GHG in the atmosphere (UNFCCC 1998). Although the Protocol's first (2008–2012) and second (2012–2020) commitments set goals for emission reductions, the global emissions have increased dramatically, as mentioned earlier. Thus, negotiations were taken, after the second commitment period ends in 2020 in the framework of the UNFCCC Climate Change Conferences, which resulted in the adoption of the Paris Agreement.

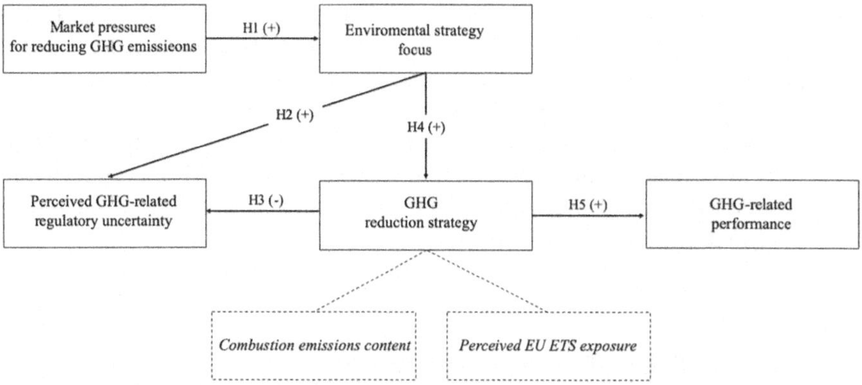

Fig. 2 Conceptual model of carbon reduction strategy (Scheme taken from Cadez et al. 2018). EU ETS: the European Union Emissions Trading Scheme

The Paris Agreement was negotiated in Le Bourget, France, on December 2015. Currently, it has been signed by 189 parties, with the exception of Iran and Turkey. The Paris Agreement long-term goal is to maintain the increase of global temperature to well below 2 °C above pre-industrial levels, to limit the increase to 1.5 °C and to increase the ability of parties to adapt to the adverse impacts of climate change. This last one in order to make "finance flows consistent with a pathway towards low greenhouse gas emissions and climate-resilient development." Also, this agreement also states that to reduce emissions is necessary to "achieve a balance achieve a balance between anthropogenic emissions by sources and removals by sinks of greenhouse gases." According to the main long terms under the Paris Agreement, each country has to develop plans and report the contribution, if it has been made, to mitigate the global warming. However, it was not set a specific emission target by a specific date.

European schemes aim to increase the understanding of the climate change mitigation strategies of GHG-intensive firms, particularly their antecedents and effects. Some authors have proposed a conceptual model (Fig. 2), which includes three antecedents that affect directly or indirectly which affect GHG reduction strategies, two exogenous (market pressure and perceived GHG-related regulatory uncertainty) and one endogenous (environmental strategy focus).

This model also includes variables, the combustion emissions content, and the perceived European Union Emissions Trading Scheme (EU ETS) exposure. This model also includes variables as the combustion emissions content, and the perceived European Union Emissions Trading Scheme (EU ETS) exposure. The first variable (combustion emissions content) involves the ratio of combustion and process emissions in the total GHG emissions mix. While, the second variable (perceived EU ETS exposure) is an specific type of environmental exposure that will probably affect GHG reduction strategies. Which in matter of exposure, diverse structural characteristics are involved as the GHG-based materials (large emitters are more exposed than small emitters) (Damert and Baumgartner 2017). As mentioned earlier, the nature of emission in a particular firm is crucial to reach GHG reduction strategies of a firm (Cadez and Czerny 2010, 2016), while in the European Union Emissions Trading Scheme, the GHG-intensive firms are disproportionally exposed to regulatory requirements even if they are operating within the same regulatory framework (Hoffmann and Busch 2008).

This model led to find out that market pressures for reducing GHG emissions, perceived GHG-related regulatory uncertainty, and environmental strategy focus are important determinants of corporate GHG reduction strategies. However, the results varied depending on the type of emissions.

4 Public Policies of Developed Countries for the Implementation of Biogas Systems

The bioenergy production has gained in most of the developing countries a bio-based economy. Biogas technology is a way to contribute on a green low carbon market. The production of biogas provides the opportunity of developing a new chain of residues from agriculture management (Scarlat et al. 2018). The biogas technology has an important economic, environmental, and social significance; however, there is a scarce acceptance in some rural areas of the developing countries.

While biogas for rural energy is affordable, the construction, the operation, and the high investment present complication, and it has led the farmers to develop cheaper systems. Therefore, the technology installation is one of the barriers, and the investment is out of the financial budget from developing countries. Nevertheless, the installation costs of the biogas plants could be reduced by the government support providing subsides and programs. The market is potentially higher when countries have positive policy frameworks, programs, and financial support when biogas is produced for its general use (Scarlat et al. 2018; Surendra et al. 2014; Rajendran et al. 2012).

The countries developed in policies for the regulation of renewable energies from the AD process are mainly those with greater agricultural performance, where their main objective is the reduction of GHG. Policy implementation improves the biogas plant application by enabling the use of new materials for the plants, the process optimization, and the increase of biogas generation and gas uses. Policies can be classified according to the implementation of the AD as shown in Fig. 3. Table 2 shows some policies and their implementation in developed countries (Hoo et al. 2017).

4.1 United States

The development of the implementation of AD in the USA as renewable energy has been very slow over the years; however, by 2015, 2100 plants were operating. In this country the Renewable Energy for America Program provides loans for AD systems

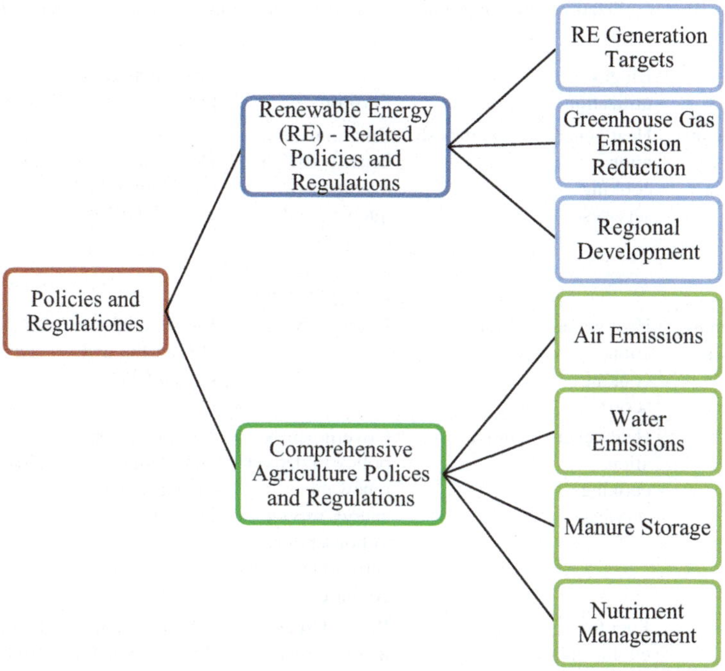

Fig. 3 Policies and regulations to the implementation of AD (Vasco-Correa et al. 2018)

in rural areas. For this sector, the USA has the Clean Water Act that provides greater promotion to the AD (Vasco-Correa et al. 2018). In addition, some activities have laws that help regulate waste; therefore, these laws promote the use of AD as an alternative for its treatment (Edwards et al. 2015).

4.2 European Countries

The different forms of regulation and policies propose goals to promote the production of energy from biogas. The European Parliament's is interested in the renewable energy production growing the policies for the investment on the generation of renewable energy offering funding for new biogas projects. Europe biogas production by the year 2030 is estimated to have the capacity of 18–20 billion m^3, which will correspond to the European consumption decreasing the GHG emissions as the main concern (Lara Grando et al. 2017). European governments mostly offer attractive initiatives to incentive the biogas sector, and the policy on each country is decisive. The tariff for biogas production depends on its capacity and the end use of biogas generation, but it certainly favors the production. For example, in Austria the biogas production from agriculture ranges from 0.13 €/kWh to 0.185 €/kWh (Ferreira et al. 2012).

Table 2 Biogas utilization and policy enforcement based on countries' socioeconomics (Hoo et al. 2017)

Category	Biogas utilization	Countries	Purpose	Project/program/policy	References
Least developed countries	Heat generation, cooking services	Bangladesh	Environmental benefits (63%) average frequency) and economic (59% average frequency) benefits	National Domestic Biogas and Manure Program (NDBMP)	Kabir et al. (2013)
Developing countries	Heat generation, cooking services	China	Energy saving	Biogas digesters at Lianshui and Guichi China	Xiaohua et al. (2007)
	Heat generation, cooking services	India	Government's policy to deliver renewable energy services to households across the countries	National Biogas and Manure Management Program	Raha et al. (2014)
	Fuel for engine and electricity generation	Malaysia	Policy target to increase renewable energy share to 11% by 2020	Fifth Fuel Policy 2000 National Renewable Energy Policy 2010 Small Renewable Energy Power (SREP) Program Renewable energy incentives Feed-in-Tariff (FiT)	Hashim and Ho (2011)
	Fuel for electricity generation	Thailand	To achieve 14% of all energy needs from renewable resources by 2022	Energy Conservation Promotion Act Renewable Energy Development Plan	Aggarangsi et al. (2013)
Developing countries	Natural gas grid injection	European countries	Contribution on reducing greenhouse gases (GHG) emission; policy target to increase renewable energy share to 20% by individual country members	Intelligent Energy for Europe (IEE)	Strauch and Singhal (2013)

(continued)

Table 2 (continued)

Category	Biogas utilization	Countries	Purpose	Project/program/ policy	References
	Vehicle fuel	Sweden	To impose profound societal structural change in combating climate change	Swedish Transport Policy	Fallde and Eklund (2015)
	Food waste management	Japan	To reduce greenhouse gases emission	Future Energy Policy Feed-in Tariff Scheme for Renewable Energy	Matsuda et al. (2012)

4.3 Asian Countries

The Asian sector has many strategies for the increasement on the investment for the renewable energy sector to promote the construction of biogas plants in rural areas. In the period from 2001 to 2012, China provided 4.86 billion dollars from government funds for biogas development. In 2009, the "Renewable Energy Law" was amended and was estimated to have the capacity to develop 15% of the energy by the year of 2020, investing 180 billion in renewable energy from this renewable energy policy. The investment by subsidies and law implementations increase 10%, and the installations of biogas were 2.2% higher (Wang et al. 2016).

A Sino-German project (Resource Recovery of Bio-organic Municipal Waste) which is a cooperation platform between China and Germany to integrate an efficient biogas generation focused on the urban and agricultural area estimating the generation of 6 GW_{tot} of BMW-derived biogas generated until 2020 and 25 GW_{tot} until 2050 (Yousuf et al. 2016).

In India the concern on waste management and climate change held to develop new policy initiatives enabling strategies to set up more institutions and laws related to renewable electricity and its enhancement. The creation of the Indian Renewable Energy Development (IREDA), as a Non-Banking Financial Institution, under the administrative control of Ministry of New and Renewable Energy (MNRE) impulse the way for the biomass sector along with the renewable sector to get strong institutional support in terms of finance (Singh and Setiawan 2013). The 40% of the biogas plants, which is five million of plants approximately, were installed under the biogas development program by the MNRE. Moreover, 400 biogas plants were installed for the capacity of 5.5 MW (Mittal et al. 2018).

In 1992, Thailand created the Energy Conservation Promotion Act and programs on energy issues, adding renewable and rural energy. With this plan, energy savings of 33.4Mt are expected in 2030 (Aggarangsi et al. 2013). Based on funding by the Energy Conservation Fund, Thailand promoted biogas generation with the biogas project to promote power generation on livestock farms, which started in 1995 and were divided into four phases until 2013, where the government provided subsidies for the construction of biogas plants on farms for the use of its waste, extending to the use of wastewater (Aggarangsi et al. 2013).

5 Public Policies of Developing Countries for the Implementation of Biogas Systems

The developing countries have sources of energy as the main activities for their economic growth. It has been shown that power generation and its consumption can improve the competition with other countries and the economic growth of developing countries. However, obtaining different types of energy has effects on social, economic, and environmental aspects (Fig. 4) (Amigun et al. 2011; Jan and Akram 2018).

The manufacture of fuels as a source of energy is one of the largest services that provide a greater economic gain in the developing countries. However, this can cause severe environmental damage for the GHG emissions. The production of biofuels is one possible solution that is taking advantage of agricultural waste and reduces the energy production costs, in comparison with other fuels. One of the well-known processes for generating electricity is the implementation of biogas systems (Amigun et al. 2011).

The biogas production from agriculture residues, industry, and municipal residues are attractive options to rise the global economy for developing countries. Some developing countries as Pakistan and African countries, among others, have implemented the installation of biogas systems as a new alternative to take advantage of disused agro-industrial waste and reduce energy production costs.

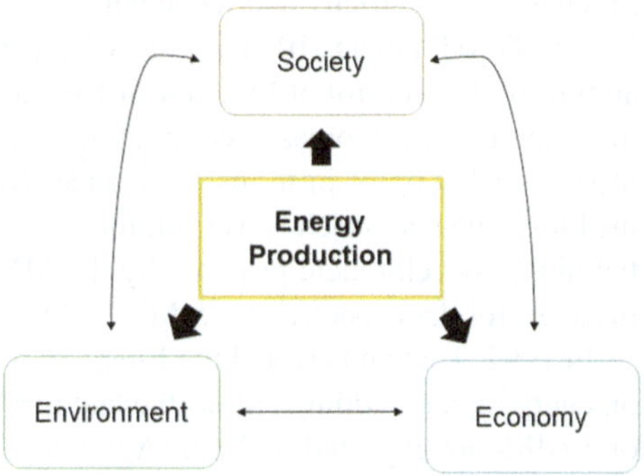

Fig. 4 Relation between effects by energy production in developing countries (Amigun et al. 2011)

5.1 African Countries

Africa is one of the developing countries with a large livestock population and agricultural areas. Because of this, it makes it a highly competitive country for the use of waste as the main source of sustainable energy generation. However, in 2014 only 38% of its population had access to electricity service (Bos et al. 2018). For this reason, the implementation of biogas systems for the use of waste and energy generation has been developed in many of their countries. However, the biogas installations launched so far are family sized (Cheng et al. 2014).

The government of the African continent, over time, has implemented new policies and support for the generation of renewable energies. Of the 56 countries that make up this continent 45 manage to have some law corresponding to the support of the implementation of these energies (Renewable Energy Network 2018). The South African government in 2009 managed to introduce a plan for renewable energy projects. However, this was not carried out due to opposition from the state power company (Becker and Fischer 2013).

Ghana has a large number of policies for the implementation of sustainable energies. One of them is the National Electrification Scheme (NES) implemented in 1989 which was extended until 2020. This policy deals with the purchase of renewable energy to promote the development of biogas systems (Kemausuor and Ackom 2017).

Nowadays Nigeria country keeps a strictly energy polices named National Renewable Energy and Energy Efficiency Policy (NREEEP), implemented since 2014 by the Energy Commission of Nigeria. A part of this policy includes the manufacture of biogas systems of various designs to support domestic, industrial, and institutional energy. So far with this law, 500 short-, 6000 medium-, and 8000 long-term digesters have been implemented (NREEEP 2014).

However in East Africa, Tanzania is the only country that have policies to support biogas systems through phase I investment subsidy in 2015 (Clemens et al. 2018).

5.2 Latin America

Countries within Latin America have great potential in biogas generation. However, the development of these systems has been slow compared to other developing countries. In 2009 the Network for Biodigester in Latin America and Caribbean (RedBioLAC) was created. This Network has been of great use, as it has promoted the installation of domestic biogas systems in Latin countries such as Bolivia, Costa Rica, Ecuador, Mexico, Nicaragua, and Peru. Bolivia leads these countries with 1000 functional biogas plants installed (Alemán-Nava et al. 2015; Garfí et al. 2016).

As for the installation of industrial biogas systems, 127 plants have been installed in Brazil, generating 584 billion m^3 of biogas per year. In addition, some other countries such as Colombia, Honduras, and Argentina have started with the construction of large-scale biogas plants due to RedBioLAC (Vögeli et al. 2014).

Also, global biogas production in 2017 as an electricity harvest was 31 million m^3, which 65% tested from the Asian continent and about 33% was produced in Latin America (Zervous 2019).

In México, the policies implemented by the federal government are expected to produce 4.8% of the renewable electricity used in the country in 2028. Mainly, the highest waste for biogas production in the states of Mexico is cow manure, organic waste, and wastewater (Díaz-Trujillo and Nápoles-Rivera 2019). Also, in 2050 the Mexican government foresaw greenhouse gas emissions to decline approximately 50% with the implementation of appropriate biogas systems. In addition, México is the only country in Latin America to provide credits to renewable transport fuel suppliers in order to use fuels from the generation of biogas, by the process of AD to gas-powered vehicles (Global Methane Initiative 2014). Although these countries possess a high capacity for the production of biogas systems, the lack of government support has restricted the developed of fully biogas systems.

Currently, some Mexican public policies related to the implementation of biodigesters for methane mitigation are carried out. The "Programa Especial de Cambio Climático" (PECC 2014–2018) has an advance of 76.12% and is based on an inclusive sustainable development model that incorporates the transition to a low carbon economy. Especially in the strategy 4.2, it is mentioned how to reduce methane emissions in wastewater treatment plants, sanitary landfills, and the oil and agriculture sectors through the construction of biodigesters. Nowadays, the implementation of the PECC 2020–2024 is beginning; however no progress has been reported (SEMARNAT 2014).

Other public policy is the "Programa Sectorial del Medio Ambiente y Recursos Naturales" (PROMARNART 2019–2024), which promotes the construction of biodigestion plants of organic waste (PROMARNART 2019). The "Estrategia Nacional de Cambio Climático vision 10-20-40" is another national policy and projects long terms to face the effects of climate change and to move towards a low carbon economy. One of its strategies is to implement energy efficiency actions such as promoting biodigesters (INECC 2015). However, although this country possesses a high capacity for the production of biogas systems, the lack of government support has restricted the developed of fully biogas systems.

5.3 *Pakistan*

Pakistan is the sixth most populous country in the world. According to the World Bank Group database in 2018, it had 212.22 million people, and more than 50% of the population live in rural areas. For this reason, the government implemented new policies to improve the growth and its economy in this sector. This country has about 175 million head of cattle, which produce about 650 million kilograms of manure a day. Farmers began to use these wastes, burning them for domestic purposes and by creating electricity and caloric energy from biogas systems. However, there is still lack of knowledge about the exploitation of natural carbon sources from waste due this amount of manure, used properly, which could generate about 16.3 million m^3 of biogas per day, being a great opportunity for the country's economic increase (Amjid et al. 2011).

In 2000, the government of Pakistan initiated a Biogas Support Program (BSP), thus installing 1200 biogas units. With a future perspective to implement around 10,000 more biogas units in the next years, it will be possible to take advantage of approximately 30% of the emissions formed by livestock waste. The objective of this program is to reduce the deforestation and to increase the agricultural production through biogas systems (Ilyas 2006). This program has only been implemented for animal waste; however there is also knowledge of food waste and paper industry, which are not still included by the government support (Ilyas 2006; Amjid et al. 2011; Jan and Akram 2018).

5.4 *Biogas Systems Challenges in Developing Countries*

Unfortunately, in developing countries there are a wide range of challenges for the implementation of biogas systems (Fig. 5) (Patinvoh and Taherzadeh 2019). Inadequate infrastructure and poor technical training for the operation of biogas systems are one of the main causes, making these systems have low biogas production (Gebreegziabher et al. 2014). The scarce knowledge of the performance of

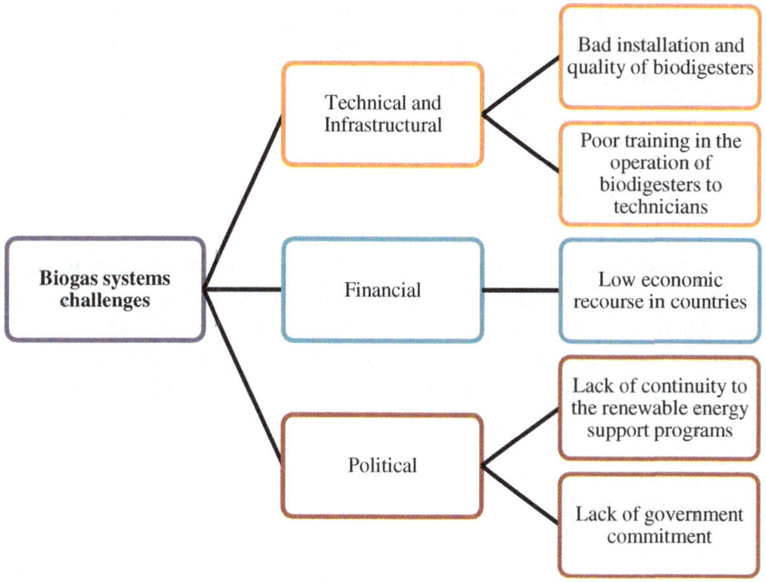

Fig. 5 Biogas systems challenges on developing countries

biodigesters and the poor installation result in the digester malfunctioning, as well as the zero domestication and industrial service of the energy generated by the industrial services (Surendra et al. 2011; Patinvoh and Taherzadeh 2019).

Countries with low economic resource are another challenge. Domestic biodigesters may be somewhat of low quality, and a treatment for biogas purification is required. Only a domestic biodigester comes to cover about $1500 per day and those of industrial scale about $500,000–$1,000,000 (Morgan et al. 2018). Policies about the implementation of biogas systems, transport, and waste collection by this system are null or not carried out in developing countries. In addition, the lack of government commitment and the intermitent continuity to renewable energy support programs have contributed to the poor current level of biogas technology (Akinbomi et al. 2014).

6 Perspectives

In many ways, biogas is a serious alternative to other fossil resources and a complement to other renewable energy sources as wind and solar. However, biogas production technology still has a high potential to improve its efficiency in the production process. This is due to the fact that biogas and other end products have a significant disadvantage: production costs remain relatively high, despite of a wide range of progress (Bahrs and Angenendt 2019).

Although there is already a several applications of biogas technologies worldwide, the industry is still in its early stages of development. World Biogas Association proposes that the biogas industry can be analyzed in three broad categories: micro-digesters that use biogas, scale digesters that generate electricity, and large-scale digesters that produce biomethane (Jain 2019).

6.1 Micro-digesters

Micro-digesters play a very important role in rural areas of developing countries, where they are an integral part of agriculture, waste management, and energy security. There is a total of about 50 million microscale digesters that operate worldwide. The biogas from these digesters is most often used for cooking or heating, displacing high-emission solid fuels such as firewood and coal.

6.2 Digesters to Scale

Scale digesters are mainly used for electricity generation. This is a technology that is widely used throughout the world. Operators of biogas plants at scale are trying to maximize efficiency and input flows by increasing heat utilization. The global generation of electricity from scale digesters reached a growth of 90% compared to 2010, generating 87,500 GWh in 2016.

6.3 *Medium- to Large-Scale Digesters*

Obtaining biomethane from biogas is a relatively new technology and is mainly used in local and national networks as fuel. While CO_2 is also used, in greenhouses and in the food industry, it is estimated that, worldwide, there are 700 plants that upgrade the biogas to biomethane.

6.4 *Biogas Production in the World*

In developing countries, biogas production is mainly on a domestic scale, to obtain fuel used in the kitchen or as lighting, compared to developed countries where it is focused on large-scale biogas plants, based on farms, electricity, and heat (Scarlat et al. 2018).

6.4.1 Asia

Several countries in Asia, such as China, Thai, India, Nepal, Vietnam, Bangladesh, Sri Lanka, and Pakistan, have programs for national biogas production, where support is given to develop domestic systems to provide the population with alternative energy sources.

Nepal has one of the most successful biogas programs, with more than 330,000 domestic biogas plants installed under the Biogas Support Program. Bangladesh opened the National Biogas and Domestic Manure Program in 2006 for rural areas that resulted in 36,000 biogas digesters at the end of 2014, the installation of 130 commercial digesters in 2017 and the construction of 100,000 small biogas plants by 2020 (Adib et al. 2015). China had an estimated 100,000 modern biogas plants in 2014, generating approximately 15 billion m^3 of biogas, equivalent to 9 billion m^3 of biomethane. Thus, with the Medium and Long Term Renewable Energy Development Plan, it plans to reach around 80 million biogas plants in homes by 2020, 8000 large-scale biogas projects, and an annual biogas production of 50 billion m^3 (Scarlat et al. 2018).

6.4.2 Africa

In Africa, despite being a poorly developed region in biogas production, the Biogas Association Program, with the support of the Netherlands Ministry of Foreign

Affairs, aimed at developing national biogas programs in Ethiopia, Kenya, Tanzania, Uganda, and Burkina Faso, for the construction of 100,000 domestic digesters. Currently, the African "Biogas for a better life" initiative aims to provide two million domestic biogas digesters by 2020. With this support, it has been estimated that the technical potential of biogas in Africa allows the construction of 18.5 million plants biogas domestic (Austin and Morris 2012; Marro and Bertsch 2015).

6.4.3 America

In 2017, the USA had more than 2100 biogas plants: 250 use cattle manure, 654 were biogas recovery plants from landfills and about 1240 wastewater treatment plant operated anaerobic digesters producing biogas (Scarlat et al. 2018). On the other hand, Latin America has both agricultural and domestic biogas plants. It has the Network for Biodigesters in Latin America and the Caribbean that promotes the development of small biodigesters in countries such as Mexico, Peru, Cosa Rica, Ecuador, Nicaragua, and Bolivia, where the latter has 1000 domestic biogas plants installed (Vögeli 2014; Persson and Baxter 2015).

6.4.4 Europe

In 2015, the European Union had more than 17,400 biogas plants, ranging from small anaerobic digesters on farms to large co-digestion plants, where total biogas production reached more than 18 billion m^3. Biogas production in Europe has grown significantly in recent years, mainly due to the support schemes established by the member countries of the European Union. The greatest contribution of biogas comes from anaerobic digestion, which is carried out mainly in countries such as Germany, Italy, Czech Republic, and France, followed by biogas from the recovery of landfill gases, where the UK, Italy, France, and Spain are the main producers. Meanwhile, biogas from wastewater treatment predominates in a few countries, such as Sweden, Poland, and Lithuania (Van Foreest 2012; Scarlat et al. 2018). The Green Gas Grids Project expects by 2030 a production of between 48 and 50 billion m_3 per year of biomethane (EBA 2013).

Güsewell et al. (2019) raised future perspectives for Biogas Plants determined mainly by the following aspects:

- Existing biogas plants continue to show a high cost of electricity even after depreciation of the main components. This is mainly due to the high cost of continuous capital for the replacement of technical components.
- High costs of biogas production caused by the cultivation of energy crops. A considerable reduction in costs seems unlikely due to competition in the use of biomass in different sectors and the increase in the means of production.
- The constant need for adjustments and modifications of biogas plants is due to new functions such as flexible energy generation (flexibilization) to balance the increasing quotas of fluctuating renewable electricity and the implementation of

new legal requirements in the agricultural sector such as the ordinance of fertilization.
- Continuous replacements and optimization measures are required due to the breakdown, the projected end of life and the technical progress of the components of the biogas plants.

These adjustments and measurement are considered under the term "repowering," which is defined as the replacement of old power plants or central components to increase the efficiency rate or capacity of the plant and reduce greenhouse emissions. Although there is already a wide application of biogas technologies worldwide, the industry is still in its early stages of development.

Economics and governmental policies are the key determining factor that affects the development of biogas production. However, the ability to produce energy resources while treating waste streams and mitigating greenhouse gas emissions makes anaerobic digestion technology very popular. Biogas can play an important role in the future as an energy carrier because it is flexible in use and is storable, which makes it very valuable for balancing energy networks. With the constant increase in the price of crude oil in recent years, process integration could be an important area to make it economically more attractive. Biogas production could increase from prolonged use of various flows of organic waste, such as food waste, crop residues, sewage sludge, or microalgae sludge.

7 Conclusions

Government policies in terms of AD are directly influenced by the factors of interest in each country, mainly agricultural and environmental factors that lead to the production of biogas, taking advantage of the opportunity of its conversion to energy as an economical strategy. In some countries, it is possible to produce up to 90% of energy for consumption; in addition, this alternative presents the use of agricultural residues, mainly reducing the environmental impact. Thus, policies play a major role in planning strategies for energy generation in the energy sector, which also impacts in reducing the GHG emissions in the atmosphere. The implementation of the policies promotes integrated technologies, turning AD as an alternative to energy generation and boosting the economic growth in developed and developing countries.

References

Abbasi T, Tauseef SM, Abbasi SA (2012) A brief history of anaerobic digestion and "biogas". In: Biogas energy. Springer, New York, NY, pp 11–23

Adib R, Murdock HE, Appavou F, Brown A, Epp B, Leidreiter A, et al (2015) Renewables 2015 global status report. REN21 Secretariat, Paris

Aggarangsi P, Tippayawong N, Moran JC, Rerkkriangkrai P (2013) Overview of livestock biogas technology development and implementation in Thailand. Energy Sustain Dev 17(4):371–377

Akinbomi J, Brandberg T, Sanni S, Taherzadeh M (2014) Development and dissemination strategies for accelerating biogas production in Nigeria. Bioresources 9(3):5707–5737

Alemán-Nava G, Meneses-Jácome A, Cárdenas-Chávez D, Díaz-Chavez R, Scarlat N, Dallemand J, Parra R (2015) Bioenergy in Mexico: status and perspective. Biofuels Bioprod Biorefin 9(1):8–20

Amigun B, Musango J, Stafford W (2011) Biofuels and sustainability in Africa. Renew Sust Energ Rev 15(2):1360–1372

Amjid S, Bilal M, Nazir M, Hussain A (2011) Biogas, renewable energy resourse for Pakistan. Renew Sust Energ Rev 15(6):2833–2837

Austin G, Morris G (2012) Biogas production in Africa. In: Bioenergy for sustainable development in Africa. Springer, Dordrecht, pp 103–115

Bahrs E, Angenendt E (2019) Status quo and perspectives of biogas production for energy and material utilization. GCB Bioenergy 11(1):9–20

Becker B, Fischer D (2013) Promoting renewable electricity generation in emerging economies. Energy Policy 56:446–455

Boiral O (2006) Global warming: should companies adopt a proactive strategy? Long Range Plan 39(3):315–330

Bond T, Templeton MR (2011) History and future of domestic biogas plants in the developing world. Energy Sustain Dev 15(4):347–354

Bos K, Chaplin D, Mamun A (2018) Benefits and challenges of expanding grid electricity in Africa: a review of rigorous evidence on household impacts in developing countries. Energy Sustain Dev 44:64–77

Cadez S, Czerny A (2010) Carbon management strategies in manufacturing companies: an exploratory note. J East Eur Manag Stud 15(4):348–360

Cadez S, Czerny A (2016) Climate change mitigation strategies in carbon-intensive firms. J Clean Prod 112:4132–4143

Cadez S, Czerny A, Letmathe P (2018) Stakeholder pressures and corporate climate change mitigation strategies. Bus Strateg Environ 28(1):1–14

Cheng S, Li Z, Mang P, Huba M, Gao R, Wang X (2014) Development and application of prefabricated biogas digesters in developing countries. Renew Sustain Energy Rev 34:387–400

Clemens H, Bailis R, Nyambane A, Ndung'u V (2018) Africa biogas partnership program: a review of clean cooking implementation through market development in East Africa. Energy Sustain Dev 46:23–31

Damert M, Baumgartner RJ (2017) Intra-sectoral differences in climate change strategies: evidence from the global automotive industry. Bus Strateg Environ 27(3):265–281

Díaz-Trujillo L, Nápoles-Rivera F (2019) Optimization of biogás supply chain in Mexico considering economic and environmental aspect. Renew Energy 139:1227–1240

Dubois O, Pirelli T, Peressotti A (2019) Biomass anaerobic digestion and gasification in non-OECD countries—an overview. In: Substitute natural gas from waste. Academic, pp. 343–387

Edwards J, Othman M, Burn S (2015) A review of policy drivers and barriers for the use of anaerobic digestion in Europe, the United States and Australia. Renew Sust Energ Rev 52:815–828

European Biogas Association (2013) Green gas grid. Proposal for a European bio- methane roadmap. Eureopean Biogas Association, Brussels. https://www.europeanbiogas.eu/green-gas-grids-project-proposal-european-biomethane-roadmap

Fallde M, Eklund M (2015) Towards a sustainable socio-technical system of biogas for transport: the case of the city of Linköping in Sweden. J Clean Prod 98:17–28
Ferreira M, Marques IP, Malico I (2012) Biogas in Portugal: status and public policies in a European context. Energy Policy 43:267–274
Freitas FF, De Souza SS, Ferreira LRA, Otto RB, Alessio FJ, De Souza SNM, Venturini OJ, Junior OA (2019) The Brazilian market of distributed biogas generation: overview, technological development and case study. Renew Sust Energ Rev 101:146–157
Garfí M, Martí-Herrero J, Garwood A, Ferrer I (2016) Household anaerobic digesters for biogas production in Latin America: a review. Renew Sust Energ Rev 60:599–614
Gebreegziabher Z, Naik L, Melamu R, Balana B (2014) Prospects and challenges for urban application of biogas installations in Sub-Saharan Africa. Biomass Bioenergy 70:130–140
Global Methane Initiative (2014) A global perspective of anaerobic digestion polices and incentives. Agriculture Subcommittee, November 2014. https://www.globalmethane.org/documents/tools/A-Global-Perspective-of-AD-Policies-Incentives.pdf
Güsewell J, Härdtlein M, Eltrop L (2019) A plant-specific model approach to assess effects of repowering measures on existing biogas plants: the case of Baden-Wuerttemberg. GCB Bioenergy 11(1):85–106
Hashim H, Ho WS (2011) Renewable energy policies and initiatives for a sustainable energy future in Malaysia. Renew Sust Energ Rev 15(9):4780–4787
Hoffmann VH, Busch T (2008) Corporate carbon performance indicators. J Ind Ecol 12(4):505–520
Hoo PY, Hashim H, Ho WS, Tan ST (2017) Successful biogas implementation – a mini-review on biogas utilization, energy policies and economic incentives. Chem Eng Trans 61:1495–1500
IEA (2019) Global energy & CO2 status report 2019. Flashing report, March 2019
Ilyas S (2006) Biogas support program is a reason for its success in Pakistan. Am Eurasian J Sci Res 1(1):42–45
INECC (2015) Estrategia Nacional de Cambio Climático Vision 10-20-40. Diciembre 2015. https://www.gob.mx/inecc/documentos/estrategia-nacional-de-cambio-climatico-vision-10-20-40
IRENA (2019) Renewable energy statistics 2019. The International Renewable Energy Agency, Abu Dhabi
Jain S (2019) Global potential of biogas. World Biogas Assosiation, June 2019. https://www.worldbiogasassociation.org/wp-content/uploads/2019/09/WBA-globalreport-56ppa4_digital-Sept-2019.pdf
Jan I, Akram W (2018) Willingness of rural communities to adopt biogás sistema in Pakistan: Critical factors and policy implications. Renew Sust Energ Rev 81:3178–3185
Jeswani HK, Wehrmeyer W, Mulugetta Y (2008) How warm is the corporate response to climate change? Evidence from Pakistan and the UK. Bus Strateg Environ 17(1):46–60
Kabir H, Yegbemey RN, Bauer S (2013) Factors determinant of biogas adoption in Bangladesh. Renew Sust Energ Rev 28:881–889
Kemausuor F, Ackom E (2017) Toward universal electrification in Ghana. Wiley Interdiscip Rev Energy Environ 6(1):e225
Kolk A, Pinkse J (2005) Business responses to climate change: identifying emergent strategies. Calif Manag Rev 47(3):6–20
Lara Grando R, de Souza Antune AM, Da Fonseca FV, Sánchez A, Barrena R, Font X (2017) Technology overview of biogas production in anaerobic digestion plants: a European evaluation of research and development. Renew Sust Energ Rev 80:44–53
Marro P, Bertsch N (2015) Making renewable energy a success in Bangladesh: getting the business model right. Asian Development Bank
Matsuda T, Yano J, Hirai Y, Sakai S (2012) Life-cycle greenhouse gas inventory analysis of household waste management and food waste reduction activities in Kyoto, Japan. Int J Life Cycle Assess 17(6):743–752
Mittal S, Ahlgren EO, Shukla PR (2018) Barriers to biogas dissemination in India: a review. Energy Policy 112:361–370

Morgan H, Xie W, Liang J, Mao H, Lei H, Ruan R, Bu Q (2018) A techno-economic evaluation of anaerobic biogas producing systems in developing countries. Bioresour Technol 250:910–921

NREEEP (2014) Energy Commission of Nigeria. http://www.energy.gov.ng/energy_laws.php

Patinvoh R, Taherzadeh M (2019) Challenges of biogas implementation in developing countries. Curr Opin Environ Sci Health 12:30–37

Persson T, Baxter D (eds) (2015) IEA bioenergy task 37: country reports summary 2014. IEA bioenergy

PROMARNART (2019). Programa Sectorial de Medio Ambiente y Recursos Naturales. https://www.biodiversidad.gob.mx/media/1/planeta/internacional/files/post2020/5PND_CesarRodriguez.pdf

Raha D, Mahanta P, Clarke ML (2014) The implementation of decentralised biogas plants in Assam, NE India: the impact and effectiveness of the National Biogas and Manure Management Programme. Energy Policy 68:80–91

Rajendran K, Aslanzadeh S, Taherzadeh MJ (2012) Household biogas digesters—a review. Energies 5(8):2911–2942

Ramanathan V, Feng Y (2008) On avoiding dangerous anthropogenic interference with the climate system: formidable challenges ahead. Proc Natl Acad Sci U S A 105(38):14245–14250

REN21 R (2015) Global status report, Paris, 2017

Renewable Energy Network of the 21st Century. Renewables 2018 Global Status Report. 2018. http://www.ren21.net/status-of-renewables/global-status-report

Rittmann BE, McCarty PL (2001) environmental biotechnology: principles and applications. McGraw-Hill, New York, NY

Scarlat N, Dallemand JF, Fahl F (2018) Biogas: developments and perspectives in Europe. Renew Energy 129:457–472

SEMARNAT (2014) Programa Especial de Cambio Climático 2014–2018. Secretaria de Medio Ambiente y Recursos Naturales. http://apps3.semarnat.gob.mx/pecc/pagina.php

Singh R, Setiawan AD (2013) Biomass energy policies and strategies: harvesting potential in India and Indonesia. Renew Sust Energ Rev 22:332–345

Strauch SJK, Singhal A (2013) Biomethane guide for decision makers - policy guide on biogas injection into natural gas grid. www.greengasgrids.eu/fileadmin/greengas/media/Downloads/Documentation_from_the_GreenGasGrids_project/Biomethane_Guide_for_Decision_Makers.pdf. Accessed 05.02.2017

Surendra K, Khanal S, Shrestha P, Lamsal B (2011) Current status of renewable energy in Nepal: opportunities and challenges. Renew Sust Energ Rev 15(8):4107–4117

Surendra KC, Takara D, Hashimoto AG, Khanal SK (2014) Biogas as a sustainable energy source for developing countries: opportunities and challenges. Renew Sust Energ Rev 31:846–859

UNFCCC (1992) United Nations framework convention on climate change. Article 2. Retrieved 23 May 2016. United Nations. http://unfccc.int/files/essential_background/background_publications_htmlpdf/application/pdf/conveng.pdf

UNFCCC (1998) Report of the conference of the parties on its third session, held at Kyoto from 1 to 11 December 1997. Addendum Part two. United Nations Framework Convention on Climate Change. FCCP/CP/1997/7/Add.1. https://unfccc.int/resource/docs/cop3/07a01.pdf#page=28

Van Foreest F (2012) Perspectives for biogas in Europe. Oxford Institute for Energy Studies

Vasco-Correa J, Khanal S, Manandhar A, Shah A (2018) Anaerobic digestion for bioenergy production: global status, environmental and techno-economic implications, and government policies. Bioresour Technol 247:1015–1026

Viancelli A, Michelon W, ElMahdy EM (2019) Current efforts for the production and use of biogas around the world. In: Improving biogas production. Springer, Cham

Vögeli Y (2014) Perspectives for biogas in Europe. Oxford Institute for Energy Studies. Anaerobic digestion of biowaste in developing countries: Practical information and case studies. Eawag-Sandec

Vögeli Y, Riu C, Gallardo A, Diener S, Zurbügg C (2014) Anaerobic digestion of biowaste in developing countries. Practical Information and Case Studies. Swiss Federal Institute of Aquatic Science and Technology (Eawag), Dübendorf

Wang X, Lu X, Yang G, Feng Y, Ren G, Han X (2016) Development process and probable future transformations of rural biogas in China. Renew Sust Energ Rev 55:703–712

WBA, World Bioenergy Association (2019) WBA Global bioenergy statistics 2019

Weiland P (2010) Biogas production: current state and perspectives. Appl Microbiol Biotechnol 85 (4):849–860

Xiaohua W, Chonglan D, Xiaoyan H, Weiming W, Xiaoping J, Shangyun J (2007) The influence of using biogas digesters on family energy consumption and its economic benefit in rural areas—comparative study between Lianshui and Guichi in China. Renew Sust Energ Rev 11 (5):1018–1024

Yousuf A, Khan MR, Pirozzi D, Ab Wahid Z (2016) Financial sustainability of biogas technology: barriers, opportunities, and solutions. Energ Source Part B Econ Plann Policy 11(9):841–848

Zervous A (2019) Renewables 2019 global status report; GSR 2019. Chair, REN21. https://www.ren21.net/gsr-2019/

The manufacturer's authorised representative in the EU is Springer Nature Customer Service Centre GmbH, Europaplatz 3, 69115 Heidelberg, Germany. If you have any concerns regarding our products, please contact ProductSafety@springernature.com

Printed and bound by CPI Group (UK) Ltd, Croydon, CR0 4YY
28/09/2026
02224065-0011